678·5·01

LIBRARY
No. B 5618
26 OCT 1972
R.P.E. WESTCOTT

Rocket Propulsion Establishment
Library

Please return this publication, or request renewal, before the last date stamped below.

Name	Date
~~Mr H J Russell~~	~~14.9.73~~
G.F. HAYES	~~1/10/79~~
~~G. RAMSDEN~~	5.10.79
Mr D Young	16.1.88

RPE Form 243 (revised 6/71) 739490

RPE Form 243 (revised 6/71) 739490

~~G R Ramsden~~

Handbook of Plastics Test Methods

EDITOR ENGINEERING SERIES

R. M. OGORKIEWICZ
M.SC.(ENG.), A.C.G.I., D.I.C., C.ENG., F.I.MECH.E.

EDITOR CHEMISTRY SERIES

PROFESSOR P. D. RITCHIE
PH.D., B.SC., F.R.I.C., F.R.S.E., F.P.I.

Handbook of Plastics Test Methods

G. C. IVES,
B.SC., F.R.I.C., F.P.I., F.I.E.I., F. INST. PET.,

Caleb Brett and Son Ltd.
(Formerly of Yarsley Testing Laboratories Ltd.)

J. A. MEAD, B.SC., and
M. M. RILEY, B.SC.,

Yarsley Testing Laboratories Ltd.

Published for The Plastics Institute

LONDON
ILIFFE BOOKS

THE BUTTERWORTH GROUP

ENGLAND
Butterworth & Co (Publishers) Ltd
London: 88 Kingsway, WC2B 6AB

AUSTRALIA
Butterworth & Co (Australia) Ltd
Sydney: 586 Pacific Highway
Chatswood, NSW 2067
Melbourne: 343 Little Collins Street, 3000
Brisbane: 240 Queen Street, 4000

CANADA
Butterworth & Co (Canada) Ltd
Toronto: 14 Curity Avenue, 374

NEW ZEALAND
Butterworth & Co (New Zealand) Ltd
Wellington: 26-28 Waring Taylor Street, 1
Auckland: 35 High Street, 1

SOUTH AFRICA
Butterworth & Co. (South Africa) (Pty) Ltd
Durban: 152-154 Gale Street

First published in 1971 by
Iliffe Books, an imprint of
the Butterworth Group

Published for The Plastics Institute,
11 Hobart Place, London, S.W.1

© The Plastics Institute, 1971

ISBN 0 592 05449 7

Filmset by Photoprint Plates Ltd
Rayleigh, Essex

Printed in England by
Fletcher & Son Ltd., Norwich

Preface

A Plastics Institute publication on the general subject of testing has not appeared since 1954*. In the intervening years the picture has changed radically. The industry itself has grown at a very rapid rate, thereby not only increasing the range of uses of plastics materials but also involving the introduction of many novel polymer compositions with improved properties. These two factors have inevitably required an increase in the number of test methods available to assess the performance of plastics, both for 'type approval' and 'quality control' purposes.

Again, since 1954 the standards and specifications picture has altered out of all recognition. Taking the United Kingdom and United States as examples, there are now volumes specifically devoted to the testing of plastics—British Standard 2782:1970 and American Society for Testing and Materials book of Standards, Part 27. On the international level progress is being made with the production of I.S.O. Recommendations, which are being written into national standards as time and opportunity allow. It might well be asked, therefore, 'Is there any need for a handbook of plastics test methods?' In answer to this, it is hoped that three justifications have been fulfilled:

1. The need to compare various standard methods and draw attention to their differences (and the differences of results obtained).
2. The necessity of providing background information to the tests available, to demonstrate what the results thereof will and will not provide.
3. The desirability of describing some useful tests which do not, at least at the time of writing (see below) appear in standards and specifications.

Again it might reasonably be questioned why elastomers should not be included, when the dividing line between them and plastics is becoming ever more diffuse; however, quite apart from any question of remit to the authors, there is already in existence an excellent textbook on the subject†.

Even within the compass of the monographs series which has so far been published by The Plastics Institute, there now exists two of considerable relevance to the subject generally:

1. GORDON, M. *High Polymers. Structure and Properties,* 2nd Ed. (1963) and
2. RITCHIE, P. D., Ed. *Physics of Plastics* (1965).

Frequent reference has been made in this handbook to these two

*COLLINS, J. H., *Testing and Analysis of Plastics, Part II: The Testing of Plastics*, 2nd (Revised) Edn.

†SCOTT, J. R., *Physical Testing of Rubbers*, Maclaren & Sons, Ltd. (London) 1965.

books in particular and, for this reason, acoustical properties of plastics, for instance, have not been discussed (as they form Chapter 9 of 'Physics of Plastics'). Likewise, subjects already well covered by existing textbooks, and particularly those on the fringe of plastics testing when interpreted literally, have not been considered in detail; however, references to textbooks and technical articles have been given wherever possible, in an endeavour to make the present volume self-contained in some degree. As a corollary, in the case of Chapter 8 a simple description of some mechanical test machines is included. In the authors' experience, these are often used by people who have no conception of the machines' capabilities, and particularly their limitations, and there does not seem to be a suitable up-to-date textbook on the subject.

Chemical analysis has not been deemed to fall within the compass of the subject in any sense (see Chapter 17) and, by and large, attention has been restricted to the general testing of plastics as such, i.e., the many methods which exist for evaluating specific shapes, forms and even some unique types of plastics have not been given more than a passing reference (again, see Chapter 17). The testing of compounding ingredients and aids to processing—solvents, plasticisers, lubricants, stabilisers, pigments and the like—has not been covered except in so far as their evaluation logically forms part of the testing of a plastics material per se.

As far as reference to national standards is concerned, attention has been restricted to British, United States and German documents, not only for reasons of space but also because there is no doubt that standardisation in these three countries is currently more advanced than elsewhere; international (I.S.O.) standards have not been described in detail because their practical value emerges only when adopted nationally. The reader interested in other sources is recommended to study Chapter 2 and particularly the appendices thereto. Many of the standard tests are described in considerable detail, *not* as a substitute for studying the official document when necessary, but to try to drive home the absolute necessity of working precisely to instructions—particularly for quality control—in such mundane matters as dimensions of tolerances; in many instances, chosen at random the actual specified tolerances have been quoted to try to drive the point home. In over twenty years of testing plastics, the authors have many times seen time wasted and effort needlessly expended in arguments between supplier and customer, for example, over disagreements in results when post mortem examination has shown that one or other party, or perhaps both, has been 'cutting corners' or deviating from the standard method in some way or another.

I was originally asked in 1963 to write this book, but over-optimism on speed of output combined with family and business affairs has delayed its completion. I am therefore particularly grateful to my colleagues, Messrs. J. A. Mead and M. M. Riley, for so readily agreeing to undertake the writing of five of the more specialised chapters when it looked as though the project would never be completed. We are all greatly indebted to our colleague, Mr. D. B. S. Berry, for his helpful comments and criticisms. And whilst thanks are being expressed, the gratitude of all three of us is due to our colleagues in the information and library department for tracing all the textbooks and journals to which reference has been made (and every one of which has been examined). Full titles are given to references so that the reader may be able to identify particular articles or books which may interest him beyond that matter relevant to the subject under discussion. Generally speaking, references are restricted to articles describing or discussing test methods, but where properties are being described which are not commonly quoted, sources of information are included. The more commonly available data are to be found in trade literature of the more reputable manufacturers or, for instance, in *Design Engineering Handbook. Plastics*, (Product Journals Ltd., Summit House, Glebe Way, West Wickham, Kent. BR4 OSL—1968).

When at last it seemed possible that this book would be finished in the foreseeable future, a date stop was applied of early 1970. This means, *inter alia*, that standard methods have been taken from:

1. B.S. 2782:1970, 'Methods of Testing Plastics'
2. A.S.T.M. Parts dated 1969 (Particularly Parts 26 and 27).
3. DIN standards up to that date.

Anyone specifically interested in the most up-to-date description of a standard test method, and many are altered quite frequently, *must* examine the latest edition of B.S. 2782 and its amendments, and the latest A.S.T.M. manual, etc. All contractual arrangements, and particularly those involving government departments should call up the very latest document. For product specifications in particular, where the British Standards Institute publications are not so helpful as those of their counterparts in the U.S.A. and Germany in providing a comprehensive list in one or two documents, Appendix 2 of Chapter 2 has been prepared. To be brought right up-to-date every issue including and after June 1970 of 'B.S.I. News' should be studied—a suitable short cut is, however, possible by the judicious study of the latest issue of the British Standards Yearbook (the current coverage includes the 1970 issue).

Currently many of those countries, including the U.K., which have traditionally used Imperial units, or non-metric units generally, are in a state of transition. Internationally, a system of units known as SI (Système International d'Unités) has been agreed and is in course of being adopted in many countries; this is not the c.g.s., or metric technical, scheme but is somewhat akin thereto. Hence official standards, literature sources and information sheets may or may not have yet been 'metricated' at this moment; where they have, some give the equivalents in the traditional units, others do not. In this handbook, the policy has been generally to quote the units as given in the source of information being cited.

The subject of plastics testing could have been treated in a number of ways, but it is hoped that the one selected will be found logical and convenient to two classes of reader in particular:

1. The student of plastics technology
2. The person in industry who has either to test plastics to earn his living or who is indirectly involved in their evaluation

An advanced treatise on the subject has not been attempted, as it is considered that there are sufficient misconceptions of the subject at the relatively simple level to be ironed out, without delving too deeply into the more esoteric realms of, for instance, design data prediction.

Chapters 1 and 17 contain what could not conveniently be fitted into the other fifteen, the former describing all the matter which is common to the remainder and the latter covering the subjects left out after the first sixteen had been written. Chapter 1 is also intended to provide a gentle introduction to the subject, an aim which I am encouraged to think has been achieved at least to some degree by the comment of my (then) secretary: after typing the first few chapters she said 'I enjoyed Chapter 1 and understood it—pity you had to go on!'

Acknowledgements are due to:
All authors, publishers and companies mentioned for their kind permission to reproduce diagrams, photographs etc. supplied and to the British Standards Institution, American Society for Testing and Materials and the Deutscher Normenausschuss for permission to reproduce extracts and diagrams from their standards, copies of which may be obtained from the addresses given in Chapter 2.

G. C. IVES

Contents

1

Introduction to the Testing of Plastics

Test: 'That by which the existence, quality or genuineness of anything, is or may be determined' *(The Shorter Oxford English Dictionary)*.

1.1 PHILOSOPHY—WHY TEST?

Why test plastics or, for that matter, why test anything? Why not rely on experience and good workmanship, backed up by sound judgement? The answer to these apparently simple questions is not so obvious as might be first thought, particularly when it will be seen, throughout what follows, that practically every test has limitations, the data it yields being only applicable with discretion to the everyday conditions of service.

When dealing specifically with plastics materials, however, the questions posed above are already part answered by the very case history of the industry, or rather the lack of it! Very few plastics in common use today were available ten years ago in precisely the same form; even if the polymer is basically the same—and there may be refinements of processing which have led to a more pure product, for instance—the plasticiser has probably changed or a new stabiliser system has found favour. The polymer molecules of many other current plastics compositions were not even known ten years ago! Thus, there is not much experience upon which to rely. (These circumstances make it very difficult to promote the use of plastics in, for instance, structural applications where the designer may insist upon a guaranteed 50 year performance.)

Testing, then, may be undertaken to assess the performance of a material in relation to the duty which it is to perform, i.e. to establish *suitability for purpose;* the mechanical properties will be measured and correlated with the calculated stresses, the electrical properties with the known circuitry, and so on. Judicious use of such measurements ideally will obviate the inspired guess, or gross wastage because of the necessity to 'play safe', and in this way the necessary experience will be assembled with a minimum of headaches.

Not that the Plastics Industry can aspire ultimately to an idyllic future, free from tiresome and profit reducing testing, any more than the more traditional industries do today. All men and machines are fallible and liable to vary in performance for a variety of reasons; the things they produce will alter in quality correspondingly. Thus we find the second principal reason for testing, as a *control of quality*. Under this broad heading come regular quality control schemes, production batch testing by random sample, purchasers' tests and control sample examination in the case of dispute.

1.2 HOW TEST?

If it is accepted that we must test, then surely that is the end of the matter. Why write a whole book on the unfortunate necessity? After all, everyone knows that 'tensile strength' is the resistance of a material to breaking by stretching, so where can there be room for argument? In a limited sense there is little in this particular case; it is true that, for instance, vendor and purchaser would both have the same broad picture in mind in discussing the tensile strength of a material, though there are many cases where different interpretations can be put on one property name and many more where a number of names apply to the same property. More generally, however, the case of tensile strength is no exception and well illustrates why testing needs such a lengthy description.

Let us suppose that a new moulding composition has been offered, of unknown origin and with a plain statement on its container:

'Tensile strength: 8000 force units/area', say '8000X'.

It is a penny per unit weight cheaper than the material we are at present using, which is satisfactory for our application; this, we will assume for convenience, is one (highly fictional!) where only tensile forces come into play and that there are no other considerations to bother us*. Our current material also has, we have been told by our suppliers, a tensile strength of '8000X'. Why should we not change materials and save money—the same strength for a penny a pound cheaper!

In fact one good reason for looking into the matter further has already become evident. Moulding compositions are usually in the form of granules, pellets, chips or powders, but tensile strength is determined on a moulding of some form or other; therefore we must produce the appropriate test specimens and we must ensure that although small they are representative of the product as a whole, in other words our *sampling procedure* must be sound. Any moulder or fabricator will know just how much the appearance and strength of a moulding can be altered simply by varying the cycle applied to a given moulding composition. Thus our two figures of '8000X' can only be compared if the *conditions of any preheat, moulding temperature, cycle time, etc.*, are comparable or, better still, if the figures are realistic optima which can be obtained using conditions (especially of time) that are commercially feasible. (Obviously a material which only attains an adequate tensile strength after an uneconomic cycle cannot be considered.)

So far we have achieved the moulding of our test pieces and already we have run into two good reasons why the quoted tensile strengths might not be comparable: however, we have hardly started yet! Most plastics materials are affected by quite small changes in *temperature;* the stiffening of some

*For the perspicacious—the specific gravities are the same too!

types of highly plasticised PVC is a classic example at low temperatures, but many other materials, especially in the thermoplastics class, are affected in a similar if not so spectacular way[1]. How can we know that our two alternative materials were both tested at the same temperature? Perhaps the cheaper product was examined at a lower temperature—and the tensile strength of plastics generally shows a negative coefficient with respect to temperature change! A limited number of plastics are influenced by the *relative humidity* (or measure of moisture content) of the surrounding air; cellulose acetate is a classic example. If our materials are like cellulose acetate then perhaps the cheaper was tested in a drier condition, which again will tend to give it a higher tensile strength.

For the mechanically minded, it will not be difficult to envisage that *shape* of the test specimen could be important; the stress distributions in the two shapes shown in Figure 1.1 will be quite different because (b) has undesirable

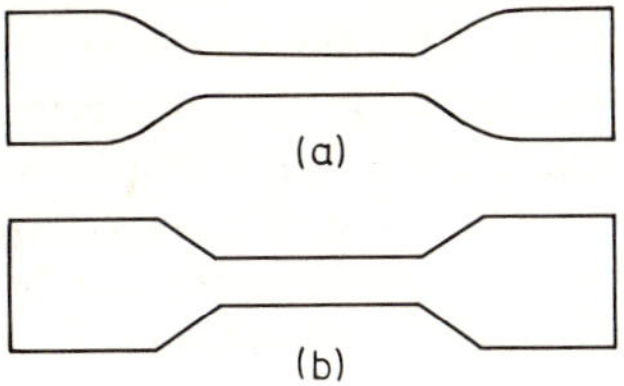

Figure 1.1. Possible shapes for tensile test specimens

'stress raisers' at the sharp 'shoulders'*. It is not difficult to imagine that these will tend to weaken the specimen and this effect may be so significant as to cause (b) to break at a lower load than (a), thus yielding a lower tensile strength. Can we be certain that our two alternative materials were examined using comparable test specimens?

Again, many materials behave rather differently according to how quickly they are stressed; the classical example is 'bouncing putty', a silicone composition which behaves like conventional putty when 'worked' (slow stressing), but which rebounds like an elastic solid when dropped on to the floor (quick stressing). Although this particular response to the time factor may be unique, in general plastics materials react to changes in *speed of stressing*[3] and therefore will give different strengths when the rate of increase of load of tensile specimens is altered. Were our two materials tested under identical stressing conditions?

There are even more influences which could affect the measured value of tensile strength of a moulding composition, but enough has now been said to make the point that the bald statement 'Tensile strength: 8000X' is meaningless without detailed qualification of how the figure was obtained. Almost without exception, analogous doubts could be raised for the other properties commonly quoted with abandon.

1.3 THE NEED FOR STANDARDS AND SPECIFICATIONS

We can see now that, even if a property is quoted with all the experimental test details to put the data into correct perspective, the situation is still far

*For a very readable account of this subject, see Chapters 4 and 5 of Reference 2.

from satisfactory. Two materials, to be compared as before, may have (known) tensile strengths measured by different (known) techniques; we may know the qualitative effects of the changes of temperature, loading rate, etc. between the two techniques, but our state of knowledge at the moment is certainly not generally sufficient to be able to reduce both sets of data to a common and comparable level or 'norm', not even to reduce one variable, still less a whole host of them which as likely as not interact amongst themselves. Thus, knowledge of the test conditions will probably indicate that the data are not comparable, which is a pretty negative contribution, and means that we will have to do tests ourselves on one material, in a manner identical with the details of moulding, etc., cited for the other; even then we may not feel absolutely confident of our ability to read the correct interpretation into the conditions given for the first set of tests and so we may decide to test both materials by our own preferred method, just for safety.

How can we overcome this sorry state of affairs? Quite simply and obviously by insisting that everyone uses the same methods of test. In this way supplier and user can be guaranteed to speak the same language, thus facilitating trade by giving some definite meaning to sales contracts. This is not only a national matter, for if agreement on methods can be achieved continentally or better still universally, then international trade is also benefited. The organisation of official standards and specifications is described in Chapter 2.

1.4 MISCELLANEOUS PITFALLS IN TESTING

In the discussion above on the ambiguity of a simply stated tensile strength, attention has already been drawn to the number of pitfalls; throughout this book, when dealing with practically every test method described, it will be necessary to mention pitfalls specific to that technique and peculiar limitations of the data so obtained. There are, however, a number of points which are so general that it is desirable and convenient to deal with them right from the start—and where better than in the 'language' of the subject, i.e. the units?

1.4.1 Units

Except for a few isolated instances, such as relative density (specific gravity) and dielectric constant, physical properties have dimensions and therefore must be expressed in units. The United Kingdom, parts of the Commonwealth and the United States, in certain fields, regrettably do not yet use the 'metric' system universally. In fact, the metric system is itself out of date and a new system of international units (S.I.*) has been worked out and is in the process of being adopted in the U.K.[4–6] However at the moment a variety of alternatives, based on the metre, inch or foot, gram, pound or ton and second, minute or hour is used, with all the confusion

*Système International d'Unités.

imaginable. Fahrenheit degrees are still frequently retained, though this will not cause too many troubles as most people will remember the conversion to degrees Centigrade (more correctly 'Celsius') from their schooldays and likewise compensation for different time units is only tiresome; however few people carry round with them a head full of length and weight equivalents. Some *approximate* conversion factors might therefore be useful at this stage (S.I. units, recommended decimal multiples and submultiples underlined):

Length: $1\ \text{in} = \underline{25{\cdot}4\ \text{mm}} = 2{\cdot}54\ \text{cm}.\quad \underline{1\ \text{mm}} = 0{\cdot}0394\ \text{in}$
$1\ \text{cm} = 0{\cdot}394\ \text{in}$
$1\ \text{ft} = \underline{0{\cdot}3048\ \text{m}}\quad \underline{1\ \text{m}} = 3{\cdot}281\ \text{ft}$

Area: $1\ \text{in}^2 = \underline{6{\cdot}452\ \text{cm}^2} = \underline{645{\cdot}2\ \text{mm}^2}\quad \underline{1\ \text{cm}^2} = 0{\cdot}1550\ \text{in}^2$
$\underline{1\ \text{mm}^2} = 0{\cdot}001550\ \text{in}^2$
$1\ \text{ft}^2 = \underline{0{\cdot}09290\ \text{m}^2}\quad \underline{1\ \text{m}^2} = 1{\cdot}196\ \text{yd}^2$

Volume: $1\ \text{in}^3 = \underline{16{\cdot}39\ \text{cm}^3} = \underline{16387\ \text{mm}^3}\quad \underline{1\ \text{cm}^3} = 0{\cdot}06102\ \text{in}^3$
$\underline{1\ \text{mm}^3} = 0{\cdot}00006102\ \text{in}^3$
$1\ \text{ft}^3 = \underline{28317\ \text{cm}^3} = \underline{0{\cdot}028317\ \text{m}^3}$
$= 28{\cdot}316\ \text{litre}\quad 1\ \text{litre} = 61{\cdot}025\ \text{in}^3$

Mass: $1\ \text{oz} = \underline{28{\cdot}35\ \text{g}}\quad \underline{1\ \text{g}} = 0{\cdot}03527\ \text{oz}$
$1\ \text{lb} = \underline{0{\cdot}4536\ \text{kg}}\quad \underline{1\ \text{kg}} = 2{\cdot}205\ \text{lb}$
$1\ \text{ton} = \underline{1016\ \text{kg}} = \underline{1{\cdot}016\ \text{tonne}}$

Density: $1\ \text{lb/ft}^3 = 0{\cdot}01602\ \text{g/cm}^3 = \underline{16{\cdot}02\ \text{kg/m}^3}$
$1\ \text{g/cm}^3 = 62{\cdot}43\ \text{lb/ft}^3\quad \underline{1\ \text{kg/m}^3} = 0{\cdot}06243\ \text{lb/ft}^3$

Force: $1\ \text{lb force} = \underline{4{\cdot}448\ \text{Newton (N)}}\quad 1\ \text{ton force} = \underline{9{\cdot}964\ \text{kN}}$
$1\ \text{kg force} = \underline{9{\cdot}807\ \text{N}}\quad \underline{1\ \text{N}} = 0{\cdot}2248\ \text{lb force}$

Stress: $1\ \text{lb force/in}^2 = 0{\cdot}07031\ \text{kg force/cm}^2 = \underline{6895\ \text{N/m}^2}$
$= \underline{6{\cdot}895\ \text{kN/m}^2}$
$1\ \text{ton force/in}^2 = 157{\cdot}5\ \text{kg force/cm}^2 = \underline{15{\cdot}44\ \text{MN/m}^2}$
$1\ \text{kg force/cm}^2 = 14{\cdot}22\ \text{lb force/in}^2 = 9{\cdot}807\ \text{N/cm}^2 = \underline{0{\cdot}09807\ \text{N/mm}^2}$
$\underline{1\ \text{N/m}^2} = 0{\cdot}0001450\ \text{lb force/in}^2$
$\underline{1\ \text{kN/m}^2} = 20{\cdot}89\ \text{lb force/ft}^2$
$1\ \text{bar} = \underline{10^5\ \text{N/m}^2}$

(For more accurate values, and many others, see for instance Refs 6–9)

Many other equivalents will be mentioned throughout the succeeding chapters. Luckily there is no dispute over electrical units; volts, amperes and ohms are fairly universal.

1.4.2 Sampling

Although the statistical concept of a sample will be mentioned towards the end of this chapter, it is appropriate to consider the physical aspect of sampling amongst the pitfalls. The majority of standard test pieces are quite small, weighing of the order of 50 g or less, and usually five is the maximum number used in any one property measurement. Therefore the combined mass of test pieces will be very small in relation, say, to the contents of the normal sized sack, perhaps 25 kg or $\frac{1}{2}$ cwt, and far smaller in relation to the production batch from which the selected sack was taken for sampling and

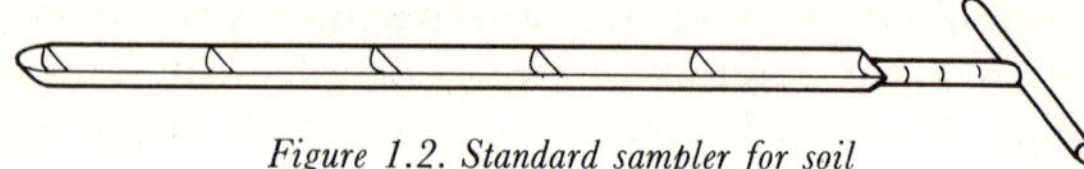

Figure 1.2. Standard sampler for soil

testing. Again, small test pieces cut from a large sheet represent a very small fraction of the total area. It is obviously very important to ensure that the essentially small quantity of material tested is truly representative of the whole.

The ordinary fabric filled phenol-formaldehyde moulding powder, for instance, is a mixture of 'fines' and coarse fibrous agglomerates and in the vibration of transportation, the particles will tend to segregate, the finer ones to the bottom. There are devices for taking representative samples over the whole depth, for instance a type of soil tester which consists of a half cylinder, open down its length and fitted with steps: see Figure 1.2.

Another type, specified at one time for sampling cellulose acetate moulding materials[10], is a hollow cylinder with an Archimedean screw running down the centre (see Figure 1.3).

A general technique for sampling powders, pellets and granules is that of 'quartering'. The sack contents, for instance, would be tipped on to a clean

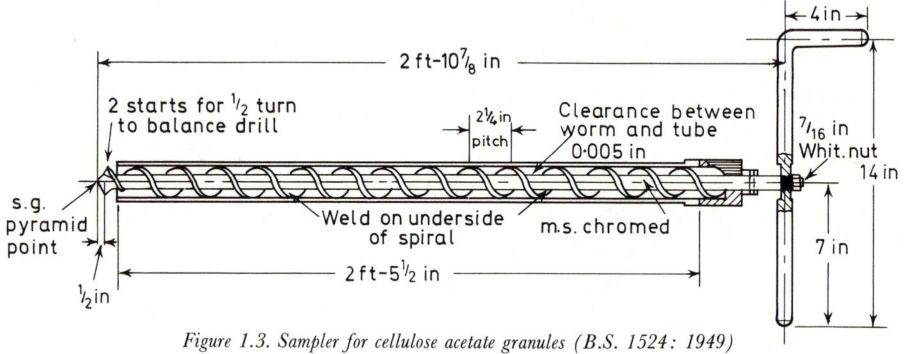

Figure 1.3. Sampler for cellulose acetate granules (B.S. 1524: 1949)

flat surface to form a conical pile, which is then flattened from the centre outwards and divided into quarters. A pair of opposite quarters are discarded and the remainder formed into a new conical pile, by taking shovels from each quarter; care must be taken in emptying the shovel that the materials run uniformly down the surface of the cone. The quartering process is repeated as many times as are necessary to yield a sample of the requisite size.

American practice[11] favours the division of such materials into arbitrary smaller units which are numbered, for example the contents of a drum into horizontal strata, followed by selection of the units to be used in testing with the aid of a table of random numbers (a set of tables designed to pick numbers out randomly without the risk of any unconscious bias).

Sheets, film, rod, tube and pipe should also be sampled randomly by division into a number of small areas or lengths as appropriate; again a table of random numbers may be used with advantage to avoid any bias.

If, for any reason, random treatment is not feasible, and some specified system has to be adopted, care should be taken to avoid coinciding the frequency of sampling with some regular occurrence, e.g. at identical times of each day (when some external influence may come into play, such as perhaps a tendency to low temperature on a compression moulding press at the start of a shift) or at intervals on calendered sheet corresponding to the same area of one of the bowls (which might be less polished than the rest). In these cases, test results obtained from samples taken at such corresponding intervals would not be representative of the whole, i.e. they would show bias because the unsampled fractions may have different properties due to the variation in press temperature or calender bowl finish.

The overall frequency of sampling should be governed by statistical principles, which are discussed later, and by commercial considerations.

Many standard test methods include sampling techniques appropriate to the particular product in question.

1.4.3 Preparation of Test Specimens

Reference has already been made to the influence on test results of fabrication conditions; when specified, these must be followed rigorously and quoted with the property data obtained. There are further complications, however, particularly in any test where the specimens are examined in other than the complete moulded form. If they have to be cut, sawn, drilled, routed or etc. from sheet or large mouldings, care must be taken to avoid excess heating; this could change the degree of cure of a phenol formaldehyde moulding or laminated sheet, for instance, and might advance it sufficiently to give specimens of different physical properties. Another possible effect is the driving off of absorbed moisture which again could materially affect certain characteristics, particularly electrical insulating properties. Severe overheating will tend to degrade the plastic polymer with detrimental effects this time, but again yielding unrepresentative test data.

Many properties are influenced by the condition of the surface of the specimen. Notches and scores act as 'stress raisers' and weaken mechanical test specimens, sometimes seriously. Most plastics are hydrophobic or water-resistant and it is not surprising that the finishing of specimens for water absorption tests must be very precisely specified. The importance of polishing the surface when assessing the light transmission characteristics of a plastics material will be obvious, surface fissures will upset resistance to high voltage ('electric strength') and so on.

By no means all plastics articles have identical properties along their principal axes, in other words there are some which are anisotropic. It is easy to envisage this in a fabric reinforced laminate because woven fabrics are usually themselves rather stronger in the warp direction than the weft, unless a special weave is incorporated. However, more homogeneous materials, with respect to composition, can also be anisotropic. An injection-moulded bar, with the gate at one end, will probably show pronounced orientation of the polymer molecules in the length direction, unless special precautions are taken, and the tensile strength, for instance, will be higher along the bar than across it. It is important therefore, to state the direction of testing when relevant or, preferably, to examine properties in two, or some cases three, orthogonal directions.

The subject of the influence of preconditioning and test atmospheres has been briefly mentioned and is dealt with more fully in Chapter 3. Quite apart from these ambient influences, however, the very age of a test specimen may be important, albeit that it may have been stored carefully at normal ambient temperatures. This is not to allude to the obvious hazards of degradation of light-unstable materials by sunlight nor to take the hypothetical case of one which is so unstable as to be attacked by the oxygen of the air, but to consider rather more subtle cases. Some materials crystallise very slowly at room temperature and, since polymer properties are often significantly different in the crystalline and amorphous forms,* it is necessary to test at equilibrium or when the rate of crystallisation is so slow as to be without significance. Similarly, the process of plasticiser/polymer gelation may proceed slowly, even at room temperature, after the normal processing cycle of mixing and pressing, so that plasticised PVC, for instance, must be examined only after some specified minimum time.

1.4.4 Size and Shape of Specimens

It might be thought that, since most properties are reduced to units of length, area or volume, to yield the basic data of the material, the precise size of test specimens cannot be of importance—it should all come out in the calculation. In actual fact, official test methods generally specify dimensions very closely and one reason for this, in the case of moulding materials, is not hard to see. A given material moulded in one case in thin section and in the other in very thick section, may have quite different properties. In a thermosetting material, the degree of cure is likely to be less in the thicker section whilst, with both thermosets and thermoplastics, the extent of locked-in strain will be greater in the thicker specimen because of the slower cooling of the centre with respect to the skin. Clearly the measured properties may vary between thin and thick mouldings due to effects other than the ratio of the cross-sectional areas.

There are a host of other reasons, however, and whilst it is not appropriate to enumerate them at this early stage, a few illustrations will serve as a warning. Reverting to the example of water absorption, which we have seen is mainly a surface phenomenon, it might be thought that logically the property would be calculated per unit surface area. In fact very often it is stated as a per cent w/w! Thus rigorous control of test piece size is necessary, so that the surface area is standard and the fictitious percentages may be compared. The mechanical property of bending (flexural) strength is calculated from the force required to break a bar at its mid-point, the breadth and thickness of the bar and the distance between the outer supports (see Figure 1.4).

$$\text{Bending strength} = \frac{3Fs}{2bt^2} \tag{1.1}$$

Why, then, worry about the precise values of s, b and t as long as they are measured accurately? In fact the classical bending formula above only

*For instance in density, rigidity, softening point, strength, gas permeability and resistance to solvents.

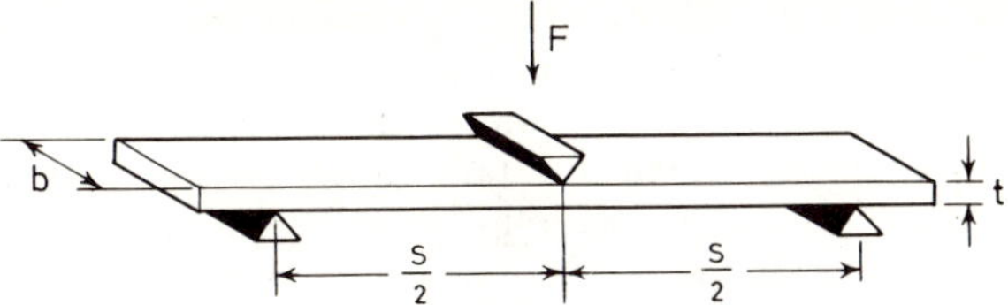

Figure 1.4. Bending strength measurement—general arrangement

holds in the case of ideal 3-point bending for certain ratios of s/t. Thus, whatever the absolute merits of the property, for comparison of data the test specimens must be essentially identical.

Electric strength is calculated from the failure voltage (breakdown voltage) of a flat specimen and its thickness, often in V/mil where 1 mil = 0·001 in. The thickness must be carefully controlled, however, as this

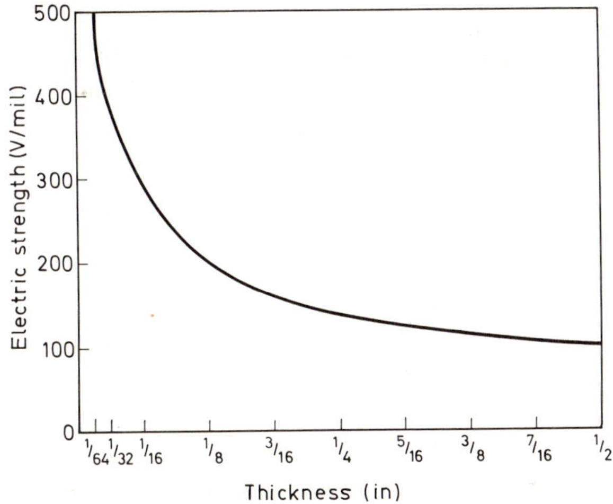

Figure 1.5. Electric strength (proof voltage) of synthetic resin bonded paper sheets (N.P.L. Notes on Applied Science No. 3)*

calculated value is by no means itself independent of thickness (see Figure 1.5).

These are just typical examples of the influence of test specimen shape on measured properties; there are many more (and many more outside influences which affect the properties considered above), but they will serve to illustrate the extreme care that is necessary in checking all instructions concerning the execution of any given test.

1.4.5 Thickness Measurement

Most mechanical properties and many others, thermal, electrical, optical, etc., depend on thickness and the test values obtained must be reduced to

*Published by Her Majesty's Stationery Office. (Out of Print).

unit thickness (often with other similar reductions in length and breadth) to give the inherent characteristics of the material. Owing to the requirements of accuracy, and in view of the values of thickness most often encountered, simple measurement by tape, ruler or dividers is generally unsuitable. Optical means, as with a travelling microscope, are hardly conducive to rapid working and recourse is often made to micrometers and dial gauges; the requisite qualities of such instruments are adequately specified, for instance References 12 and 13. It is not any basic deficiency in the equipment which creates any difficulty, however, but the very nature of many *plastics* materials, i.e. their tendency to yield under stress. Ratchet micrometers are normally screwed down to a certain pressure before the ratchet slips; dial gauges operate under a certain spring force on to their anvil area.

Ratchet micrometers are thus only suitable for measuring very rigid materials such as phenol-formaldehyde mouldings; their indefinite pressures will lead to variable results with, for instance, highly plasticised PVC or low density expanded polystyrene.

Dial gauges can more readily be specified, for easy checking, in terms of spring force and anvil area and, by this means, reproducible results within and between testing laboratories are feasible, although in the case of extremely soft materials, the thickness value recorded, with a finite force bearing on a finite area, may have no absolute significance. Many test methods for plastics materials include a technique for thickness measurement based on this principle.

A more fundamental value, not so easy to determine, is *gravimetric thickness*. The area A of a sheet is determined accurately by mensuration, its mass m taken and the density D measured by, say, a displacement method (see Chapter 5). The thickness t is given by:

$$t = \frac{m}{DA} \qquad (1.2)$$

provided the units are consistent, e.g. if A is in cm^2, m in g, and D in g/cm^3 then t will be in cm. For products with uneven surfaces, e.g. embossed sheeting, this is a convenient method for obtaining a meaningful average value of t. With more uniform sheeting, the value obtained is clearly not biased by external influences such as gauge pressure, though it is influenced by any heterogeneities within such as voids and inclusions.

There are other, refined, techniques for determining thickness such as β-ray gauges, but these are not normally employed for testing properties; their principal use is in monitoring production.

1.5 THE CONCEPT OF STATISTICS

'A judicious man looks at statistics, not to get knowledge, but to save himself from having ignorance foisted on him'.

Carlyle.

No responsible person even remotely connected with science or technology should be without a working knowledge of statistical principles; anyone ignorant of the basic ideas of this subject runs the risk of undertaking his work inefficiently and of being unable to draw the correct conclusion from the results he obtains. An example in plastics testing will illustrate the second point quite simply.

Suppose that two laminated sheet materials, *A* and *B*, of equal thickness, are to be compared for impact strength (see Chapter 8 Section 8.10). The edgewise Izod values are determined in the usual way, in quintuplicate, with the following results:

	A	*B*
	1·65 J	1·09 J
	1·37 J	1·48 J
	1·41 J	1·38 J
	1·26 J	1·26 J
	1·60 J	1·39 J
Mean:	1·46 J	1·32 J

On the face of it, by looking at the two mean values *A* may be selected as the material with the better impact strength, but a little pause for reflection will indicate that the picture may not be so clear cut. Impact strength tests on laminated sheet, by this method, invariably show a high scatter of results, that is the spread of individual values, and this is certainly true of the figures above. It is instructive to compare the ranges of each set:

A: 1·26–1·65 J
B: 1·09–1·48 J

There is seen to be a region, of 1·26–1·48 J, within which a test figure taken at random might be placed in either set of figures; three out of five of the individual values for *A* fall in this region and four out of five of Material *B*! The essential difference, therefore, is two high figures, of 1·60 and 1·65 for *A* and one low figure, of 1·09, for *B*. Are these the all important values, pointing out the relative merits of the two materials, or are they the 'odd men out', and not within the common region due to some chance happening; say test pieces cut from a fabric reinforced laminate in such a way that they contain one more or one less thread than the normal average? We can tell from these figures whether there is likely to be a real or significant difference between the two sets of data only by *statistically analysing* them to find out whether there is a sound reason for thinking that all ten figures have been drawn from one set of data (i.e. there is no difference between the materials), rather than from two different sets (i.e. *A* is more impact resistant than *B*). The greater part of the study of statistics is to learn to assess the magnitude of chance error in relation to the particular effect under examination.

In any group of data, if the *spread* of values is *normal* and the measurements are sufficient, a histogram plot of frequency (of measured value) against the value itself will approximate to the diagram, known as a *Gaussian* curve (see Figure 1.6).

This represents the distribution of data about the mean ($\bar{x}$) when variations therefrom are due only to chance, with no bias or skewness. The shape and position can be defined by $\bar{x}$ and a measure of the spread, usually the standard deviation σ, defined by

$$\sigma = \left\{\frac{\Sigma(x-\bar{x})^2}{n}\right\}^{\frac{1}{2}}$$

where n is the number of observations. Thus, in Figure 1.6, approximately 68% of all values lie within the range $\pm\sigma$ of the mean, 95% within $\pm 2\sigma$ of the mean and 99·7% within $\pm 3\sigma$. Clearly one can go on ad infinitum, widening the range to increase the percentage of measurements embraced by the normal curve, but for most practical purposes the *one-in-forty* probability limits ($2\frac{1}{2}$% below and $2\frac{1}{2}$% above, i.e. ±95% included) are most often used as a guide to checking statistical significance. Various standard tables have been computed, relating such probabilities to the relevant circumstances of a number of measurements, spreads of values, etc.

It would be undesirable, and indeed impossible, to try to give even a summary of simple statistical principles of use to the plastics tester as a small section of this handbook. Fortunately there are many readable accounts,

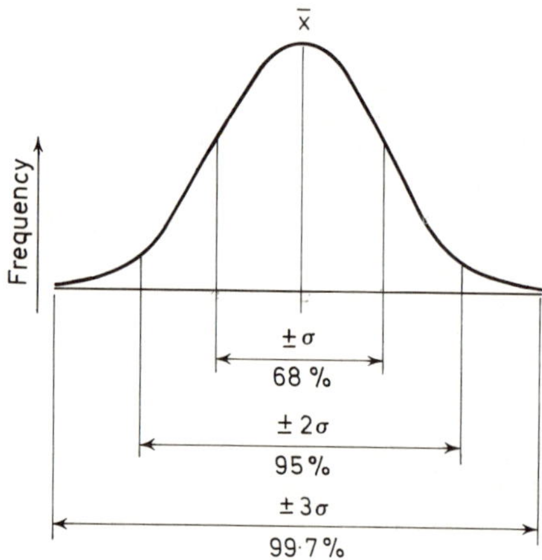

Figure 1.6. Normal frequency distribution curve

ranging from books for the layman [14, 15], to the more detailed accounts which still make no serious mathematical demands[16–19]. There are also standards[20, 21]. It will be sufficient here to indicate some of the problems which can be examined by simple statistical techniques:

1. The calculation of a meaningful figure for the spread of a set of individual values such as those for impact strength for materials A or B above—*the standard error* (related to standard deviation).
2. Comparison of two sets of data, such as A and B above, to see if one shows a significantly higher scatter than the other—in this case to ascertain if one is a more variable product than the other.
3. To check that a finite number of values can be assumed to be in line with a mean level of quality established over a virtually infinite number of tests in the past.
4. To establish whether one set of values differs significantly from another, or others—such as the data on A and B above.
5. To establish if there is any significant relationship between two sets of data which might be thought to be related and to draw the most probable straight line or curve through the various points on a graph of the plotted data—or to create the most probable relationship equating the two variables.

For example, it might be desirable to check the modulus (stiffness) of a plastics material by the more simply and quickly measured property of (Vickers diamond) hardness if, as we suspect, there is a relationship between the two properties. Accordingly, we would carry out both tests on a number of samples and use the techniques of *regression* and *correlation* to ascertain if there is any significant relationship and, if so, what is its nature and how reliable it is.

6. The design of investigations to yield the most reliable data with the smallest number of experiments.
7. The organisation of sampling procedures.
8. The establishment of satisfactory quality control.

1.6 QUALITY CONTROL

Any manufacturer with a modicum of conscience, or just plain sound business sense, will take precautions to ensure that the quality of his products is consistently meeting a certain standard without, preferably, too much variation; in other words he will institute some form of *quality control*[22, 23]. Quite apart from keeping his customers happy and meeting any contractual obligations, the use of regular quality control will confirm or otherwise that the manufacturing process is operating satisfactorily; a sudden change in a measured property, or a slow deterioration, may be symptomatic of something radically wrong along the production line. Well organised quality control will help to identify this at the earliest possible moment and, quite apart from keeping inferior goods away from the customer, will assist in the early rectification of the fault.

The frequency and comprehensiveness of quality testing must be governed, to a large extent, by commercial considerations. In the ultimate, the only way to be absolutely certain that nothing below standard gets past inspection is to examine 100% production. This may be feasible if a simple visual inspection is adequate, but, at the other end of the scale, if the only usable test were destructive in nature, the cost could be very high indeed! Statistical principles, of the type referred to in the previous section, point the way to achieving the best check of quality for a given rate of expenditure and indicate what risk is being taken of allowing inferior quality products through.

A particularly useful method of recording production property data is by the use of *quality control charts*[24, 25]. For example, with any one property regularly checked, a graph is constructed, with the property as ordinate and time of sampling as abscissa. With this visual aid, almost immediately it will become possible to identify any trends or gross variations. The greatest value, however, is derived when the scheme has been operating long enough for reliable estimates to be made of the grand mean (i.e. the average figure derived from all measurements to date) and the standard deviation of results, i.e. a statistical measure of the overall variation between samples. From the latter and published tables, the limits can be calculated between which 95%, 99%, 99·9%, etc., of all data should lie providing everything is under control. A horizontal line is drawn on the quality control chart at the value of the grand mean and pairs of lines are drawn, one either side of the mean line, to indicate the 95%, etc., limits; the 95% lines for example are called the one-in-forty limits because, on average, not more than one in forty measurements should stray above or below the limits. By carrying on the plotting with newly acquired day-to-day data it will be obvious how the charts can be used to identify a real production failing from the chance low result that is within normal statistical variability.

A number of variants may be operated along these general lines; for example, the range of results for each set of readings, such as the values pertaining to a particular shift, may be plotted as ordinates and limits calculated to indicate whether the variability within, say, each shift is under control or a particular shift or other fixed period is exhibiting excessive variation.

The selection of the actual tests to employ for quality control must be a compromise between scientific interest and economic necessity. If the control can be exercised adequately with a non-destructive test or tests, only the actual cost of the latter is relevant for no production is actually destroyed. This, unfortunately, is rarely the case. It is, however, often possible to reduce the number of different types of tests carried out regularly to a reasonable minimum, for it will often be found that a number of them give data which correlate (see previous section, point 5).

1.7 PRESENTATION OF DATA

In addition to the laying out of quality control charts, there is quite an art in the general manner of recording and reporting test data. It should be obvious by now that sufficient details of test conditions and other variables must be given. Again, an adequate case has been made for indicating the reliability of the data either by quoting the individual test values and the mean, or the latter with the calculated standard error or at least the range of results, as appropriate. The units should be stated, and correctly!

The format of any report should be as brief as possible, commensurate with the recording of all the relevant facts. Tabulation is desirable when comparisons are the main object of the work. Opinions differ on the ideal layout of a technical report, but one recommended order of sections is as follows:

Summary
(Contents)
1. Introduction
2. Materials/Products Examined
3. Test Method Details
4. Results
5. Conclusions
6. Recommendations
(Appendices)

The summary is for the busy managing director who has not the necessary time to digest all the report; it is not identical with the conclusions because the summary is a précis of the report as a whole. It should not, therefore, contain any data or opinions which are not in the body of the report. The introduction essentially gives the reasons why the work was done and the various happenings which led up to it. It is always important to record precisely what was examined, not just in terms of a type description, but also batch numbers and other relevant data. Sections 3 and 4 will be self evident and so should section 5, a statement of the conclusions to be drawn from the work in the light of the data of the report and any other information available from experience. A section giving recommendations, for example for further investigations, is often desirable.

If the work being reported contains, for instance, a mass of numerical data, it is advisable to collect this together into an appropriate number of appendices suitably cross-referenced in the text, so that the main body of the report keeps within reasonable proportions; section 4 can be restricted then to summary tables of mean values, for instance. Graphical representation of results should always be considered as a concise way of reporting data and one which will often assist the drawing of the correct conclusions. Graphs are usually best placed with the appendices.

In certain cases, a section of bibliography or literature cited should be included, between section 6 and any appendices. With large reports, a contents list immediately after the Summary is very useful.

The actual mode of expression of numerical results is a subject which deserves some study. Whilst it is not desirable to dispense with individual style, the needs of clarity, elegance and uniformity should take precedence. Thus, a quarter of a metre should be written 0·25 m and not ·25 m; 2 500 000 $\mathrm{lbf/in^2}$ is inferior to $2{\cdot}5 \times 10^6$ $\mathrm{lbf/in^2}$. For guidance on this and the rounding of quoted values, reference should be made to Refs. 26 and 27 and for guidance on the use of symbols, signs and abbreviations to Ref. 28. The B.S. glossary of plastics[29] terms is somewhat out-of-date (though it is currently under revision)[30] and there is also a recommended list of common names and abbreviations[31].

1.8 LIMITATIONS OF TEST DATA

When we have taken all these precautions, heeded all these warnings—not mixed our units, have prepared our test specimens correctly, used the correct technique and reported all these facts—what is the value of the data we have obtained? Essentially the figures for strength, resistance, etc., derived from our measurements only relate to *conditions which simulate precisely those under which we performed our tests*. Much has been said already, and much more will be stated throughout this book, on just how often property values vary according to the manner in which they were determined; even more regrettably, there is very rarely a user condition which in fact comes anywhere near to copying normal laboratory test practice. A little thought will indicate that this is not necessarily a damning criticism of the practice for there are factors of economics and time in particular to take into account.

It follows, therefore, that the test data obtained from the routine type of test, whilst admirable for quality control and perhaps as an indication of service performance if interpreted carefully, will rarely give the designer the values upon which to base his calculations. The largest gap between most published property data and performance behaviour lies in the time scale, just how long will a given component withstand a certain stress without failing; trouble-free service over a period of years may be essential, yet tests have lasted but a few seconds or minutes. (One of the greatest drawbacks in the young plastics industry is this lack of case history, though in some traditional industries the necessity of designing more competitively has raised some doubts as to the validity of many time honoured canons!) Accelerated tests will often provide very useful guides and frequently are the only solution to such demands. However, it must always be remembered that, to produce this very acceleration, some test variable or variables have had to be intensified, for example the temperature raised, the nature of the environment changed or the frequency of stressing increased. These necessary changes may in themselves induce effects which would *never* occur at the usual ambient temperature, etc., and thus misleading data may result.

It cannot be emphasised too forcefully that all measured properties should be most critically assessed to establish their true significance and applicability.

REFERENCES

1. NIELSEN, L. E., *Mechanical Properties of Polymers,* 108 et seq., Chapman & Hall Ltd., London (1962)
2. GORDON, J. E., *The New Science of Strong Materials,* Penguin Books Ltd., Harmondsworth, Middx (1968)
3. NIELSEN, L. E., *Mechanical Properties of Polymers,* 111, Chapman & Hall Ltd., London (1962)
4. B.S. 3763: 1964, *The International System (SI) Units* and P.D. 5686, *The Use of SI Units,* British Standards Institution (1969)
5. ANDERTON, P., and BIGG, P. H., *Changing to the Metric System. Conversion Factors, Symbols and Definitions,* Her Majesty's Stationery Office, London (1969)
6. DRYDEN, I. G. C., *SI and Related Units: Quick-Reference Conversion Factors,* British Coal Utilisation Research Association, Leatherhead (1968)
7. B.S. 350: Part 1: 1959, *Conversion Factors and Tables, Part 1, Basis of Tables. Conversion Factors,* British Standards Institution, London
8. B.S. 350: Part 2: 1962, *Conversion Factors and Tables, Part 2, Detailed Conversion Tables,* British Standards Institution, London
9. P.D. 6203, *Additional Tables for SI Conversions,* British Standards Institution, London (1967)
10. B.S. 1524: 1949, *Cellulose Acetate Moulding Materials,* Appendix K, 17 (Now replaced by B.S. 1524: 1955) British Standards Institution, London
11. A.S.T.M. Designation D.1898–68, *Recommended Practice for Sampling of Plastics,* American Society for Testing and Materials (Philadelphia)
12. B.S. 870: 1950, *External Micrometers,* British Standards Institution, London
13. B.S. 907: 1965, *Dial Gauges for Linear Measurement,* British Standards Institution, London
14. MORONEY, M. J., *Facts from Figures,* Penguin Books Ltd., Harmondsworth, Middx (1951)
15. GOODMAN, R., *Teach Yourself Statistics,* E.U.P., London (1957)
16. DAVIES, O. L. (ed.) *Statistical Methods in Research and Production, 3rd edn,* Oliver and Boyd, London (1957)
17. *Symposium on Application of Statistics, Special Technical Publication No. 103,* American Society for Testing and Materials, Philadelphia (1950)
18. HOEL, P. G., *Introduction to Mathematical Statistics, 2nd edn,* John Wiley and Sons Inc., London (1954)
19. MANDEL, J. and LASHOF, T. W., 'Processing of Numerical Test Data', *Testing of Polymers,* Vol. 2 (Edited by J. V. Schmitz), Interscience Publishers, New York (1966)
20. B.S. 600: 1935, *The Application of Statistical Methods to Industrial Standardisation and Quality Control,* British Standards Institute, London
21. A.S.T.M. Designation E.177–68T, *Tentative Recommended Practice for Use of the Terms Precision and Accuracy as Applied to Measurement of a Property of a Material,* American Society for Testing and Materials, Philadelphia
22. *Statistical Quality Control, Special Technical Publication No. 66,* American Society for Testing and Materials, Philadelphia (1946)

23. GRANT, E. L., *Statistical Quality Control,* 2nd edn, McGraw-Hill Book Company Inc., New York (1952)
24. B.S. 600 R: 1942, *Quality Control Charts,* British Standards Institution (Now withdrawn from sale), London
25. *Manual on Quality Control of Materials, Special Technical Publication No. 15-C,* American Society for Testing and Materials, Philadelphia (1951)
26. B.S. 1957: 1953, *The Presentation of Numerical Values,* British Standards Institution, London
27. A.S.T.M. Designation E.29–67, *Recommended Practice for Indicating which Places of Figures are to be Considered Significant in Specified Limiting Values,* American Society for Testing and Materials, Philadelphia
28. B.S. 1991: Part 1: 1967, *Letter Symbols, Signs and Abbreviations, Part 1, General,* British Standards Institution, London
29. B.S. 1755: 1951, *Glossary of Terms used in the Plastics Industry,* British Standards Institution, London
30. B.S. 1755: Part 1: 1967, *Glossary of Terms used in the Plastics Industry. Polymerisation and Plastics Materials,* British Standards Institution, London
31. B.S. 3502: 1967, *List of Common Names and Abbreviations for Plastics and Rubbers,* British Standards Institution, London

2

Standards and Specifications

2.1 INTRODUCTION

In Chapter 1, Section 1.3, the need for official standardisation of test methods .was explained, that is the reason for the setting up of official *standards* and *specifications*. However, before going further we should settle what is meant by these two terms. It is probably true to say that there is no universal agreement, but the B.S.I. (see below) of the United Kingdom now terms all its documents *British Standards,* but only those which specify minimum requirements of properties, or the composition, of materials and products are termed 'Specifications'. Thus 'Glossaries', 'Recommended Practices' and 'Methods of Test' are readily identified.

Many countries have a national organisation for the preparation of their internal standards; a list of names and addresses of the majority appears in Appendix 1 to this Chapter. Most of these countries are members, through their standardising body, of the International Organisation for Standardisation (I.S.O.) the general concept of which will be self-evident. However, despite the prime importance of this international body, the national scene will be discussed first because of the natural order of events in this field. Briefly, this is that each country's standards picture was more or less well advanced before the I.S.O. was created, even in the relatively young plastics industry, and not surprisingly the progress in international agreement is very slow, hampered by preconceived ideas and sometimes national prejudices.

It is again impractical to describe the work of all the national standardising bodies; therefore discussion will be restricted to the United Kingdom, the United States of America and more briefly, Germany, probably the three most advanced bodies in the field of plastics standards.

2.2 UNITED KINGDOM STANDARDS

There are many bodies producing standards in the United Kingdom, with sponsorship from industrial associations, armed forces, and nationalised industries in particular, but nowadays these groups generally speaking only resort to such activity if their needs are not satisfied by the existing coverage of the British Standards Institution. The B.S.I. is the approved body for the preparation and promulgation of national standards (in the U.K.).

2.2.1 British Standards Institution

THE STRUCTURE

The B.S.I. started in 1901 with a predominantly engineering interest, but it now covers every branch of industry. It is financially supported by the Government, but draws a significant fraction of its income from membership fees and the sale of its publications. The members are principally manufacturing companies, user firms, and other bodies and individuals interested in standardisation. Membership is voluntary, though it is difficult to keep fully abreast of the current position without being a member; a limited number of standards are obligatory, i.e. manufacturers of goods described in these standards *must* only market goods meeting the requirements laid down therein.

The B.S.I. has a permanent staff comprising essentially a Director General, administrative staff at various levels, secretarial staff and editorial staff. A General Council, consisting of senior personnel from industry and Government departments, decides general policy. Various divisional councils deal with policy matters relating to various broad sections of industry; a typical example is the Chemical Division Council. These councils, and the various subsidiary committees, are comprised of appointed members from industry, who give their services voluntarily, with the B.S.I. providing the secretariat. Under these divisional councils, industry standards committees operate and these are the policy making bodies within their defined fields. Typical are the Plastics Industry Committee and the Rubber Industry Committee, respectively known as PLC/- and RUC/- (PL = Plastics, RU = Rubber and C = Chemical (Division)). These industry committees appoint technical committees to deal with defined fields where it is considered that there is a need for standardisation. Again the secretariat is supplied by the B.S.I. with the members appointed from industry, the armed forces, the Government ministries, the universities and research associations and stations (the last two financed by industry, industry/Government or Government alone). As far as the industry representation is concerned, it is not usual to appoint members from specific companies, but rather to take representatives from the appropriate manufacturing and user trade associations* (which representatives derive from the said companies of course) though experts are sometimes co-opted. A Chairman is elected from within the members, preferably one independent of manufacturing or user interests. A technical committee of particular interest in the present context is known as PLC/17 and deals with methods of testing plastics, set up rather belatedly to rationalise the many test methods for plastics that had been developed (or just arisen) and were causing just the type of confusion referred to in Section 1.2, and to standardise new methods for properties not yet covered. There are many other PLC/committees, concerned principally with specifying materials

*The British Plastics Federation, for example.

and products (ideally using the standardised test methods where these exist); these other committees are not of such immediate interest in this book. An important exception is PLC/36 dealing with recommendations for the presentation of design data for plastics; the first document from this committee has just been published (1970).

For general information, the current edition of the B.S.I. annual report should be consulted.

HOW A B.S. IS PREPARED

The technical committees prepare draft standards, usually based on work already done by one or more of the bodies represented, for circulation for comment. Any work required to prove a test or establish a standard is carried out voluntarily in the laboratories or works of the members represented. The B.S.I. does not perform this function usually and the Institution laboratories (sited at Hemel Hempstead, Hertfordshire) exist essentially to carry out approval testing to *existing* standards, of certain ranges of manufactured goods, especially those associated with human safety, for example car safety belts.

The draft standards, which have no standing whatsoever at this stage, are circulated for comment to known experts in the field and to any member of the B.S.I. who cares to apply for a copy (free); announcement of availability is contained in the journal *B.S.I. News* issued monthly to all members. All comments so collected are considered by the technical committee and a final draft is prepared for editing by the B.S.I. staff; this is not a technical function, but ensures consistency of layout and expression. This draft is submitted for approval to the industry committee after which it is printed and published, its availability again being announced in *B.S.I. News* and often in the technical press. The final standards are *not* issued free!

The average time taken from the inception of the committee work to publication is about two years, this long period of time being due mainly to the facts that the work of the members is voluntary, unanimity is sought, and the B.S.I. are currently short of staff.

THE USE OF A B.S.

Generally speaking, British Standards are documents for *voluntary* use, though in a few cases legislation exists to make their use mandatory, as for example with the manufacture of motor-cyclists' crash helmets. In these cases, such goods *must not be sold* unless they are known and stated to be up to the quality demanded by the appropriate B.S.

As it is not possible to quickly assimilate all the British Standards relating to plastics from any B.S.I. publication, e.g. the British Standards Yearbook, a list of those currently* available is given in Appendix 2, but in view of the constantly changing position it should be borne in mind that this list will perforce omit the most recently issued documents and perhaps the latest

*See Preface.

versions of those that are mentioned (see below). The B.S.I. does not classify its standards in any way, and an attempt has been made in the appendix to place them into various groups according to the subject matter; the choice is sometimes arbitrary and entirely the authors'. In addition there are many 'fringe' standards, which do or may relate to the plastics industry, that have been omitted for one reason or another. Thus, for instance, no mention has been made of those standards which relate to liquids (and some solids) of possible utility as plasticisers, as it is considered that these are just outside the scope of this book.

When proposing to use a B.S. the first thing to do is to check that (a) it is the relevant or the most relevant available, (b) it is the current version and (c) it has a full complement of the issued appendices; how to do this is explained below. Next examine the 'scope' clause (and the 'introduction' preferably) to reinforce your selection of the standard selected before noting if any other B.S. documents are referred to, which should be available in order that the original may be fully understood. Components parts, test equipment or analytical chemicals, for example, may be specified by reference to other relevant documents, and there is a general policy of grouping test methods for a given type of material into one standard, with the materials specifications simply specifying quality and cross-referencing to the standard for the test methods. Thus the committee PLC/17 referred to above has produced B.S.2782, *Methods of Testing Plastics,* which is a standard devoted solely to describing routine methods of test for plastics; the more recent plastics quality specifications 'call up' B.S.2782 for most of the methods to be used in evaluating the material or product in question—the exceptions being those techniques which are considered to be specific to the product in question. B.S. 2782 is currently issued as a single bound volume; it is added to or amended (see below) not more than twice a year and revised about once every three or four years. At the time of writing, the latest version is B.S. 2782: 1970, which is almost entirely composed of test methods originally drafted in Imperial units and which have been simply converted precisely into S.I. units (with metric technical and Imperial unit equivalents). At a later stage, B.S. 2782 will be revised into solely S.I. units.

After gathering together all the necessary documents, there is no adequate substitute, when using British Standards, for a complete and careful reading of them. Regrettably they are not all identical in their layout, though uniformity is slowly being achieved. The usual and preferred format, following the clauses mentioned above, is a series of requirements clauses containing the minimum (or maximum) test values with reference to another B.S., or an appendix in the same standard, where the requisite test method is to be found. Sometimes this information may be in tabular form and occasionally the tests may be split into those that should be frequently carried out and those that need only occasional attention (perhaps a difficult or costly test or one for a property that is unlikely to be unsatisfactory if all the frequently checked ones are up to scratch). Some standards give recommendations for quality control schemes and statistical limits relating to standard deviations and corresponding quality levels according to the number of tests carried out or the numbers of batches examined. Exceptionally, appendices may be included of an advisory nature, e.g. on how to organise a quality control scheme.

A particularly important point in connection with using a standard is to ensure that there are no general clauses which refer to each property clause and that may not be mentioned specifically in each of the latter. Such clauses may cover 'sampling' and 'conditioning of test pieces' in particular, and are often sited immediately before the property requirement clause.

A manufacturer who makes a product that meets the appropriate British Standard is likely to want to capitalise on this fact and he may do this by labelling his product 'Conforming to B.S. XXXX'. In the eyes of the law, this means that every single batch, packet, bottle, model or whatever, is up to this quality; it is perhaps more than a little unfair that a manufacturer whose general quality of produce may be well above the minimum set by the B.S. he is quoting, could be prosecuted because one batch tested happened to be below par due to some statistical chance (he would probably be only too willing to replace the offending goods, but this is by the way).

A far more satisfactory arrangement all round is that represented by the British Standards Mark, or 'Kite Mark' from its design (see Figure 2.1). This is a registered certification trade mark owned by the B.S.I. and it may be used in connection with certain goods covered by British Standards (a list of these standards, including several for plastics materials and goods, is

Figure 2.1. British Standards Mark

to be found in the B.S. Yearbook—see below). A licence is issued to the successful applicant to use this mark on his products, providing he submits to a routine of inspection, sampling and testing prescribed by the B.S.I. and related to the particular product and its mode of manufacture. Thus the consumer is guaranteed that his purchase is being made under controlled conditions, checked independently and regularly, and complies to a certain minimum quality, whilst the manufacturer is protected against the vagaries of statistical chance, as the licence schemes are based on statistical principles.

HOW A B.S. IS KEPT UP TO DATE

After a B.S. is issued, the technical committee remains in being—indeed it may have other standards to prepare. However, as far as the B.S. in question is concerned, the duties of the committee extend to keeping a constant watching brief on changes in the scientific or commercial world which may necessitate alterations to the standard in the form of additional tests, additional properties to be specified or modification of qualities already specified. Relatively minor changes are dealt with by amendment slips (carrying a PD reference up to the middle of 1968 and an AMD reference thereafter, and usually issued free) which also are issued to cover any errors in original drafting or printing (amendment slips are announced in *B.S.I. News*). Major changes are covered by a complete revision, which goes through essentially the same procedure as a new standard. Each B.S. is given a separate number to which is added the year of publication; new issues generally carry the same numbers but the new date. Thus B.S. 771 for phenolformaldehyde moulding materials was first published in 1938 and has been revised in 1948, 1954 and 1959. All standards are considered for revision after a period not exceeding ten years.

A complete list of British Standards in numerical order is contained in the British Standards Yearbook which gives a brief precis of each. The index is alphabetical only and thus there is no comprehensive list of plastic standards for instance. The yearbook is issued to members and is available for purchase by the public at large; any standard ordered by reference to the yearbook should be the current version with all the issued amendment slips (given in the yearbook) attached therein. In view of the somewhat unsatisfactory way in which the amendments are prepared and printed, it is as well to immediately edit the main text with reference to the amendments before use. From time to time standards are reissued with amendments actually printed into the text.

Necessarily the yearbook may be up to 18 months out of date, and the already mentioned *B.S.I. News* (only issued to members) gives monthly information on new standards published and new amendment slips issued. The B.S.I. does *not* automatically send the slips to all holders of the appropriate B.S.—they have to be obtained by the standards user—and thus it is almost imperative that anyone needing to be up-to-date with his B.S. documents should be a B.S.I. member. Simple reference to a B.S. by number, with no qualifying date, implies the latest version with all issued amendment slips therein. Again, when the amendments are received, the editing exercise should be carried out.

OTHER SERVICES OF THE B.S.I.

The B.S.I. library at the headquarters in London is available for consultation by the public at large and many foreign standards and related documents may be borrowed, or purchased through the sales office.

2.2.2 Other Official U.K. Standards Issuing Bodies

Despite the existence of the B.S.I., there are many occasions when the production of standards by other bodies is justified, though it must be admitted that much overlap still exists as a result of parochial interests, ignorance and lack of coordination. The average delay in preparing a B.S., and the evolution of the general need for it in the first place, may force organisations with specific or urgent needs to act in advance of the national body and produce their own documents. A particular end use may demand a certain property and special test quite inappropriate to a document intended for use by industry at large.

Thus, particular industries or even individual companies, with a need for new materials such as plastics and at the same time having to be very function conscious, frequently produce their own standards. A good example is the motor car industry with its needs for reliability in the face of extreme competition, but much the same is true of other large industries such as aircraft, coal, telecommunications and, it is to be hoped, building.

Likewise, the Government service departments undertake much of their own standardisation whilst at the same time contributing to the work of the B.S.I. The justification may be one of urgency, but more likely it results from the specific and/or unique interests of the armed forces. There is a considerable amount of overlap between the various departments issuing standards which can be very confusing, though currently a serious attempt is being made to rationalise procedure, particularly through the Ministry of Defence with the DEF specifications series. However, when the sources have been traced most at least have itemised lists of their publications and thus it is not difficult to identify the plastics documents. For this reason no such lists are given here, but Appendix 3 of this chapter summarises the most important bodies issuing such specifications and the addresses from which they may be obtained.

It should be mentioned that, with the current reorganisation of the Ministry of Defence and consequent changes in the individual armed services, and the creation of the Quality Assurance Directorate, the general picture of these bodies' standards is altering radically at the moment.

2.3 U.S.A. STANDARDS

The scheme of national standardisation in the U.S.A. differs in several respects from that in the U.K.[1,2]; the most widely known body in this field is certainly the American Society for Testing and Materials (A.S.T.M.) but it is not the nearest counterpart to the B.S.I. This is in fact the U.S.A. Standards Institute (U.S.A.S.I.) which is, inter alia, the U.S.A. member of the I.S.O.

2.3.1 U.S.A. Standards Institute

The U.S.A.S.I. was in fact originally formed by five engineering societies, of which the A.S.T.M. was one, and three government departments; until recently it was known as the 'American Standards Association' (A.S.A.). Its members are technical societies, trade associations and similar groups and companies. A monthly publication, *Magazine of Standards*, gives details of documents approved as American Standards, but the U.S.A.S.I. does *not* write standards—it considers whether those produced by the A.S.T.M. and other official bodies are suitable to be deemed American Standards. For this reason, greater attention will be paid to the A.S.T.M. and its activities, though the importance of U.S.A.S.I. as the premier standardising body in the U.S.A. must not be overlooked[3,4].

2.3.2 American Society for Testing and Materials

The A.S.T.M., as its name implies, is primarily concerned with tests and specifications for materials, though semi-finished products such as moulded sheet and pipe are tackled as well as some consumer goods. It has a permanent staff and its membership is drawn essentially from the same types of interests as the B.S.I. The work of standardisation is carried out by a multitude of technical committees with carefully prescribed terms of reference in order to reduce duplication of effort to the minimum. The A.S.T.M., incidentally, enters into other activities such as running a 'Division of Materials Science', organising technical meetings which are one of the premier sources of material for the special technical publications—often detailed monographs—it publishes at frequent intervals, and it also publishes the journals *Materials Research and Standards* (replacing the *A.S.T.M. Bulletin*) and *The Journal of Materials*.

Each technical committee produces recommendations for standards falling within its jurisdiction, which are submitted for approval and publication either at Annual Meetings or through action by the Administrative Committee on Standards. If approved, the specification or test method or whatever is published as a tentative (identified by a capital T after the number and abbreviated date) for at least one year. Any criticisms or comments that are forthcoming are borne in mind by the technical committee before it recommends that the document be adopted by the A.S.T.M. as a formal standard. Every standard before adoption is submitted to a ballot by letter vote of the entire Society membership and a favourable two thirds vote is necessary.

A complete list of numbers and titles of A.S.T.M. standards is contained in the annually published *Index to A.S.T.M. Standards**. This covers all the individual *Books of Standards* which deal with fairly narrow sections of industry and generally each contains only an index to its own contents (unfortunately test methods specified in one book may be described in another). Beginning in 1964, the Books of Standards have been issued, more or less simultaneously, annually—the latest (1969) set runs to 32 volumes of which the following are of principal interest to the reader:

1. Part 14. 'Thermal and Cryogenic Insulating Materials; Acoustical Materials; Fire Tests; Building Constructions'.
2. Part 16. 'Structural Sandwich Constructions; Wood; Adhesives'.
3. Part 26. 'Plastics—Specifications; Methods of Testing Pipe, Film, Reinforced and Cellular Plastics'.
4. Part 27. 'Plastics—General Methods of Testing, Nomenclature'.
5. Part 28. 'Rubber; Carbon Black; Gaskets'.
6. Part 29. 'Electrical Insulating Materials'.

Prior to 1964, the standards books were issued triennially and were brought up to date in each of the two intervening years between publication by *supplements*, one for each book. The supplements contained completely new standards and revised versions of previously issued ones; the presence of the latter, and any minor editorial alterations to standards appearing in the bound parts, was readily identified by the provision of stickers, one for each standard, that were attached to the relevant pages of the book of standards. It is considered that this regularised issue of amendments (and their mode of presentation) was a distinct improvement on the corresponding B.S.I. procedures, though conversely the degree of cross-referencing between

*In 1970 this is to be replaced by Part 33, 'Glossary of Definitions and Index to Standards'.

one A.S.T.M. standard and another gets rather out of hand on occasions.

In past years, the Society has published special compilations of standards of interest to a particular industry, but perforce these have been available only some time after the publication of the original books of standards. The special compilations thus have tended to be out of date, added to which there was no amendment slip arrangement to bring these rather infrequently published documents up-to-date. It is relevant to point out that up to 1961 the Books of Standards were available only in rather large volumes (11 in 1961 and fewer still before) each covering a much wider range of subjects than the 1969 versions. There was then a logical need for special (smaller) compilations of standards, but with the new arrangement this need probably no longer exists.

The A.S.T.M. standards are available as individual items and as a whole they cover the field of plastics materials more comprehensively than do the British Standards. In view of the availability of the above-mentioned special books of standards, each with an adequate index, no list of A.S.T.M. standards for plastics is given here. (For address see Appendix 4 of this Chapter.)

2.3.3 Other Official U.S.A. Standards Issuing Bodies

For similar reasons as those given for the U.K. in Section 2.2.2. above, there are a number of other bodies in the U.S.A. besides the A.S.T.M. which issue standards, notably commercial organisations, government departments, and the armed forces. Some of these have extremely comprehensive lists of standards available and complement the A.S.T.M. to a considerable extent on the finished plastics products side. Useful summaries of the more important standardising bodies and their functions will be found in References 5, 6 and 7.

One of the most useful sources of U.S.A. standards for plastics other than the A.S.T.M. is the Federal Standards list[8]. Other useful summaries will be found in References 9–13.

A list of the addresses of the sources of various important U.S.A. standards is given in Appendix 4.

2.4 GERMAN STANDARDS

The Deutscher Normenausschuss (D.N.A.) dates back to 1917 and, after World War II, was authorised to continue its work for the whole of Germany. It is the I.S.O. member for that country and operates from West Berlin with branch offices in Cologne and East Berlin. D.N.A. is financed by members' subscriptions and sales of documents, though for particularly large sections, for example the electrical industry, additional finance is obtained from the industry itself.

Standards are drafted by working committees at the direction of the D.N.A. Council which generally prescribes the composition of the committees and their spheres of duty. Drafts are submitted to a Standards Examining Board for checking coordination with the efforts of related working committees and style of format. The so-produced draft (Entwurf) DIN standard is published for comment by members of the Council and by the public; draft standards for plastics appear in the journals *Kunststoffe* and *Plaste und Kautschuk* for instance. These comments are duly heeded and a final manuscript submitted to the Standards Examining Board.

DIN standards are published individually and also grouped into various handbooks; one such handbook, *Taschenbuch* 21, *Kunststoffnormen* contains or refers to all the standards on plastics. A considerable number of individual standards are available as English translations of which a list of titles is published as a separate booklet. *(DIN. English Translations of German Standards.)* There is also a yearbook covering all DIN standards *(DIN Normblatt–Verzeichnis)*.

2.5 I.S.O. STANDARDS

The desirability of all nations working to common standards will be obvious but from what has been said in Chapter 1, the chances of this ideal being achieved between individual pairs of trading nations must be limited. Clearly an international forum is required.

Such a concept was met by the International Standards Association, which was set up in 1926. However, it did not achieve a great deal, not the least stumbling block being the fact that the U.S.A. was not represented, and it died a natural death in World War II.

A fresh impetus was provided by the United Nations and, in 1946, one of its groups became the International Organisation for Standardisation (I.S.O.) which has its headquarters at 1, Rue de Varembé, 1211 Geneva 20, Switzerland, where the secretariat, under a Secretary-General, coordinates the work of all the committees. There are currently about 60 member countries of I.S.O., represented normally by their national standardising bodies.

The technical work of I.S.O. is carried out by technical committees (T.C.s) relating to various industries and there are now about 125 of these. The secretariat of each technical committee is held by a member body, that for plastics (ISO/TC.61) being held by the U.S.A. (The secretariat for rubber—ISO/TC.45—is held by the U.K.).

Each member country is given the option of joining every technical committee, as a 'participating member' (P) with voting rights at committee meetings, or 'observer member' (O); there is no obligation to join. Since I.S.O. work is costly and time consuming, and many countries may have little or no interest in certain industries, the reasons for the various choices will be readily seen. At the time of writing ISO/TC.61 has 26 'P' members:

Australia	Finland	Italy
Austria	France	Japan
Belgium	Germany	Netherlands
Brazil	Hungary	Poland
Bulgaria	India	Rumania
Canada	Iran	S. Africa
Czechoslovakia	Israel	Spain

Sweden	United Kingdom	U.S.S.R.
Switzerland	U.S.A.	

and 18 'O' members:

Argentina	Korea, R.	Peru
Chile	Mexico	Portugal
Denmark	Morocco	Singapore
Greece	New Zealand	Turkey
Ireland	Norway	U.A.R.
Korea, D.	Pakistan	Yugoslavia

Representation by the actual delegates is decided by the national standards bodies, in the case of the U.K. nominated by B.S.I. Committee PLC/-/1 and approved by PLC/-, the industry standards committee.

ISO/TC.61 held its first meeting in 1951 in New York and has held its Plenary Sessions at approximately yearly intervals since, at various centres around the world. It can well be imagined that the amount of work is formidable and to expedite matters various sub-committees and/or 'working groups' are set up. There are ten working groups in ISO/TC.61:

WG1 Equivalent Terms.
WG2 Strength properties
WG3 Standard laboratory atmospheres and conditioning procedures
WG4 Thermal properties
WG5 Physical-chemical properties
WG6 Chemical and environmental resistance
WG7 Preparation of test methods
WG8 Electrical properties
WG9 Specifications
WG10 Cellular materials.

WG9 is split into 'task groups' for various polymers and materials; there are currently eleven 'TGs'.

Normally the various working groups prepare documents for consideration at the Plenary Sessions of the Technical Committee. If 50% of the 'P' members approve, then the document is circulated from Geneva to all I.S.O. members as a Draft Recommendation. If, as a result of a postal ballot, 60% of members approve, it is submitted to the I.S.O. council; if this body approves, the document is published as an 'I.S.O. Recommendation', which, it will be appreciated, is not binding on each member country to adopt as its national standard (though it is hoped that this will happen and B.S.I. Committees must these days have a good reason for not adopting I.S.O. methods). If 100% of I.S.O. members approve a document, it becomes an 'I.S.O. Standard' and is obligatory. This ideal has not yet been achieved!

Progress in I.S.O. is inevitably slow, though ISO/TC.61 is amongst the most successful of the committees, because of distance, national prejudice and language difficulties—the official languages are English, French and Russian. There are about twelve other technical committees with related interests, probably the most relevant being ISO/TC.45 for rubber.

A list of I.S.O. Recommendations and Draft Recommendations for plastics is given in Appendix 5 of this chapter, derived from the I.S.O. Catalogue. The B.S.I. publishes a supplement to its yearbook, listing I.S.O., I.E.C., C.E.E., and other international publications (see below).

The structure of I.S.O., its committees and progress to date are described in Reference 14.

2.6 OTHER INTERNATIONAL STANDARDS

Because of the importance of plastics materials in electrical insulation, mention must be made of:

1. The International Electrotechnical Commission (I.E.C.) a world wide organisation with 40 members and headquarters at 1 Rue de Varembé, 1211 Geneva 20, Switzerland, and
2. The International Commission on Rules for the Approval of Electrical Equipment (C.E.E.) with membership restricted to European countries and currently 19 members. Address:
   Central Office,
   Utrechtsweg,
   Arnhem, Holland.

2.7 WHEN AND WHEN NOT TO USE STANDARDS

Relevant specifications (standards for materials and products) should be used whenever possible in the execution of trade contracts for the reasons outlined in Chapter 1. The trouble is that, in the plastics industry, new products are being developed so rapidly that the process of standardisation cannot keep pace. Since specifications are based on standard methods, a case for the latter has also been made.

However, by the time a contract has been placed it follows that the materials or products required have already been chosen and the question is on what basis? Assuming that the use demands some modicum of performance, even if only stability of appearance, and the choice has been dictated by other than only financial considerations, what faith can be placed in the data provided by the supplier or obtained by the research department of the user? Here published standard methods *may* be appropriate, indeed will be if they simulate the use conditions, but very often this is not the case. Many standard tests are designed for quick completion as quality control measures and it therefore follows inevitably that the conditions cannot approach those of the vast majority of uses. The complex behaviour of plastics materials under mechanical stress, in particular their tendency to creep, makes the usually measured ultimate tensile strength quite valueless as a source of design data and likewise the apparently more reliable modulus. The dependence of breakdown voltage on section thickness is a complex matter; the effects of time, temperature, the presence of voids and a host of other variables mean that the standard electric strength data must be treated with extreme caution. The dependence of apparent viscosity (the ratio of shear stress to shear rate) on shear rate in polymer melts means that very probably the standard melt flow test for polythene (with a low rate of shear) will give data quite irrelvant to, say, the calculation of shear forces obtaining in paper coating by extrusion.

Nevertheless this is not to decry the value of standard tests in any way, but rather to draw attention to the limitations of the existing range of standard tests. There are some which are aimed at obtaining what might be termed more 'fundamental' data and current thinking is to extend the coverage in this field. B.S.I. Committee PLC/36 has been set up for just this purpose.

Finally, it is not possible to envisage the happy state of affairs when standard tests will be available to cope with every situation. This is particularly true of the use of materials in specialist applications and such cases are best dealt with by evaluation of the performance of a prototype or at least by design of an appropriate 'ad hoc' test to simulate the use conditions.

APPENDIX 1

NATIONAL STANDARDS ORGANISATIONS

Those marked with an asterisk(*) are members of the International Organisation for Standardisation (ISO).

ALBANIA

Bureau de Standardisation auprès de la Commission,
du Plan d'Etat de la Republique Populaire d'Albanie,
Tirana, Albania

ARGENTINA

Institute Argentino de Racionalización de Materiales.
Chile No. 1192,
Buenos Aires, Argentina

AUSTRALIA*

Standards Association of Australia,
Science House,
80-86 Arthur Street,
North Sydney, NSW-2060,
Australia

AUSTRIA*

Oesterreichisches Normunginstitut (ON),
Borsegasse 18,
A-1010 Wien,
Austria

BELGIUM*

Institut Belge de Normalisation,
29 Avenue de la Brabançonne,
Bruxelles 4, Belgium

BRAZIL*

Associaçao Brasileira de Normas Técnicas,
Caixa Postal No. 1680,
Rio de Janeiro, Brazil

BULGARIA*
Institut de Normalisation,
Mesures et Appareils de Mesure,
8 Rue Sveta Sofia,
Sofia, Bulgaria

BURMA
Department of Standards,
Union of Burma Applied Research Institute,
Junction of Kaba Aye Pagoda-Kanbe Roads,
Rangoon, Burma

CANADA*
Canadian Standards Association,
77 Spencer Street,
Ottawa 3, Canada

CENTRAL AMERICA *(Costa Rica, El Salvador, Guatemala, Honduras, Nicaragua, Panama)*
Instituto Centroamericano de Investigación y Tecnologia Industrial,
4A, Calle y Avenida la Reforma,
Zona 10,
Apartado Postal 1552,
Guatemala

CEYLON*
Bureau of Ceylon Standards,
141/2 Vajira Road,
Colombo 5,
Ceylon

CHILE*
Instituto Nacional de Investigaciones Technologicas y Normalizacion,
Plaza Bulnes 1302, Of. 62,
Casilla de correo 995,
Santiago, Chile

COLOMBIA*
Instituto Colombiano de Normas Tecnicas,
Apartado Aereo: 14237,
Bogota D.E., Colombia

CUBA*
Direccion de Normas y Metrologia,
Ministerio de Industrias,
Reina 408,
Habana, Cuba

CZECHOSLOVAKIA*
Uřad pro normalisaci a meřeni,
Václavské náměstí 19
Praha 1- Nové Mešto,
Czechoslovakia

DENMARK*
Dansk Standardiseringsraad,
Aurehøjvej 12,
DK 2900 Hellerup, Denmark

FINLAND*
Suomen Standardisoimisliitto r.y.,
Bulevardi 5, A.7,
Helsinki, Finland

FRANCE*
Association Française de Normalisation,
23 Rue Notre-Dame des Victoires,
75 Paris 2°, France

GERMANY*
Deutscher Normenausschuss,
4-7 Burggrafenstrasse,
1 Berlin 30, Germany

GHANA*
The Director,
National Standards Board,
c/o Ministry of Industries,
P.O. Box M39,
Accra, Ghana

GREECE*
Ministry of Industry,
Direction of Standardization,
Kanigos Street,
Athens, Greece

HUNGARY*
Magyar Szabványügyi Hivatal,
Ulloi-ut 25,
Budapest IX, Hungary

INDIA*
Indian Standards Institution,
'Manak Bhavan',
9 Bahadur Shah Zafar Marg,
New Delhi 1,
India

INDONESIA*
Jajassan 'Dana Normalisasi Indonesia',
Djalan Braga 38 Atas,
Bandung, Indonesia

IRAN*
Institute of Standards and Industrial Research of Iran,
Ministry of Economy, P.O. Box 2937,
Teheran, Iran

IRAQ*
Iraqi Organisation for Standardisation,
Ministry of Industry,
P.O. Box 11185,
Baghdad, Iraq

IRELAND*
Institute for Industrial Research and Standards,
Glasnevin House,
Ballymun Road,
Dublin 9, Ireland

ISRAEL*
The Standards Institution of Israel,
University Street,
Tel Aviv, Israel

ITALY*
Ente Nasionale Italiano di Unificasione,
Piazza Armando Dias 2,
120123 Milano, Italy

JAPAN*
Japanese Industrial Standards Committee,
Agency of Industrial Science and Technology,
Ministry of International Trade and Industry,
3-1 Kasumigaseki, Chiyodaku,
Tokyo, Japan

KOREA, Dem. P. Rep. of*
Committee for Standardisation of the Democratic People's Republic of Korea,
Pyongyang, N. Korea

KOREA, Rep. of*
Korean Bureau of Standards,
Ministry of Commerce and Industry,
1-19, Chong-Ro,
Seoul, S. Korea

LEBANON*
Lebanese Standards Institution,
P.O. Box 2806,
Beirut, Lebanon

MEXICO*
Dirección General de Normas,
Av. Cuahtémac No. 80,
Mexico 7, D.F.

MOROCCO*
Service de Normalisation Industrielle Marocaine,
Direction de l'Industrie,
Sous-Secretariat d'Etat a l'Industrie et aux Mines,
Rabat, Morocco

NETHERLANDS*
Nederlands Normalisatie-instituut,
Polakweg 5,
Rijswijk (ZH), Netherlands

NEW ZEALAND*
Standards Association of New Zealand,
Private Bag,
Wellington, C.I.,
New Zealand

NORWAY*
Norges Standardiserings-forbund,
Haakon VII's gt. 2,
Oslo 1, Norway

PAKISTAN*
Pakistan Standards Institution,
39 Garden Road,
Saddar,
Karachi-3, Pakistan

PARAGUAY
Instituto Nacional de Tecnologia y Normalización,
Brasil y José Borges,
Asunción, Paraguay

PERU*
Instituto Nacional de Normas Técnicas Industriales y Certificación,
Apartado No. 145,
Av. Republica de Chile 698,
Lima, Peru

PHILIPPINES
Philippines Standards Association,
P.O. Box 3719,
Manila, Philippines

POLAND*
Polski Komitet Normalisacyjny,
Ul. Swietokrzyska 14,
Warszawa 51, Poland

PORTUGAL*
Repartiçào de Normalisaçào,
Avenida de Berna 1,
Lisboa-1, Portugal

RHODESIA
Standards Association of Rhodesia,
9th Floor, Pax House,
87 Union Avenue,
Salisbury, Rhodesia

ROMANIA*
Oficiul de Stat pentru Standarde,
Str. Edgar Quiret 6,
Bucarest 1, Romania

SINGAPORE*
Singapore Industrial Research Unit,
P.O. Box 2611,
Singapore

SOUTH AFRICA, Rep. of*
South African Bureau of Standards,
191 Private Bag,
Pretoria, South Africa

SPAIN*
Instituto Nacional de Racionalización del Trabajo,
Serrano 150,
Madrid 6, Spain

SWEDEN*
Sveriges Standardiseringskommission,
Box 3 295, S-10366,
Stockholm 3, Sweden

SWITZERLAND*
Association Suisse de Normalisation,
Kirchenweg 4,
8032 Zurich, Switzerland

THAILAND*

Centre for Thai National Standard Specifications,
Applied Scientific Research Corporation of Thailand,
196 Phahonyothin Road, Bangkhen,
Bangkok, Thailand

TURKEY*

Türk Standardlari Enstitüsü,
Necatibey Caddesi,
Ankara, Turkey

UNITED ARAB REPUBLIC*

Egyptian Organisation for Standardisation,
2 Latin America Street,
Garden City, Cairo, Egypt

UNITED KINGDOM*

British Standards Institution,
British Standards House,
2 Park Street,
London, W1Y 4AA

Sales Branch:
Newton House,
101-113 Pentonville Road,
London, N.1 9ND

URUGUAY

Instituto Uraguayo de Normas Tecnicas,
Agraciada 1464, Piso 9,
Montevideo, Uruguay

U.S.A.*

United States of America Standards Institute,
10, East 40th Street,
New York 10016, N.Y., U.S.A.

U.S.S.R.*

Komitet Standartov, Mer i Izmeritel 'nyh,
Priborov pri Sovete Ministrov SSSR,
38 Kvartal Jugo-Zapada,
Korpus 189-a,
Moskva V-421, U.S.S.R

VENEZUELA*

Comisión Venezolana de Normas Industriales,
Dirección de Industrias,
Ministerio de Fomento,
Caracas, Venezuela

WEST INDIES
Director of Standards,
Federal Ministry of Trade and Industry,
Port of Spain,
Trinidad, West Indies

YUGOSLAVIA*
Jugoslovenskizavod za Standardizaciju
Cara Uroša ul 54, Post pregradak 933,
Beograd, Yugoslavia

APPENDIX 2

BRITISH STANDARDS FOR PLASTICS

N.B. This list may be brought up to date by studying every issue of 'B.S.I. News' from June 1970 onwards.

B.S. 1755: 1951	Glossary of terms used in the Plastics Industry
B.S. 1755:	
Part 1: 1967	Glossary of terms used in the Plastics Industry. Polymerisation and plastics materials
B.S. 3502: 1967	Schedule of common names and abbreviations for plastics and rubbers

Group 1. Test Methods for Plastics Materials

B.S. 476:	Fire tests on building materials and structures
Part 1: 1953	Fire tests on materials and structures
Part 4: 1970	Non-combustibility test for materials
Part 5: 1968	Ignitability test for materials
Part 6: 1968	Fire propagation test for materials
B.S. 874: 1965	Methods of determining thermal properties, with definitions of thermal insulating terms
B.S. 1416: 1962	Methods for the sampling and analysis of rennet casein
B.S. 1417: 1963	Methods for the sampling and analysis of acid casein
B.S. 2067: 1953	Determination of power factor and permittivity of insulating materials
B.S. 2782: 1970	Methods of testing plastics
B.S. 2880: 1957	Methods of testing cellulose acetate flake
B.S. 2918: 1957	Electric strength of solid insulating materials at power frequencies
B.S. 3177: 1959	Permeability to water vapour of flexible sheet materials
B.S. 3424: 1961	Methods of test for coated fabrics and Addendum No. 1: 1967 (PD 6227)
B.S. 3544: 1962	Methods of test for polyvinyl acetate adhesives for wood
B.S. 3667: 1963–1966	Methods of testing flexible polyurethane foam
B.S. 3715: 1964	Concentration gradient density columns
B.S. 3755: 1964	Methods of test for the assessment of odour from packaging materials used for foodstuffs

B.S. 3781: 1964 Method for determining the comparative tracking index of solid insulating material
B.S. 4066: 1969 Flame-retardant characteristics of electric cables
B.S. 4370: 1968 Methods of test for rigid cellular materials
B.S. 4443: 1969 Methods of test for flexible cellular materials
B.S. 4542: 1970 Method for the determination of loss tangent and permittivity of electrical insulating materials in sheet form (Lynch Method)
B.S. 4601: 1970 Electroplated coatings of nickel plus chromium on plastics materials*
B.S. 4584: 1970 Metal-clad base materials for printed circuits
Part 1: Methods of test
B.S. 4597: 1970 General requirements and methods of test for multilayer printed wiring boards using plated through holes
B.S. 4618: 1970 Recommendations for the presentation of plastics design data

Group 2. Compositions and Materials for Processing

B.S. 771: 1959 Phenolic moulding materials
B.S. 1133: — Packaging Code
Section 16: 1953. Adhesives for packaging
B.S. 1203: 1963 Synthetic resin adhesives (phenolic and aminoplastic) for plywood
B.S. 1204: Synthetic resin adhesives (phenolic and aminoplastic) for wood
Part 1: 1964: Gap-filling adhesives
Part 2: 1965: Close-contact adhesives
B.S. 1322: 1956 Aminoplastic moulding materials
B.S. 1444: 1970 Cold-setting casein adhesive powders for wood
B.S. 1493: 1967 Polystyrene moulding materials
B.S. 1524: 1955 Cellulose acetate moulding materials
B.S. 1539: 1949 Moulded electrical insulating materials for use at high temperatures
B.S. 1540: 1949 Moulded electrical insulating materials for use at radio frequencies
B.S. 2487: 1969 Denture base polymer
B.S. 2571: 1963 Flexible PVC compounds
B.S. 3126: 1959 Toughened polystyrene moulding materials
B.S. 3168: 1959 Rigid PVC extrusion and moulding compounds
B.S. 3241: 1960 Toughened polystyrene for sheet extrusion
B.S. 3396: Woven glass fibre fabrics for plastics reinforcement
Part 1: 1966: Loom-state fabrics
Part 2: 1966: Desized fabrics
Part 3: 1970: Finished fabrics for use with polyester resin systems
B.S. 3412: 1966 Polythene materials for moulding and extrusion

*Includes methods of test.

B.S. 3496: 1962	Glass fibre chopped strand mat for the reinforcement of polyester resin systems
B.S. 3532: 1962	Unsaturated polyester resin systems for low pressure fibre reinforced plastics
B.S. 3534:	Epoxide resin systems for glass fibre reinforced plastics. Part 1: 1962: Wet lay-up systems Part 2: 1964: Pre-impregnating systems
B.S. 3691: 1969	Glass fibre rovings for the reinforcement of polyester and of epoxide resin systems
B.S. 3749: 1964	Woven glass fibre rovings fabrics for the reinforcement of polyester resin systems
B.S. 3815: 1964	Epoxide resin casting systems for electrical applications
B.S. 3816: 1964	Cast epoxide resin insulating material for electrical applications at power frequencies
B.S. 3840: 1965	Polyester dough moulding compounds
B.S. 4045: 1966	Epoxide resin pre-impregnated glass fibre fabrics
B.S. 4071: 1966	Polyvinyl acetate (PVA) emulsion adhesives for wood

Group 3. Fabricated Articles in Plastics

Section 1. Semi-finished Products including Laminates in All Forms

B.S. 1133:	Packaging code Section 7: 1952. Paper and board wrappers, bags and containers including films, foils and laminates Section 21: 1964. Transparent cellulose films, plastics films, metal foil and flexible laminates
B.S. 1137: 1966	Phenolic resin bonded paper sheets for electrical purposes at power frequencies
B.S. 1314: 1946	Synthetic resin bonded paper tubes for use as electrical insulation for power circuits
B.S. 1763: 1967	Thin PVC sheeting, (flexible, unsupported)
B.S. 1885: 1952	Synthetic resin bonded paper insulating tubes (rectangular cross-section) for electrical power circuits up to 1000 volts
B.S. 1951: 1953	Thermosetting synthetic resin-bonded paper round tubes for use at radio frequencies
B.S. 2076: 1954	Thermosetting synthetic resin-bonded paper insulating sheets for use at radio frequencies
B.S. 2572: 1955	Phenolic laminated sheet
B.S. 2601–2: 1963	Coated fabrics for upholstered furniture (PVC and N.C. types)
B.S. 2739: 1967	Thick PVC sheeting (flexible, unsupported)
B.S. 2848: 1957	Flexible insulating sleeving for electrical purposes
B.S. 2966: 1956	Phenolic resin bonded cotton fabric sheets for electrical purposes
B.S. 3186: 1970	Cellulose acetate sheet for spectacle frames

B.S. 3217: 1963	Air-permeable PVC coated fabrics for upholstered furniture
B.S. 3253: 1960	Phenolic resin-bonded asbestos paper sheets for electrical insulation at power frequencies
B.S. 3290: 1960	Toughened polystyrene extruded sheet
B.S. 3546: 1962	PVC-coated fabrics (cotton and rayon staple) for foul weather clothing
B.S. 3695: 1963	Polythene monofilament twines
B.S. 3757:	Rigid PVC sheet Part 1: 1964: Pressed sheet Part 2: 1965: Calendered and extruded sheet
B.S. 3784: 1964	Polytetrafluoroethylene sheet
B.S. 3794: 1964	Decorative laminated plastics sheet
B.S. 3835: 1964	Rigid PVC profiles for fitting sheet lining materials
B.S. 3837: 1965	Expanded polystyrene board for thermal insulation purposes
B.S. 3869: 1965	Rigid expanded polyvinyl chloride for thermal insulation purposes and building applications
B.S. 3878: 1965	Flexible PVC sheeting for hospital use
B.S. 3888: 1965	Copper-clad synthetic-resin bonded laminated sheet for use in telecommunication and allied electronic equipment
B.S. 3927: 1965	Phenolic foam materials for thermal insulation and building applications
B.S. 3953: 1965	Synthetic resin-bonded woven glass fabric laminated sheet
B.S. 4021: 1966	Flexible polyurethane foam sheeting for use in laminates
B.S. 4023: 1966	Flexible cellular PVC sheeting Part 1: 1966: Physically blown material Part 2: 1967: Chemically blown material
B.S. 4203: 1967	Extruded rigid PVC corrugated sheeting
B.S. 4216: 1970	Coated knitted fabrics for upholstered furniture (PVC types)
B.S. 4375: 1968	Unsintered PTFE tape for thread sealing applications
B.S. 4541: 1970	Polyurethane interior foam cores for domestic mattresses for adults
B.S. 6234: 1969	Polyethylene insulation and sheath of electric cables
B.S. 6746: 1969	PVC insulation and sheath of electric cables
B.S. 6746C: 1969:	Colour chart for PVC insulation and sheath of electric cables

Section 2. Finished Products in Plastics

B.S. 1226: 1945	Draining boards
B.S. 1254: 1954	W.C. seats (plastics)
B.S. 1321: 1946	Plastics picnic type tableware
B.S. 1348: 1949	Measuring cups and spoons for cookery and medicine
B.S. 1651: 1966	Industrial gloves
B.S. 1774: 1961	Rainwear from PVC sheeting
B.S. 1776: 1951	Fabrication of lightweight articles (other than rainwear) from polyvinyl chloride sheeting

B.S. 1882: 1966	Flexible tubing for medical use
B.S. 1972: 1967	Polythene pipe (Type 32) for cold water services
B.S. 1973: 1970	Polythene pipe (Type 32) for general purposes including chemical and food industry uses
B.S. 2026: 1953	Tolerances for mouldings and thermosetting materials
B.S. 2038: 1953	Rolled sheet metal screw threads and associated threads in moulded plastics and die cast materials for general purposes
B.S. 2050: 1961	Electrical resistance of conductive and anti-static products made from flexible polymeric material
B.S. 2456: 1954	Floats for ball valves (plastics) for cold water
B.S. 2515: 1954	Reflex reflectors for vehicles, including cycles
B.S. 2552: 1955	Polystyrene tiles for walls and ceilings
B.S. 2581: 1955	Plastics trays
B.S. 2766: 1956	Moulded plastics ashtrays
B.S. 2906: 1957	Aminoplastic mouldings
B.S. 2907: 1957	Phenolic mouldings
B.S. 2919: 1968	Low and intermediate density polythene rod for general purposes
B.S. 3012: 1958	Low-density polythene sheet
B.S. 3094: 1959	Domestic articles made from low-density polythene Part 1: Circular washing-up bowls Part 2: Buckets
B.S. 3104: 1970	Polyamide (nylon) mountaineering ropes
B.S. 3167: 1959	Melamine plastics tableware
B.S. 3184: 1959	Plastics fire buckets
B.S. 3221:	Medicine Measures Part 3: 1966: Plastics medicine measures of 50 ml capacity Part 4: 1969: Plastics medicine measuring spoons of 5 ml capacity Part 5: 1966: Plastics medicine measures of 10 ml capacity
B.S. 3260: 1969	PVC (vinyl) asbestos floor tiles
B.S. 3261: 1960	Flexible PVC flooring
B.S. 3284: 1967	Polythene pipe (Type 50) for cold water services
B.S. 3330: 1961	Polythene splints and appliances for external orthopaedic use
B.S. 3501: 1962	Dinghy buoyancy equipment Part 1: Buoyancy bags made from unsupported PVC sheeting Part 3: Buoyancy equipment made from expanded polystyrene
B.S. 3505: 1968	Unplasticised PVC pipe for cold water services
B.S. 3506: 1969	Unplasticised PVC pipe for industrial uses
B.S. 3746: 1964	PVC garden hose
B.S. 3796: 1970	Polythene pipe (Type 50) for general purposes including chemical and food industry uses
B.S. 3858: 1965	Binding and identification sleeves for use on electrical cables and wires

B.S. 3867: 1965	Dimensions of pipes of plastics materials (outside diameters and pressure ratings)
B.S. 3873: 1965	Polytetrafluoroethylene moulded basic shapes
B.S. 3887: 1965	Regenerated cellulose and unplasticised PVC pressure-sensitive closing and sealing tapes
B.S. 3897: 1965	Plastics (polythene) containers
B.S. 3912: 1965	Polythene filament ropes (hawser laid)
B.S. 3924: 1965	Pressure-sensitive adhesive tapes for electrical purposes
B.S. 3932: 1965	Expanded polystyrene tiles and profiles for the building industry
B.S. 3943: 1965	Plastics waste traps
B.S. 3944: 1965	Colour filters for theatre lighting and other purposes
B.S. 3977: 1966	Polyamide (nylon) filament ropes (hawser laid)
B.S. 3990: 1966	Acrylic resin teeth
B.S. 4041: 1966	Polystyrene mouldings
B.S. 4042: 1966	Specification for a system of dimensional tolerances for small injection mouldings in rigid and semi-rigid thermoplastics materials
B.S. 4077: 1966	The sequence of measurements of film bags
B.S. 4135: 1967	Sinks for domestic purposes made from cast acrylic sheet
B.S. 4154: 1967	Corrugated plastics translucent sheets made from thermosetting polyester resins (glass fibre reinforced)
B.S. 4159: 1967	Colour marking of plastics pipes to indicate pressure ratings
B.S. 4213: 1967	Polyolefin or olefin copolymer moulded cold water storage cisterns. Metric units
B.S. 4271: 1968	Polytetrafluoroethylene (PTFE) rod
B.S. 4514: 1969	Unplasticised PVC soil and ventilating pipe, fittings and accessories
B.S. 4576: 1970	Unplasticised PVC rainwater goods
	Part 1: Half-round gutters and circular pipe

Section 3. Materials and Products Containing Plastics as an Important Part Thereof

As Insulation

B.S. 1862: 1959	Cables for vehicles
B.S. 2316: 1963	Radio-frequency cables
	Part 1: General requirements and tests
	Part 2: British Government Services requirements
B.S. 2401: 1966	Polyester film dielectric capacitors for direct current for use in telecommunication and allied electronic equipment
B.S. 3040: 1958	Radio-frequency cables for use with domestic television and VHF receiving aerials
B.S. 3289: 1960	Conveyor belting for underground use in coal mines
B.S. 3485: 1962	PVC-covered conductors for overhead power lines
B.S. 3573: 1962	Polythene-insulated telephone distribution cables
B.S. 4553: 1970	PVC-insulated split concentric cables with copper conductors for electricity supply

B.S. 6004: 1969	PVC-insulated cables (non-armoured) for electric power and lighting
B.S. 6195: 1969	Insulated flexible cables and cords for coil leads (Metric revision of B.S. 4195)
B.S. 6231: 1969	PVC-insulated cables of switchgear and control gear wiring
B.S. 6346: 1969	PVC-insulated cables for electricity supply
B.S. 6500: 1969	Insulated flexible cords

As Adhesive

B.S. 1088: 1966	British-made plywood for marine craft. Marine plywood manufactured from selected untreated tropical hardwoods
B.S. 1455: 1963	Plywood manufactured from tropical hardwoods
B.S. 2604: 1963	Resin-bonded wood chipboard
B.S. 3444: 1961	Blockboard and laminboard
B.S. 4079: 1966	Plywood for marine craft. Plywood made for marine use and treated against attack by fungi or marine borers

APPENDIX 3

SOME SOURCES OF U.K. STANDARDS OTHER THAN B.S.I.

1. *Defence Standards, C.S. and T.S. Specifications*

DGW(N): Naval Ordnance Quality Assurance Directorate
M.O.D.(N), Ensleigh, Bath, BA1 5AB
DQA (Mats): Quality Assurance Directorate (Materials),
Headquarters Building,
Royal Arsenal East,
Woolwich, London, S.E.18 6TD

2. *DEF Specifications*
Her Majesty's Stationery Office,
Various branch offices

3. *DTD Specifications*
Her Majesty's Stationery Office,
Various branch offices

4. *D. Eng. R.D. Specifications*
Ministry of Aviation Supply,
Eng. R.D.I.,
Room 163A
St. Giles Court,
London, W.C.2

5. *N.C.B. Standards*
   National Coal Board,
   Hobart House,
   Grosvenor Place,
   London, S.W.1

6. *Various branches of the Post Office*

APPENDIX 4

SOME SOURCES OF U.S.A. STANDARDS OTHER THAN U.S.A.S.I.

1. *A.S.T.M. Standards*
   American Society for Testing and Materials,
   1916 Race Street,
   Philadelphia,
   Pa. 19103, U.S.A.

2. *A.M.S. Standards*
   Society of Automotive Engineers, (S.A.E.)
   485 Lexington Avenue,
   New York 17,
   N.Y., U.S.A.

3. *Federal and Mil Standards*
   Superintendent of Documents,
   U.S. Government Printing Office,
   Washington 25,
   D.C., U.S.A.

4. *C.S. Standards*
   Commodity Standards Division,
   Office of Technical Services,
   U.S. Department of Commerce,
   Washington 25,
   D.C., U.S.A.

5. *S.P.I. Standards*
   Society of Plastics Industry Inc.,
   250 Park Avenue,
   New York 17,
   N.Y., U.S.A.

6. *N.E.M.A. Standards*
   National Electrical Manufacturers' Association,
   155 E.44th Street,
   New York 17,
   N.Y., U.S.A.

APPENDIX 5

I.S.O. RECOMMENDATIONS FOR PLASTICS*

N.B. This list may be brought up to date by studying every issue of *B.S.I. News* from June 1970 onwards.

ISO/R.59	Determination of the percentage of acetone soluble matter in phenolic mouldings
ISO/R.60	Determination of apparent density of moulding material that can be poured from a specified funnel
ISO/R.61	Determination of apparent density of moulding material that cannot be poured from a specified funnel
ISO/R.62	Determination of water absorption
ISO/R.62/A1	Amendment 1 to ISO Recommendation R.62
ISO/R.75	Determination of temperature of deflection under load
ISO/R.117	Determination of boiling water absorption
ISO/R.118	Determination of methanol-soluble matter in polystyrene
ISO/R.119	Determination of free phenols in phenol-formaldehyde mouldings
ISO/R.120	Determination of free ammonia and ammonium compounds in phenol-formaldehyde mouldings
ISO/R.161	Pipes of plastics materials for the transport of fluids Part 1: Metric series
ISO/R.171	Determination of bulk factor of moulding materials
ISO/R.172	Detection of free ammonia in phenol-formaldehyde mouldings (qualitative method)
ISO/R.173	Determination of the percentage of styrene in polystyrene with Wijs solution
ISO/R.174	Determination of viscosity number of polyvinyl chloride resin in solution
ISO/R.175	Determination of the resistance of plastics to chemical substances
ISO/R.176	Determination of the loss of plasticisers from plastics by the activated carbon method
ISO/R.177	Determination of migration of plasticisers from plastics
ISO/R.178	Determination of flexural properties of rigid plastics
ISO/R.179	Determination of the Charpy impact resistance of rigid plastics (Charpy impact flexural test)
ISO/R.180	Determination of the Izod impact resistance of rigid plastics (Izod impact flexural tests)
ISO/R.181	Determination of incandescence resistance of rigid self-extinguishing thermosetting plastics
ISO/R.182	Determination of the thermal stability of polyvinyl chloride and related copolymers and their compounds by the Congo Red method
ISO/R.183	Determination of the bleeding of colourants from plastics

*ISO/R denotes an ISO Recommendation: ISO/DR denotes a draft ISO Recommendation.

ISO/R.194 List of equivalent terms used in the plastics industry. (English, French, Russian)

ISO/R.194/P1, /Cs, /D and /NL Corresponding Polish, Czech, German and Dutch terms (respectively)

ISO/R.264 Pipes and fittings of plastics materials. Socket fittings for pipes under pressure. Basic dimensions. Metric series

ISO/R.265 Pipes and fittings of plastics materials. Socket fittings with spigot ends for domestic and industrial waste pipes. Basic dimensions. Metric series

ISO/R.291 Standard atmospheres for conditioning and testing

ISO/R.292 Plastics: Determination of the melt flow index of polyethylene and polyethylene compounds, 2nd Edition

ISO/R.293 Compression moulding test specimens of thermoplastic materials

ISO/R.294 Injection moulding test specimens of thermoplastic materials

ISO/R.295 Compression moulding test specimens of thermosetting materials

ISO/R.305 Determination of the thermal stability of polyvinyl chloride and related copolymers and their compounds by the discoloration method

ISO/R.306 Plastics: Determination of the Vicat softening point of thermoplastics, 2nd Edition

ISO/R.307 Determination of the viscosity number of polyamide resins in dilute solution

ISO/R.308 Determination of the acetone soluble matter (resin content of material in the unmoulded state) of phenolic moulding materials

ISO/R.330 Pipes of plastics materials for the transport of fluids (outside diameters and nominal pressures). Part II: Inch series

ISO/R.458 Determination of stiffness in torsion as a function of temperature

ISO/R.462 Recommended practice for the determination of change of mechanical properties after contact with chemical substances

ISO/R.472 Plastics: Definition of terms, 2nd Edition

ISO/R.483 Plastics: Methods for maintaining constant relative humidity in small enclosures by means of aqueous solutions

ISO/R.489 Plastics: Determination of the refractive index of transparent plastics

ISO/R.527 Plastics: Determination of tensile properties

ISO/R.537 Testing of plastics with the torsion pendulum

ISO/R.554 Standard atmospheres for conditioning and/or testing. Standard reference atmosphere. Specifications

ISO/R.558 Conditioning atmosphere. Test atmosphere. Reference atmosphere. Definitions

ISO/R.580 Oven test for moulded fittings in unplasticised polyvinyl chloride (PVC) for use under pressure

ISO/R.584	Plastics. Determination of the maximum temperature and the rate of increase of temperature during the setting of unsaturated polyester resins
ISO/R.585	Plastics. Determination of the moisture content of non-plasticised cellulose acetate
ISO/R.599	Plastics. Determination of the percentage of extractable materials in polyamide resins
ISO/R.600	Plastics: Determination of the viscosity ratio of polyamides in concentrated solution
ISO/R.604	Determination of compressive properties of plastics
ISO/R.727	Socket fittings for pipes under pressure. Unplasticised polyvinyl chloride (PVC) fittings with plain sockets. Metric series
ISO/DR.748(2)	Determination of tensile creep of plastics
ISO/DR.749	Plastics. Determination of the melt flow index of polypropylene and polypropylene compounds
ISO/DR.751	Plastics. Supplement to list of equivalent terms (ISO/R.194)
ISO/DR.754(2)	Plastics. Determination of viscosity index of polyethylenes and polypropylenes in dilute solution
ISO/R.800	Plastics: Basis for specification for phenolic moulding materials
ISO/DR.820	Determination of the water vapour transmission rate of plastics films and thin sheets. Dish method
ISO/DR.823	Methods for determining the density and specific gravity of non-cellular plastics
ISO/DR.824	Plastics. Method of test for the viscosity number of methyl-methacrylate polymers and copolymers in dilute solution
ISO/DR.826	Plastics. Determination of the chlorine in vinyl chloride polymers and copolymers
ISO/DR.827	Plastics. Determination of vinyl acetate in vinyl chloride-vinyl acetate copolymers
ISO/R.844	Plastics. Compression test of rigid cellular plastics
ISO/R.845	Plastics. Determination of apparent density of rigid cellular plastics
ISO/R.846	Plastics. Recommended practice for the evaluation of the resistance of plastics to fungi by visual examination
ISO/R.868	Plastics. Determination of indentation hardness of plastics by means of a durometer (Shore hardness)
ISO/R.869	Plastics. Preparation of specimens for optical tests on plastics materials. Moulding method
ISO/R.870	Plastics. Preparation of specimens for optical tests on plastics materials. Casting method
ISO/R.871	Plastics. Determination of the temperature of evolution of flammable gases from plastics
ISO/R.872	Plastics: Determination of ash of unplasticised cellulose acetate
ISO/R.877	Plastics. Determination of resistance of plastics to colour change upon exposure to daylight

ISO/R.878	Plastics. Determination of resistance of plastics to colour change upon exposure to light of the enclosed carbon arc
ISO/R.879	Plastics. Determination of resistance of plastics to colour change upon exposure to light of the xenon lamp
ISO/R.899	Determination of tensile creep of plastics
ISO/R.922	Plastics. Determination of soluble matter of crystalline polypropylene in boiling n-heptane
ISO/R.960	Plastics. Determination of the water content in polyamides
ISO/R.974	Plastics. Method of determining the brittleness point by impact
ISO/DR.999	Bending test for rigid cellular plastics
ISO/R.1000	Rules for the use of units of the international system of units and a selection of the decimal multiples and submultiples of the S.I. Units
ISO/DR.1000	Determination of flammability of plastics in the form of bars
ISO/DR.1002(2)	Plastics. Determination of viscosity number of polyethylene terephthalate in dilute solution
ISO/R.1043	Abbreviations (symbols) for plastics
ISO/R.1060	Plastics. Designation of polyvinyl chloride resins
ISO/R.1061	Plastics. Determination of free acidity of unplasticised cellulose acetate
ISO/R.1068	Plastics. PVC resins. Determination of the compacted apparent bulk density
ISO/R.1110	Plastics. Accelerated conditioning of test specimens of polyamide 66, 610 and 6
ISO/R.1133	Plastics. Determination of the melt flow rate of thermoplastics
ISO/R.1147	Plastics. Aqueous dispersions of polymers and copolymers. Freeze–thaw cycle stability test
ISO/R.1148	Plastics. Aqueous dispersions of polymers and copolymers. Determination of pH
ISO/R.1157	Plastics. Determination of the viscosity number and viscosity ratio of cellulose acetate in dilute solution
ISO/R.1184	Plastics. Determination of tensile properties of films
ISO/DR.1218	Plastics. Determination of the 'melting point' of polyamides
ISO/DR.1253	Plastics. Second supplement to list of equivalent terms (ISO/R.194)
ISO/DR.1262	Plastics. Specification for unplasticised vinyl compounds
ISO/DR.1263	Plastics. Guide for preparing specifications for polyamides
ISO/DR.1264	Plastics. PVC resins. Determination of pH of aqueous extract
ISO/DR.1265	Plastics. PVC resins. Determination of impurities and foreign matter
ISO/DR.1267	Appendix to ISO Recommendation R.295. Preparation of compression moulded test specimens from phenolic moulding materials

ISO/DR.1268	Plastics. Preparation of glass fibre reinforced, resin-bonded, low-pressure laminated plates or panels for test purposes
ISO/DR.1269	Plastics. PVC resins. Determination of volatile matter (including water)
ISO/DR.1270	Plastics. PVC resins. Determination of ash and sulphated ash
ISO/DR.1300	Plastics. Method of test for the determination of the behaviour of plastics in a ventilated tubular oven
ISO/DR.1301	Revision of ISO Recommendation R.182. New Title: Plastics. Determination of the thermal stability of polyvinyl chloride and related copolymers and their compounds as determined by splitting off hydrogen chloride
ISO/DR.1325	Plastics. Determination of the electrical properties of thin sheet and film
ISO/DR.1326	Plastics. Flammability and burning rate of plastics in the form of film
ISO/DR.1330	Plastics pipes for the transport of fluids. Unplasticised polyvinyl chloride pipes. Tolerances on outside diameters
ISO/DR.1331	Plastics pipes for the transport of fluids. Polyethylene pipes. Tolerances on outside diameters.
ISO/DR.1332	Plastics pipes for the transport of fluids. Unplasticised polyvinyl chloride pipes. Tolerances on wall thicknesses up to 6 mm
ISO/DR.1333	Plastics pipes for the transport of fluids. Polyethylene pipes. Tolerances on wall thicknesses up to 6 mm
ISO/DR.1334	Plastics pipes for the transport of fluids. Determination of the burst strength
ISO/DR.1383	First supplement to ISO Recommendation R.472-66. Plastics. Definition of terms
ISO/DR.1419	Accelerated ageing and simulated service tests of fabrics coated with rubber or plastics
ISO/DR.1420	Determination of the impermeability to water of fabrics coated with rubber and plastics
ISO/DR.1421	Determination of breaking strength and elongation at break of fabrics coated with rubber or plastics
ISO/DR.1422	Second supplement to ISO/R.472—Definitions of plastics terms
ISO/DR.1597	Plastics. Determination of acetic acid yield of unplasticised cellulose acetate
ISO/DR.1598	Plastics. Determination of insoluble particles in cellulose acetate
ISO/DR.1599	Plastics. Determination of viscosity loss on moulding of cellulose acetate
ISO/DR.1600	Plastics. Determination of light absorption of cellulose acetate before and after heating
ISO/DR.1601	Plastics. Appendix II to ISO Recommendation R.295—Preparation of compression moulded test specimens from aminoplastic (moulding materials)

ISO/DR.1602	Plastics. Appendix III to ISO Recommendation R.295—Preparation of compression moulded test specimens from polyester and epoxy resin (moulding materials)
ISO/DR.1621	Amendment I to ISO Recommendation R.264—Pipes and fittings of plastics materials. Socket fittings for pipes under pressure. Basic dimensions. Metric series
ISO/DR.1622	Plastics. Requirements for polystyrene moulding and extrusion materials
ISO/DR.1624	Plastics. Polyvinyl chloride resins. Sieve analysis in water
ISO/DR.1625	Plastics. Polymer and copolymer water dispersions. Determination of dry solid content at 105°C
ISO/DR.1627	Plastics. Migration of plasticisers from polyvinyl chloride to polyethylene (measurement of migration by electrical means)
ISO/DR.1628	Plastics. Directives for the standardisation of methods for the determination of the dilute solution viscosity of polymers
ISO/DR.1633	Plastics. Determination of the loss on ignition of glass-reinforced plastics
ISO/DR.1642	Plastics. Specification for industrial laminated sheets based on thermosetting resins
ISO/DR.1663	Plastics. Determination of water vapour transmission of rigid cellular plastics
ISO/DR.1674	Plastics. Determination of flexural properties of rigid plastics (Revision of ISO Recommendation R.178)
ISO/DR.1675	Plastics. Liquid resins. Measurement of density by the pycnometer method

REFERENCES

1. CAMPBELL, C. 'Co-operation, the key to voluntary standards,' *S.P.E.J.*, **19,** No. 5, 465 (May 1963)
2. REINHART, F. W., 'Who establishes plastics standards?', *S.P.E.J.*, **16,** No. 2, 175 (February 1960)
3. AINSWORTH, C., 'Why ASA?', *Mater. Res. Stand.*, **3,** No. 6, 492 (June 1963)
4. MOWBRAY, A. Q., 'Who Will Write the Standards?', *S.P.E.J.*, **24,** No. 12, 74 (December 1968)
5. ADAMS, C. HOWARD, 'Plastics Standards in the U.S.', *Plast. Technol.* **7,** No. 12, 40 (December 1961)
6. KAIDANOVSKY, S. P., 'Guide to Materials Standards and Specifications, Part 4—Plastics and Rubber', *Mater. des Engng*, **47,** No. 6, 113 (June 1958)
7. REINHART, F. W., 'Plastics Materials Standardisation', *S.P.E.J.*, **24,** No. 4, 100 (April 1968)
8. REINHART, F. W., 'Standards for Plastics, Federal Test Method Standards', *S.P.E.J.*, **17,** 12, 1291 (December 1961)
9. BEACH, N. E., *Government Specifications and Standards for Plastics, Covering Defense Engineering Materials and Applications, Document No. AD.410401,* Department of Commerce, Office of Technical Services, Annex 1, Washington 25, D.C., U.S.A.
10. BEACH, N. E., *Defense Specifications and Standards for and Relating to Reinforced Plastics, Document No. AD.402225,* Department of Commerce, Office of Technical Services, Annex 1, Washington 25, D.C., U.S.A.

11. SKOW, N. A., 'Specifications for Thermosetting Industrial Laminates', *S.P.E.J.*, **19,** No. 12, 1269 (December 1963)
12. REINHART, F. W., 'Standards for Plastic Pipe and Fittings', *S.P.E.J.*, **17,** No. 2, 159 (February 1961)
13. REINHART, F. W., 'Standards for Vinyl Chloride Plastics', *S.P.E.J.*, **18,** No. 3, 308 (March 1962)
14. *I.S.O. Momento 1969,* International Organisation for Standardisation, 1 Rue de Varembé, 1211 Geneva 20, Switzerland

3

Preconditioning and Test Atmospheres

3.1 THE NEED

Plastics materials, generally, are affected by ambient conditions and, in the normal test laboratory, these variables may be limited to varying temperature and humidity; hazards of irradiation, be they ultraviolet from sunlight, gamma rays from nuclear power or heat from rocket propulsion, are more logically treated under the general heading of 'Ageing' (see Chapters 11 and 13).

Sensitivity to the effects of temperature is primarily a consequence of the fact that most present-day plastics materials are based on polymers built upon a backbone chain consisting solely or predominantly of covalently bonded carbon atoms. Cross-linking, secondary intermolecular forces and crystallinity can influence reaction to temperature quite significantly, but by comparison with most other 'materials of construction' (in the broadest sense of the phrase) plastics are *temperature sensitive*. A number of them contain hydroxyl groups, e.g. cellulose acetate, polyvinyl acetate and polyvinyl alcohol, whilst others contain amino groups, such as the nylons and casein, and these groups have a distinct affinity for water molecules, such as are present in air of any finite moisture content or humidity; the water acts as a plasticiser and brings about a change of properties*. Even if the polymer itself is not subject to the influences of temperature or humidity, the other materials present in the product may well be—particularly to humidity. Wood flour or chopped rag filler and cotton, paper or glass fibre reinforcement are well known examples.

It will be obvious therefore, that to obtain results of any meaning, from the point of view of defining the physical condition of the material when tested and providing data comparable with others obtained similarly, specimens should be subjected to standard preconditioning to bring them into an equilibrium state with a specified atmosphere; this, of course, is quite apart from allowing to come to equilibrium any physical change, such as described in Section 1.4.3, which might be loosely described as a continuation of the processes induced in the fabrication of the test specimen. The need for preconditioning (hereafter 'conditioning') clearly depends

*One of the most spectacular cases is the increase in impact strength of about twenty-fold which may be achieved by allowing dry nylon-6 to absorb moisture to saturation.

on the nature of the materials under test; in many cases control of humidity is unimportant because the materials are to all intents and purposes completely hydrophobic.

The duration of conditioning, i.e. that minimum time necessary to bring the test specimens into equilibrium with their surroundings, will depend on the nature of the materials, the physical shape and size of the specimens and just how far removed from the equilibrium temperature and moisture content they were when first placed in the conditioning atmosphere. Although plastics materials are poor conductors of heat, again by virtue of their 'organic' nature, they will not take very long to equilibriate with surrounding temperature providing their cross-section is not too large. Humidifying is another matter, however, for here it is not just a question of transference of energy from one bond to another and one molecule to another as in heat conduction, but the diffusion of water molecules through entangled polymer molecules to the centre of a specimen, perhaps a path of half an inch or so, from the nearest external surface, some many million molecules away. This can readily be imagined to be a slow process and it becomes progressively slower as the specimen approaches the equilibrium state. For water sensitive materials, therefore, conditioning may take a very long time.

By the same argument, it is often necessary to actually conduct the tests in atmospheres of controlled temperature and humidity but, as this is an altogether more difficult and costly business, 'in situ' testing is dispensed with as much as possible. Whether this is feasible or not depends again on the nature of the material under examination and the type of test used. A material which is not very sensitive to the influence of temperature or humidity, or at least changes its condition only slowly when, say, placed in 'ordinary ambient conditions', may often be examined in the open laboratory, *providing the test is conducted and completed* within some appropriate time of removal from the conditioning chamber. Likewise, a quick acting test, such as that for impact strength by a pendulum method, and/or one which uses a specimen of large cross-section, will often make it unnecessary to conduct the test in a defined atmosphere. Even if only the need for humidity control, during the actual testing operation, can be dispensed with this is a distinct help as humidity control of test laboratories is generally far more complicated and costly than temperature control.

3.2 THE SELECTION OF CONDITIONS

On the national level, the actual levels of temperature and humidity laid down in standards and specifications will naturally tend to simulate average ambient weather conditions in the country in question, to facilitate the actual conditioning and testing operations and produce data relevant to the likely environment of use of the plastics materials. For practical reasons, the tendency is to stipulate a temperature above normal for most times of the year, simply because it is easy to raise room temperatures a little above ambient but difficult and costly to reduce them below.

Owing to the manner in which the plastics industry has developed, at least in the U.K., there has been a tendency in the past for each specification for each different material to dictate its own, often different, conditions. This is unfortunate and inefficient; on the face of things, only a variation in duration is logical, depending on the nature of the material in question. However, these anomalies are being removed.

Internationally, standardisation is even more difficult. 'Ambient' in North Canada means something rather different from what it does in the Sahara desert. There is not only the problem of settling conditions acceptable practically to both regions, but the fact that, even if an equitable compromise were reached, the data obtained in such conditions would probably have little practical meaning in either! It follows, therefore, that a limited number of alternative standard conditions must be settled, to be representative broadly of the various important trading areas of the world.

3.3 DEFINITIONS

Before discussing the various standard conditions proposed, it would be as well to define the terms to the used.

There can be little dispute about temperature except, perhaps, to arbitrate in the argument over Fahrenheit versus Centigrade or Celsius ($F = \frac{9}{5}C + 32$). Generally the latter is the preferred unit although the S.I. scheme specifies Kelvin.

On the water content of the air, there is more confusion. The most frequently used term is *relative humidity* (r.h.) but very often without the correct definition in mind. B.S. 1339:1965, 'Definitions, Formulae and Constants Relating to the Humidity of the Air', gives the following: 'The ratio of the actual vapour pressure to the saturation vapour pressure over a plane liquid water surface at the same (dry bulb) temperature, expressed as a percentage'.

A.S.T.M. E.337-62, 'Standard Method for Determining Relative Humidity', gives a similar phraseology.

Percentage Saturation is defined as (B.S.1339): 'The ratio of the actual mixing ratio to the saturation mixing ratio at the same temperature and pressure expressed as a percentage', where mixing ratio is given by the expression: 'The ratio of the mass of water vapour to the mass of dry air with which the water vapour is associated'.

Percentage saturation and relative humidity do not differ by more than 2%, however, at temperatures below 100°F.

For further definitions the reader is referred to B.S.1339 and A.S.T.M. E.41-63, 'Definitions of Terms Relating to Conditioning'.

3.4 I.S.O. RECOMMENDATIONS

The International Organisation for Standardisation has produced a recommendation, reference R.291, 'Plastics. Standard Atmospheres for Conditioning and Testing' (1st Edition, February 1963). This proposes the following three alternatives where both temperature and humidity control are necessary:

1. 20°C, 65% r.h., or
2. 23°C, 50% r.h., or
3. 27°C, 65% r.h.

Where only temperature control is required:

4. 20°C, or
5. 23°C, or
6. 27°C

If control of neither is necessary, it is stated that prevailing room conditions may be used.

The normal tolerance on temperature recommended is ±2°C, but provision is made for a closer tolerance of ±1°C where required. Likewise, for humidity the normal tolerance stated is ±5%, with a closer one of ±2%.

If the reaction of a given material to change in temperature and humidity is known such that test data may be corrected to a standard reference atmosphere, this latter is laid down as 20°C and 65% r.h. This seems to be a recommendation of doubtful value.

For conditions (1), (2) and (3), a standard conditioning time of 88 to 94 h is recommended, for 4, 5 and 6 not less than 3 h, although it is admitted that there may be cases where these periods may not be suitable and must accordingly be adjusted.

The actual testing operation is carried out in the same atmosphere as used for conditioning, unless it is known that, for reasons given in Section 3.1, (4), (5) or (6) may be substituted for (1), (2) or (3) respectively or the prevailing room conditions for any of the six alternatives.

ISO/R.554; 'Standard Atmospheres for Conditioning and/or Testing. Standard Reference Atmosphere. Specifications', (1st edition, January 1967), does not conflict with R.291, but being a more general document requires atmospheric pressure between 860 and 1060 mbar, with 1013 mbar at the standard reference atmosphere.

ISO/R.558 simply defines various terms such as 'conditioning atmosphere', 'test atmosphere' and 'reference atmosphere'.

3.5 U.K. PRACTICE

There is no one standard condition for plastics testing, for example even in B.S.2782, and the relevant individual materials and products specifications must be consulted for details of the levels of temperature and humidity to be used in each case. As stated previously, at the moment these may vary from one document to another, but, as new standards are produced or old ones revised, the levels are generally being reconciled with I.S.O. recommendations.

When a material or product is being tested for which there is as yet no specification, and there is a suspicion that conditioning might be important, this exercise should be carried out (and the test performed in the same atmosphere if necessary) using the most appropriate standard atmosphere, e.g. that relevant to another material (for which there is a specification) with which the new product is to be compared.

3.6 U.S.A. PRACTICE

Using A.S.T.M. as the logical example, D.618–61, 'Conditioning Plastics and Electrical Insulating Materials for Testing', goes into the subject very thoroughly. The following definitions are given:

Standard Laboratory Temperature: A temperature of 23°C (73.4°F) with a normal tolerance of ±2°C (±3.6°F) and a closer tolerance ±1°C (±1.8°F) if required.

Standard Laboratory Atmosphere: An atmosphere having a temperature of 23°C (equivalents and tolerances as above) and a relative humidity of 50%, the latter with a normal tolerance of ±5% and a closer tolerance of ±2% when required.

Room Temperature: A temperature in the range 20°–30°C (68°–86°F)

The standard lays down a neat shorthand for identifying the type of conditioning used. A first number indicates the number of hours of conditioning, a second the temperature and a third the relative humidity; the last mentioned may be replaced in certain circumstances by words indicating special treatments. Thus:

1. 96/23/50 —96 h at 23°C and 50% r.h.
2. 48/50/Water —48 h at 50°C in water
3. 48/50+96/23/50—48 h at 50°C followed by 96 h at 23°C and 50% r.h.

For the actual test condition, the conditioning data as above is followed by a colon, then a capital T, a number indicating the test temperature, and after it one indicating the relative humidity if controlled. e.g.:
96/23/50:T – 35/90 Condition as 1, above, and test at 35°C and 90% r.h.

A.S.T.M. D.618 lays down minimum conditioning times according to the thickness of test specimens and for the most commonly used atmosphere (*t*/23/50), *t* is 40 h, for specimens 7 mm or under, and 88 h for specimens above 7 mm. This is known as *Procedure A,* but there are five other alternatives to meet the needs of rapidity or to simulate special atmospheres or use environments. Corresponding instructions for test conditions are given.

This standard, in fact, is referred to in a more general document, A.S.T.M. E.171–63, 'Tentative Specification for Standard Atmospheres for Conditioning and Testing Materials', which lays down conditions for a wide range of non-metallic products including adhesives, surface coatings, paper products and sandwich constructions.

3.7 GERMAN PRACTICE

In German DIN standards, the practice is often to quote the preconditioning atmosphere for each material (frequently 20±2°C and 65±5% r.h.). However later standards cross-reference to DIN.50014, 'Normal Climates', which specifies either:

20°±2°C and 65±3% r.h.—coded 20/65
or 23°±2°C and 50±3% r.h.—coded 23/50

both at a pressure of 800–1060 mbar.

3.8 CONDITIONING CHAMBERS AND TEST ROOMS

It is not feasible, in a small section of this book, to describe the construction of conditioning chambers and the design of laboratories fitted with humidity and temperature control; this is a science in itself which, particularly in the case of the laboratory, is best left to the relevant expert. What will follow, therefore, is little more than a few hints and comments which might have been included in Section 1.4, 'Miscellaneous Pitfalls in Testing', had it not meant upsetting the general order of the treatment of the subjects in the first three chapters.

There has been a tendency to treat *conditioning chambers* in the 'do-it-yourself' manner—a suitable box, a dish of salt solution, a circulating fan (perhaps!), low wattage heater with controller, a thermometer and some form of dial hygrometer slung together. This is not to decry salt solutions as a means of controlling humidity; they can provide a cheap and accurate method if used properly and, in particular, if the salt solution has a large surface area. The often overlooked factors are in the design of the chamber: lack of thermal insulation, doors that when opened immediately cause an almost complete interchange of atmosphere with that of the surrounding room, shelves and corners which prevent adequate air circulation, fans which do not move the air sufficiently, heaters which radiate to test specimens and hygrometers which do not indicate r.h. correctly.

British Standard practice is covered by B.S. 3718: 1964, 'Laboratory Humidity Ovens (Non-injection Type)', and B.S. 3898 : 1965, 'Laboratory Humidity Ovens (Injection Type)'.

Much use is made of 'mechanical' types of hygrometer such as those which depend on the change of dimensions of hair or paper with change in humidity; they are certainly easy to use, but need very careful attention. They are prone to hysteresis, may be affected by mould growth in high humidity and fragile to the extent that strictly limited air speeds should be used (which may conflict with the maintenance of uniform r.h. in the chamber!). They are only correct at the temperature of calibration at the best of times and may have significant negative temperature coefficients. It is definitely wise to use each hygrometer over only a very restricted range of relative humidity.

General guidance on the use of hygrometers will be found in 'N.P.L. Notes on Applied Science, No. 4 Measurement of Humidity' (Published by Her Majesty's Stationery Office (London) 1958) and it will be found that the wet and dry bulb type is still amongst the most accurate. Likewise in A.S.T.M. E.337–62 (see above), the preferred instrument is the wet and dry bulb sling hygrometer or psychrometer, the use of which is carefully described.

In the N.P.L. notes (Appendix V), there is a table showing the relative humidities obtainable from a range of salts (in the form of saturated solutions) at various temperatures (see Table 3.1).

ISO Recommendation 493, 'Plastics. Methods of Maintaining Contact Relative Humidity in Small Enclosures by Means of Aqueous Solutions', gives two techniques: (a) by means of saturated salt solutions (the Table given in Annex B is virtually identical to the N.P.L. data) and (b) by using glycerol solutions (see below, A.S.T.M. E.104). The necessary requisites for constructing suitable enclosures are described.

A.S.T.M. E.104–51, 'Recommended Practice for Maintaining Constant Relative Humidity by Means of Aqueous Solutions', gives the following three methods:

1. Using glycerol solutions and checking the composition thereof by measurement of the refractive index. A table gives expectable relative

humidities (at 0°C, 25°C, 50°C and 70°C) between about 40% and 98%. 0.1% copper sulphate is added to avoid fungal growth.

2. Using aqueous sulphuric acid. Control of composition is by density and r.h. values between about 1% and 100% are obtainable over a similar temperature range.
3. Using saturated salt solutions. A similar table to that above is provided.

The A.S.T.M. recommends small airtight containers, but admits that small vents may be needed under certain conditions. It makes the very pertinent point that the chamber should not be overloaded with test specimens!

A.S.T.M. D.618 specifies that measurements of humidity (and temperature) should always be made as close as possible to the specimen being conditioned (or tested), but in no case more than 60 cm (2 ft) from it.

The measurement of temperature needs little comment in the present context because extreme accuracy is not required. The ordinary mercury in glass thermometer is quite adequate, suitably shielded from any radiation.

Test laboratories are obviously more of a problem because of their very size; they must be large enough to take the necessary test equipment and

Table 3.1 RELATIVE HUMIDITIES FROM SATURATED SALT SOLUTIONS

Saturated salt solution	*Temperature* (°C)								
	5	10	15	20	25	30	35	40	50
	Relative humidity (%)								
Potassium sulphate	98	98	97	97	97	96	96	96	96
Potassium nitrate	96	95	94	93	92	91	89	88	85
Potassium chloride	88	88	87	86	85	85	84	82	81
Ammonium sulphate	82	82	81	81	80	80	80	79	79
Sodium chloride	76	76	76	76	75	75	75	75	75
Sodium nitrite	—	—	—	66	65	63	62	62	59
Ammonium nitrate	—	73	69	65	62	59	55	53	47
Sodium dichromate	59	58	56	55	54	52	51	50	47
Magnesium nitrate	58	57	56	55	53	52	50	49	46
Potassium carbonate	—	47	44	44	43	43	43	42	—
Magnesium chloride	34	34	34	33	33	33	32	32	31
Potassium acetate	—	21	21	22	22	22	21	20	—
Lithium chloride	14	14	13	12	12	12	12	11	11
Potassium hydroxide	14	13	10	9	8	7	6	6	6

operator(s) and not to suffer any significant changes due to the running of electric motors, body temperature, humid breath, etc. Double windows and double doors, and good insulation generally, are pre-requisites. Humidity control by saturated salt solutions is hardly feasible and it is usual to employ a humidifier (e.g. an evaporator) or a dehumidifier (e.g. a refrigerator), to add water vapour or remove it, depending on whether atmospheric ambient conditions contain more water per unit volume or less than those required. Control of these units is by some form of humidistat which may be little more than an elaborate paper hygrometer used in a different way. Both with control of humidity and temperature (the latter basically simple except where refrigeration is needed to produce temperatures below ambient), careful design of the room is essential to avoid stationary pockets of air and produce adequate circulation. (See B.S.

4194 : 1967, 'Recommendations on the Design Requirements and Testing of Controlled Atmosphere Laboratories'.)

With both conditioning chambers and test laboratories, installation of recorders is essential for providing a permanent record of the humidities and temperatures prevailing throughout the conditioning and testing periods.

FURTHER READING

1. 'Symposium on Conditioning and Weathering', *Special Technical Publication No. 133, American Society for Testing and Materials,* Philadelphia (1953)
2. GAVAN, F. M. and JOY, F. A., 'Conditioning Equipment for Polymer Testing', *Testing of Polymers,* Ed. J. V. SCHMITZ, Interscience Publishers, New York (1965)

4

Characterisation of Polymer Structure and Size (Particularly by Viscosity in Solution)

4.1 INTRODUCTION

The most profound influence on the properties of a polymer is exerted by the chemical constitution of the polymer molecule itself. However, this handbook is concerned with testing and accordingly it will be assumed that the chemical constitution is known, either by reason of awareness of the starting materials used or, better, by detailed analysis. After this, properties are influenced to varying degrees by (a) molecular weight (i.e. size of molecules) and (b) crystallinity (i.e. relative positioning of molecules one to another)—though the range of values of the latter that may be encountered is very dependent on the chemical nature of the polymer. Since the result of polymerisation is invariably not a number of polymers of identical molecular weight, but rather polymers of a range of molecular weights, the influence of molecular weight on properties may be due to the average value and/or the range of values in the sample under consideration.

Generally speaking, *crystallinity* will influence the following parameters:

Density Rigidity Short time tensile strength Softening Point Solvent Resistance	(positive correlation)
Permeability to gases Shock (impact) resistance	(negative correlation)

and *molecular weight*, the following:

Melt viscosity
Shock (impact) resistance
Load bearing properties (long term)

It must be emphasised that these are broad generalisations and by no means exhaustive; thus as far as molecular weight is concerned, the comments apply to the normal commercial range of polymer size and not at low degrees of polymerisation where the influence of the variable is very significant indeed.

4.2 CRYSTALLINITY

Methods which may be used to measure the degree of crystallinity are:

X-ray diffraction
Nuclear magnetic resonance
Infra-red absorption
Density

The first mentioned is the fundamental and most widely employed technique; it and N.M.R. are, however, beyond the scope of this book and the reader is referred to References 1—13, for instance.

Infrared techniques are applicable to certain polymers, e.g. polypropylene, where the absorptions of the crystalline and amorphous regions differ significantly. It is not a direct measurement and the relationship of absorption, or ratio of absorptions at two frequencies (one selected as markedly dependent on crystallinity and the other essentially independent thereof, to act as a test variable reducing quotient), with crystallinity has to be established by calibration[14, 15].

The most simple method is by density measurement, which again is dependent on accurate calibration in the first place. It is not difficult to envisage that the more ordered (crystalline) the packing of polymer molecules the greater will be the density of the mass polymer; usually the relationship between density and percentage crystallinity is linear. Nevertheless there are exceptions and with poly(4-methyl pentene-1), for instance, the crystalline polymer is *less* dense than the amorphous variety.

The measurement of density is described in Chapter 5 and any method accurate to 0.0001 g/cm^3, and appropriate to the test specimen in question, should be suitable.

4.3 MOLECULAR WEIGHTS GENERALLY

Unless the molecular weights of all the polymer species are identical in a given sample, i.e. it is a *homodisperse* system, where the average value has a unique meaning, the expression of average molecular weight can be made in several different ways—that is for the normal *heterodisperse* system.

The number average molecular weight, $\bar{M}_N$, results from 'counting' the number of molecules present and is obtained from such methods as end-group analysis, electron microscopy and the measurement of colligative properties such as osmotic pressure, lowering of freezing point, elevation of boiling point and depression of vapour pressure. Light scattering techniques yield a weight average molecular, $\bar{M}_W$, whilst solution viscosity measurements result in an average, $\bar{M}_\alpha$, generally somewhere between the two. Higher still is z-average, $\bar{M}_Z$, approximately that obtained by sedimentation equilibrium (ultracentrifuge). Progressing through $\bar{M}_N$, $\bar{M}_\alpha$, $\bar{M}_W$ to $\bar{M}_Z$ so the average becomes more dependent on the higher molecular weight species present in the sample.

Mathematically,

$$\bar{M}_N = \frac{\sum_i N_i M_i}{\sum_i N_i}$$

$$\bar{M}_\alpha = \left[\frac{\sum_i N_i M_i^{\alpha+1}}{\sum_i N_i M_i}\right]^{1/\alpha}$$

$$\bar{M}_W = \frac{\sum_i N_i M_i^2}{\sum_i N_i M_i}$$

$$\bar{M}_Z = \frac{\sum_i N_i M_i^3}{\sum_i N_i M_i^2}$$

where α is derived from the Mark-Houwink formula:

$[\eta] = KM^\alpha$ (see below, Section 4.4.)

[N_i represents the number of the ith polymer species with molecular weight M_i].

Methods for determining molecular weight may be divided into two classifications: primary and secondary. The former includes osmotic pressure and the other colligative techniques, electron microscopy, gel permeation chromatography, ultra-centrifugal and end group analysis methods; of these, osmotic pressure is the most important, being useful up to $\bar{M}_N$ values greater than 1 000 000 in contrast to the other methods which are generally limited to a maximum of 20 000 or thereabouts.

In the secondary method class, solution viscosity is very widely used by virtue of simplicity of technique and apparatus (but *not* theory!), followed by light scattering.

The majority of these methods demand text books in their own right and the reader is referred to the standard works[16, 21] for detailed consideration. Useful information will also be gained from References 22–26 and the literature of Reference 27.

However, the technique of solution viscosity merits detailed consideration here first because (for what it is worth) the method as used for obtaining a measure of molecular weight is simple and quick and, second, because it is widely used as a process control method for molecular size by simply recording the solution viscosity, i.e. not converting the latter to molecular weight.

4.4 MOLECULAR WEIGHT FROM SOLUTION VISCOSITY

It has already been stated that this is a secondary (non-absolute) method of measuring molecular weight; considerable thought has been given to the theory of solution viscosity, but as yet no fundamental relationship has been worked out completely to make it a primary one. For detailed considerations of the current thoughts, see pages 220–237 of Reference 1, and References 17, 20 and 28; a very brief treatment follows, starting with definitions of the essential terms.

The International Union of Pure and Applied Chemistry has approved the following definitions:

$\frac{\eta}{\rho}$ = Viscosity/density ratio ('kinematic viscosity') of solution. η = dynamic or absolute viscosity (usually measured in poises or centipoises) and ρ = density.

$\frac{\eta}{\eta_0}$ = Viscosity ratio (formerly 'relative viscosity') of solution.

$\frac{\eta_0}{\rho_0}$ = Viscosity/density ratio for solvent at same temperature (for dilute solutions, $p \approx p_0$).

[$\eta_{sp} = \frac{\eta - \eta_0}{\eta_0}$ (formerly specific viscosity')]

$\frac{\eta_{sp}}{c}$ = Viscosity number $= \frac{\eta - \eta_0}{\eta_0 c}$ (formerly 'reduced viscosity' of solution).

c = concentration of solute in g/ml of solution.

$\frac{\ln \frac{\eta}{\eta_0}}{c}$ = Logarithmic viscosity (formerly 'inherent viscosity') of solution.

$[\eta]$ = Limiting viscosity number, L.V.N. (formerly 'intrinsic viscosity') of solution.

L.V.N. is produced by plotting viscosity number or logarithmic viscosity against concentration, which usually produces a straight line at least at lower concentrations, and extrapolating to the $c = 0$ axis either by eye, or preferably statistically by a 'least squares' method. Alternatively, if the particular polymer species, solvent and temperature have been reliably investigated, L.V.N. may be calculated from a single solution viscosity determination at a known concentration by the Huggins equation:

$$\frac{\eta_{sp}}{c} = [\eta] + k'[\eta]c^2$$

where k' is the Huggins constant and is assumed to have been reliably determined previously.

The L.V.N. having been so determined, to convert it to an average molecular weight figure involves the use of an appropriate mathematical relationship previously established between solution viscosity, for the polymer, solvent and temperature in question, and molecular weight as measured by a primary method—most often that of osmotic pressure—using monodisperse samples as far as possible.

For many years the empirical Mark-Houwink equation:

$$[\eta] = KM^{\alpha},$$

where K and α are constants and M = molecular weight, has been the most favoured relationship, and though more recent theoretical approaches, notably that of Fox and Flory, have produced far more complicated equations, they can often be reduced by approximation and certainly for convenience to the above form.

The literature abounds with papers dealing with practical investigations establishing values of K', K and α for all manner of polymeric species. In using these values, it is imperative to satisfy oneself as to the reliability of the reported work in respect of:

1. The efficiency of fractionation of the polymer studied—i.e. does it approach monodisperse?
2. The accuracy of the primary method of measuring molecular weight, e.g. in osmosis the efficiency of the semi-permeable membrane, the temperature control of the experimental set-up, the attainment of equilibrium, etc.
3. The general accuracy of the viscosity measurements with reference to such factors as temperature control, completeness of solution, cleanliness of solutions and of viscometer, and purity of solvent.

If all these factors seem in order, then the user of the reported data must be certain he is operating with *precisely* the same polymeric species, the same solvent, the same temperature and using the same concentration units (there is much confusion and indefinition in the literature about this last-mentioned). Put very simply, solution viscosity assesses molecular weight because the larger the polymer molecule dissolved the greater the friction in flow, i.e. the greater the viscosity; again, over simplified, a 'good solvent will encourage the polymer chain to stretch out, whereas a 'poor' one will cause it to coil up, so that a given polymer will yield a lower viscosity solution in a 'poor' solvent than in a 'good' solvent, even though the viscosities of the solvents themselves be identical. Thus a given polymer could yield different solution viscosities, even though at the same concentration and temperature, in a pure solvent and in the same solvent contaminated with a small portion of a different one, if the 'polymer solvent' interactions were different. Again, in a given pure solvent, though two polymer species may have the same molecular weights, the solution viscosity of a homopolymer could be quite different to that of a copolymer even though the latter is based primarily on the same monomer unit. Both these effects will be seen to give different calculated (apparent) molecular weights from solution viscosity measurements on polymers of identical molecular weight.

When all is said and done, it is very rare that a knowledge of absolute molecular weight is needed; most often the *trend*, qualitative or quantitative, is of prime interest, be it to determine the effect of polymerisation conditions on molecular weight (decrease or increase and in what ratio) or the dependence of some physical property—say softening point—on molecular weight and particularly where increase in the latter ceases to affect the former. Frequently, therefore, a guide to molecular magnitude, as from L.V.N. or even η_{sp}/c, will be quite sufficient.

4.5 MEASUREMENT OF SOLUTION VISCOSITY

Standard methods in the U.K. generally follow the procedures laid down in B.S. 188[29] which is a suitable manual for all such measurements on low viscosity liquids. The glass capillary viscometers described are almost invariably used; the other technique, falling sphere viscometry, is more suitable for 'dopes' because of the much higher viscosities involved.

Three types of capillary viscometer are described (see Figures 4.1–4.3).

In the ultimate, every viscometer is calibrated using a liquid of known viscosity, though this calibration may be conveniently achieved using an

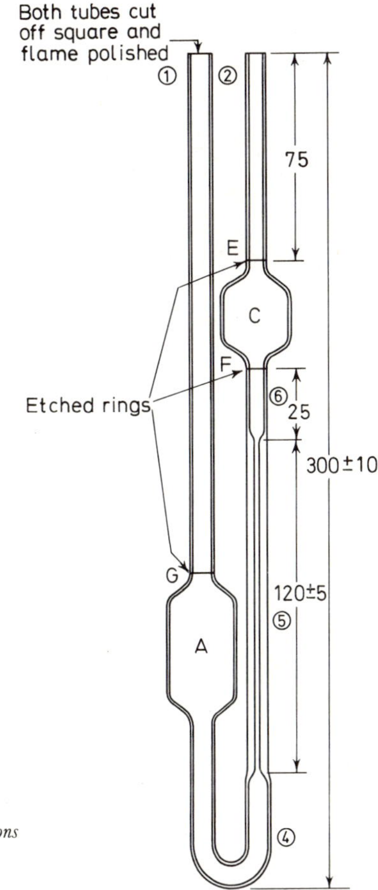

Figure 4.1. U-tube viscometer. All dimensions in millimetres (B.S. 188)

unknown, homogeneous and stable liquid run through both the uncalibrated viscometer and a calibrated 'master' viscometer.

The kinematic viscosity of a liquid $\nu(=\eta/\rho)$ and its flow time t through a capillary viscometer are connected by a formula of the type

$$\nu = Ct$$

where ν is most often expressed in centistokes, t in seconds and C is the viscometer constant.

This equation is applicable in all cases except where the viscosity is so low, and the rate of efflux so great, that a correction must be made to allow for the effects of kinetic energy of efflux. In this case,

$$\nu = Ct - \frac{B}{t}$$

B may be determined experimentally or calculated from the known dimensions of the viscometer. However, it is usually possible to use the simpler equation by choosing a viscometer to give a sufficiently large value of t. Under these conditions, it will be noted that, for the determination of such values as viscosity ratio, viscosity number and logarithmic viscosity, the value of c need not be known (since it cancels out)—that is the viscometer can be uncalibrated; however, it follows that one and the same viscometer must be suitable for both the solution and the (less viscous)

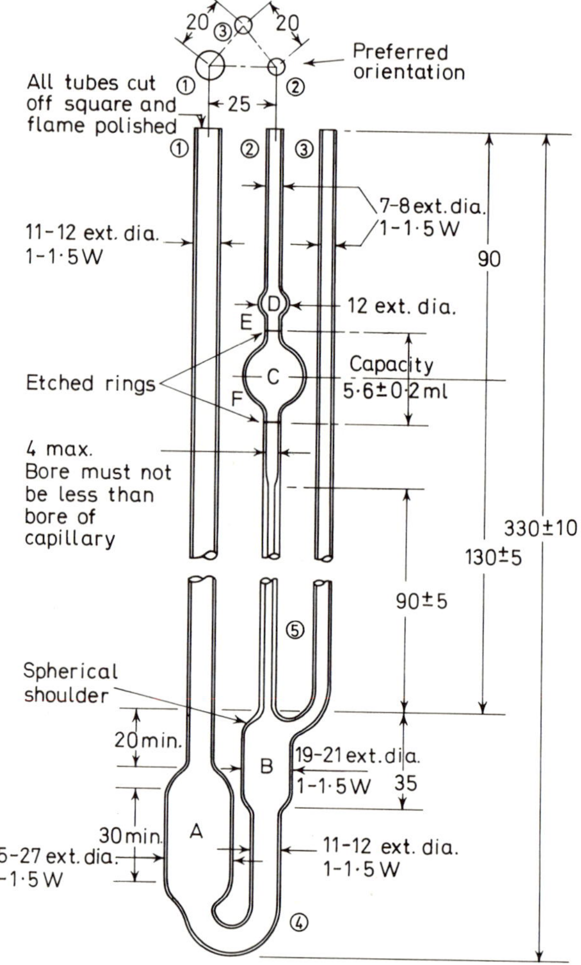

Figure 4.2. Suspended level viscometer. All dimensions in millimetres (B.S. 188)

solvent and this is often not easy to arrange. Table 3 of B.S. 188 describes a series of viscometers of different sizes, in each type, and their usable ranges using the simple equation, and the kinetic energy correction factor.

The use of the viscometers is fully described in the above mentioned standard; it suffices to confirm here the efflux time t is the period of flow of the liquid between two etched marks (e.g. *E* and *F* in the U-tube viscometer) under the action of gravity. B.S. 188 lays down requirements for temperature measurement and control of the thermostat bath ($\pm 0.015°C$ in range 15°C to 100°C and $\pm 0.03°C$ outside this), verticality of the viscometer and accuracy of the timing device (0.07% over a period of not less

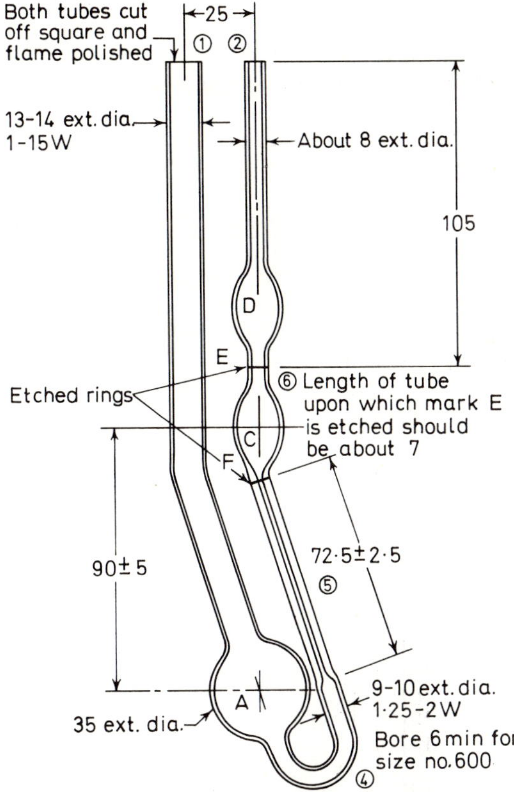

Figure 4.3. Cannon-Fenske viscometer. All dimensions in millimetres (B.S. 188)

than 15 min); it is essential to check stopwatches to ensure that they meet this requirement. The B.S. also draws attention to the need for extreme cleanliness in undertaking such measurements.

U.S. practice is laid down in A.S.T.M. D.445–65[30] which again covers all the important variables mentioned above and allows the use of some nineteen types of viscometers including those of B.S. 188. For general guidance on German practice, see Reference 31.

Viscous solutions may conveniently be examined by the 'falling sphere' method, which is fully described in B.S. 188 for instance. The kinematic viscosity ν (in centistokes) of the liquid is given by:

$$\nu = Mg\frac{(\delta-\rho)}{0{\cdot}03\pi v d\rho\delta}\times F$$

$$\text{or}\quad \nu = \frac{d^2 g(\delta-\rho)}{0{\cdot}18 v\rho}\times F$$

Where M is the mass of the sphere in g
d is the diameter of the sphere in cm
δ is the density of the sphere in g/cm^3
ρ is the density of the liquid in g/cm^3
v is the velocity of fall of the sphere in cm/s
g is the local acceleration due to gravity in cm/s^2
and F is known as the Faxén term and is a correction for the effect of the wall of the fall-tube on the motion of the sphere
If D is the diameter of the fall tube in cm

$$F \approx 1{\cdot}00 - 2{\cdot}104\frac{d}{D} + 2{\cdot}09\frac{d^3}{D^3}$$

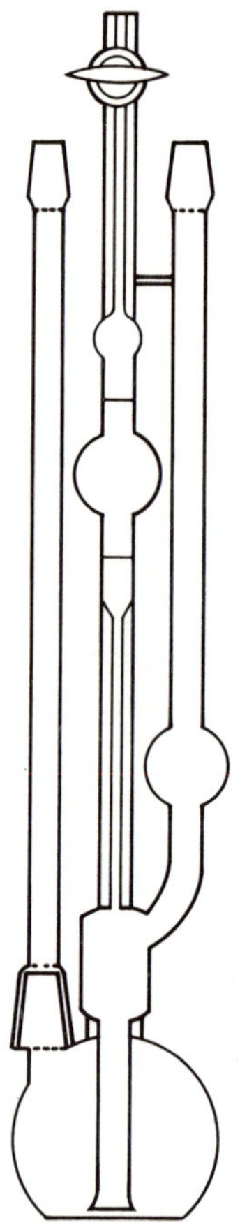

Figure 4.4. Suspended level dilution viscometer (filter type)

Temperature control requirements are as before and an accurate timing device is again essential. Other than this, carefully machined steel balls $\frac{1}{16}, \frac{3}{32}, \frac{1}{8}, \frac{5}{32}$ in diameters) and a 'fall tube', with parallel sides and horizontal timing marks, are all that are required.

4.6 USE OF DILUTION VISCOMETERS

For the determinations of L.V.N. ($[\eta]$), the viscosities of solutions of various concentrations must be determined or, if the efflux times are long enough and the same viscometer is used throughout, just the flow times. Depending on the molecular weight of the polymer, the most usual concentrations to examine are in the range 0·05 to 0·5 g/100 ml, i.e. 0·0005—0·05 g/ml in the I.U.P.A.C. units.

To obviate the tedium of making up these solutions, however, commercial viscometers are available[27] which permit the determination of the efflux time of a precisely known volume of the most concentrated solution to be examined, followed by successive dilutions, with known volumes of solvent, and viscosity determinations at each stage; a final measurement with pure solvent is of course, necessary. The feasibility of such a 'dilution viscometer' results from the fact that the suspended level viscometer, Figure 4.2 above, does not require the use of a carefully controlled volume of liquid, because of the 'free fall' of the liquid at the lower end of the capillary. Thus, by careful selection of the dimensions of the bulbs, etc., dilutions of up to 10-fold can be made without upsetting the functioning of the viscometer. As a further refinement, a sintered glass filter has been built in (see Figure 4.4).

These so-called 'Modified Ubbelohde Viscometers' are described in Reference 32.

4.7 STANDARDISED PROCEDURES FOR DETERMINING SOLUTION VISCOSITY

Table 4.1 summarises some of the currently (see Preface) nationally standardised methods for measuring the solution viscosities of polymers:

Table 4.1 STANDARD PROCEDURES FOR SOLUTION VISCOSITY DETERMINATION

Polymer type	*Solvent*	*Concentration*	*Temperature*	*Expression of Results*	*Source*
PVC	Cyclohexanone	0·005 g/1 ml	25°C	Viscosity Number	B.S. 2782: Method 404A: 1970
Polystyrene	Toluene	0·02 g/1 ml	25°C	Viscosity in Ns/m^2 or cP	B.S. 2782: Method 404B: 1970 B.S. 1493: 1967
Polyamide	Formic acid (90%)	0·005 g/1 ml	25°C	Viscosity Number	B.S. 2782: Method 404C: 1970
Polyamide	Metacresol	0·005 g/1 ml	25°C	Viscosity Number	B.S. 2782: Method 404D: 1970
Cellulose Acetate	Acetone	0·06 g/1 ml	25°C	Percentage change in viscosity on moulding	B.S. 1524: 1955 Appendix H
Cellulose Acetate	90 parts dichloro-methane 10 parts methanol	0·01 g/1 ml	25°C	Viscosity Ratio ('Relative Viscosity')	B.S. 2880: 1957 Clause 8*

Table 4.1—*continued*

Polymer type	*Solvent*	*Concentration*	*Temperature*	*Expression of Results*	*Source*
Cellulose Acetate	Various	0·0026 g/1 ml	25°C	L.V.N. ('Intrinsic Viscosity') calculated from given Baker-Philippoff equation	A.S.T.M. D.871-63
Cellulose Acetate Butyrate					A.S.T.M. D.817-65
Cellulose Acetate Propionate	Various	20 parts to 100 parts of soln. —by weight		Viscosity in poises by falling sphere method	
Cellulose Nitrate	Ethanol Toluene Ethyl Acetate (Various proportions acc. viscosity)	12·2 parts, 20 parts or 25 parts to 100 parts of soln. (wt.)	25°C	Seconds to fall by falling sphere method	A.S.T.M. D.301-56
Ethyl Cellulose	Various	5 parts to 100 parts of soln. (wt.)	25°C	Viscosity in cP	A.S.T.M. D.914-50
PVC	Cyclohexanone	0·002 g/1 ml	30°C	Inherent Viscosity	A.S.T.M.D.1243-66
Polyvinylidene chloride	Orthodichloro-benzene (96–99%)	'2%' (0·02663 g/ml solvent)	120°C	Viscosity in cP	A.S.T.M. D.729-57 Para. (c)
Polyamide ('nylon')	Formic acid (90%)	0·11 g/1 ml solvent	25°C	Viscosity ratio ('Relative Viscosity')	A.S.T.M. D.789-66 Para. (d)
Polyamide ('nylon')	Metacresol	0·0944 g/1 ml solvent	25°C	Viscosity ratio ('Relative Viscosity')	
Polyethylene	Decalin	Approx. 0·001, 0·002, 0·003 and 0·004 g/ml	130°C	Viscosity ratio, Logarithmic Viscosity Number, L.V.N.	A.S.T.M.D.1601-61
PVC	Cyclohexanone	0·005 g/1 ml	25°C	k Number' where $\log\frac{\eta}{\eta_0} = \frac{75k^2}{1+1\cdot5kC}+kC$ where C = concentration in g/100 ml	DIN.53726
Polystyrene	Benzene	0·01 g/1 ml	25°C	'K value' calculated from viscosity ratio	DIN.7741

Table 4.1—*continued*

Polymer type	*Solvent*	*Concentration*	*Temperature*	*Expression of Results*	*Source*
Polycarbonate	Methylene chloride	0·005 g/1 ml	25°C	'*K* value' calculated from viscosity ratio	DIN.7744
Polymethyl-methacrylate	Chloroform	~ 250 mg/ 100 ml	25°C	Viscosity Number	DIN.7745
Polyamide	See below	ISO/R.307			DIN.53727
Cellulose Acetate	See below	ISO/R.1157			DIN.53728

*Also includes a method for testing a 25% solution in 95% acetone by falling sphere technique.

†This method derives from Reference 33 and the reader is warned that many commercial manufacturers quote values using ethylene dichloride as solvent; results are somewhat different to those of solutions in cyclohexanone[34]. Results for a wide range of vinyl chloride polymers in various solvents, and expressed in a variety of forms, are given by Matthews and Pearson[35].

The above table gives only some of the nationally standardised methods and there are many other methods produced by other bodies mentioned in Chapter 2. The whole picture well illustrates the need for international standards where the current (as before) position is as shown in Table 4.2.

Table 4.2

Polymer type	*Solvent*	*Concentration*	*Temperature*	*Expression of Results*	
PVC	Cyclohexanone	0·005 g/1 ml	25°C	Viscosity Number	ISO/R.174
Polyamide	Formic acid (90%) or metacresol	0·25 g/50 ml	25°C	Viscosity Number	ISO/R.307
Polyamide	Formic acid (90%)	5·5 g/50 ml	25°C	Viscosity Ratio ('Relative Viscosity')	ISO/R.600
Polyethylene and Polypropylene	Decahydro-naphthalene	0·05 g/50 ml	135°C	Viscosity Number	ISO/DR.754
Methylmeth-acrylate polymers and copolymers	Chloroform	0·5 g/100 ml	20°C	Viscosity Number	ISO/DR.824
Cellulose acetate (⩾ 50% acetic acid yield)	Dichloromethane (90%) and Methanol (10%)	0·5 g/100 ml	25°C	Viscosity Number and Viscosity Ratio	ISO/R.1157
Poly (ethylene terephthalate)	o-chlorophenol	0·01 g/ml	25°C	Viscosity Number	ISO/DR.1002
Directives for the standardisation of methods for the determination of the dilute solution viscosity of polymers.					ISO/DR.1628

REFERENCES

1. GORDON, M., *High Polymers. Structure and Physical Properties,* 2nd edn, Plastics Institute Monograph, Chapter 6, Iliffe Books Ltd., London (1963)
2. FRITH, E. M. and TUCKETT, R. F., *Linear Polymers,* Chapter 1, Longmans, Green and Co. London (1951)
3. WEISSBERGER, A., *Physical Methods of Organic Chemistry,* Volume I, Part II, Interscience Publishers Inc., New York (1949)
4. KLINE, G. M., ed, 'Analytical Chemistry of Polymers, Part II, Analysis of Molecular Structure and Chemical Groups', *High Polymers,* Volume XII, Interscience Publishers Inc., New York (1962)
5. KE, BACON, 'Newer Methods of Polymer Characterisation', *Polymer Reviews,* Volume 6, Interscience Publishers Inc.: John Wiley & Sons Inc., New York (1964)
6. MILLER, M. L., *The Structure of Polymers,* Reinhold Publishing Corporation, New York (1966)
7. STATTON, W. O., 'The Use of X-Ray Diffraction and Scattering in Characterisation of Polymer Structure', *Int. Symp. Plast. Test. Stand.,* Special Technical Publication No. 247, American Society for Testing and Materials, Philadelphia (1959)
8. BUNN, C. W., 'Crystallinity in Polymers: Occurrence, Measurement and Influence on Properties', *The Physical Properties of Polymers, S.C.I. Monograph No. 5,* Society of Chemical Industry (London), 3 (1959)
9. ALLEN, G., 'Applications of N.M.R. Spectroscopy in Polymer Science', *Rev. pure appl. Chem.* **17,** 67 (June 1967)
10. BOVEY, F. A., 'The High Resolution Nuclear Magnetic Resonance Spectroscopy of Polymers', *Polymer Engng Sci.,* **7,** No. 2, 128 (April 1967)
11. SLICHTER, W. P., 'High Resolution N.M.R. in Polymers', *J. chem. Educ.,* **45,** No. 1, 10 (January 1968)
12. KAVESH, S., and SCHULTZ, J. M., 'Meaning and Measurement of Crystallinity in Polymers: A Review', *Polymer Engng Sci.,* **9,** No. 5, 331 (September 1969)
13. KAVESH, S., and SCHULTZ, J. M., 'Meaning and Measurement of Crystallinity in Polymers: A Review', *Polymer Engng Sci.,* **9,** No. 6, 452 (November 1969)
14. POTTS, W. J., 'The Use of Infra-red Spectroscopy in Characterisation of Polymer Structure' in *Int. Symp. Plast. Test. Stand.,* Special Technical Publication No. 247, American Society for Testing and Materials (Philadelphia) (1959)
15. ELLIOTT, A., 'Some Applications of Infra-Red Spectroscopy to Problems of Polymer Structure', *Symp. Tech. polymer Sci.,* R.I.C. Monograph No. 5, The Royal Institute of Chemistry (London), 26 (1956)
16. WEISSBERGER, A. (ed.) *Physical Methods of Organic Chemistry,* Volume 1, Interscience Publishers Inc., New York Part I (1949) and Part III (1954)
17. TANFORD, C., *Physical Chemistry of Macromolecules,* John Wiley & Sons Inc., London (1961)
18. ROBB, J. C., and PEAKER, F. W., *Progress in High Polymers,* Heywood and Company Ltd., London (1961)
19. JONYIIAN, CH'IEN, *Determination of molecular weights of high polymers,* translated by Dr. J. Schmorak, Israel Program for Scientific Translations Ltd., Jerusalem (1963)
20. ALLEN, P. W., *Techniques of Polymer Characterisation,* Butterworths (Publishers) Ltd., London (1959)
21. SMITH, D. A., ed., *Addition Polymers. Formation and Characterisation,* Chapter 5, Butterworths (Publishers) Ltd., London (1968)
22. MCINTYRE, D., 'The Accurate Determination of Molecular Weights of Macromolecules' in *Int. Symp. Plast. Test. Stand.,* Special Technical Publication No. 247, American Society for Testing and Materials , Philadelphia (1959)
23. ONYON, P. F., 'Osmometry and Light Scattering', *Symp. Tech. Polymer Sci.,* R.I.C. Monograph No. 5, 4, The Royal Institute of Chemistry, London (1956)
24. ZICHY, E. L., 'Ebulliometric Determination of Molecular Weight', *Tech. Polymer Sci.,* S.C.I. Monograph No. 17, Society of Chemical Industry, London, 122 (1963)
25. QUALE, D. V., 'Molecular Weight Determinations on Polymers by Electron Microscopy', *Brit. Polymer J.,* **1,** No. 1, 15 (January 1969)
26. DETERMANN, H., *Gel Chromatography,* translated by E. Gross, Springer-Verlag (New York) Inc. (1968)
27. WHITEHEAD, ALAN D., The Ancient House, Ardleigh, Nr. Colchester, Essex.
28. FLORY, P. J., *Principles of Polymer Chemistry,* Cornell University Press, Ithaca (1953)
29. B.S. 188: 1957, *Determination of the Viscosity of Liquids in C.G.S. units,* British Standards Institution, London

30. A.S.T.M. D.445–65, *Standard Method of Test for Viscosity of Transparent and Opaque Liquids (Kinematic and Dynamic Viscosities)*, American Society for Testing and Materials, Philadelphia
31. DIN.51550, *Bestimmung der Viskositat,* Deutscher Normenausschuss, Berlin (1960)
32. A.S.T.M. D.1601–61, *Standard Method of Test for Dilute Solution Viscosity of Ethylene Polymers,* American Society for Testing and Materials, Philadelphia
33. FIKENTSCHER, H., 'Systematik de Cellulosen auf Grund ihrer Viskosität in Lösung', Cellulose-Chem. **13,** 58 (1932)
34. PENN, W. S., *PVC Technology,* 2nd edn, Maclaren & Sons Ltd., London, 9 (1966)
35. MATTHEWS, G. A. R., and PEARSON, R. B., 'Molecular Weight Characteristics of PVC', *Plastics, Lond.,* **28,** No. 307, 98 (May 1963)

5

Density, Relative Density (Specific Gravity) and Dimensions

5.1 DENSITY AND RELATIVE DENSITY (SPECIFIC GRAVITY)

5.1.1 Definitions

Density: Mass per unit volume (at defined temperature).
Metric units: g/cm^3 (frequently, but not exactly identical, in units of g/ml).

Relative Density:* Mass (of substance) compared to the mass of an equal volume of a specific (reference) substance—most often water. In this instance the temperature of both the substance and the reference substance must be stated. Being a ratio, the property is dimensionless.

5.1.2 Standard Tests

Recommendation 5.1 of B.S. 4618 from Committee PLC/36 lays down the form of presentation of density data.

DISPLACEMENT METHOD

Method 509A of B.S. 2782:1970 is for solid plastics at 20 ± 2°C or 23 ± 2°C; it measures density relative to water, but assumes the error is negligible in quoting this numerical value as equal to the density in g/ml (Density of water $\approx$ 1 g/ml).

A specimen, with smooth surfaces free from crevices and dust and preferably of mass not less than 5 g, is weighed in air (W_1) and then in freshly boiled distilled water (W_2) at the requisite temperature, after allowing sufficient time for the specimen to reach the latter.

If the density of the material under examination is less than that of water, a sinker is attached and the mass of this in water (W_3) and of it plus the specimen also in water (W_4) are taken. Fine wire is used for the suspension of the specimen (and sinker) and it is counterbalanced by a similar length of wire—the upthrust on part of the wire on the specimen side of the balance

*The more usual term is *specific gravity* which, however, is not employed here in view of the recently announed I.S.O. and B.S.I. rulings.

is sufficiently small to be ignored. All air bubbles must be removed, for example by use of a minute quantity of detergent.

If no sinker is required:

$$\text{Relative density} = \frac{W_1}{W_1 - W_2}$$

When a sinker is used:

$$\text{Relative density} = \frac{W_1}{W_1 + W_3 - W_4}$$

Specific gravity balances are available commercially which are based on this method[1].

A.S.T.M. D 792–66, Method A–1, is essentially the same but determines the mass of the wire under partially immersed conditions and gives the actual density of water at the test temperature (23°C) for a more accurate result to be calculated. It is for specimens of 50 g or less, as is Method A–2 which is for materials affected by water and uses a liquid without effect on the test specimen. Method A–3 is simply an extension to cover heavier specimens.

PYKNOMETER METHOD

Method B of the same A.S.T.M. is for moulding powders, pellet and flake. A pyknometer (see Figure 5.1) is accurately weighed (W_1), a suitable quantity of the material (1–5 g) is added and the whole reweighed (W_2). The material is covered with freshly boiled distilled water and the pyknometer

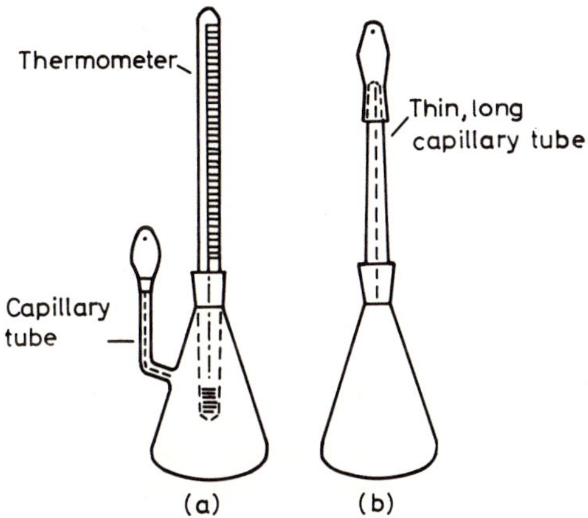

Figure 5.1. Typical pyknometers

is then placed in a vacuum dessicator and all air removed. The pyknometer is filled with water and placed in a constant temperature bath at 23°C (±0·1°C). When the temperature has been achieved the capillary is filled or emptied as necessary to the appropriate level and reweighed (W_3).

The pyknometer is emptied and filled with freshly boiled distilled water by repeating the procedure. The mass (W_4) is recorded.

$$\text{Relative density } 23/23°\text{C} = \frac{(W_2 - W_1) \times d}{(W_3 - W_4) + (W_2 - W_1)}$$

For water $d = 1{\cdot}000$, but if another displacement liquid is used, for example when the material is not inert to water, d is the relative density of this liquid.

$$\text{Density } D^{23°\text{C}} \text{ (g/ml)} = \text{Relative density } 23/23°\text{C} \times 0{\cdot}9976$$

DENSITY GRADIENT METHOD

If two miscible liquids, of significantly differing densities, are carefully mixed, a uniform density gradient from bottom to top can be set up. If this gradient is calibrated, then a specimen allowed to sink into the mixed liquids will come to rest at a point where its density equals that of the immediately surrounding liquid.

This principle is used in A.S.T.M. D.1505–68 as a general test method for plastics and in Method 509B of B.S. 2782:1970 for polyethylene film. The latter makes reference to B.S. 3715:1964, 'Concentration Gradient Density Columns', and, like the A.S.T.M., recommends liquid mixtures for total density ranges of 0·8—2·9 g/ml (see Table 5.1). Two methods of

Table 5.1 SUITABLE LIQUID MIXTURES (B.S. 3715)

Liquids	*Density range* (g/ml)
Ethanol and water	0·8–1·0
Ethanol and carbon tetrachloride	0·8–1·6
Water and calcium nitrate	1·0–1·6
Carbon tetrachloride and bromoform	1·6–2·9

making density gradient columns are described of which one is as shown in Figure 5.2.

Two solutions are made (A and B), of which A has a density lower than the lowest density of interest by about 20% and B a density higher than the highest of interest by about 30%. One litre of A is placed in vessel 1 and one litre of B in vessel 2, the taps being closed.

The magnetic stirrer in vessel 1 is started (to give thorough mixing but *not* bubble formation) and the connecting tap opened when probably some of solution B will flow into solution A. When hydrostatic equilibrium has been established the right hand tap is opened so that, in conjunction with the capillary bore, the column takes about 2 h to fill. Whilst this happens, solution B flows into, and is mixed with, solution A in vessel 1 which therefore increases in density; hydrostatic equilibrium between the two vessels must be maintained throughout and, for this, a wide-bore connecting tube is necessary with slow filling. Finally the filling tube, with the upper end closed, is removed carefully and slowly (see also Reference 2).

In use, careful control of temperature is of course essential. Total density gradients of more than 0·2 g/ml or less than 0·02 g/ml are not recommended. Needless to say, the mixed liquids must be without effect on the material under test, be transparent, of low volatility and low viscosity. The tube is calibrated with spherical glass floats, about 5 mm in diameter, the densities of which may, for instance, be obtained by a method such as that described in the next Section.

These floats are placed gently in the column and after at least 12 h the heights of each above a reference level is determined by a cathetometer. The measured heights are plotted against density and a smooth curve,

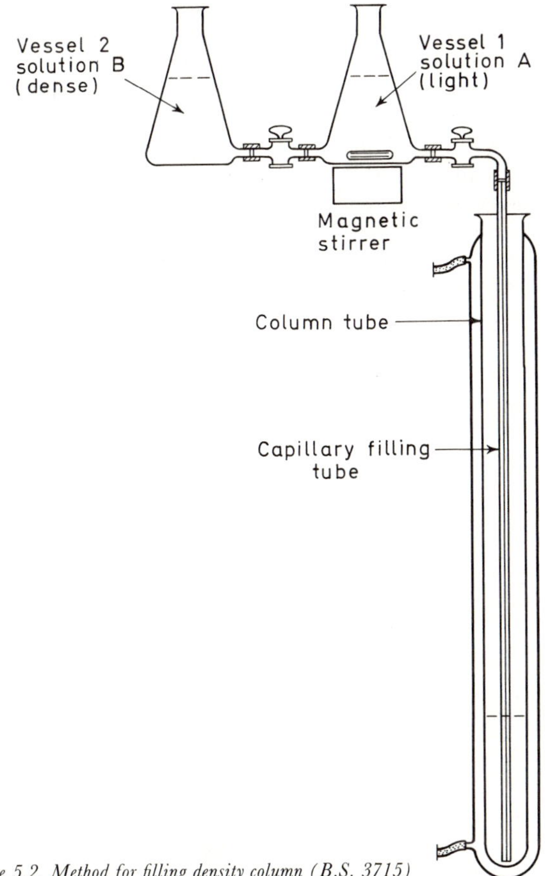

Figure 5.2. Method for filling density column (B.S. 3715)

without discontinuities, should result. With careful use, a column should last several months.

Test specimens should be of easily identifiable shape and size, the latter being such that the centres of their volumes can be estimated to within 1 mm. At least two are placed gently in this column and their heights recorded after not less than 12 h. The densities are read off the calibration graph and, in the case of polyethylene film for instance, should be within 0·0005 g/ml. Care must be taken to avoid air bubbles, dirt and interference between specimens and floats.

The column is cleared as and when necessary by a wire gauze basket, of the widest practicable mesh, drawn upwards through the column at a uniform rate not exceeding 3 cm/min after which the basket is returned to the bottom of the column which is then recalibrated.

FLOTATION METHOD

Method 509C of B.S. 2782:1970 describes a method suitable for polythene film and B.S. 3715:1964, in Method 1 of Appendix A, employs the technique for the standardising of the floats used in the density gradient tube. The latter well illustrates the basis of the method.

First, two clear miscible liquids are chosen, with densities one either side of the range of interest. The floats are placed in a 250 ml measuring cylinder and about 150 ml of the less dense liquid added; if this is chosen correctly, all floats will sink to the bottom. The cylinder and contents are brought to the specified temperature and the more dense liquid is run in from a burette, with stirring, until the lowest density float barely sinks after allowing the liquid to stand for at least three minutes. Another

Figure 5.3. Typical density gradient apparatus. (Courtesy Davenport (London) Ltd.)

drop of liquid is added from the burette, the mix is stirred and left for three minutes. If the float rises, a sample of the liquid is removed; if not more drops are added until it does, when the sample is removed. The procedure is then repeated successively for each float.

The densities of the liquid samples, which are taken as the densities of the corresponding floats, are measured by density bottle or pyknometer technique.

In Method 509C of B.S. 2782:1970, the densities are taken of two liquids, the more dense in which the sample floats and the less dense in which it sinks, so that limits of density may be quoted. Ethanol/water mixtures are recommended, with 0·1% of detergent added; zinc chloride may be used to increase the density of the liquid. Figure 5.3 shows a typical density gradient apparatus.

5.2 THICKNESS (AND WIDTH AND LENGTH)

It is a relatively straightforward matter to measure the thickness of sections (and their width and length if these are small enough to need sensitive techniques) of most conventional materials of construction—e.g. wood, metal and stone; the same applies to rigid plastics such as the phenolics, aminoplastics, polystyrene, unplasticised PVC, polymethylmethacrylate, polycarbonate, etc. Simple specification of the accuracy of the measuring instrument will suffice.

A useful review will be found in Reference 3.

5.2.1 Use of Micrometers and Dial Gauges

External micrometers are too well known to need description here; they are the most frequently employed instruments for the purpose, with calibrations of 0·001 in usually. They are accurately standardised in B.S. 870[4] and their use described in Methods A and B of A.S.T.M. D.374[5] for instance. Likewise dial gauges (strictly 'comparator gauges') are frequently used; B.S. 907[6] covers these in smallest units of 0·001 in, 0·0001 in, and 0·01 mm and Method C of A.S.T.M. D.374[5] details requirements for U.S. practice. (See also DIN 863 and DIN 878.)

The measurement of soft materials which readily yield, particularly highly plasticised PVC and cellular plastics, presents difficulties in defining what the thickness (or width) really is. A figure under no compression force is difficult to obtain, for optical means are hardly suitable for routine rapid working. Micrometers without ratchets are clearly open to considerable subjectiveness in how hard the spindle is screwed down. Ratchet micrometers are little better, slipping at some arbitrary force. Many product specifications require the use of dial gauges, specifying the force on the plunger (which will include any spring loading) and the dimensions of the bearing faces ('anvils'), and there is much variation from one document to another. Each must be carefully studied and the instrument selected accordingly; a general method is given in Method 512B of B.S. 2782:1970 for flexible sheet, where the gauge must read to 0·0025 cm (0·0001 in) for sheet up to 1·8 mm (0·070 in) thick, the foot of the anvil must be not less than 6·3 mm (0·25 in) diameter and the pressure 10–20 kN/m^2 (0·1–0·2 kgf/cm^2; 1·5–3·0 lbf/in^2). For thicker sheet, a gauge reading to 0·025 mm (0·001 in) may be used.

5.2.2 Gravimetric Thickness

A method which does not rely upon the arbitrary selection of bearing pressure is that by which gravimetric thickness is determined. It is more or less indispensable as a means of obtaining a measure of the thickness of embossed sheeting.

In Method 512A of B.S. 2782:1970 an area of not less than 16 cm^2 (2·5 in^2) is cut with a punch, razor, or sharp knife. The mass (W) of the specimen in grammes, its area (A) in square centimetres and its relative density (as in Section 5.1.2 above) are all determined to an accuracy of at least 1 per cent, at 23 ± 1°C.

$$\text{Gravimetric thickness} = \frac{10{\cdot}0W}{A \times \text{relative density}}\ \text{mm}$$

$$\frac{0{\cdot}394W}{A \times \text{relative density}}\ \text{in}$$

5.2.3 Miscellaneous Methods

Accurate measurement of large dimensions may be made by vernier caliper. Continuous monitoring of calendered sheet, extruded film and the like is conveniently effected by instruments based on the absorption of beta radiation which is dependent on the mass of material it passes through, i.e. the thickness of a homogeneous product.

REFERENCES

1. Binney and Smith and Ashby Ltd., 116 Cannon Street, London E.C.4.
2. PAYNE, N., and STEPHENSON, C. E., 'Measuring the density of polyolefins with an improved gradient column', *Mater. Res. Stand.*, **4,** No. 1, 3 (January 1964)
3. KEINWATH, G., *The Measurement of Thickness,* United States Department of Commerce, National Bureau of Standards, Circular 585, Washington D.C. (1958)
4. B.S. 870: 1950, *External Micrometers,* British Standards Institution, London
5. A.S.T.M. D.374–68, *Standard Methods of Test for Thickness of Solid Electrical Insulation,* American Society for Testing and Materials (Philadelphia)
6. B.S. 907: 1965, *Dial Gauges for Linear Measurement,* British Standards Institution, London

6

Testing of Materials Before Moulding

6.1 INTRODUCTION

Whilst the majority of the properties of plastics that concern us relate to the finished article, certainly if we are users pure and simple, there are a number of important tests which are applied to plastics compositions in the unmoulded or precast state. Obviously certain of the properties of such 'raw materials' are of paramount importance to the processor and fabricator if he is to predict his machinery requirements rather than pursue an entirely ad hoc approach. Again, there are tests designed for application to polymers prior to their formulation into moulding compositions. Thus this chapter considers tests which range from a simple melting point determination on, say, a novolak resin to a comprehensive study of the rheological (flow) behaviour of a polymer in the molten state.

There are many 'tests', such as for free phenol in moulded PF articles, free monomer in polystyrene, antioxidant in polyethylene and epoxide equivalent, which are really standardised analytical techniques. These are not within the province of this book and the reader is referred to the relevant B.S. or A.S.T.M. standards, for example, relating to the product in question, to B.S. 2782 and Reference 1.

6.2 MELTING POINT

Under this heading only those tests will be considered which are suitable for materials with a fairly sharp change of state, (solid to liquid), i.e. those which behave somewhat like an ordinary low molecular weight organic compound; as in the characterisation of these latter, most of the standard methods depend on a visual observation of the transition. These melting point tests are generally *not* suitable for the majority of plastics, or more correctly thermoplastics, which usually soften slowly over a temperature range which may be quite extensive, and frequently never reach a free-flowing condition before thermal decomposition sets in. With such an indefinite change of state there is no melting point in the usual sense of the term and it is necessary, if it is required to put a temperature figure to such an indefinite property, to follow the change in some parameter—usually mechanical—with rise of temperature and ascribe a softening point to that temperature at which the property being observed reaches some prescribed value. This type of test is discussed more fully in Chapter 11.

There are several tests available for melting point determination and attention is given below to those which have been standardised. However, mention must be made of the simple and useful Durrans method, where 3 g of resin is gently melted, in the bottom of a test-tube, and the bulb of a thermometer inserted in the resin which is then cooled to solidify. 50 g of mercury is then poured on top of the resin and the whole assembly heated so that the resin temperature rises at a rate of 2 deg C/min until the resin melts and floats on the mercury. The temperature at which this first occurs is recorded as the melting point of the resin[2].

6.2.1 'Ring and Ball' Method

This method appears as Method 103A of B.S. 2782:1970 and is generally used for synthetic resins of the novolak type (see Figure 6.1).

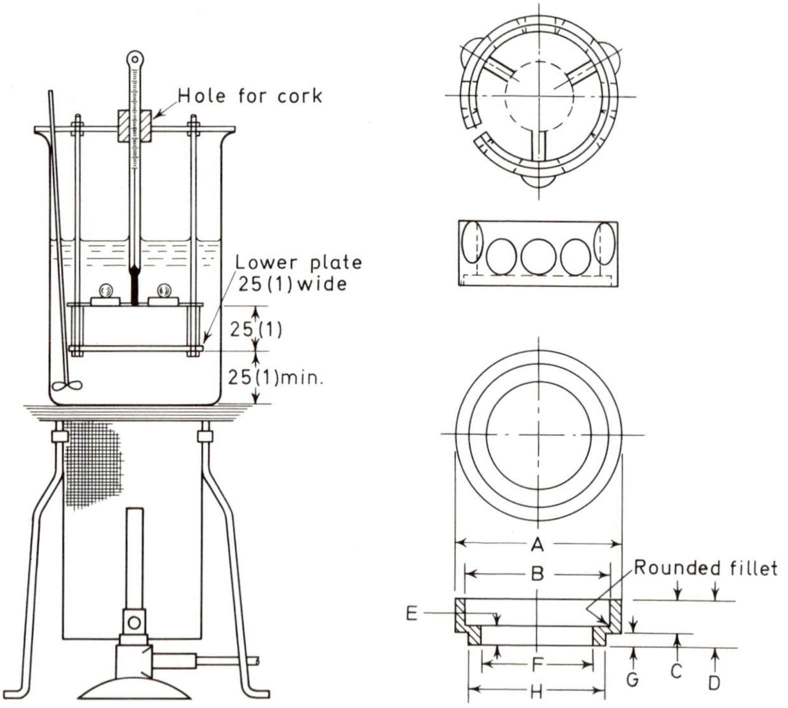

Figure 6.1. 'Ring and Ball' softening point apparatus (B.S. 2782) For (a) dimensions are in millimetres with inch equivalents in parentheses. Dimensions for (c) are as follows: A. 23·0 ± 0·1 mm (0·906 ± 0·004 in) B. 19·9 ± 0·1 mm (0·781 ± 0·004 in) C. 4·4 ± 0·1 mm (0·172 ± 0·004 in) D. 6·4 ± 0·1 mm (0·250 ± 0·004 in) E. 2·8 ± 0·1 mm (0·109 ± 0·004 in) F. 15·9 ± 0·1 mm (0·625 ± 0·004 in) G. 2·0 mm (0·08 in) H. 19·0 ± 0·1 mm (0·750 ± 0·004 in)

The sample under examination is reduced to small pieces (not powder, as bubbles may result on melting) and 1·6 g weighed into a ring (as shown) placed on a metal plate, of suitable finish to avoid sticking. This assembly is then placed in an oven at a temperature 10–20 deg C above the expected melting point until the resin has melted (time should not exceed 15 min). The assembly is then removed, placed on a flat metal surface, and any excess resin over the top of the ring immediately removed by scraping with a hot knife. After allowing to cool for at least 15 min, the ring filled with resin is removed from the plate, if necessary using gentle tapping.

A steel ball, 9·53 mm ($\frac{3}{8}$ in) diameter and mass 3·45–3·55 g, is placed centrally on the cast resin, using a centering guide (as illustrated) which rests on the ring.

Two such specimens are prepared and placed in the apparatus illustrated, using as heat transfer medium a liquid without effect on the resin (glycerol will often be suitable). The heating bath is raised in temperature by 5 deg C $\pm$0·5 per min, with constant mechanical stirring, and the melting point recorded as that temperature at which the ball or surrounding resin first touches the lower plate. The agreement between the duplicate results must be within 1 deg C.

6.2.2 Capillary Tube Method

This will be familiar to all would be organic analysts and employs the well-known Thiele tube. It is described in Method 103B of B.S. 2782:1970 (see Figure 6.2).

The sample under examination is ground to a fine powder with a pestle and mortar and filled to a depth of about 1 cm into a thin walled capillary tube, length about 6 cm, and internal diameter about 1 mm, which is sealed at one end. The sample is compacted as much as possible, by tapping the sealed end of the tube sharply on a hard surface and then the open end of the tube is sealed in a Bunsen flame.

Two such specimens are used; they are attached to the bulb of the thermometer in diametrically opposite positions by an elastic band, which is placed in the Thiele tube in the position shown in the figure, i.e. with bulb level with the upper junction of the side arm. Heating is effected by a small heater (e.g. 175 W) attached to the side arm and controlled by a variable transformer. A lens opposite the bulb provides a convenient means of viewing the behaviour of the sample, with the aid of a lamp.

The nature of the heating medium is obviously not critical as it makes no contact with the sample, but water or glycerol are specified in B.S. 2782. The test is started about 20°C below the expected melting point and the temperature rise is controlled to $1\frac{1}{2}$–2 deg C per min. As the melting point is approached, the powdered sample first contracts away from the tube wall and coalesces (the 'sinter point'). Immediately afterwards it becomes translucent and subsequently becomes transparent. Finally, if the sample truly melts, it runs down the tube.

The melting point is taken as that temperature at which the specimen

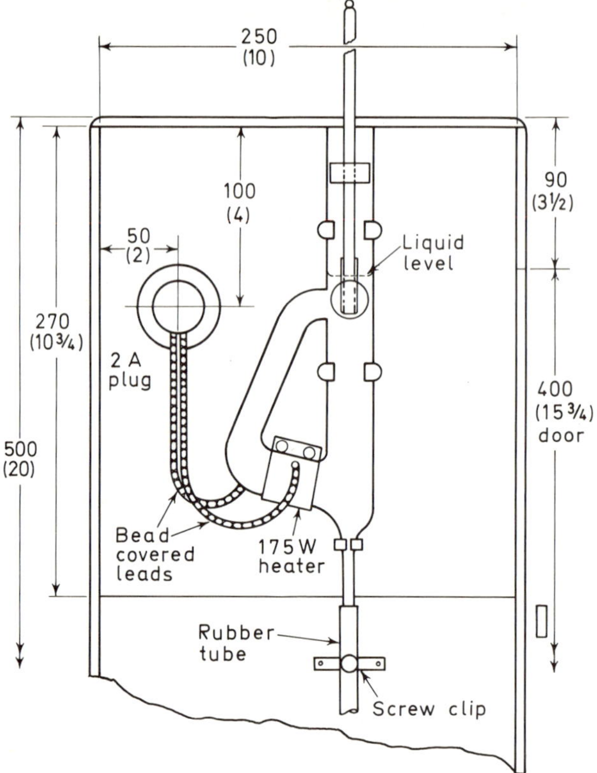

Figure 6.2. Thiele tube and general capillary tube method apparatus (B.S. 2782) All dimensions are in millimetres with inch equivalents in parentheses

first transmits light through half its bulk. Again the agreement of the two separate values must be within 1 deg C.

6.2.3 Hot Plate Method

This method is described as specifically for nylon (Method 103C of B.S. 2782:1970), but is presumably suitable for any material with similar melting characteristics.

The test specimen is a single granule of carefully defined size (related to a certain sieve size) which is placed on a microscope slide resting across the centre of a hot plate. A few drops of silicone fluid are also placed on the slide, adjacent to the granule, and the spacing is such that a cover glass placed on the latter is slightly tilted towards the fluid which makes contact with only a part of the underside of the glass (see Figure 6.3).

The temperature of the hot plate, which has a thermometer inserted in a cavity immediately below its centre, is raised at a rate of $1 \pm \frac{1}{4}$ deg C/min and the boundary line of the silicone fluid carefully observed. The melting point is taken as that temperature when the boundary line moves rapidly across the cover glass, i.e. when the granule no longer supports the glass. Two such tests are carried out.

This is the basis of the methods described in A.S.T.M. D.789–66, a specification for nylon materials, and in A.S.T.M. D.2133–66 for acetal resin materials, where the use of a commercial apparatus, the Fisher-Johns

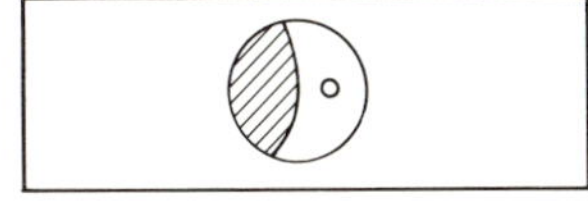

Plan view

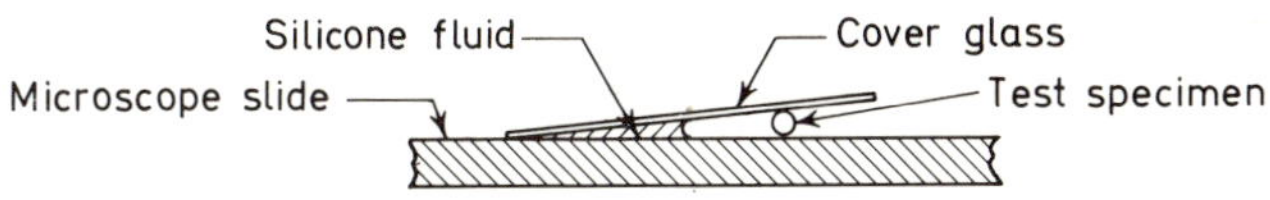

Figure 6.3. Specimen mounting for hot plate method (B.S. 2782)

melting point apparatus, is specified. Using the same apparatus, appearance changes are used to measure the melting points of polytetrafluoroethylene resins (A.S.T.M. D.1457–69) and fluorinated ethylene propylene copolymer (A.S.T.M. D.2116–66).

6.2.4 Microscope/Polarised Light Method

This method is only applicable to crystalline polymers because it depends on their birefringent, or double refracting, property. Thus when crystalline, such a material appears transparent when viewed between crossed Nicol prisms, i.e. it is visible against the dark background, but when molten it has become non-crystalline and becomes invisible in the 'blackness' of the crossed prisms.

The specimen is prepared as a thin layer between a microscope slide and cover glass. If it is a moulded or tabletted material that is being examined, it may simply be microtomed to 0·01–0·05 mm thickness and if it is in film form of this thickness it may be used direct. To avoid light scattering and ensure good thermal contact it is advisable to immerse the specimen in a liquid, preferably of similar refractive index, which of course must not affect the material. Powdered material, or thicker film or sheet, is placed between the microscope slide and cover glass, which together are put on a hot plate and heated to a temperature about 25 deg C above the melting point. This cover glass is pressed down to reduce the specimen to a thickness of 0·01–0·05 mm and, after keeping the elevated temperature for about 10 min, the assembly is then allowed to cool slowly, to induce optimum crystallinity. If the material is liable to thermal decomposition under these conditions the temperature may have to be lowered and the cooling started immediately.

A microscope is used, fitted with analyser and polariser and capable of giving 50–100 magnification. It is also fitted with a hot stage, which is mounted just above the microscope stage, and which consists of an insulated metal block with a central hole, for the passage of light, and with a recess close to the hole for the insertion of a thermometer or thermocouple. The hot stage is electrically heated at controlled rates and should be designed so that it can be enclosed in an atmosphere of nitrogen if necessary.

At the start of the test, the polariser and analyser are adjusted for complete extinction of light and the specimen assembly placed on the hot stage. The polariser and analyser are then rotated again, if necessary, to obtain maximum brightness of the specimen and thus overcome the effects of strain in the latter. The hot stage is then heated at a (linear) controlled rate, usually $1 \pm 0{\cdot}25$ deg C/min, and the melting point taken as that temperature when the field becomes completely dark. For high melting materials, the test may be conveniently started 50 deg C below the expected softening point.

A test of this type forms Method 103D of B.S. 2782:1970 and the above is essentially similar to that described in A.S.T.M. D.2117–64.

6.3 PARTICLE SIZE

The particle size of powders, granules or pellets may be of profound importance for it will influence the packing density of a moulding powder, and hence the charge in a fully positive mould, and the gelation characteristics of a paste making polymer, to take two common examples. The most simple method, and the most widely used on a routine basis, is that of sieving. This technique is not normally advised for particle sizes much below about 75 μm diameter (though A.S.T.M. D.1921–63 suggests 37 μm as minimum) where it is necessary to use sedimentation, optical or other methods[3–7].

A word of warning is necessary in the use of test sieves, which are usually numbered to indicate directly or indirectly the number of apertures per unit dimension. Sieve screens may be made of woven silk or fine wire mesh and it is not difficult to imagine that these two types differ quite considerably in their resistance to distortion under pressure. Wire mesh is most often specified, but again it is necessary to appreciate that a simple statement of the number of holes per unit dimension will not suffice since the diameter of the wire will influence that all important factor, the aperture size. Specification of the latter alone is not adequate either, since the diameter of the wire used to achieve that size will control the actual shape of the aperture as the weave of the mesh must of necessity yield a non-planar hole. Obviously a coarse wire will give a more distorted (and larger) hole than one obtained from a fine wire, though in the main plane of the mesh both holes may be identical. For this reason it is absolutely essential to tie sieve analysis—as particle sizing by test sieves is termed—to the particular type of sieves used, for example as specified in the appropriate British Standard (B.S. 410:1969, 'Test Sieves') which follows the recommendations of ISO Technical Committee TC24 and, in particular, ISO/R.565.

There are a variety of ways of using test sieves, well illustrated by reference to the relevant A.S.T.M. standards (D.1705–61 and D.1921–63).

DIN.53477 gives a method for sieve analysis of granular thermosetting compression moulding materials.

6.3.1 Normal Dry Sieving

An appropriate range of sieves (usually 8 in diameter), appropriate that is in the total span of aperture sizes they cover and in the intervals of sizes between them, are 'nested' together, decreasing in aperture size from top to bottom; a tray or pan is fitted under the smallest mesh sieve.

100 g of sample is weighed out (or any convenient quantity) and transferred to the top sieve and the cover placed on the top sieve. Although hand tapping of the assembled nest of sieves may be used it is both tedious and inefficient and therefore a mechanical vibrator is recommended (A.S.T.M. D.1921 specifies one with a rotary motion and tapping at 150 taps per minute). After 10 min or some appropriate period to attain equilibrium the sieves and tray are separated and the contents weighed to the nearest 0·1 g either by direct weighing in the sieves (having already weighed the empty sieves) or by transferring the contents to weighing bottles, etc.; in this latter case, extreme care must be taken to remove all particles from the walls and mesh of the sieve by a soft brush.

Normally the total recovered contents will amount to 98% or more of the original sample weight, in which case any lost is added to the figure passing through the finest sieve used, i.e. on the tray; if less than 98% is recovered the operator is examined!

An obvious variant on this method is to simplify it by employing just one sieve or a very limited number.

6.3.2 Vacuum Dry Sieving

Smaller sieves, normally of 3 in diameter, are used and, after drying in a dessicator, are weighed individually before starting as before. They are then inserted in a conical adaptor of appropriate diameter to take the nested sieves and fitted with a lead at its smaller diameter to a vertical standpipe attached to a vacuum system. The suction of the latter is adjusted so that the finest sieve is not appreciably strained and then the vibrator started. An appropriate quantity of sample is accurately weighed (1 g for very fine powders or 5–10 g for coarser ones) and carefully transferred to the top sieve; tapping of the sieve and light brushing of the sample assists. When no more powder is transferred through the top sieve, suction and vibration are ceased, the sieve removed and any powder adhering underneath is brushed on to the next sieve below, when the operation is repeated and so on. The sieves with retained powders are weighed, whence the weights of the latter are obtained by difference.

6.3.3 Wet Sieving

A method is described in A.S.T.M. D.1705–61 for powdered polymers and copolymers of vinyl chloride. The use of the wet technique eliminates static charges and troubles associated therewith such as agglomeration; (it also reduces the tendency to 'flying' of powder and hence loss of sample).

Sieves are used as in Section 6.3.1 above and are weighed prior to testing. An appropriate quantity of polymer is weighed out (25–100 g) into a beaker and approximately 300 ml of 0·5% wetting agent solution (anionic sulphate or sulphonate type) is added with stirring. This mixture is poured into the top sieve and the powder washed through this and the lower ones by more solution and then water. The sieves are dried in an oven and weighed.

In A.S.T.M. D.1457–69 for polytetrafluoroethylene materials, the apparatus is illustrated in support of the description of the test method.

6.4 APPARENT POWDER DENSITY AND BULK FACTOR

In discussing particle sizing, mention has already been made of the moulding difficulties encountered with material which has a high bulk. Two standard tests exist for measuring the apparent density of moulding material.

6.4.1 Apparent Powder Density

Firstly there is a test developed for moulding material that can be poured from a funnel. Method 501A of B.S. 2782:1970 is identical with ISO/R.60 and DIN.53468.

A funnel of the form shown in Figure 6.4 is used. It is mounted vertically with its lower orifice 20–30 mm above the top of a measuring cylinder (capacity 100 ml, internal diameter 40–50 mm) and coaxial with it. With the lower orifice closed, 110–120 ml of well mixed powder are poured into

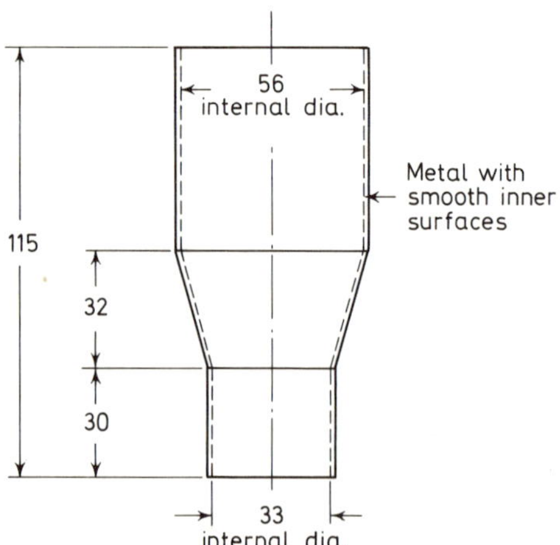

Figure 6.4. B.S./I.S.O. standard funnel. Dimensions in millimetres

the funnel and then the powder is allowed to flow into the measuring cylinder, assisted if necessary by loosening it with a rod. When the cylinder is full, a straight edged blade is drawn across the top of the cylinder to remove excess and then the contents are weighed. The mean of two determinations is taken and expressed in g/ml.

A.S.T.M. D.1895–67 offers two similar methods, but both with different sized funnels to the above, and some other, minor, variations. One of these funnels (of Method A) is used to measure pourability by timing the rate of flow of the powder out of the funnel.

Method C of this A.S.T.M. is essentially identical with ISO/R.61 which is the same as Method 501B of B.S. 2782:1970 and DIN.53467. All are for powders which cannot be poured through such funnels, which thus forms the second basic method.

A cylinder is used which is 1000 ml in capacity and internal diameter 90 ± 2 mm. Into this fits a plunger of slightly smaller diameter and with a total mass of 2300 g. A cylinder closed at its lower end and weighted with lead shot may be used.

60 g of the powder is dropped, little by little, into the cylinder so that it is evenly distributed and has a level surface. The plunger is then lowered on to the powder and allowed to remain there one minute before the height of the powder is measured with the plunger still in position. From this height of powder, the diameter of the cylinder and the weight of powder, the apparent density is thus calculated.

Apparently the A.S.T.M. D.1895 method is not suitable for polytetrafluoroethylene because of the nature of the particles and A.S.T.M. D.1457–69 gives a suitable variant of the funnel technique.

ISO/R.1068, for PVC resins, is a cylinder method for determining 'expected apparent bulk density'; a shaking machine is used to tamp down the material under a piston.

6.4.2 Bulk Factor

The bulk factor of a moulding is defined as the ratio of the volume of a given mass of moulding material to its volume in the moulded form. It is thus the ratio of the density of the moulded material to its apparent density before moulding.

Method 501C of B.S. 2782, ISO/R.171 and the method of A.S.T.M. D.1895 all require determination of apparent powder density by the corresponding techniques (as above) and moulded density by the appropriate method (see Section 5.1). The German practice is similar and is to be found in DIN.53466.

6.5 MELT FLOW BEHAVIOUR

Study of the flow behaviour of molten polymers is clearly of paramount interest to the fabricator if he is to establish the feasibility of any intended process involving melting. If his test is ideal it will forecast a variety of properties such as adequacy of thermal stability, ease of processing (is the machine powerful enough?) and quality of mouldings. The reputable manufacturer of plastics compositions will likewise be interested in these subjects in order to be able to give his customers an adequate technical service; in addition, since the manufacturer can exercise a considerable degree of control over processability by altering parameters such as average molecular weight and molecular weight range of polymer, plasticiser type and content, lubricant and stabiliser, efficient reliable laboratory scale test procedures are essential if experimentation is to be kept within reasonable bounds during the early stages. Ultimately, however, when a 'short list' has been prepared, there is no real substitute for trying out the most promising compositions on the factory scale using the actual processing equipment to be employed in manufacture.

Unfortunately, the study of the flow behaviour of polymer melts is not without its complications. In an ideal or Newtonian liquid, to which water at normal temperatures approaches very closely, there is a linear relationship between shear rate (rate of deformation) $\mathrm{d}u(\mathrm{d}r$ and shearing stress τ such that

$$\tau = \eta \frac{\mathrm{d}u}{\mathrm{d}r}$$

where η is a constant known as the coefficient of viscosity, or simply the *viscosity;* this constant, as is well known, is markedly dependent on temperature. Application of this concept to flow of a Newtonian liquid through a capillary of radius r and length l leads to the familiar Poiseuille formula:

$$\frac{Pr}{2l} = \eta \frac{4Q}{\pi r^3}$$

where P is the pressure on the melt and Q the volume rate of flow. Regrettably this formula does not apply to a great number of liquid systems, including polymer melts, all of which are non-ideal. Some liquids change their flow behaviour with time, for example one may 'thicken up' after stirring, another 'thin down'. Even ignoring these complications however, and considering only time stable phenomena, there are still a number of possibilities, some of which are illustrated in Figure 6.5.

Many polymer melts follow the 'pseudoplastic' pattern, whence it is seen that there is no constant of proportionality and hence no viscosity! It follows that, since the various common processing techniques for thermoplastics are associated with shear rates between about 100 s^{-1} and 100 000 s^{-1}, there is really no substitute for a comprehensive study of shear stress against shear rate over a range of the latter embracing all likely processing techniques, and furthermore a study of this nature throughout a suitable temperature range. Regrettably the majority if not all standard flow tests employ only one (low) shear rate, either fixed by the machine used or dictated in part by the flow behaviour of the material under examination. Clearly such a one-point observation cannot yield much of use for predicting general processability and although the test may be simply a quality control measure, it can be misleading even in this context since the precise shapes of all pseudoplastic curves are not identical but may vary (as shown in Figure 6.6).

Detailed consideration of flow behaviour of molten polymers is a science in itself and the reader is referred to Reference 8, for instance, for a detailed study of the subject in relation to common processing techniques, and to Reference 9 for a new work on the subject as a whole.

So far, we have only considered materials whose flow characteristics, if not time independent, are such that no change of state is involved. Except where heating to fluidise has been carried on for so long that significant decomposition sets in, thermoplastics fit this description. When, however,

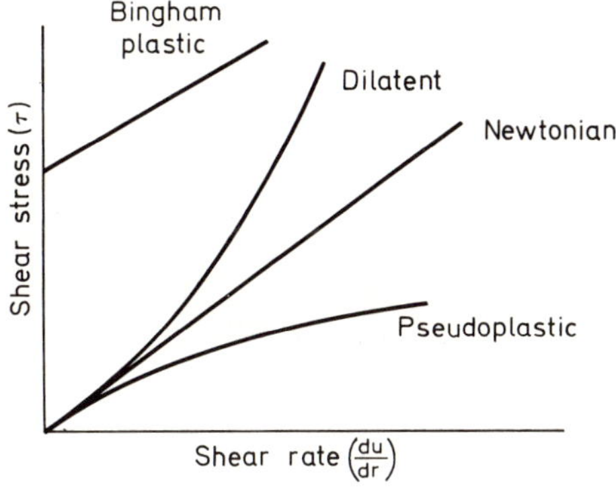

Figure 6.5. Flow relationships for various types of liquids

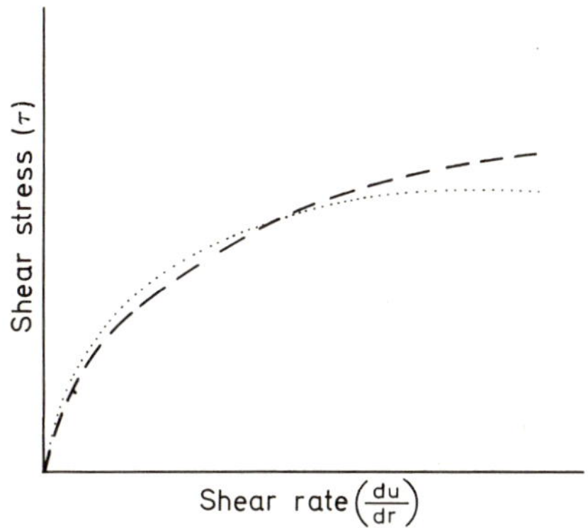

Figure 6.6. Different pseudoplastic curves

we turn to thermosets, when a fluid mass sets to the solid state whilst still hot and during processing, it can well be imagined that study of such an unstable system is very complicated.

6.5.1 Melt Flow Index Determinations

The determination of melt flow index (M.F.I.) is one of the most widely used tests, though it suffers from the advantages discussed above (in the B.S. test, for M.F.I. values in the range 0·3–20, the corresponding shear rates used are 1–73 s^{-1}); it exists in a number of variants according to polymer type and molecular weight and is described in B.S.I., A.S.T.M. and ISO documents.

In method 105C of B.S. 2782, three variants are described, corresponding to ISO/R.292 (2nd Edition). The basic principle is that of extrusion plastometry and Figure 6.7 shows a cross-section of the important parts.

The method in B.S. 2782 is restricted to polyethylene and polyethylene compounds and, in the U.K. is often referred to as 'grading', the result thus derived being termed the 'grade' of the material under test.

The cylinder is of hardened steel and is fitted with heaters, lagged for operation at 190 ± 5°C; the dimensions of the cylinder, and particularly

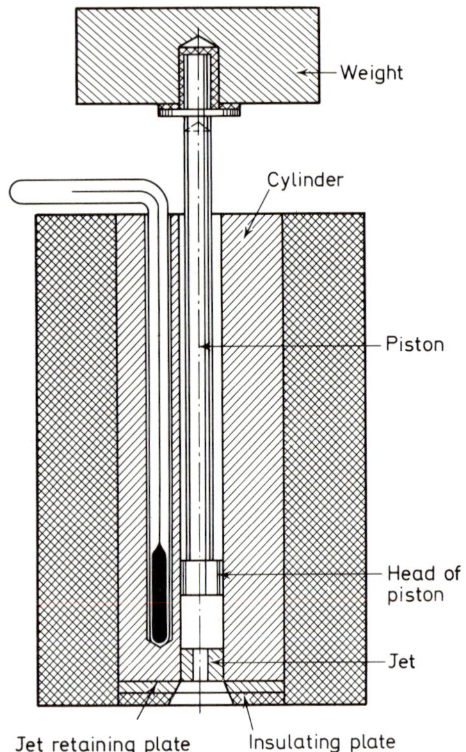

Figure 6.7. Apparatus for determination of melt flow index (B.S. 2782)

the radius, are very carefully specified (see Poiseuille formula above, for although as explained, it does not apply to polymer melts, often a power law of the type $\tau = k(\mathrm{d}u/\mathrm{d}r)^n$ does). The piston is of mild steel and the diameter of its head is $0{\cdot}075 \pm 0{\cdot}015$ mm ($0{\cdot}0030 \pm 0{\cdot}0006$ in) less than that of the internal diameter of the cylinder at all points along the working length of the latter.

Two removable weights are provided, which with that of the piston in one case total 2160 g and in the other 5000 g. Likewise two alternative jets are used, one of internal diameter 2·095 mm (0·0825 in) and the other

1·180 mm (0·0465 in); both are 8·000 mm (0·315 in) long and made of hardened steel.

The first standard procedure ('A'), uses the jet of internal diameter 2·095 mm with the load of 2160 g; it is for material of M.F.I. in the range 1 to 25 (see below). The first requisite is to ensure that the apparatus, and particularly the cylinder, piston and jet, is scrupulously clean, for example by washing with rag and cotton wool soaked in tetrahydronaphthalene or xylene whilst the apparatus is still hot. When clean, the apparatus is kept at 190° ± 0·5°C for 15 min before use, on completion of which period a charge of about 5 g is inserted into the top of the cylinder, within a period not exceeding one minute, and the piston reinserted. (For materials of M.F.I. above 20, more than 5 g may be needed.) Six minutes later (a period found to be adequate for the temperature to attain equilibrium again after adding the cold test sample) the weight is added and the molten material starts to extrude through the jet. The rate of extrusion is measured by cutting off the extrudate at the jet at suitable time intervals (indicated in the Standard). Several such 'cut offs' are taken up to 30 min after insertion of the sample into the cylinder and all when the piston head is between 50 mm and 20 mm above the upper end of the die (as marked by scribed marks on the piston). The first cut off and any containing air bubbles are ignored; the remainder, at least three, are weighed to the nearest milligramme and the average taken; the difference between the maximum and minimum weights of these so-accepted 'cut offs' must be not more than 10%.

$$\text{M.F.I. } (A) = \frac{600 \times \text{average weight of cut-off in grammes}}{\text{interval of time in seconds}}$$

The value is thus the amount extruded in ten minutes under the specified conditions.

Procedure 'B' uses the same load (2160 g) but the smaller jet, of diameter 1·180 mm. It is for materials of M.F.I. between 25 and 250 and the jet is calibrated against that of Procedure 'A' using a hot homogenised sample of M.F.I. between 7 and 10 by Procedure 'A' and containing antioxidant. The conversion factor is defined as the ratio of the average weight of cut off, using the larger jet and an interval time of 30 s, to the average weight of cut off using the smaller jet and the same intervals.

The actual test is carried out as in Procedure 'A', the result being calculated as follows:

$$\text{M.F.I. } (B) = \frac{600 \times \text{average weight of cut off in grammes}}{\text{interval of time in seconds}} \times \text{conversion factor}$$

The value has the same significance as before.

For materials of M.F.I. below 1, Procedure 'C' is used, in which the jet of 2·095 mm diameter is again employed, but this time with the load of 5000 g. The technique is as before.

$$\text{M.F.I. } (C) = \frac{150 \times \text{average weight of cut off in grammes}}{\text{interval of time in seconds}}$$

This time, the value corresponds to the amount extruded in $2\frac{1}{2}$ min under the specified conditions.

Note: There is *no* absolute correlation between M.F.I.(*A*) and M.F.I.(*C*) although the values should agree approximately.

ISO/DR.749 extends the method to the examination of polypropylene by raising the test temperature to 230°C. A.S.T.M. D.1238–65T employs only the jet of diameter 0·0825 in diameter but temperatures of 125, 150, 190, 200, 230, 235, 265, 275 and 300°C and loads of 325, 1050, 1200, 2160, 3800, 5000, 10 000, 12 500 and 21 600 g according to whether acetals, acrylics, ABS copolymers, cellulose esters, nylon, polychlorotrifluoroethylene, polyethylene, polycarbonate, polypropylene, polystyrene or 'vinyl acetal' is being examined. Except for polyethylene, the results are termed 'flow rates'. Marshall and Welford[10] on the contrary, recommend Method 105C of B.S. 2782 (see above) with a temperature of 220°C and loads equivalent to 0·8 kgf, 2·16 kgf or 5 kgf for polystyrene and ABS materials. However, B.S. 1493:1967, 'Polystyrene Moulding Materials', requires the measurement of melt flow index to be carried out using the load of '*C*' with jet '*A*' at 200 ± 0·5C.

6.5.2 Rossi-Peakes Test

In Method 105A of B.S. 2782 this test is described for determining the 'flow temperature of thermoplastic moulding material', that temperature at which 'a thermoplastic moulding material flows a specified distance under a specified pressure when tested in a standard apparatus'.

The apparatus is specified in terms of all essential dimensions, etc., and illustrated by reference to the Rossi-Peakes flow tester (see Figure 6.8).

The orifice 1 is highly polished and runs in a split cone which is clamped in heater block 3. There is a thermometer well inserted into a hole drilled into one half of the split cone. The charge chamber 2 is situated below the orifice and a heated ram 4 is so positioned that it forces material in the chamber up into the orifice. 5 is a system for applying a pressure of 10·3 MN/m^2 (105 kgf/cm^2; 1500 lbf/in^2) to the ram. The rate of flow is measured by 6, consisting of a follower rod and indicator exerting a specified pressure.

To carry out the test, the apparatus is assembled and the temperature adjusted according to the material under examination. A pellet of the latter, of size to match the charge chamber, is introduced therein and the pressure applied immediately. The length of flow after 2 min is determined and then the material is removed and the operation repeated with a fresh pellet. Two pellets are thus tested at each of three temperatures, at all of which the flow length is between 13 and 38 mm (0·5 and 1·5 in) and where at least one measurement is above 25 mm (1 in) and one below. The results are plotted and the flow temperature read off where the length of flow is 25·4 mm (1·00 in).

In A.S.T.M. D.569–59, two alternative procedures are offered, one as before and the other where the degree of flow is measured at a specified temperature. In this latter, pressures of 500, 1000 or 1500 lbf/in^2 are used and the degree of flow measured at the prescribed temperature.

6.5.3 Other Flow Tests

Although the melt flow index and Rossi-Peakes tests are the only ones generally standardised by B.S.I., A.S.T.M. or I.S.O. for thermoplastics (there is a tentative method, A.S.T.M. D.1238–65T, 'Measuring Flow Rates of Thermoplastics by Extrusion Plastometer'), there are many others of which most suffer from the same disadvantages as those described above, namely using operating conditions, especially shear rates, far removed from practice. For instance, Hayes[11] has described two extrusion plastometers, one a weight-loaded instrument, i.e. of the Rossi-Peakes or M.F.I. type,

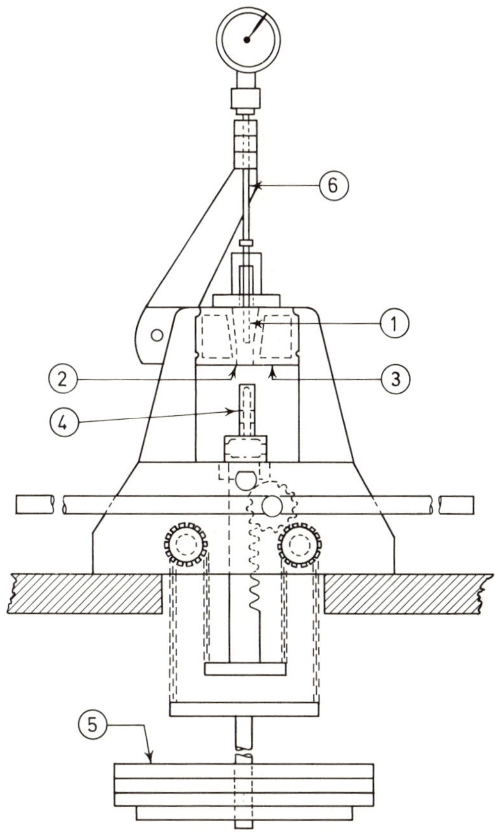

Figure 6.8. Rossi-Peakes flow tester (B.S. 2782)

the other where the piston is motor driven to give a fixed constant rate of extrusion and the pressure measured. More recently Gray, Holford and Combs[12] have developed an apparatus where the change in apparent viscosity with temperature can be followed continuously and, it is claimed, the minimum processing temperature (at low shear rate!) and maximum processing temperature can be predicted.

On the commercial scale, an extrusion rheometer is available from which any shear rate up to 10 000 s^{-1} can be obtained (see Figure 6.9).

A useful 'ad hoc' method for comparative testing is a spiral disc mould where a fair length of flow channel can be provided by a conveniently sized mould fitted to an injection machine. Campbell and Griffiths[13] describe one where the effective length is 77 in, in a mould of overall dimensions 12 in × 9 in. The flow length of the material into the mould, which is vented to atmosphere at its inner end, is determined and related to the test conditions of cycle time, temperatures of mould and barrel, pressure, etc. The Brabender Plastograph has found wide use for studying

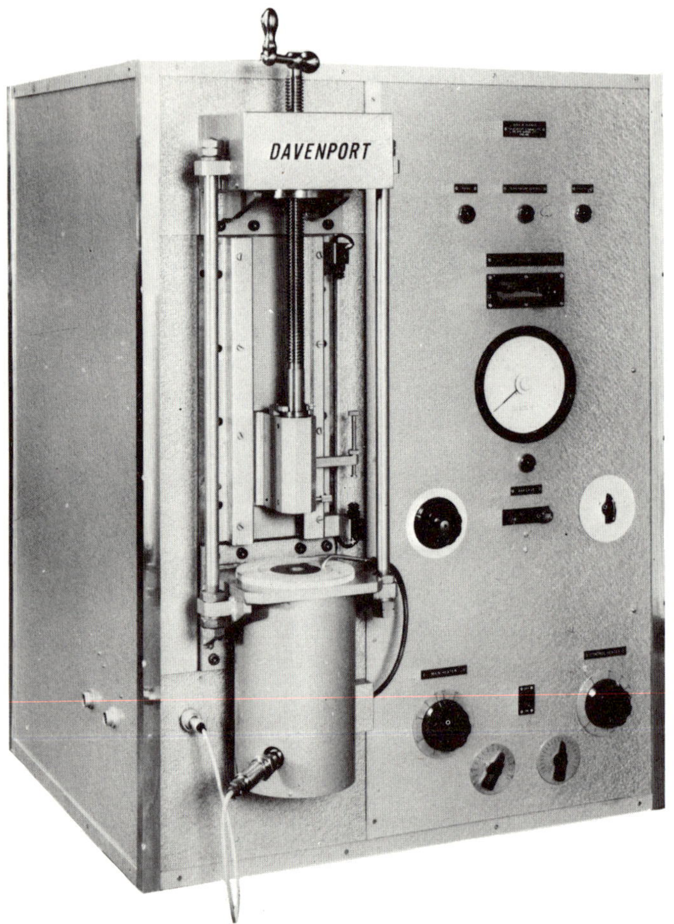

Figure 6.9. Extrusion rheometer (Courtesy Davenport (London) Ltd.)

the rheological behaviour of plastics materials (see References 14–17), the last of which contains a useful bibliography.

In recent years, much work has also been described with specially designed pieces of apparatus aimed at bridging the gap between laboratory tests and factory operation. Various workers have attached measuring instruments to actual extruders or injection moulding machines and plastometers, or as they are popularly known 'rheometers' (rheology is the study of flow); facilities for studying wide ranges of rates of shear, have been described, for example[18]. However, this subject merits a substantial volume to itself and cannot be given further space here; for a general review of the subject of capillary rheometry, the reader is referred to Reference 19. The presentation of data is covered by a recommended practice in A.S.T.M. D.1703–62.

Before leaving the subject of flow tests on 'reversible' materials, mention should be made of two A.S.T.M. Standard methods for highly plasticised compositions:

(a) A.S.T.M. D.1823–66, 'Apparent Viscosity of Plastisols and Organosols at High Shear Rates by Castor-Severs Viscometer', and
(b) A.S.T.M. D.1824–66, 'Apparent Viscosity of Plastisols and Organosols at Low Shear Rates by Brookfield Viscometer'.

6.5.4 Flow Tests for Thermosetting Materials

The M.F.I. test and those briefly mentioned in Section 6.5.3 have all been designed for the examination of thermoplastics. The Rossi-Peakes test is applied to thermoplastics but it is also used for thermosets. Von Meysenbug[20] has described a modified form; there are also other non-standard tests as for instance the spiral type mentioned previously and the flow disc, where a given charge of material is placed between two heated metal discs which are then closed at a fixed speed. The radius of the moulding formed, or its thickness, is taken as an indication of flow or stiffness respectively.

A test designed specifically for thermosetting moulding materials is the so-called 'cup flow' determination. Method 105B of B.S. 2782 describes the U.K. practice and limits applicability to phenolic and alkyd moulding materials, excluding 'fast-curing' ones. The cup flow mould is shown in Figure 6.10.

Needless to say, in a flow test such as this, the surfaces of the inner male and female members must be smooth and highly polished. The general form and dimensions (including steam coring) are very carefully specified.

The temperature of the mould is laid down as 163 ± 1°C and the thrust on the mould 100 ± 5 kN ($10\,160 \pm 500$ kgf; $10 \pm \frac{1}{2}$ ton f). The closing speed of the mould when empty must be 130 mm (5 in) in 4 to 5 s. The mass of moulding material to use is found, by trial and error, so that a flash of between 2 and 2·5 g is obtained in the test. The mould is opened just sufficiently for charging the weighed-out sample of material, at 10 to 40°C, with the aid of a scoop and the mould closed again immediately; the time between dropping the powder in the mould and the registration of full pressure must be between 5 and 10 s.

The time of flow is measured in seconds as that between the first registration of pressure on the hydraulic gauge of the press to the instant the flash is seen to cease moving. This is the 'cup flow' of the material.

The technique described in A.S.T.M. D.731–67, to determine moulding index, follows much the same pattern though the mould cavity does differ in design (see Figure 6.11).

An elaborate method is laid down for obtaining the appropriate weight of material to use. For materials of impact strength under 2·7 cm kgf/cm

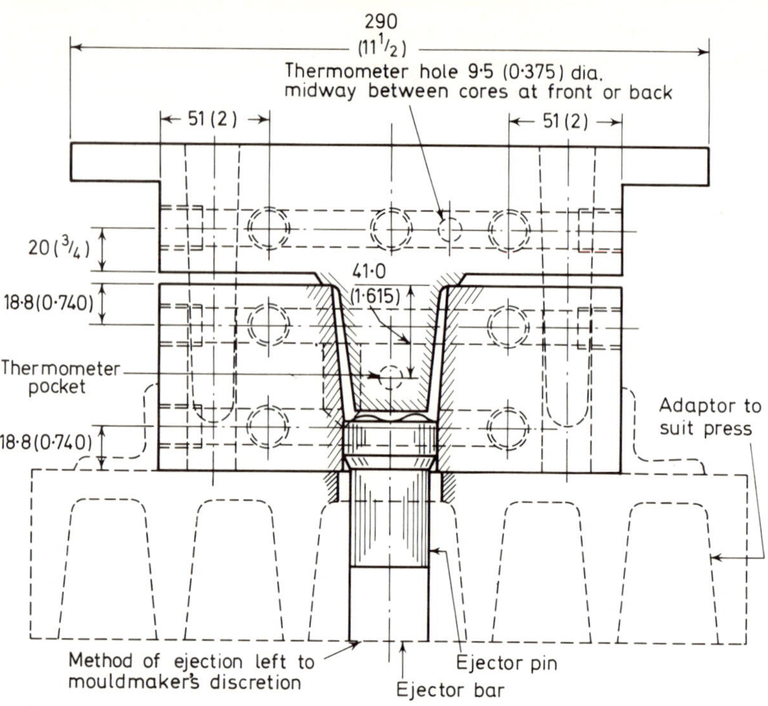

Figure 6.10. B.S. cup flow mould (B.S. 2782) Dimensions in millimetres with inch equivalents in parentheses

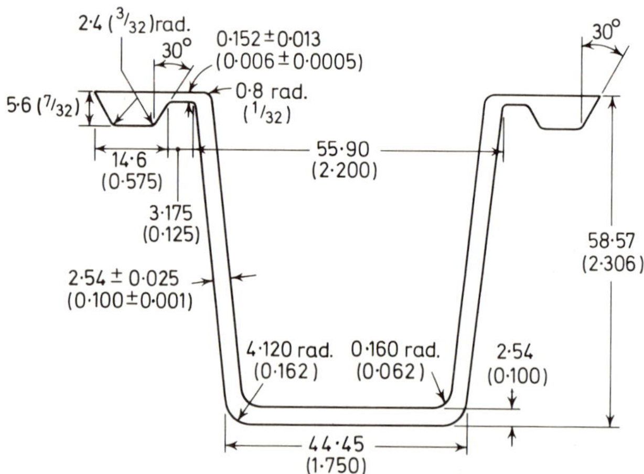

Figure 6.11. A.S.T.M. cup mould (A.S.T.M. D.731) Dimensions are in millimetres with inch equivalents in parentheses. All surfaces highly polished to no. 2 micro-finish (SPI-SPE standard for mould finish.) Rockwell C-58 steel. Tolerances on dimensions are ± 0·025 mm (0·001 in) except as noted

(0·5 ft lb/in)* a cup with flash of thickness 0·15–0·20 mm (0·006–0·008 in) is moulded, the flash removed and the cup weighed; this weight, multiplied by 1·1, is the charge to use in the test except for materials of impact strength above 2·7 cm kgf/cm (0·5 ft lb/in)* when flash of thickness 0·51–0·66 mm (0·020–0·026 in) is produced and the cup weight multiplied by 1·05. The preferred temperatures are 150, 155 or 165°C (all ± 1 deg C), according to the material under test. A load is first determined which produces a cup of the required flash thickness and then the next lowest load is selected, from a table of eight specified alternatives, for test purposes. The time of flow in seconds is recorded from the instant the hydraulic gauge registers an applied load of 454 kg (1000 lb) to when the prescribed flash thickness has been reached.

The result is expressed as the 'moulding index' the minimum force required to produce the necessary flash, with a subscript of the closing time in seconds, e.g. 6800_{18} kgf for a 6800 kg minimum load and 18 s flow time.

DIN.53465 also uses a 'cup flow' type of mould and expresses the results as 'closing time' of thermosetting moulding materials.

The behaviour of glass filled phenolic and diallyl phthalate moulding compounds has been examined by Sundstrom, Walters and Goff[21] using five flow tests and the filling of an impact test bar mould; amongst the flow tests used were the Brabender, cup flow and spiral mould methods. A general correlation of test results, for each type of compound, was noted.

(*Note:* There is a multitude of other tests which are applied to materials prior to fabrication but, except for those described above, it is considered that they are so restricted in application or so specific to one type of material as not to warrant mention.)

REFERENCES

1. HASLAM, J., WILLIS, H. A. and SQUIRREL, D. C. M., *Identification and Analysis of Plastics,* 2nd edn, Iliffe Books Ltd., London (in preparation)
2. DURRANS, T. H., 'A New Method for the Determination of the Melting Point of Resins', *J. Oil Col. Chem. Ass.,* **12,** No. 108, 173 (June 1929)
3. B.S. 3406: 1961–3 (Parts 1–4), *Methods for the Determination of Particle Size of Powders,* British Standards Institution, London
4. 'Classification of Methods for Determining Size: A Review', Particle Size Analysis Sub-Committee of the Analytical Methods Committee of the Society for Analytical Chemistry, *Analyst,* **88,** No. 1044, 156 (March 1963)
5. IRANI, R. R. and CALLIS, C. F., *Particle Size: Measurement, Interpretation and Application,* John Wiley & Sons Inc., London (1963)
6. DALLAVALLA, J. M., *Micromeritics,* 2nd edn, Pitman Publishing Corporation, London (1948)
7. ORR, C. Jr. and DALLAVALLA, J. M., *Fine Particle Measurement,* Macmillan & Co. Inc., New York (1959)
8. BERNHARDT, E. C., *Processing of Thermoplastic Materials,* Rheinhold Publishing Corporation, New York (1959)
9. LENK, R. S., *Plastics Rheology,* Maclaren & Sons, London (1968)
10. MARSHALL, B. I. and WELFORD, P. N. MCC, 'Melt Flow Index Test for Polystyrene and A.B.S.', *Brit. Plast.* **39,** No. 10, 591 (October 1966)
11. HAYES, R., 'Two Extrusion Plastometers for Use with Polyvinyl Chloride', *Chem Indy,* **44,** 1069 (1st November 1952)

*By the A.S.T.M. test! (see below, Chapter 8)

12. GRAY, T. F. Jr., HOLFORD, T. G. and COMBS, R. L., 'Processing Temperature Indexer—a Novel Rheometer,' *S.P.E.J.*, **24,** No. 9, 35 (September 1968)
13. CAMPBELL, G. and GRIFFITHS, L. L., 'Recent Advances in Injection Moulding Techniques and in the Evaluation of Moulding Materials', *Plastics Progress 1955*, Iliffe Books Ltd., London, 259 (1956)
14. MCCABE, C. C., 'Rheological Measurements with the Brabender Plastograph', *Trans. Soc. Rheol.*, **IV,** 335 (1960)
15. MCCABE, C. C., 'Rheological Measurements with the Brabender Plastograph', *Chem. Can.*, **12,** No. 10, 44 (October 1960)
16. SCHRAMM, G., 'Measuring the Fusion Rate of Rigid PVC Dry Blend', *Int. plast. Engng*, **5,** No. 12, 420 (December 1965)
17. MATTHAN, J., 'Evaluation of the Variables of the Brabender Plastograph', *Research Report No. 165*, Rubber and Plastics Research Association of Great Britain, Shawbury (1968)
18. BALLMAN, R. L. and BROWN, J. J., *Capillary Rheometry*, Instron Application Series, Instron Engineering Corporation, (Canton, Massachussetts)
19. TURNER, L. W., 'A Capillary Rheometer as a Means of Evaluating Polymers', *Tech. Polymer Sci.*, 151, S.C.I. Monograph No. 17, Society of Chemical Industry, London (1963)
20. MEYSENBUG, C. M. VON, 'An Improved Flow Tester for Thermosetting and Thermoplastic Moulding Compounds' *Int. Symp. Plast. Test. Stand.*, Special Technical Publication No. 247, American Society for Testing and Materials, Philadelphia (1959)
21. SUNDSTROM, D. W., WALTERS, L. A., and GOFF, C. S., 'Flow Tests for Thermosets', *Mod. Plast.* **46,** No. 3, 104 (March 1969)

7

Preparation of Test Specimens

7.1 INTRODUCTION

In Chapter 4 it was mentioned that the degree of crystallinity could be estimated from density measurements, the density of the amorphous and crystalline forms generally being significantly different. Since the degree of crystallinity in a specimen will depend on the manner in which it has been prepared—temperature, pressure, profile of the cavity, rate of cooling etc.—it therefore follows that the magnitude of this apparently most simple and straightforward of properties, the mass per unit volume, will vary according to the history of the specimen under examination; this comment applies to practically every property measured on the massive polymer. Reference has also been made in earlier chapters to the influences of test specimen shape and preconditioning and to these can be added such considerations as mechanical surface finish and, for certain properties, cleanliness of surface.

Whilst for the purposes of assessing merit in a specific end-use the condition of material under examination should simulate realism, for the purposes of a standard test the condition should approach, as it were, idealism—that is history, shape, condition, etc., should be carefully defined. It is a fact that much effort is wasted on producing standard test data, even within the limited confines of the value of such data, with insufficient attention to the preparation of the test specimens.

The subject as a whole has been discussed by Cohen[1].

7.2 TREATMENT OF THE RAW MATERIAL

By and large the examination of plastics 'raw materials' is outside the scope of this book but, in passing, it is worth commenting on the necessity of, for instance, predrying hygroscopic powders or flake before making up solutions for viscosity measurement. Similarly, moulding granules may need predrying, e.g. of urea-formaldehyde or nylon, if satisfactory test mouldings are to be obtained.

7.3 MOULDING OF TEST SPECIMENS

Many tests are normally performed on specially moulded specimens; in fact all the tests usually described as being on moulding materials are really on specimens moulded from the latter, with the exception of the few which may be broadly described as checking the 'mouldability' of the material—particle size, bulk factor, volatile content, cup flow and melt behaviour generally.

The first and most important decision is to settle which method of moulding is appropriate, for the orientation induced in an injection moulded specimen may yield quite different properties to those from a relatively non-orientated compression moulded or extruded specimen. Hayes[2] illustrates this well with some bending strength measurements on $5 \times \frac{1}{2} \times \frac{1}{4}$ in bars of toughened polystyrene, compression moulded in one case and injection moulded from one end in the other (see Table 7.1).

Table 7.1

	Bending strength (lbf/in²)	*Deflection at break* (in)
Compression moulded	6 570	0·71
Injection moulded	8 170	> 1·00

In the same paper, the effect of degree of orientation is demonstrated by measuring the same property on bars produced at different moulding temperatures and pressures, the greatest degree of orientation resulting from the lowest temperature and highest pressure (see Table 7.2).

Table 7.2

Temperature (°C)	*Pressure* (lbf/in²)	*Breaking strength* (lbf/in²)
180	10 000	8 170
200	8,000	7 430
220	5,000	7 460
240	4,500	7 010

Similar studies have been made inter alia by Budesheim and Knappe[3], Horsley, Lee and Wright[4], Bryant and Hulse[5], Bossu, Chatain, DuBois and Rougeaux[6], Högberg[7], Morgan and Vale[8], Williams and Mighton[9], Dasch[10, 11], Koda[12] and Malac[13]. Högberg, for instance, examined impact strength measured on bars in which, in one case, flow had occurred in the direction of the bar and, in the other, the direction of flow was perpendicular thereto. Four styrene-acrylonitrile copolymers were examined and the ratios of the values were between $2\frac{1}{2}$ and 3 to 1, in favour of the bars where flow had been down their length. Amongst the effects studied by Williams and Mighton was the influence of moulding conditions on the stability of mouldings. Koda's study, on polycarbonates, embraced abrasion resistance, hardness, density and heat shrinkage measurements.

The subject, in relation to injection moulding, has been comprehensively reviewed by Whisson[14]. Wintergerst[15] has endeavoured to anneal test pieces of high impact polystyrene back to a condition of 'basic strength' from the injection moulded condition.

On the other hand, moulding conditions may have a more macroscopic effect on the material; a filled or reinforced composition can yield rather different data depending on whether the filler is subject to fracture and whether the process is rigorous enough to cause it. Work reported by Spiwak[16] well demonstrates these points. A range of thermosetting materials was examined, using compression moulded specimens and two sets prepared by transfer moulding, one gate (B) being twice the size of the other (A). The figures shown in Table 7.3 have been selected from Spiwak's work (Courtesy of the Society of Plastics Engineers Inc.):

Table 7.3

Material	*Impact strength* (ft lbf/in)			*Bending strength* (lbf/in^2)		
	Compression moulded	*Transfer moulded (A)*	*Transfer moulded (B)*	*Compression moulded*	*Transfer moulded (A)*	*Transfer moulded (B)*
P.F. with short cotton fibre	0·44	0·45	0·44	9 200	9 930	9 210
P.F. with long asbestos fibre	2·31	1·23	1·17	12 100	6 400	7 280
D.A.P. with short acrylic fibre	0·64	0·66	0·66	10 160	11 590	9 940
D.A.P. with long glass fibre	3·68	0·68	0·88	14 800	10 220	8 210

A somewhat similar exercise has been described by Elmer and Harrington using phenolic materials in a round robin study[17].

Panov[18] investigated the effect of specimen size of the tensile strengths of a range of plastics materials and found that the strengths decreased with increasing cross-sectional area—over 20% decrease was observed for polystyrene and unplasticised PVC by increasing the area from 25 mm^2 to 400 mm^2.

Parts of a recent book edited by Ogorkiewicz[19] deal with the effect of processing variables on the properties of thermoplastics.

Standard specifications therefore tend to tie down fairly rigidly the moulding conditions to be used for preparing test specimens, endeavouring to strike a reasonable compromise between standardisation and the inevitable variations in moulding behaviour between one manufacture and the next. In British Standards, instructions on moulding are given in the appropriate product specification; B.S. 771:1959, 'Phenolic Moulding Materials', requires all specimens to be moulded in flash, positive or semi-positive moulds under a pressure of 1–3 tons f/in^2 with a mould temperature of 155°C–170°C (except in the case of the test for mould shrinkage when the temperature must be 155±1°C). Curing times are specified for each type of test specimen. B.S. 3126:1959, 'Toughened Polystyrene Moulding Materials', uses injection moulded specimens, the shapes of which (as in B.S. 771) are carefully specified, but no conditions are laid down concerning the actual technique of moulding. However, some control of the orientation induced in the moulded specimen is effected by a 'reversion' test which is carried out on typical impact and tensile strength specimens. This test involves heating measured specimens in a bath of glycerine maintained at a temperature 20 deg C above the measured softening point (see below, Chapter 11). After removing from the bath and cooling, the 'reversion' (i.e. percentage shrinkage) shall not be greater than 15% for impact specimens and not greater than 35% for tensile specimens. In B.S. 1493:1967; 'Polystyrene Moulding Materials', injection-moulded tensile specimens must not revert by more than 30%; impact strength specimens, after notching, are tested for susceptibility to 'crazing' (see Chapter 13, Section 13.5) if first results seem doubtful.

In A.S.T.M. practice, the usual policy of standardising every step and operation is followed. Thus A.S.T.M. D.647–68 recommends designs for various specimens, produced by compression, transfer and injection moulding, without reference to specific materials. The next link in the chain is provided by A.S.T.M. D.796–65, D.956–51, D.1130–63 and D.1897–68 (amongst others) for, respectively, moulding test specimens from phenolics and aminoplastics and injection moulding specimens from thermoplastics generally. All cross-reference to A.S.T.M. D.647 by selecting appropriate moulds therefrom and, in the first two mentioned, temperatures, pressures, preheating, etc., are carefully specified. DIN.53451 follows the A.S.T.M. practice in tabulating temperatures, pressures, etc., for moulding test pieces from various thermosets and thermoplastics.

I.S.O. has advanced some way by producing three general recommendations—ISO/R.293, R.294 and R.295 and draft recommendations DR.1267, DR.1601 and DR.1602 (see Appendix 5 to Chapter 2).

7.4 MACHINING OF TEST SPECIMENS

If moulding is not necessary because the product to be tested has already been fabricated in this sense, then specimens of a shape appropriate to the requisite property may have to be prepared from some 'massive form', for example calendered sheet, laminate, casting or large commercial moulding. In many ways this is a more complex problem than moulding specimens directly, for the machining of test specimens is a difficult operation to standardise especially in terms of the quality of the finishing. No simple recourse to using a stainless steel mould with mirror finish is possible here. Yet many properties are profoundly influenced by the presence of surface flaws such as nicks and scratches.

The point is well illustrated by reference to the apparently simple operation of cutting out specimens from soft thin film with a die. Patterson[20] gives the results of a correlation exercise in tensile testing of 1·5 mil polyethylene, 1 mil polyester and 4 mil polycarbonate (1 mil = 0·001 in) undertaken by four laboratories using four different techniques:

1. die cut
2. manual razor cut
3. hand driven rotary razor cut

   shear cutter.

Yield strength data showed the least scatter—10% or less covering all four techniques and laboratories. However, much greater divergencies were found with ultimate tensile strength—30% or more—and elongation at break—up to approximately 75%—with the laboratories being relatively consistent and the techniques accounting for the majority if not all the variation. Die cut specimens were clearly the worst in this exercise though this should not be taken as a total indictment of the technique; indeed, although in this work simple rectangular specimens were examined, many more complex shapes could not be conveniently prepared by any other technique. More importantly, the lesson to be learned is of the necessity to ensure that cutting dies are in good, sharp, condition and 'clean' cuts are made, for example, by a clicking press. Ennor[21] has described a method for sharpening cutting dies and Delabertauche and Dean[22] give details of one with detachable blades. Die cutting cannot be used for stamping out specimens from thick sheet since a cross-section which is not rectangular will result; the limit of thickness, however, depends on the material being examined and can only be found by trial and error.

Generally speaking, standards do not specify or even recommend methods of achieving the machining of the specimens required nor, regrettably, do they even draw attention to the care needed. In the main satisfactory techniques are only evolved by trial and error but, by way of illustration, some suggestions are made based on the findings of the authors' colleagues.

7.4.1 To Reduce Rigid and Semi-rigid Sheets and Mouldings to Thinner and Uniform Section

The 'work' is held in a simple vacuum chuck and the thickness is reduced using a swing grinder or a fly cutter of suitable form (rubber may be machined using a cup grinding wheel). The results depend on the accuracy and rigidity of the machine tool, but the technique has, for instance, been found suitable for reducing PTFE sheet to appropriate thickness (a few mm) for measurement of dielectric properties, with a maximum variation of 0·025 mm over the whole area of the specimen (53 mm diameter)—that is $\pm$0·0005 in!

7.4.2 To Prepare Shaped Specimens From Glass Fibre Reinforced Sheets

If simple shapes (for example rectangular) are to be cut, rough shaping by bandsaw or hacksaw, followed by grinding to yield smooth edges, is effective. The use of a diamond saw in a precision bench is even better. For more complex shapes one may:

1. hand file between hardened steel jigs,
2. rout to a template, using a tungsten carbide cutter or
3. rout using a pantographic cutter.

These techniques are all somewhat laborious and profligate on the tools, but no adequate alternative has yet been found. Generally, the effort involved in preparing specimens of this type is many times that of actually carrying out the test.

A grinding machine for preparing test specimens from flexible plastics, amongst other materials, has been described by Eller and Gondek[23]. Commercial equipment is available from, for instance', H. W. Wallace

& Co. Ltd., 172 St. James's Road, Croydon, Surrey, CR9 2HR and Metallurgical Services Laboratories Ltd., Reliant Works, Brockham, Betchworth, Surrey.

REFERENCES

1. COHEN, L. A., 'Influence on Properties of Specimen Shape and Preparation', *Testing of Polymers,* Schmitz, J. V. and Brown, W. E. (eds), Volume 3, Interscience Publishers, New York, 15 (1967)
2. HAYES, R., 'The Effect of Temperature and Orientation on the Mechanical Properties of Plastics', *Trans. Plast. Inst., Lond.,* **27,** No. 49, 219 (July 1954)
3. BUDESHEIM, H. and KNAPPE, W., 'Effect of Processing Conditions on Mechanical Properties of Injection-moulded Polystyrene Standard Test-pieces', *Kunststoffe,* **49,** No. 6, 257 (June 1959)
4. HORSLEY, R. A., LEE, D. J. A. and WRIGHT, P. B., 'The Effect of Injection Moulding Conditions and Flow Orientation on Properties; *The Physical Properties of Polymers,* S.C.I. Monograph No. 5, Society of Chemical Industry, London, 63 (1959)
5. BRYANT, K. C. and HULSE, G., 'The Effect of Injection Moulding Variables on the Quality of Mouldings' *Plastics Progress, Papers and Discussions at the British Plastics Convention 1955,* Iliffe Books Ltd., London (1955)
6. BOSSU, B., CHATAIN, M., DUBOIS, P. and ROUGEAUX, J., 'Fundamentals of the Mechanical Properties of Plastics', *Int. Symp. Plast. Test. Stand.,* Special Technical Publication No. 247, American Society for Testing and Materials, Philadelphia, 67 (1959)
7. HOGBERG, H., 'Anisotropic Effects in Testing Plastics' in *Int. Symp. Plast. Test. Stand.,* Special Technical Publication No. 247, American Society for Testing and Materials, Philadelphia, 95 (1959)
8. MORGAN, D. E. and VALE, C. P., 'Variation of Physical Properties of Melamine Moulding Materials with Curing Time', *The Physical Properties of Polymers,* S.C.I. Monograph No. 5, Society of Chemical Industry, London, 169 (1959)
9. WILLIAMS, J. L. and MIGHTON, J. W., 'The Effects of Moulding Conditions upon the Permanence of Plastics', *Symp. Plast. Test.—Present and Future,* Special Technical Publication No. 132, 32, American Society for Testing and Materials, Philadelphia (1953)
10. DASCH, J., 'Effect of Shape of Test Pieces on Strength Values of Thermoplastics', *Kunststoffe,* **57,** No. 2, 117 (February 1967)
11. DASCH, J. 'Effect of Specimen Shape on the Strength of Thermoplastics, Part 2, Anisotropy of Flexural Strength', *Kunststoffe,* **58,** No. 11, 769 (November 1968)
12. KODA, H., 'Effects of Moulding Conditions on Properties of Injection-Moulded Polycarbonates', *J. appl. Polymer Sci.,* **12,** No. 10, 2257 (October 1968)
13. MALAC, J., 'Properties of PVC III. Influence of the Way of Processing on the Properties of PVC Samples,' *J. appl. Polymer Sci.,* **13,** No. 8, 1767 (August 1969)
14. WHISSON, R. R., 'The Effect of Processing Conditions during Injection Moulding on the Properties of Moulded Thermoplastics', *Technical Review 42,* Rubber and Plastics Research Association of Great Britain, Shawbury (September 1967)
15. WINTERGERST, S., 'Determination of the Basic Strength in Specimens of Thermoplastic Materials', *Kunststoffe,* **57,** No. 3, 188 (March 1967)
16. SPIWAK, L., 'Effect of Transfer Moulding on Strength of Reinforced Compounds,' *S.P.E.J.,* **19,** No. 6, 557 (June 1963)
17. ELMER, C. and HARRINGTON, E. C. JR., 'Flexural and Impact Variations of Phenolic Mouldings: A Statistical Round-robin Study', *Bul. Amer. Soc. Test. Mat., No.* 249, 35 (October 1960)
18. PANOV, P., 'Influence of Absolute Dimensions of Test Specimens on the Tensile Strength of Plastics', *Plaste u. Kautsch.,* **14,** No. 7, 491 (July 1967)
19. OGORKIEWICZ, R. M., ed., *Thermoplastics: Effects of Processing,* Iliffe Books Ltd., London (1969)
20. PATTERSON, G. D. JR., 'An Interlaboratory Study of Cutting Plastic Film Tension Specimens', *Mater. Res. Stand.,* **4,** No. 4, 159 (April 1964)
21. ENNOR, J. L., 'Sharpening the Dies used in Tensile Testing of Rubber and Plastics', *J. Instn. Rubb. Ind.,* **2,** No. 2, 98 (April 1968)

22. DELABERTAUCHE, C. and DEAN, S. K., 'A Die for Cutting Rubber Test Pieces', *J. Inst. Rubb. Ind.*, 2, No. 2, 99 (April 1968)

23. ELLER, S. A. and GONDEK, W. K., 'A Laboratory Grinding Machine for Preparing Test Specimens from Rubber and Other Flexible Products', *Bul. Amer. Soc. Test. Mat.*, **No. 206,** 70 (May 1955)

8

Short-Term Mechanical Properties

8.1 INTRODUCTION

In Section 2.7 an apparently heretical statement was made to the effect that standard tests may not be the answer to all problems. Nowhere is this more true than in the evaluation of mechanical properties and particularly in stress–strain measurements. Most tests, be they in tension, compression, shear, or bending (a combination of the other three), that have been specified for plastics have been adapted from time honoured methods used for metals. Over the past two decades the metallurgists have begun to realise that the mechanical behaviour of their materials is not so uncomplicated as once thought, but the situation with plastics is far worse. Sufficient mention has been made in previous chapters of the influences of time, temperature and history on plastics to make it obvious that a simple, single, test for a mechanical property at one (arbitrary) combination of time scale and temperature and in one (arbitrary) physical condition cannot give a very useful guide to the general performance of the plastics material.

Nevertheless, illogical as it may seem, the mostly arbitrary short-term tests will be described now, leaving the more difficult subjects of the effects of (long) time, load cycling and temperature on mechanical properties until later. Quite apart from the (limited) justification of tackling the simple things initially, most readers will encounter the standard tests first in their practical work and this type of test is widely used—and generally quite suitable—as an instrument of quality control: the important thing at this stage is to remember the *limitations of the normal standard mechanical test.*

For a general resumé of 'Terms Relating to Methods of Mechanical Testing', as defined in the U.S.A., see A.S.T.M. E.6–66.

8.2 TENSILE PROPERTIES

The most common type of stress—strain measurement is made in *tension*, that is by stretching the material. A *tensile* stress is thus applied, defined for a section of uniform cross-section area, A_0 by the formula $\sigma_1 = F_1/A_0$ where σ_1 = tensile stress and F_1 = tensile force (see Figure 8.1).

If this tensile stress induces a stretch to length l_1, then the tensile strain ε_1 is defined as

$$\varepsilon_1 = \frac{l_1 - l_0}{l_0}$$

Taking the stressing operation to the ultimate, that is increasing the force until the material breaks, tensile strength (ultimate tensile stress)

$$\sigma = \frac{F}{A}$$

where F = force at failure
A = area of cross-section at failure

As the material stretches so its dimensions orthogonal to the axis of applied force will decrease and thus the area of cross-section will decrease.

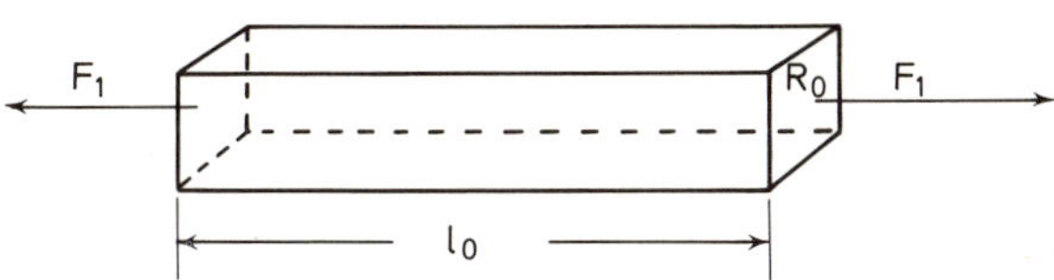

Figure 8.1. Material in tension

However, for experimental convenience, most tensile strengths are based on the *original* cross-section (A_0) since this is easily measured, before the test is started.

Ultimate elongation, or elongation at break, equals

$$l - l_0$$

where l is the length at failure.

This is usually expressed as a percentage of the original length, i.e.

$$\frac{l - l_0}{l_0} \times 100\%$$

Note on units of force:

The following is derived essentially from Reference 1.

Measurements of load or force are conveniently standardised by reference to the weights of given masses, a method capable of high accuracy. However, the weight of one pound mass, for instance, i.e. the gravitational pull on a one pound mass, varies over the earth's surface and errors of a few parts per thousand can be incurred unless this variation in gravitational acceleration is taken into account.

This difficulty in standardisation caused by varying gravitational pull is overcome by using a unit of force based on an internationally accepted standard acceleration. Such a unit of force is termed a 'technical unit of force'.

The international standard acceleration is 980·665 cm/s^2 and its equivalent in British units is 32·1740 ft/s^2. Hence the *British technical unit of force* is that force which, acting alone, will give to a one pound mass an acceleration of 32·1740 ft/s^2, i.e.

1 British technical unit of force = 1 lbf
= 32·1740 pdl

Likewise,

1 Metric technical unit of force = 1 kgf
= 980 665 dyne
= 9·80665 Newtons [N]

The Newton is the S.I. unit (see Section 1.4.1) of force and, like the dyne, is defined in magnitude to take account of international standard acceleration. A force of 1 N, when applied for second will give a mass of 1 kg a speed of 1 m/s (i.e. an acceleration of 1 m/s^2).

Note: The term 'kilopond' (kp) is sometimes used for kilogramme force. Another correction is technically required to correct for the upthrust on a mass due to the buoyancy of the air it displaces. For a mass of density about 8 g/cm^3 the correction is about 1 part in 7000 and need not concern us in the present context.

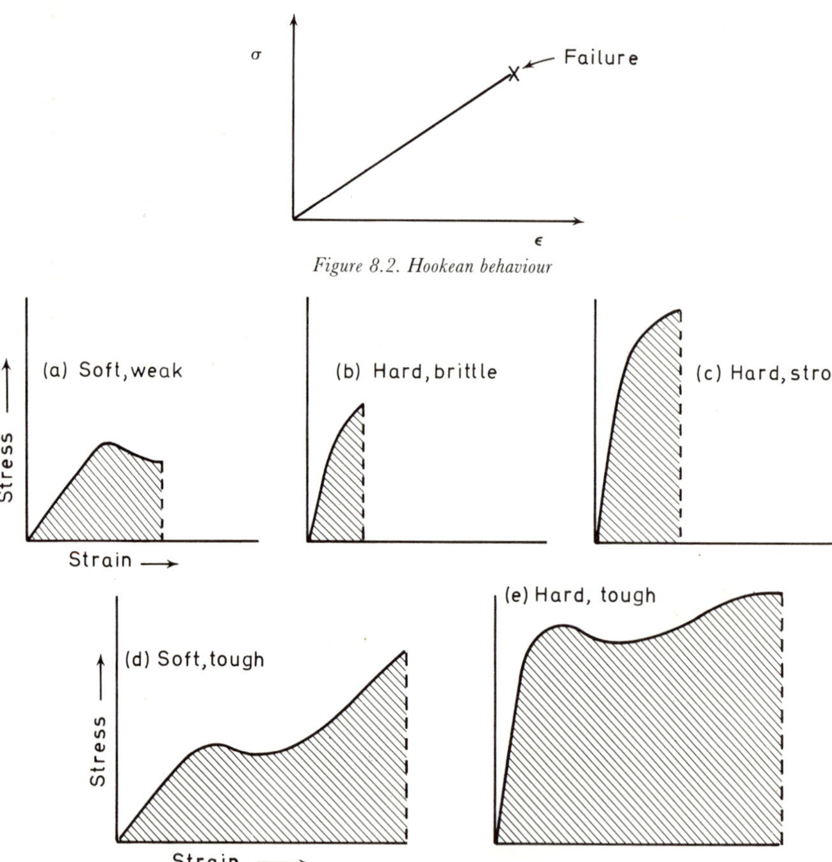

Figure 8.2. Hookean behaviour

Figure 8.3. Stress–strain behaviour of various types of plastics (after Carswell and Nason)

According to Hooke's Law, for an ideal elastic solid stress is proportional to strain (see Figure 8.2). Even ignoring the effects of temperature, and most of those of time, for the moment, the fact is that no plastics material comes very close to this ideal. Plastics exhibit a whole spectrum of behaviour which in qualitative terms it is easy to envisage deriving from such dissimilar materials as soft PVC, polystyrene, nylon and unplasticised PVC.

Expressed in the form of stress–strain diagrams, Carswell and Nason[2] have envisaged five possibilities as shown in Figure 8.3.

It will immediately be obvious that such terms as 'tensile strength' and 'elongation at break' can be misleading or gross oversimplifications even when used with the knowledge that such properties themselves have strictly limited application. For instance, in (a) above is the correct value for tensile strength that deriving from the stress at break or that at the peak of the graph? Again, what significance has an elongation at break from a behaviour such as that depicted in (d), when we have a combination of steady extension

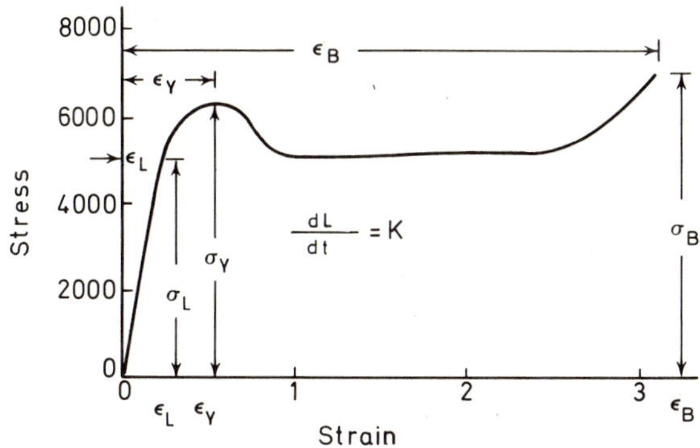

Figure 8.4. Schematic stress–strain curve (after Nielsen)

with increase in stress, increase in extension with no increase in stress (even a drop in stress) and finally a rapid extension with relatively little increase in stress?

Nielsen presents a general plot of stress–strain[3] of the type shown in Figure 8.4.

Where σ_Y = yield stress
ε_Y = elongation at yield
σ_B = ultimate or tensile strength
ε_B = ultimate elongation or elongation at break

(*Note:* The precise location of the yield point *Y* is a matter of argument and, according to different authorities, may be anywhere between *Y* and *L*.)

A Hookean material has already been defined and the constant ratio σ/ε is called the 'Youngs Modulus'. *If* a given plastics material has a stress–strain relationship which is initially linear, as in Figure 8.4 up to position *L*, then

$$\text{Young's Modulus} = \frac{\sigma_L}{\varepsilon_L}$$

A more complete description of stress–strain behaviour will be found in References 3–6 for instance. The above brief discourse is simply to illustrate what is likely to be needed in practical terms to measure tensile properties, i.e. the requisite testing equipment. With controlled temperature conditions taken as read, the equipment is some machine capable of applying a force (whence is calculated the stress), increasing from zero to the maximum required at a controlled rate. According to whether a very thin foil or a large moulding is being examined, so a maximum force of a few grams or several tons may be required; thus one machine, or a range of machines, must be available to apply these forces with equal pro-rata accuracy, and to achieve these desired forces within a convenient space of time (usually specified) whether the force be low or high and whether the material stretches only about 1% before breaking (as polystyrene) or perhaps 500–1000%, i.e. up to 10 times its original length (as soft PVC or low density polyethylene). With such a range of ultimate elongations clearly the means of following and measuring elongation (whence strain) must be many and varied.

Since the testing machines used for measuring tensile properties are almost invariably used also for compression, flexure and shear tests, it is convenient at this stage to consider the equipment, before going into details of tensile (and other) measurements.

8.3 MECHANICAL TESTING MACHINES

Put in simple terms, a mechanical testing machine is a device for applying a force on a test piece coupled to means of measuring that force; the latter is often referred to as a dynamometer, especially if self contained, though this term should really be applied to devices for measuring power output. There are a number of methods of achieving these ends, even for routine testing, and a brief description of some of them is appropriate. The subject divides itself more or less logically into two parts: firstly the testing machines themselves and secondly the means available for measuring extension, compression strain and deformation generally.

8.3.1 Machines for Measuring Force

'CONVENTIONAL' TESTING MACHINES

Over the past ten years or so, a radical change has overtaken the mechanical testing machine industry with the introduction of the proof ring/transducer type of machine, popularly known as an 'electronic testing machine', which is especially suitable for the low and relatively low loads encountered in plastics testing. However, these 'electronic' machines are only developments of the two-part device referred to above, and many conventional machines are still in use; indeed in the U.K. undoubtedly the majority of mechanical testing is still carried out on 'conventional' machines.

The most simple method of measuring breaking load is to add known weights (see Figure 8.5).

Here the added weights provide both the means of applying the force and the means of measuring it. Obvious disadvantages are the 'jerkiness' of applying the added weights—especially bearing in mind the known time-sensitivity of plastics (see Section 8.9)—and the insensitivity resulting from the necessity of adding finite increments of weight. A variant of the technique is to add lead shot at a steady rate into a canister suspended in place of the weight in Figure 8.5. The supply of lead shot is stopped when the specimen breaks and the weight of canister plus lead shot plus lower clamp gives the force required to fracture the specimen. However the method is clearly not very elegant, has obvious limitations of weight that can be added and it is not very easily controlled to give steady application of load, let alone easy to vary the rate of application of load as required.

Machines for mechanical testing have been in existence since the early nineteenth century; for a good review of early development and type of machines (some of which are still in use today!) see Chapter IV, Reference 7. Machines may be for tensile (stretching), compression or bending tests only, but it is not difficult to design for examination of all three—with shear and hardness as well—and such 'Universal' machines are commonplace

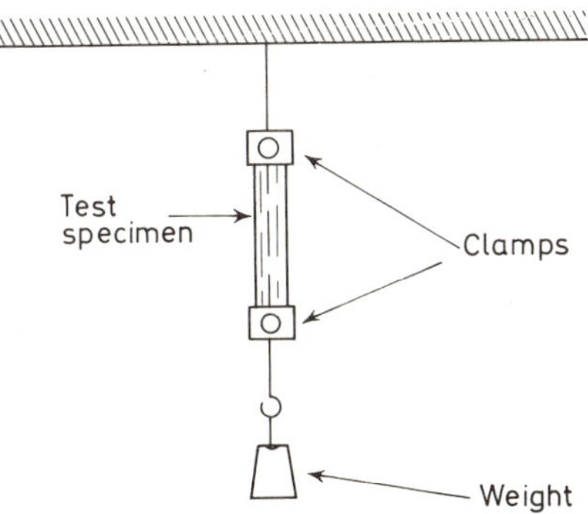

Figure 8.5. Simple tensile test

nowadays. Another and more important way of classifying machines derives from whether the strain on the test specimen is varied, the force being measured up to fracture, or the force is varied (in known fashion) and the strain increased to fracture of the specimen. *Very broadly*, the former is known as a constant rate of traverse (C.R.T.) machine and the latter a constant rate of loading (C.R.L.) machine, though this is very much an over-simplification of the state of affairs. Suffice to state at this stage that the one (C.R.T.) operates within a time factor governed by the stretch or compression of the test specimen whilst the other (C.R.L.) within a time factor governed by the breaking load of the specimen. Mention has already been made of the sensitivity of plastics materials to time—generally speaking the force to fracture decreases with increase in its time of application—and so quite obviously these two types of machine offer two different methods of controlling the period of time to break the test specimens.

C.R.L. TESTING MACHINES

One way of controlling the time to break of test specimens, within a range that represents the limits between which the breaking load is essentially unaffected by time, is to select a suitable rate of loading on a C.R.L. machine; the approximate breaking load of the specimen will probably be known and the rate of loading may be selected to give a break in a convenient space of time, typically between 15 and 45 s.

Possibly the most popular type of C.R.L. machine is based on the steelyard principles as shown in Figure 8.6.

The poise weight is driven from left to right, from a position of equilibrium about the fulcrum such that there is no force on the specimen, along the arm at a steady rate, whereby the moment of the poise weight about the fulcrum is increased steadily and thus the force on the specimen is increased at a steady rate. The poise weight is driven by a separate engaging screw running parallel to the steelyard, or by a similar mechanism.

Constant rate of loading conditions are most commonly called for in testing textiles and fabrics; they are not particularly popular in plastics testing and will not therefore be given further detailed examination. However, it is worth mentioning that, just as the steelyard type of machine may be used under C.R.T. conditions (see below), so many of the normally C.R.T. operated machines may be adapted to C.R.L. operation. For instance, a pointer driven at a steady rate over a dial indicating the force,

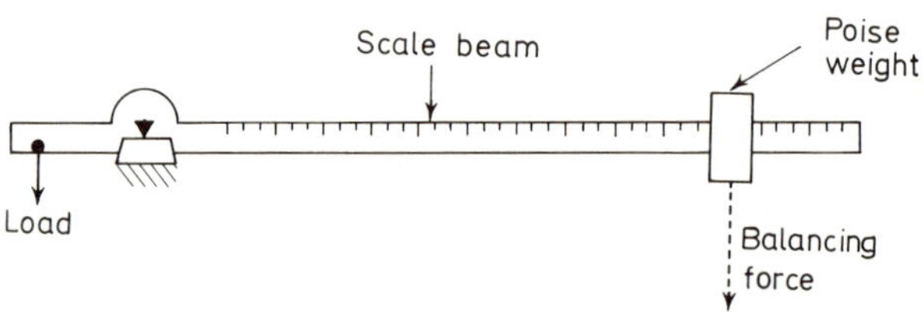

Figure 8.6. C.R.L. steelyard type testing machine

and at such a rate as to indicate the required rate of loading (a 'load pacer') may be followed by another pointer which indicates applied force and is constrained to keep pace with the 'pacer' by varying the cross-head speed. This latter can be achieved by a variable control on the speed of the drive motor. Obviously the technique lends itself to sophisticated solutions such as servomechanisms or alternatively the following of the pointer may simply be achieved by a handle driving the test machine directly.

One significant disadvantage with C.R.L. machines, and loading by weights or lead shot, is that the load increases all the time, or at least cannot decrease. Thus, stress–strain behaviour of the type shown in Figure 8.4, where there is a portion of the curve with negative slope, cannot be detected. However the same criticism may also be levelled at some C.R.T. machines.

C.R.T. TESTING MACHINES

Under this heading will be considered those types of machines where the test is controlled by the straining of the specimen and the resulting force is measured. This is a somewhat crude description because very few machines actually achieve the neat and tidy ideal of constant rate of extension of the specimen. In the first place, as will be seen in Section 8.4, the usual type of tensile specimen does not extend uniformly over its whole length, so that even if constant rate of separation of the jaws holding the ends of the specimen is achieved, extension will only be increased at a constant rate as an average over the whole length of the specimen. Secondly, the rate of separation of the driven jaws of the test machine is the sum of the deformation of the specimen *plus* movements in the jaws themselves and in the means of attaching the jaws to the machine, so that the rate of jaw separation, whether constant or not, is related to the specimen deformation (stretching, bending or compression) by a variable and probably unknown amount. Only by taking a gauge length of the specimen (see below), over which straining properties are uniform, and controlling the rate of separation of the jaws with reference to the extension of the gauge length, will true constant rate of extension be achieved. Thus constant rate of traverse, i.e. constant rate of separation of the crosshead holding one jaw from the other crosshead holding the other jaw, usually means no more than just that, i.e. 'C.R.T.' Moreover, many machines fondly thought to be 'C.R.T.' are not this by any manner of means. Thus the 'dynamometer' or force measuring device may 'yield' significantly under the load it is measuring and thus contribute to the apparent extension of a tensile specimen for instance: machines are termed 'soft' if this yield is high relative to the deformation in the specimen and 'hard' if low. Finally, the mechanism driving or separating the jaws may alter in performance significantly according to the force it is acting against in the specimen.

After drawing attention to all these shortcomings it is only fair to comment that with many machines the deviation from constancy of rate of transverse is not so serious as to significantly affect mechanical test results on plastics within practical limits, providing the load is increased without 'surges' and the time to break is achieved within a defined period of time. Thus many specifications for mechanical tests speak in terms of 'extending the test specimen (or separating the jaws of the machine) at a rate which is *essentially* constant'.

Starting with the first requisite of the testing machine, the means of applying the load, many commercial machines employ a motor-driven screw gear mechanism driving a movable 'head' which transmits the load through the test specimen to a fixed 'head' (the test specimen is attached to the 'heads', or crossheads, by the jaws or chucks which are secured thereto)—see Figure 8.7.

The fixed 'head' is attached directly to a plate or via a fixed crosshead which is connected to the means of measuring the force. It will be seen that depending where the specimen is put in the machine so tensile or compressive (or bending) forces may be applied. Control of speed of separation of the crossheads or platens is effected by appropriate gearing of the driving motor; hand winding may also be used although speed is obviously not very controlled under these conditions.

There are many variants on this theme, with the screws rotating within bearing nuts mounted in the movable head as shown or the screws fixed to the movable head and the bearing nuts in the platen for instance. Alternatively in a tensile testing machine, a single rotating screw may be used as in the well known Hounsfield Tensometer as shown in Figure 8.8. In the Model 'E' Tensometer the straining head is moved by twin vertical recirculating ball screws.

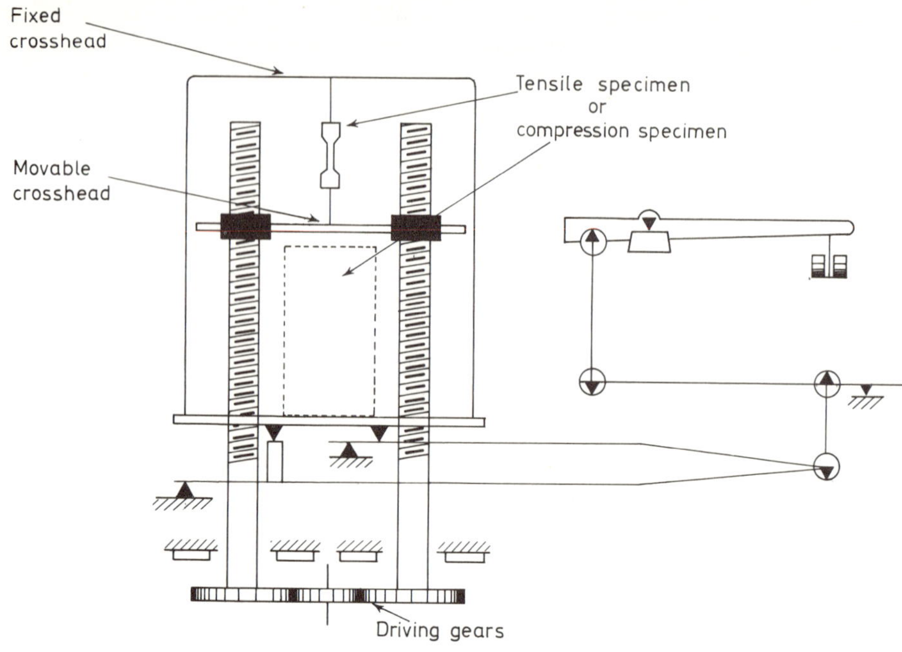

Figure 8.7. A screw gear testing machine

Figure 8.8. Single screw machine (Courtesy Tensometer Ltd.)

Hydraulic rams can also be used to apply the load which may be indicated by a suitably calibrated Bourdon gauge which actually responds to the pressure developed within the hydraulic cylinder; however errors due to friction are difficult to overcome and a separate load measuring device is preferable—see Figure 8.9.

Such machines are essentially compressive in character and strictly speaking are not C.R.T. (but neither are they C.R.L. though the loading is increased by the hydraulic system).

Coming now to the devices for measuring the force applied to the test specimen, the earliest test machines used the steelyard principle referred to above. Here, however, the poise weight is not used to apply the load but

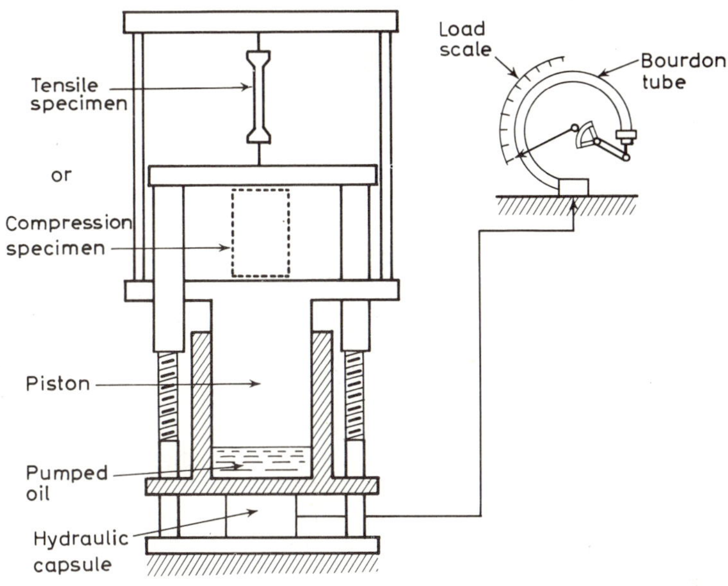

Figure 8.9. Hydraulic testing machine

is moved along the arm to balance out the force on the specimen, and thus the position of the poise measures the force. The balancing may be achieved by hand driving of the poise and visual observation of the position of the end of the steelyard arm, or more elegantly by a motor controlled by the levelling of the arm for instance. Different scales of force range may be obtained by altering the size of the poise weight.

Other types of simple machines use a range of calibrated springs, the linear responses of which to applied force may be indicated on a suitable dial gauge; a variant of this is to combine springs with a lever arm (Figure 8.10). In the Hounsfield Tensometer referred to above, force is transmitted through the test specimen via the fixed crosshead to the centre of a calibrated steel beam, the deflection of which is linearly proportional to the force applied. This deflection also moves a piston in a reservoir of mercury so that the mercury flows into a glass tube of uniform cross-section, calibrated in a force scale. The mercury meniscus is followed visually or may be connected to a pen recorder. A range of beams is available from 62 lbf to 2 tonf maximum capacity.

All these types of machine are relatively 'soft' in response to applied force, but even more so is the pendulum type of machine popular for elastomer testing and inherited to some degree by the testers of soft plastics (see Figure 8.11).

If the rate of movement of the pendulum is low,

$$FR = Mgd$$

where g = gravitational acceleration. Thus, since R is constant and so

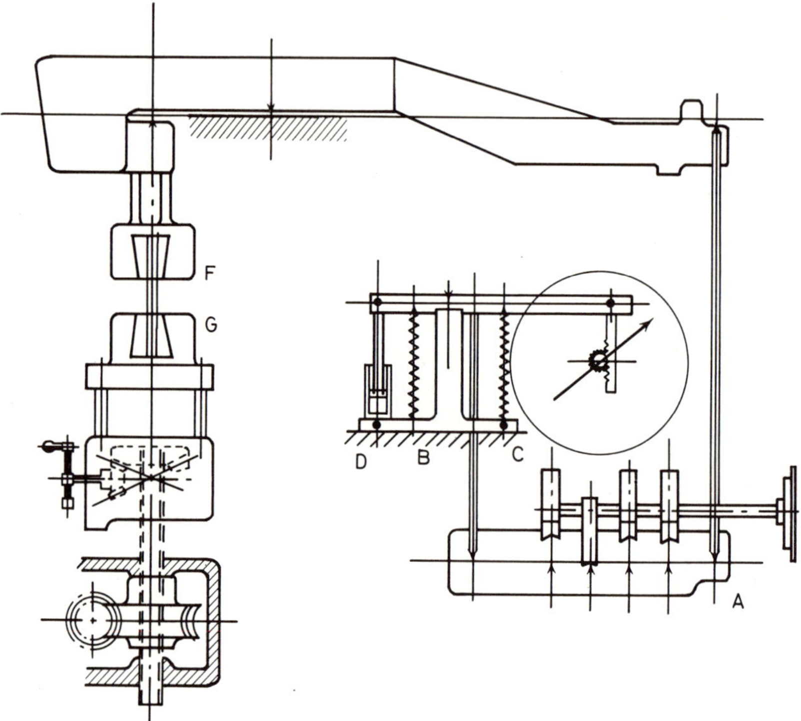

Figure 8.10. Load weighing mechanism used in Denison machine. A: capacity change lever, fulcrum selected by hand operated cam. B: major weighing spring. C: minor weighing spring. D: shock absorbing dashpot. F: weighing wedgebox. G: straining wedgebox. (Reproduced by kind permission of Pergamon Press Ltd. from an article by Dr. J. H. Lamble[8])

is M, for a given size of 'bob' (changing these alters the force ranges on the machine), the distance d may be used to measure the force F. However, more correctly

$$FR = Mgd + I\,\frac{d\omega}{dt}$$

where I is the moment of inertia of the pendulum about the pivot and

$$\frac{d\omega}{dt} = \text{angular acceleration of pendulum}$$

This second term can be quite significant, especially at the rates of jaw separation normally used for soft rubber or highly plasticised PVC for instance. Thus there are large errors inherent in the pendulum type of machine used this way and further, since a pendulum has a natural period of oscillation, the arm may move in a series of jerks rather than uniformly.

Mention has already been made of hydraulic machines and direct reading of force from a Bourdon gauge connected to the pump system; other weighing devices may be used such as pendulums (Figure 8.12). Direct reading may be inaccurate however, because if the piston fits too loosely in the cylinder there will be leakage of oil and if too tightly (e.g. with packing) there will be errors due to friction. A controlled leak of oil, without

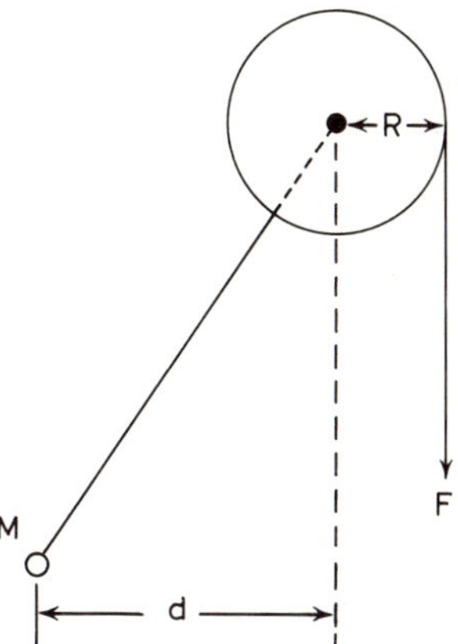

Figure 8.11. Principle of pendulum testing machine

packing, and rotating cylinders overcome these troubles. Thus in Figure 8.12, the Avery machine Model 7110 CCJ, the main ram works without packing in an accurately bored cylinder with a controlled oil leak; the latter ensures that the ram is fully lubricated and the oil is returned to the oil tank by means of a suction pump. Load 'weighing' is achieved by a pendulum connected to a dial gauge—the horizontal displacement of the pendulum is linearly proportional to the load (see above) and thus to the horizontal movement of the connected arm. This therefore moves the cam on which the dial gauge pointer is mounted by a linearly proportional amount of its circumference which in turn moves the pointer over the circumferential scale in the same direct proportionality. The purpose of the dashpot is to dampen the return of the pendulum when the specimen fractures. A 'slave' pointer records maximum load. The capacity change wheel selects five alternative fulcra to give five alternative capacity ranges.

Alternatively, load 'weighing' in hydraulic machines may be achieved by use of a hydraulic capsule (Figure 8.9), an example of which consists of a thin, flexible metal diaphragm filled with a liquid. This latter is virtually incompressible and thus the load on the platen is transmitted to the Bourdon gauge or other weighing device with scarcely any movement of the diaphragm. Much more popular currently, however, are the dynamometers based on proof rings and transducers or similar systems, the so-called 'electronic' machines. Further, since their response to force is reversible

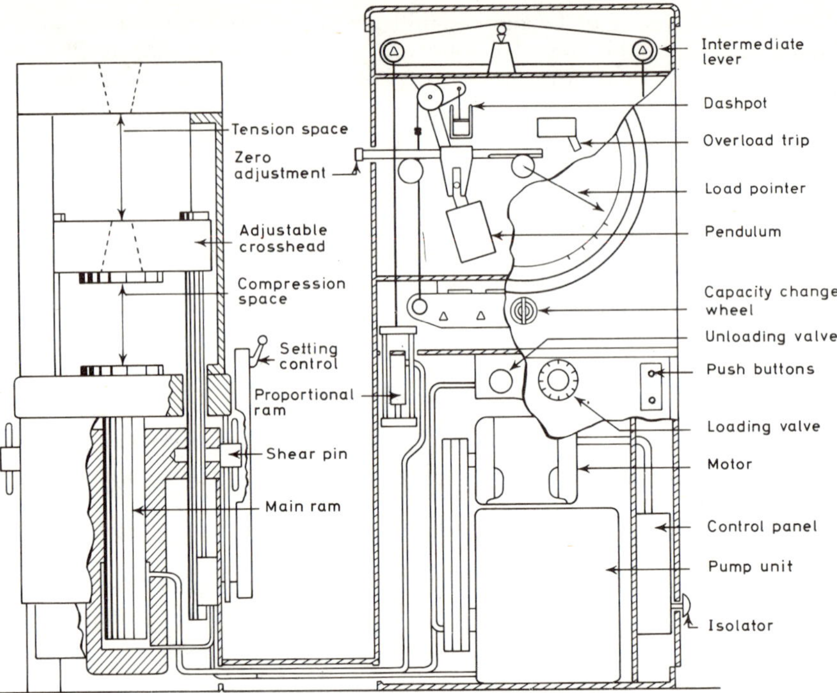

Figure 8.12. Principles of operation of the Avery 7110 CCJ universal testing machine

they will follow stress–strain behaviour as illustrated in Figure 8.4 whereas many of the above C.R.T. machines cannot.

'ELECTRONIC' TESTING MACHINES

Virtually any of the methods described above for applying load to the test specimen may be combined with the proof ring transducer type of load measuring device. Proof rings are stiff but perfectly elastic metal rings or other hollow shapes whose linear deformation, albeit extremely small, is proportional to load. One such type of proof ring is shown in Figure 8.13.

The deformation is followed by an electrical device of which there are many types, but typical and widely used are the differential transformer and differential inductance transducers; in the former movement is linearly converted to an electrical parameter by displacement of the coils of a variable transformer and in the latter by movement of a steel core within an electrical coil. A description will be found in Reference 9 of an early inductance transducer/proof ring dynamometer where the latter is fitted to a pendulum machine (Hounsfield) of the type generally illustrated in Figure 8.11, the dynamometer being in 'series' with the driven crosshead and the test specimen, and attached between the jaw holding the fixed end of the specimen and the fixed crosshead (which is connected to the pendulum pivot). The electrical signal is usually fed to a chart recorder—see Section 8.3.2 below.

A commercial testing machine based on the principle is the already-mentioned Tensometer Type 'E', the layout of the instrument up to the transducer, but excluding the chart recorder and control gear, being shown in Figure 8.14.

The transducer is housed in the 'load cell'. Instruments operating on similar principles are marketed by Goodbrand and Co. Ltd., Samuel Denison and Son Ltd., Instron Engineering Corporation, Tinius Olsen Testing Machine Company, Scott Testers Inc., Wiedemann (Baldwin) Machine Company, Thwing-Albert Instrument Company, Karl Frank G.m.b.H., and Contraves A. G., amongst others, whilst the Riehle Testing Machines Model F-400 retains the poise weight balancing principle (see the description of the steelyard machines above), but drives the poise along the arm from a control actuated by a differential transformer. H. W.

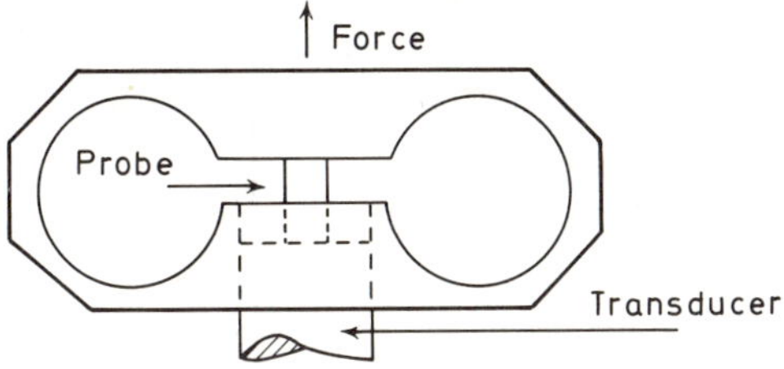

Figure 8.13. A proof ring design

Wallace and Co. Ltd., offer a self-contained proof ring/transducer dynamometer to be fitted to existing test machines or any suitable apparatus applying force.

Different capacity force scales can obviously be obtained from different sized proof rings; in addition, however, using one and the same proof ring, the electrical circuitry may be switched to vary the scales within the overall range of the proof ring (see Reference 9). This offers a great advantage for it means that a test may be started using a sensitive scale, so that the initial stress–strain behaviour may be followed in detail, and then by switching (the electrical circuit) to a large force scale the test may be completed without any discontinuity, providing the same proof ring is used. In addition, this type of dynamometer is relatively 'hard' so that the 'fixed' head and jaw or grip is essentially fixed, there are no inertia errors and a very large range of force may be covered by the same machine (many tons down to 1 g or so). Finally this type of machine is readily adaptable to chart recording, is time saving and very suitable for routine testing, where for instance the amplification of the transducer signal can be set to suit the dimension of the type of test piece under examination (see Section 8.4.2 for instance) so that stress is read directly.

A comparison of the accuracies and speeds of operation of a conventional pendulum machine and an electronic dynamometer machine is described in Reference 10. In an examination involving five natural rubber compounds (of different hardnesses), quite different tensile strengths, ultimate elongations and stress–strain curves were obtained in many cases from the two machines and the electronic machine was found to be much more speedy in operation as well as the more accurate.

It is now possible to obtain electronic testing machines with sufficiently stable electrical equipment and accurate recorders so that they comply with the requirements for a Grade A machine to B.S. 1610[11]. (The corresponding standard governing accuracy of testing machines are, in the U.S., A.S.T.M.E.4[12] and in Germany DIN.51221. See also Reference 1.) They may be adapted to C.R.L. conditions by using a signal from the load

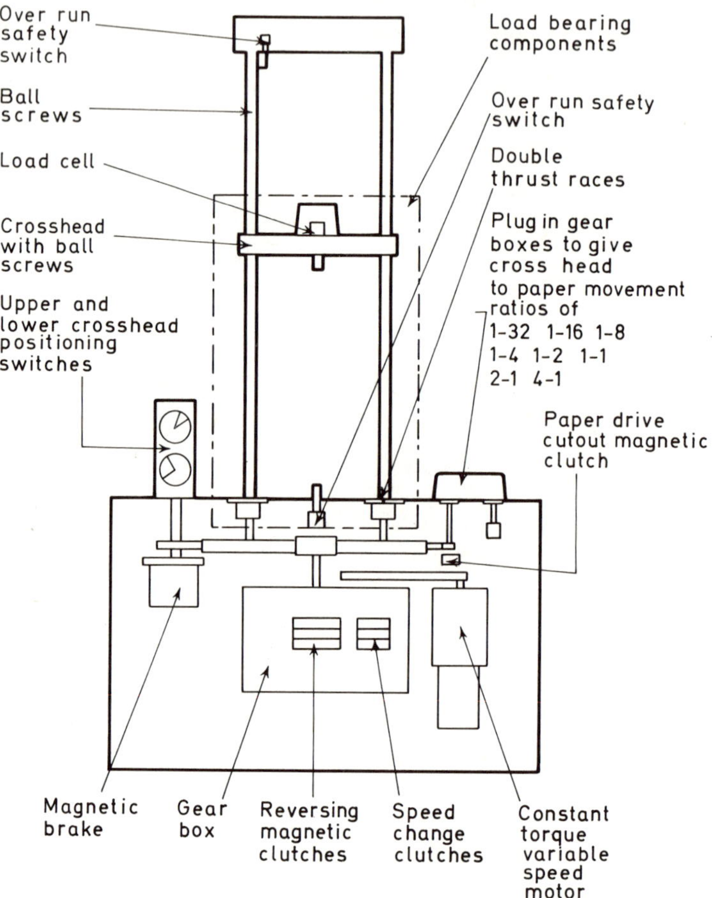

Figure 8.14. Partial layout of 'E' type tensometer (Courtesy Tensometer Ltd.)

transducer to control the motor driving the crosshead and to true constant rate of extension conditions by taking a signal from an extensometer (see below).

CHECKING AND RECALIBRATING OF TESTING MACHINES

Mention has just been made of three standards relevant to the checking of test machines; it is fair to state that generally such equipment is not recalibrated frequently enough and quite large errors in force measurement are tolerated unknowingly. Checking is usually carried out with the aid of standard dead weights, or steelyards connected thereto, but two convenient methods based on springs have been suggested[13].

8.3.2 Extensometers and Elongation Indicators

Many mechanical testing machines offer some means of following or measuring extension or elongation and, from this, strain. Most popular are those which plot, on a pen chart recorder or similar device, the force on the Y axis and the movement of the movable crosshead on the X axis (see Section 8.3.1). The chart may be wound on a drum which is driven by gearing to the screw driving the crosshead, or whatever driving means is used, and the pen moved vertically over the paper by link attachment to the weighing device, for example a pendulum. In the Hounsfield Tensometer (Figure 8.7), the drum is driven as just described but the plot is obtained by manually following the meniscus of the mercury indicator with a cursor, on the arm of which is a pin which is depressed into the graph paper (which is separated from the metal drum by a soft rubber sleeve). The cursor is depressed as frequently as is necessary to give an accurate representation of the force yield behaviour, and at the end the 'pin pricks' are joined up by pen.

There are objections to this principle of following deformation however. In the first place the movement of the crosshead is in part due to the yield of the stress indicator (e.g. the spring steel beam in the Hounsfield Tensometer) and this must be compensated for; as it is often linear with respect to load this need not be too difficult, though for a 'soft' machine it may change the force deformation curve quite appreciably and involve a significant correction to elongation or deformation figures. In the second place the separation of the jaws on the crosshead may initially be increased in taking up 'slack' in the specimen/jaw and jaw/lug connections, but again this can be overcome by applying a slight load before test, insufficient to strain the specimen significantly but enough to straighten out all links and take up all 'looseness'.

Finally, the true separation of the jaws may not give an accurate measure of the elongation or deformation of a definable length of the specimen. This is particularly true in tensile measurements and is discussed more fully in Section 8.4.5 below. As a result, sensitive units for actually measuring deformation directly are required, either for simply assessing elongation or strain, for translating on to stress–strain diagrams or for feeding on to chart recorders to give direct stress–strain plots as described above.

It may readily be imagined that the problems involved in measuring strains of 2% or less on the one hand, and of upwards of 1000% on the other, are quite different and so are the solutions. Yet, as already stated, these are the ranges of *breaking* strain encountered in, respectively, rigid plastics such as polystyrene and soft plastics such as low density polyethylene; furthermore the strain behaviour of most interest, as far as rigid plastics are concerned, stops well short of the breaking point. However, all such extensometers follow the relative movement of *gauge marks*, representing the limits of the gauge length or central section of the test specimen over which elongation or strain is to be measured (see below, Section 8.4).

For small extensions there are many types of extensometer, based on micrometers for instance, where the micrometer is used to read the distance between arms fixed on the gauge marks of the test specimen and running out normal from it (see Figure 8.15).

The two arms are affixed to the test piece at gauge marks a and b on a plane running through its axis, by set screws; the lower arm carries a micrometer screw e and on the hardened point of this the upper frame rests. Thus if the arms are at right angles to the test specimen the micrometer screw measures the variation in distance ab. Levels c and f are provided to set the arms normal to the test specimen by first adjusting the lower

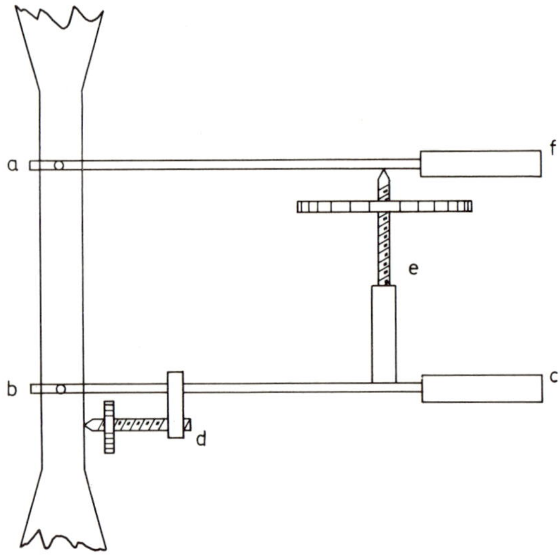

Figure 8.15. Micrometer extensometer

one c by the screw d, after which the upper one f is set by the micrometer screw e. The arms are then clamped to the specimen.

Movement of the gauge marks with respect to one another is followed discontinuously by screwing up the micrometer until the hardened point just makes contact with the upper arm. As in all extensometry involving devices attached to gauge marks on the test specimen, care must be exercised to ensure that any pressure thus executed on the gauge mark areas does not influence the stress–strain characteristics of the specimen.

There are many variants of this simple type of mechanical extensometer, using arms to magnify the extension, electrical circuits to show contact, and dial gauges to read distance directly for instance. The technique is suitable for tensile or compressive measurements.

In another classical technique, mirrors are attached to the gauge marks, e.g. on rollers which rotate as the gauge marks separate, so that the extension may be followed by telescope, light reflection on to scales or whatever, with magnification more or less chosen at will by selection of the details of the optical system. Alternatively, a travelling microscope or cathetometer may be used directly to follow the separation of simple marks drawn on the test piece, but although the danger of damage to the test specimen is virtually removed, the method is slow and tedious and discontinuous, like the micrometer technique.

A useful review of these techniques for small extension measurements will be found in Chapter XI of Reference 7 and in Chapter XV of Reference 14. Some extensometers are fragile and care must be taken to prevent damage from the shock of the fracture of the test specimen.

For extensions of what might be termed the 'intermediate range' (up to about 50%), Kinloch and Waters[15] have described an extensometer where steel rods connected indirectly to the gauge marks move independently of each other through polytetrafluoroethylene bearings fitted into the ends of a central brass electrode; the rods are insulated from the electrode by a liner of suitable dielectric constant and the whole assembly is encased in a brass screen with holes to allow the steel rods to pass freely. The movement of the rods varies the capacitance of what is essentially a coaxial capacitor in a manner such that the variation of the capacitance is virtually directly related to the separation of the rods, i.e. of the gauge marks. The method lends itself to continuous reading and plotting of stress–strain diagrams (see below); the extensometer is stated to be robust and capable of withstanding the shock caused by fracture of the test specimen.

For highly extensible materials, Farlie, Hawkes and Waters[16] have described an extensometer which uses the pinpricking chart marking technique to which brief reference was made above in connection with the Hounsfield Tensometer. In the same company's pendulum machine for testing elastomers and soft, extensible, plastics, the chart on the drum is moved in the Y axis direction according to the load applied and the corresponding elongations are 'pricked' by the operator who follows the displacement of the gauge marks by eye on a scale running parallel to the test specimen. Farlie et alia have retained the marking technique but the pin is depressed by electromagnetic operation of the hammer which strikes the head of the pin (Figure 8.16). Very briefly, a pulley M is made to rotate as the gauge mark separation increases (two other pulleys P on the other side help this) and in direct relation thereto; a spring S keeps the mechanism slightly taut. This rotation drives a concentric sprocket N which has 20 teeth on its circumference and one rotation is equal to 100% elongation. The teeth cause a contact (P) to be made in a relay circuit which actuates the hammer depressing the pin and thus extension is recorded every 5% increase in elongation. One variant of this works off a roll of film with alternating light transmitting and non-light transmitting areas taking the place of the teeth of the sprocket wheel, the light pulses falling on a photoelectric cell. Eagles and Payne[17] describe a similar technique where the teeth of the wheel, filled with transparent resin, provide the pulsed light beam and the output of the relay is fed to the X axis of an electrically driven chart recorder, with the dynamometer feeding the Y axis. A commercial version of this instrument has been evaluated by Meardon[18] and found to give improved repeatability both between operators and between times, compared with that obtained visually from a scale, though the instrument gave no better reproducibility for a given operator in any one series of tests.

Instron Ltd. market an extensometer based on optical reflection of gauge marks on to photocells; when the distance between the gauge marks changes, the photocell heads follow by a servo system. Influence of the gauge marks on the specimen is thus kept to an absolute minimum.

Mention has already been made, in connection with the Kinloch and Waters extensometer[15], of the utility of extensometers giving a change of electrical parameter to provide the drive on the *X* axis of a chart recorder, the *Y* axis being plotted from the corresponding load. With 'electronic' test machines in particular such extensometers are now finding wide use and are standard ancillaries, based on change in capacitance, inductance

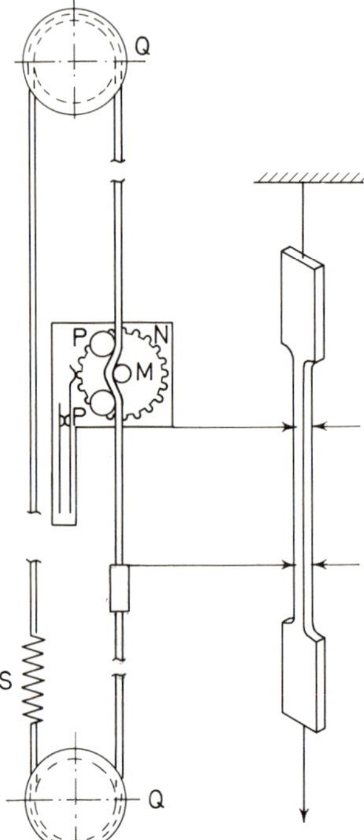

Figure 8.16. Extensometer for highly extensible materials (Farlie, Hawkes and Waters) (Reproduced by courtesy of Rubber Journal)

or resistance of appropriate transducers, differential transformers or the pulsed light technique, generally to drive a servomechanism which rotates the recorder drum around the *Y* axis.

A very useful review of stress–strain recorders has been made by Bouche and Tate[14].

8.4 TENSILE STRESS–STRAIN, TENSILE STRENGTH, ELONGATION AND MODULUS MEASUREMENTS

This subject has been considered in some detail—theory, effect of variables, test equipment etc.—by Harris[20].

8.4.1 Definitions

In Section 8.2 an attempt was made to define some of the various parameters encountered in a tensile test. In a *general* specification for tensile properties of plastics, A.S.T.M. D.638–68 (which has no precise counterpart in the B.S. series), the appendix defines inter alia the following:

(a) Tensile Stress (nominal): The tensile load per unit area of minimum original cross-section*, within the gauge boundaries**, carried by the test specimen at any given moment.

(b) Tensile Strength (nominal): The maximum tensile stress (nominal) sustained by the specimen during a tension test.

(c) Tensile Strength at Break (nominal): The tensile stress (nominal) is so designated if the maximum stress occurs at break.

Note: (i) If a stress–strain behaviour is imagined such as that in Figure 8.2 without the final (r.h.s.) upward trend, the significance of (b) in relation to (c) is readily grasped.

(ii) The use of the word 'nominal' is to allow for the fact that true tensile stress, etc., would be based on the minimum cross-section during test.

(d) Gauge Length: The original length of that portion of the specimen over which strain or change in length is determined.

(e) Elongation: The increase in length produced in the gauge length of the test specimen by a tensile load.

(f) Percentage Elongation: The elongation of a test specimen expressed as a percentage of the gauge length.

(g) Percentage Elongation at Break: The percentage elongation at the moment of rupture of the specimen.

(h) Strain: The ratio of the elongation to the gauge length of the test specimen.

(i) Yield Point: The first point on the stress–strain curve at which an increase in strain occurs without an increase in stress.

Note: Reverting to Figure 8.3 only variants *a*, *d* and *e* show a 'yield point' by this definition; in the latter two, strictly speaking, there are two points of zero slope, but it is the first (l.h.s.) in each case which is the 'yield point'. For cases of types *b* and *c*, see (k) below.

(j) Yield Strength: The stress at which a material (sic) exhibits a specified limiting deviation from the proportionality of stress to strain. Unless otherwise specified, this stress will be the stress at the yield point.

(k) Offset Yield Strength (see Figure 8.17). The stress at which the strain exceeds by a specified amount (the offset) on extension of the initial proportional portion of the stress–strain curve.

(l) Proportional Limit: The greatest stress which a material (sic) is capable of sustaining without any deviation from proportionality of stress to strain (Hooke's Law).

(m) Elastic Limit: The greatest stress which a material (sic) is capable of sustaining without any permanent strain remaining upon complete release of the strain.

*See below, the consideration of test specimens.

**See definition (d).

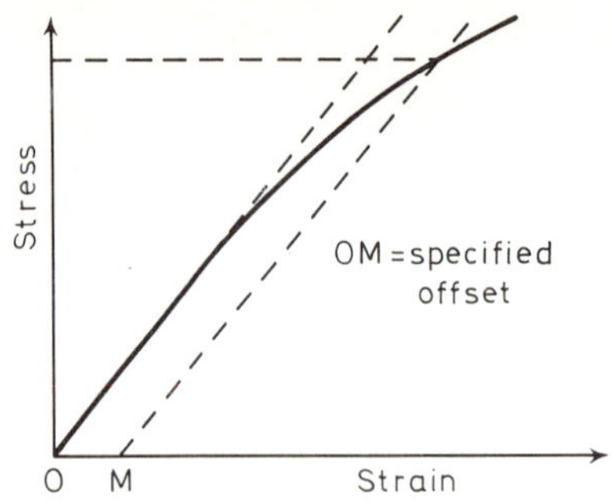

Figure 8.17. Offset yield strength (A.S.T.M. D. 638)

Definitions (c) and (g) are the most important in what follows and (l) and (m) are somewhat academic as far as plastics materials are concerned.

8.4.2 Test Specimens

In earlier chapters we have talked of preconditioning our test specimens and even how to mould or machine them, but there remains the question of the shape and size of pieces on which we are going to perform the measurements. It probably will not come as a surprise to learn that different tests (for different properties) demand different test specimens—hence the reason for leaving the subject until discussion of each individual method—but it may be less obvious that, for instance, tensile tests on different materials employ specimens of different sizes; the effect of specimen size on tensile strength has been studied by Panov[21], for instance. However, reference again to the widely varying types of materials embraced in the term 'plastics', with quite different ultimate tensile strengths and percentage elongations at break, will justify this complication.

To carry out a tensile test, i.e. a stretching test, some form of elongated specimen capable of being gripped at both ends is needed. The simple rectangular bar illustrated in Figure 8.1 is not suitable and even if reduced in thickness to facilitate gripping it is still often unsatisfactory since the fracture point could occur anywhere along its length and most probably would occur within one or other of the gripped portions because of the weakening effect of the compression exerted by the gripping chucks (see below). Thus a failure load may be obtained which is lower than the characteristic tensile value for the material in the specimen cross-section and, additionally, the failure may occur outside the portion of the test piece being examined for elongation (the gauge length). In order to select the portion in which failure takes place and obtain a breaking load on a cross-sectional area unbiassed by gripping mechanisms, a 'dumb-bell' specimen is normally employed. For metals this is a fairly accurate description since test specimens thereof are usually produced by lathe turning as shown in Figure 8.18.

Many of the more rigid plastics may be machined on a lathe and so it is possible to produce similar specimens out of cast epoxide or polymethyl methacrylate, for example; however in the majority of instances, specimens have to be cut out from (thin) flat sheet or moulded from powder or granules and the orthodox dumb-bell specimen is not used as a standard test piece, not even when machining from castings. For sheet materials, the specimens are 'two dimensional dumb-bells', retaining the flat profile of the sheet. Typical types are as shown in Figure 8.19.

For mouldings, at least when compression moulded from thermosetting materials, the 'tab' ends are thickened as shown in Figure 8.20.

A.S.T.M. D.638–68, already mentioned, describes two general shapes of specimens, as shown in Figure 8.21.

Four types are recommended, related to the above and selected according to thickness range. The choice of specimen size is also governed by the relevant product specification, but a different specimen is used for thin film sheeting generally (A.S.T.M. D.882–67); two of the A.S.T.M. D.638 specimen types are used for 'Testing Rigid Sheet and Plate Materials Used for Electrical Insulation' (A.S.T.M. D.229–69). A specimen similar to that in Figure 8.20 is specified in 'Tensile Strength of Moulded Electrical Insulating Materials' (A.S.T.M. D.651–48). If insufficient material is available for the standard tests, and only if, the use of 'micro-tensile' specimens is allowed in A.S.T.M. D.1708–66 (Figure 8.22).

In B.S. 2782:1970, the Method 301 series of tests describes eleven tensile methods, each associated with a particular type or class of materials,

Figure 8.18. Metal dumb-bell tensile test specimen

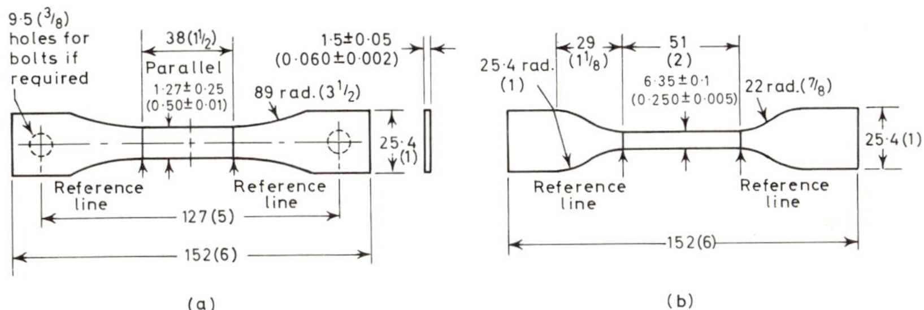

Figure 8.19. Typical sheet plastics tensile test specimens (B.S. 2782) Dimensions in millimetres with inch equivalents in parentheses. (a) Variation in width of parallel portion not to exceed 0·1 (0·005). (b) Variation in width between reference lines not to exceed 0·1 (0·005)

employing nine different shapes or sizes of dumb-bell specimen. Eight are of the types described above and are selected according to the strength and/or extension characteristics of the material under test, its physical form (e.g. thickness of sheet) and mode of moulding. According to Method 301L, however, certain reinforced plastics, e.g. glass rovings bonded with polyester resin, can be tested successfully using a rectangular specimen. A rectangular strip is used, not less than 230 mm (9 in) long and 25 ± 0·5 mm (1·0 ± 0·02 in) wide with four rectangular end-pieces, 45 mm long or more and not less than 3 mm ($\frac{1}{8}$ in) thick; the thicknesses must be equal. The end pieces are of similar material to that of the test strip and are bonded

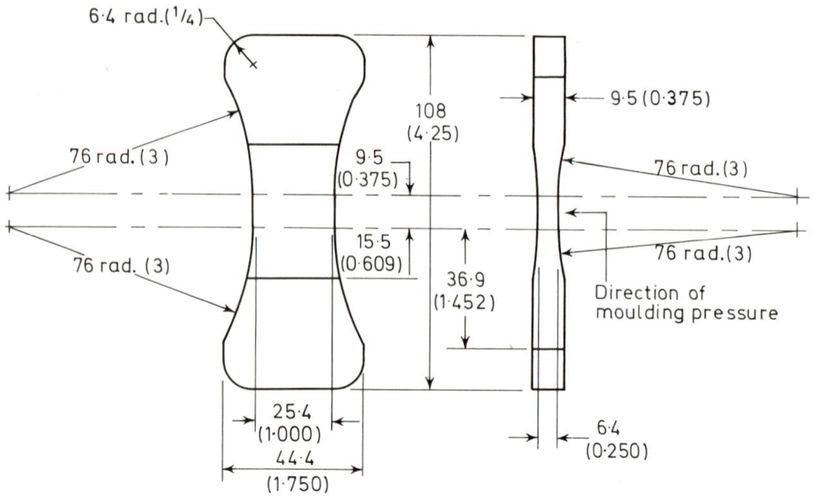

Figure 8.20. Compression moulded test specimen (B.S. 2782). Dimensions in millimetres with inch equivalents in parentheses. Tolerance on all dimensions ±4%

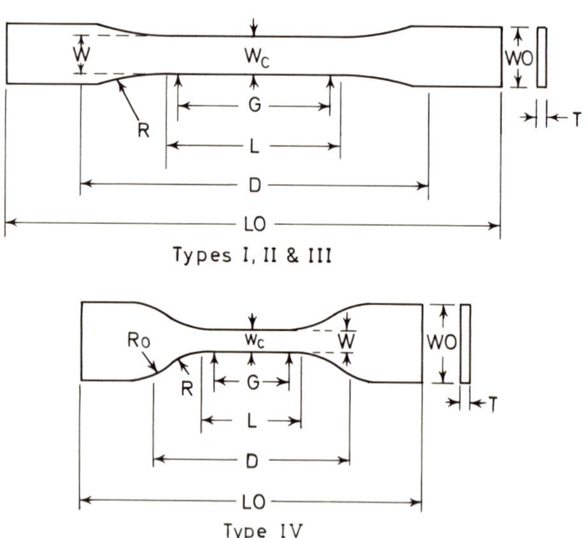

Figure 8.21. A.S.T.M. D.638 tensile test specimens

to the strip, two each end and one either side thereof, by a cold hardening epoxide resin (see Figure 8.23 for a diagram showing a sheet of such a reinforced plastics material, with end pieces bonded, from which strips are cut for test). The use of rectangular specimens where possible simplifies and reduces the amount of machining required. 'Where possible' is dictated

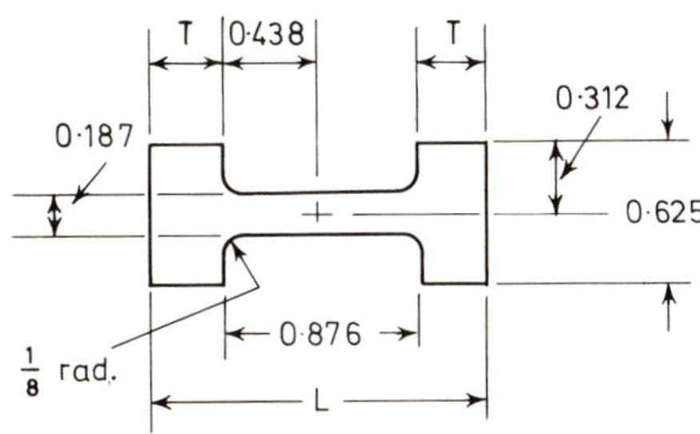

Figure 8.22. Micro-tensile test specimen (A.S.T.M. D.1708) All dimensions are in inches. Tolerance: ± 0·002 in. Thickness = thickness of sheet. Minimum tab length, T = 0·312 in (larger tabs shall be used wherever possible). Minimum length, L = 1·50 in

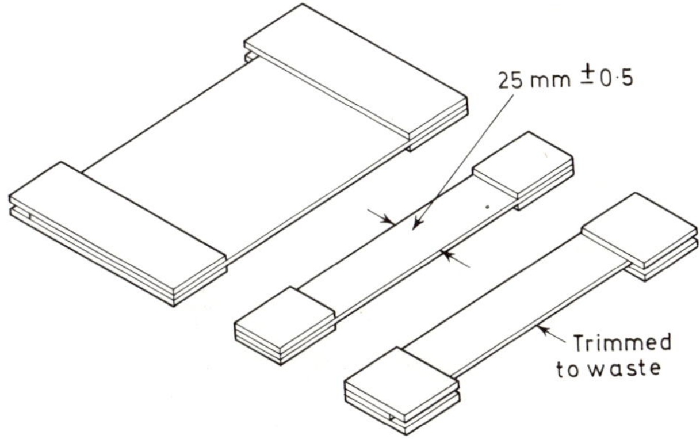

Figure 8.23. Rectangular strip specimen preparation (B.S. 2782)

by the requirement that results from strips breaking within the areas of the end-pieces are disregarded.

DIN 53455 follows a similar pattern in offering a range of specimens etc.

8.4.3 Gripping Test Specimens

Thus far we have the testing machine and the specimens; there remains the very important matter of holding the latter in such a way that they do not slip out nor get crushed at the gripped ends before the central portion of the specimen fractures; in addition the design of the grips or chucks and their mode of attachment to the testing machine must be such as to ensure alignment of test piece in the direction of strain, without any bending or shearing component.

There are many designs of grips to achieve this aim, some complementary but mostly having special applications according to the nature of the material under test and its physical form, be it strong or weak, soft or hard, wide or narrow, thick or thin. Some designs are described below.

Probably the most widely used is the 'wedge action' type, suitable particularly for rigid flat sheeting (Figure 8.24).

Some control over the friction between the test specimen and the gripping faces of the wedges can be obtained by having different surface patterns

Figure 8.24. Wedge action grips (Courtesy Tensometer Ltd.)

on the latter, varying from serrated patterns to perfectly smooth; this last mentioned may be further modified by insertion of emery paper between jaw face and test specimen, with the rough paper surface facing the test specimen. The final choice depends on trial and error.

Another type, for the moulded thermoset type of tensile specimen, is illustrated in Figure 8.25.

For thin brittle films, vice chucks are often suitable (see Figure 8.26), or windlass chucks (see Figure 8.27a and b).

The film is gripped, vice fashion, in between the 'flat' on the transverse rod of the chuck and the plate bolted thereto, and then the film and the rod/plate assembly are rotated as shown by the arrow so that the film has to wind round almost half a circumference of the rod before leaving the chuck to form a similar assembly at the other end with an identical chuck.

Very strong laminates and sheet materials are best gripped by strip chucks (Figure 8.28) where the test piece is first drilled with a hole at each tab end (see l.h.s. of Figure 8.20) to suit the bolts which are screwed down

Figure 8.25. Grips for moulded test pieces (Courtesy Tensometer Ltd.)

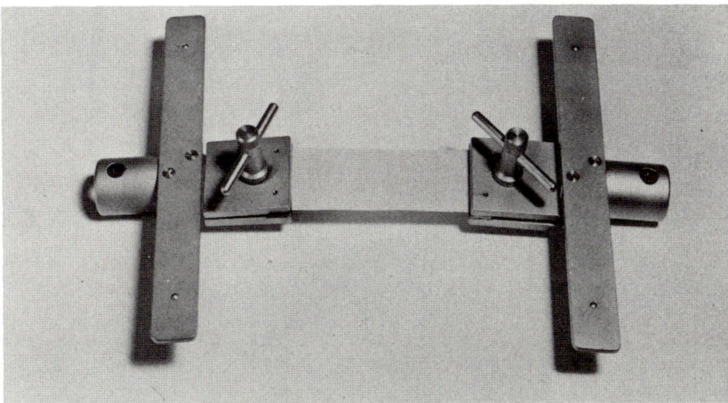

Figure 8.26. Vice type chucks (Courtesy Tensometer Ltd.)

hard on stout serrated washers which fit into the chucks, themselves hollow to fit round the free floating lugs of the test machine and be gripped by pins fitted in the holes shown matching holes in the lugs. (This method of attaching the chucks to the test machine lugs is widely used.)

Finally, in this by-no-means exhaustive list, may be mentioned self tightening jaw chucks used widely for soft rubber and plastics sheeting, such as highly plasticised PVC (see Figure 8.29).

(a)

(b)

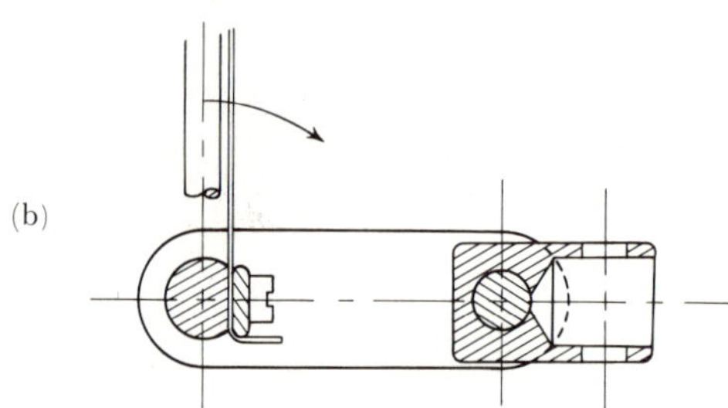

Figure 8.27. Windlass chucks (Courtesy Tensometer Ltd.)

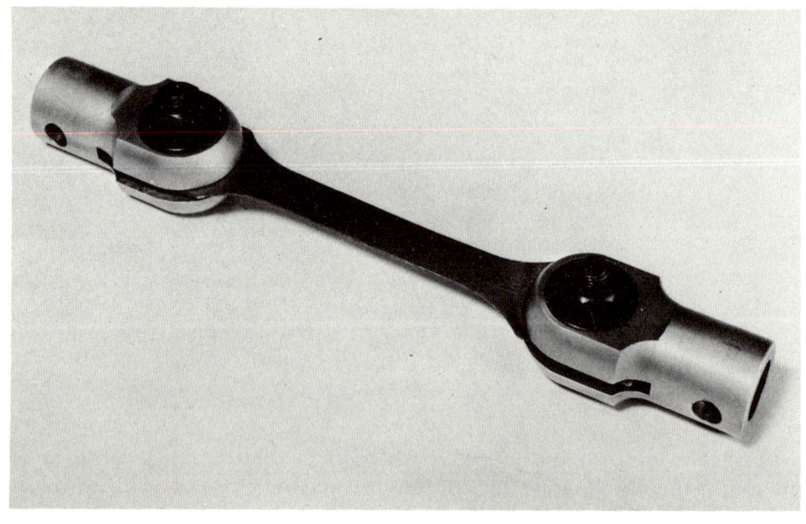

Figure 8.28. Strip chucks (Courtesy Tensometer Ltd.)

Figure 8.29. Self tightening jaws (Courtesy Instron Ltd.)

A self aligning grip system for Instron machines has been described by Holland[22].

8.4.4 Some Details of Specific Tests

With the conditioned tensile specimen suitably mounted in the test machine, in the necessary surrounding atmosphere, the appropriate force scale selected, and the extensometer device attached to follow strain if required, there remains one variable to settle—the rate of stressing or straining to be applied, which will control the time scale of the experiment. In tests such as those being described in Sections 8.4–8.8, i.e. quality control tests yielding reproducible data without necessarily having application to design considerations, choice of rates of stressing or straining are for experimental convenience, to strike a reasonable compromise between a speed so fast that the test is completed before it can be accurately followed with ordinary routine test equipment and one so slow that the test takes an inordinate time to complete. Most plastics tests are conducted on (nominally) constant rate of transverse machines and thus low speeds are used for virtually inextensible materials and fast speeds for highly extensible materials.

Some idea of the variation necessitated by these considerations is gained by reference to the Methods 301 of B.S. 2782: 1970 (see Table 8.1). Likewise the replication of the test varies, the choice having been governed by known variability of the material in question.

Table 8.1 TENSILE TESTS IN B.S. 2782: 1970

Test ref.	*Material*	*No. of test specimens*	*Speed of test*	*Temperature of test*
Method 301A	Thermosetting moulding material	3	Break in ½–1½ min	20±5°C
Method 301B	Cellulose acetate moulding material	5	5·1–6·3 mm/min (0·20–0·25 in/min)	23±1°C
Method 301C	Laminated sheet (thermosetting)	3*	Break in ½–1½ min	20±5°C
Method 301D	Flexible PVC extrusion compound	4	460±75 mm/min (18±3 in/min)	23±1°C
Method 301E	Flexible unsupported PVC sheet	5*	280±25 mm/min (11±1 in/min)	23±1°C
Method 301F	Polythene	4	460±75 mm/min (18±3 in/min)	23±2°C
Method 301G	Rigid PVC compound	4	25±6 mm/min (1±0·25 in/min)	23±1°C
Method 301H	Toughened polystyrene for sheet extrusion	5	5·1–6·3 mm/min (0·20–0·25 in/min)	23±2°C
Method 301J	Toughened polystyrene moulding material	5	5·1–6·3 mm/min (0·20–0·25 in/min)	23±2°C
Method 301K	PTFE (excluding rod)	5*	50±5 mm/min	35±1°C†
Method 301L	Reinforced plastics	≮5*	Break in ½–1½ min	20±5°C

*If the property differs with the direction in the plane of the sheet, sets of specimens are cut in two directions at right angles to each other, usually with their major axes in one case parallel and the other perpendicular to some feature of the sheet, e.g. direction of cloth reinforcement or calendering.

†PTFE has several transition points (see Chapter 11), including two near ordinary ambient temperatures (at 19 and 30°C), where mechanical properties change significantly (e.g. see Koo and Andrews[23]); the aim of testing at 35°C is to conduct measurements significantly above these critical points and avoid the inevitable scatter of results which would occur if the tensile strength were measured in the region of either of these transition points.

Note: In all the tests, the 'rate of traverse of the loading grip' or the 'rate of separation of the grips' must be 'substantially constant'.

The test specimens used are specified (see Section 8.4.2). The temperatures of test are generally related to I.S.O. conditions (see Chapter 3), with tolerance in particular selected according to the degree of sensitivity of tensile strength of the material under test to variation in surrounding temperature. All eleven methods require the calculation of the individual tensile strength of each test specimen by dividing the load at break by the original cross-section and expressing the results as MN/m^2, or as an interim measure, as kgf/cm^2 or lbf/in^2, except 301G, H and J which use the maximum load recorded; Method 301F includes tensile stress at yield also and some of the methods specify elongation measurements. All place some control on the accuracy of the test machine to be used, generally that the force shall be recorded to an accuracy of 1%, $1\frac{1}{2}$% or 2% (according to method) or, if the values are very low, to ±1·1 N (113 gf or 0·25 lbf). In all cases, the mean value of tensile strength, etc., is also calculated.

A.S.T.M. D.638–68 requires at least five specimens, or two sets of five if there are directional effects, and gives a choice of four speeds, ranging from 2·5 mm/min (0·10 in/min) to 510 mm/min (20 in/min). Accuracy of load measurement must be ±1% or better and results are expressed in kilograms per square centimetre (or pounds per square inch). Average values are calculated and so are standard deviations*. As with B.S. 2782, but rather more comprehensively, a list of items to be reported is laid down.

In B.S. 903, covering tests for vulcanised rubber, tensile values are derived from the median (central) value because the distribution of individual results is rarely normal or Gaussian*. The use of 'order statistics' to obtain a more accurate estimate of the true value from a very limited number of individual results has been described by Heap[24] in which, very briefly, far less significance is given to the lower or lowest values, which are the ones most likely to be spurious, than in the calculation of a mean value or selection of a median value.

DIN.53455 specifies the official German practice and gives two shapes of dumb-bell specimen, one for moulding materials and the other for sheet materials. A rectangular shape is also included. (See, in addition, ISO/R.527.)

8.4.5 Strain and Elongation Measurements

It might be thought that the easiest way to measure elongation (extension) and therefore strain would be simply to follow the relative movement of the jaws, i.e. the 'jaw separation'; if the machine is relatively 'hard', i.e. the fixed jaw moves insignificantly because the loading device has a small displacement response to force, then only the moving jaw should need to be watched. However, as stated in Section 8.3.2, it is virtually impossible to measure elongation accurately this way, and still less possible to obtain strain.

The movement of the jaws will depend on the extension of the test specimen, certainly, but other factors that may contribute are the taking up of slack in the lugs in the mechanism of jaws and, if the material under test is soft, the in-line movement of the jaws resulting from their compressing the gripped portions. For hard materials, where the last mentioned effect is insignificant, it is true that the 'slack' can be taken up by moving the crosshead until a force just registers and the test started from there, but then the second difficulty is encountered, that is just to what length of the test piece does the measured extension relate? If it is a dumb-bell specimen, obviously the waisted portion will stretch, but will the tab ends and, if so, by how much pro-rata? Again, be it a dumb-bell or even simple strip specimen, the portions under compression in the chucks will obviously be

*See Chapter 1, Section 1.5.

constrained to some degree from responding to the tensile (stretching) force. Thus no reliable figure for extension in relation to a known length of specimen, in other words strain, is likely to result from following jaw separation in a tensile test.

There is no substitute for the use of *gauge marks* to define a *gauge length* (see Section 8.4.1 above) and following the separation of these marks, which thus yields the elongation in relation to a fixed and known length. If the material is very extensible, say low density polyethylene with an elongation at break of about 400%, a 25 mm (1 in) gauge length would extend by 100 mm (4 cm) and if this is measured to 2·5 mm (0·1 in) the elongation at break will be obtained to the nearest 10%, which is sufficiently accurate for most practical purposes. Such accuracy is obtainable by a skilled operator using a ruler graduated in divisions of 1 mm, moving the zero end to coincide with the 'fixed' gauge mark (which will move as the portion of test piece between the gauge mark and the fixed chuck elongates) and watching the position of the 'moving' gauge mark. This is the technique suggested in Methods 301D, E and F of B.S. 2782: 1970.

More refined techniques are available for extensible materials, as mentioned in Section 8.3.2.

For the relatively inextensible materials such as most thermosets, polystyrene, polymethylmethacrylate, etc., where elongation at break may be 2% or less, clearly such visual observation methods are impracticable, and mechanical or electrical extensometers must be used. If elongation at break as such is to be measured, as opposed to say extension over the initial elastic range (see below), care must be exercised to ensure that the extensometer is not of such a type that it will be damaged when the test piece breaks or shatters. However, despite the fact that elongation at break is often specified, it has little or no practical significance.

Force–elongation, and hence stress–strain, plots may be obtained by manual plotting of observed measured force values and the corresponding elongation figures or, better, from extensometers fed to a chart recorder on the *X* axis, with force forming the *Y* axis (see Section 8.3.2).

8.4.6 Tensile (Young's) Modulus

A.S.T.M. D.638–68 contains the following definition:

(a) Modulus of Elasticity Elastic Modulus Young's Modulus	The ratio of stress (nominal) to corresponding strain below the proportional limit of the material.

It is pointed out that many plastics do not conform to Hooke's Law, of constant proportionality of stress to strain, throughout the elastic range, indeed many scarcely show any Hookean behaviour at all. As an alternative, therefore, for such materials it is usual to take the tangent to the stress–strain curve at a low stress. In the A.S.T.M. there is the definition:

(b) Secant Modulus: The ratio of stress (nominal to corresponding strain at any specified point on the stress–strain curve).

Note: It is worth mentioning here, to avoid a possible confusion, that in the U.K. it is common, especially in relation to elastomers, to quote '100% modulus', '200% modulus' and so on. These are not true moduli, but simply the tensile stresses at, respectively, 100% extension, 200% extension etc. Obviously only in the case of the 100% value will the value be identical with 'true' modulus (making the rash assumption of Hookean behaviour up to this extension!).

Since strain is a ratio of two length measurements, and is therefore dimensionless, the units of modulus are those of stress, that is most usually MN/m^2 (kgf/cm^2 or lbf/in^2). It will also be readily appreciated that since modulus is essentially the resistance of a material to stretching (tensile strength) divided by the stretch induced (the extension), the lower the extension for a given stress, the higher the modulus, i.e. the modulus indicates the *stiffness* of the material.

The measurement of tensile modulus has been covered essentially in the sections above dealing with test machinery, tensile stress measurement, extension measurement, the production of stress–strain diagrams and the definition of modulus. As far as standardised techniques go, A.S.T.M. D.638 requires that the initial linear portion of the load extension curve be extended and the difference in the stress (using the initial specimen cross-section) corresponding to some section of the straight line be divided by the corresponding difference in strain (extension divided by original gauge length). In B.S. 2782, Method 302A is given a technique for *secant modulus* of rigid materials, the value being derived from the tensile stress at 0·2% strain. (Because of the general lack of linear stress–strain relationship in plastics materials, the B.S. sets itself *against* determining modulus from the initial part of such a curve.)

Two forms of specimen are allowed. One is a somewhat strange rectangular bar not less than 150 mm long, 10–25 mm wide and 1–10 mm thick; this dates back to the time when the method was described in detail in B.S. 771 for phenolic moulding materials and, for convenience, used the same specimen as for plastic yield (see Chapter 11). The second type of specimen is dumb-bell in form and identical with one of those specified for the measurement of tensile strength. Two specimens are used; the gauge length is 50 mm and the length of the specimen between the chuck jaws is 115 ± 5 mm. The extensometer must be capable of measuring to an accuracy of at least 0·0025 mm over the 50 mm gauge length.

An initial tensile force w is applied to straighten out the jaws and test piece; this force is conveniently about 10% of that expected to be required to induce 0·2% strain. With this force applied, the extensometer is set to zero or the reading noted and the force is steadily increased by separating the jaws at a rate of 1 mm/min ± 25% until the gauge mark separation, as indicated by the extensometer, has increased by 0·10 mm, at which point the force is noted (W).

$$\text{Elastic (secant) modulus in tension} = \frac{W-w}{0{\cdot}002A}$$

where A = initial cross-sectional area of test specimen.

The units are stipulated as MN/m^2, or kgf/cm^2 as an interim measure.

If 0·2% strain is likely to lead to fracture of the specimen, 0·1% strain may be used, for which

$$\text{Elastic (secant) modulus in tension} = \frac{W-w}{0{\cdot}001A}$$

8.5 COMPRESSIVE PROPERTIES AND MEASUREMENT

8.5.1 Definitions, Test Specimens and Techniques

If two equal and opposite colinear forces are applied to a material a compressive stress is set up and compressive strain results, in every way analogous to the conditions of tensile stress and strain described in Section 8.1, but tending to crush rather than stretch. To distinguish compressive parameters from their tensile counterparts, the former are taken as negative in value.

A.S.T.M. D.695–68T defines compressive stress (nominal), compressive strength, compressive strength at failure, compressive deformation, compressive strain, percentage compressive strain, compressive yield point, compressive yield strength, offset compressive yield strength, proportional limit and modulus of elasticity in similar terms to their tensile analogues (see Section 8.4.1). Crushing load is also defined, relating compressive force to a particular object under test and it is worth noting here that the adjectives 'compressive' and 'crushing' may be used synonymously—B.S. 2782:1970, in Methods 303A, B and C, employs the term 'crushing strength'.

More important still, however, is the definition included in A.S.T.M. D.695:

> Slenderness Ratio: The ratio of the length of a column of uniform cross-section to its least radius of gyration. For specimens of uniform rectangular cross-section, the radius of gyration is 0·289 times the smaller cross-sectional dimension. For specimens of uniform circular cross-section, the radius of gyration is 0·250 times the diameter.

Although the precise mechanical significance of this ratio need not concern us here, the definition serves to indicate what different sizes and shapes of test specimens will give comparable results, for compressive property data are influenced by the specimens used.

Cubes, prisms and rods seem obvious choices for compression test specimens, with perhaps a preference for the last, as it should make for more uniform distribution of load. There is certainly no problem of gripping as there is with tensile tests and therefore no complications with waisted portions and tab ends. On the other hand there are difficulties of a different type such as the following:

1. The need to apply a truly axial force
2. The tendency of the specimen to bend or buckle
3. Friction between the flat plate compressing chucks and the ends of the test specimens, which will tend to prevent lateral expansion of the specimen as it is compressed.

(1) is serious because in compression tests it is not possible to have knuckle-type links which have their slack taken up under (tensile) stress. (2) will

be obviated by using specimens of large cross-section, but this will lead to greater friction in (3). The answer, such as it is, lies in compromise between the two, and hence the importance of stipulating the slenderness ratio already mentioned.

As far as apparatus is concerned, many machines are made as compression machines (Section 8.3.1) compression being effected between two approaching flat surfaces; those of tensile type can easily be made to apply compressive forces, for instance by use of chucks of the type shown in Figure 8.30.

Because of the relatively large cross-sectional areas of standard test specimens (see below). the forces to be measured are somewhat higher than

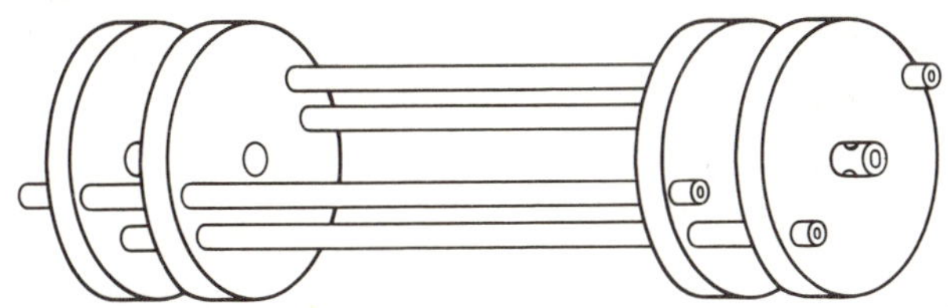

Figure 8.30. Compression chucks for tensile machine

in tensile tests and thus higher capacity machines, or larger scales on the same machines, have to be used. By and large compressive strain measurement causes little difficulty because it is usually allowable to relate the approach of the chuck plates to the change in length of the specimen and this is followed readily by dial gauge either directly or via a lever mechanism (see Section 8.3.2).

8.5.2 Some Details of Standard Tests

B.S. 2782:1970, Methods 303A, B and C cover crushing strength tests on thermosetting moulding material, thermosetting sheet and casting and laminating resin systems respectively. The first and last employ a cylinder 9·53 mm (0·375 in) long and 9·53 mm diameter and Method 301B a cube of side 12·7 mm (0·500 in) (which may be built up from sheet of thickness less than 12·7 mm thick). The test, which is performed in duplicate, must be completed within 15 and 45 s of first applying load.

$$\text{Crushing strength} = \frac{W}{A}\ \text{MN/m}^2,\ \text{kgf/cm}^2 \text{ or lbf/in}^2$$

W = load at fracture
A = cross-sectional area of test specimen

A.S.T.M. D.695–68T specifies the use, where applicable, of right cylinders or prisms of length twice their principal width or diameter, preferably 12·7 × 12·7 × 25·4 mm (0·5 × 0·5 × 1·0 in) prisms or 12·7 mm dia. × 25·4 mm long cylinders. For moduli and offset yield stress data (i.e. where the test is not to fracture) a slenderness ratio of between 11 and 15:1 is employed. Plying up is again allowed but for thin (under 3·2 mm ($\frac{1}{8}$ in) thick) glass-reinforced plastics in particular a dumb-bell specimen is used, which is compressed along its length and prevented from buckling by being clamped as a flange (loose) between two suitable, polished, steel plates. Five specimens are used, or ten if directional effects in the material are suspected. Full details of speed of testing, extensometry etc. are given.

The A.S.T.M. requires that loading shall be axial within 1 part in 1000

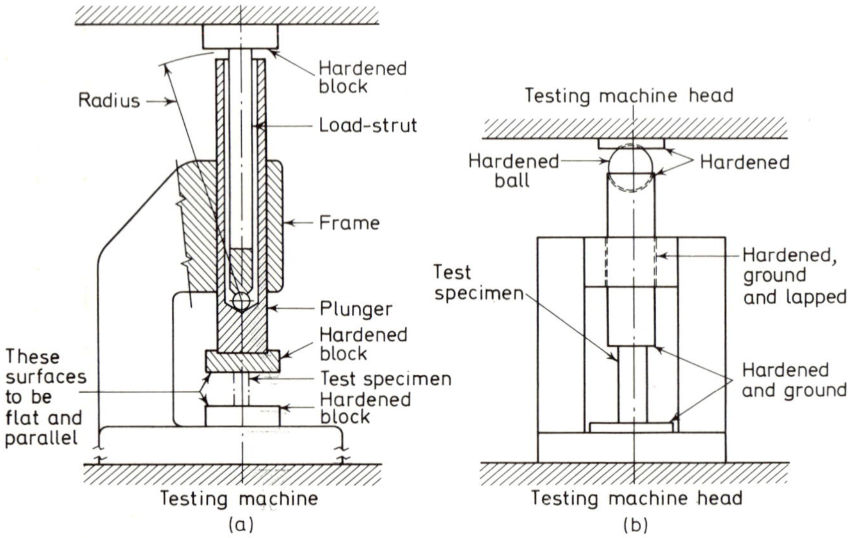

Figure 8.31. Compression tools (A.S.T.M. D.695)

and applied through surfaces that are flat within 0·025 mm (0·001 in) and parallel. Two compression tools are illustrated in Figure 8.31.

Another compression test is contained in A.S.T.M. D.621–64, but as it is most often used as a measure of heat resistance its description will be found in Chapter 11.

DIN.53454 gives the equivalent German practice.

8.5.3 Plain Strain Compression Tests

Williams and Ford[25], using the fact that strain is so much easier to measure in compression tests than over gauge lengths, have evaluated the 'plain strain compression test', first developed for metals, as a means of determining the permanent and total deformation curves for plastics up to high levels of strain such as may be encountered in engineering applications. The outline of the experimental set up is shown in Figure 8.32.

By this technique, the area under load remains constant. The specimen faces can be lubricated with a number of materials: molybdenum disulphide grease was used by Williams and Ford.

The application of the technique to the examination of polymers at large strains has been further described by Williams[26] and the reader is referred

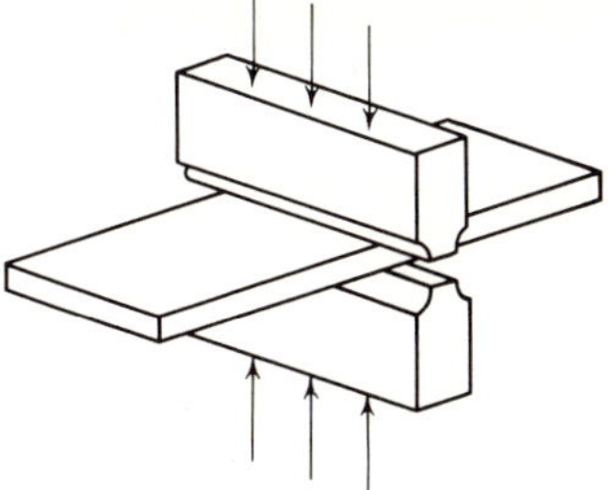

Figure 8.32. Plain strain compression test (after Williams and Ford)

to these two papers in particular for the full experimental details, mathematics and some results.

8.6 SHEAR PROPERTIES AND MEASUREMENT

The third type of force is that of shear which, unlike tensile and compressive forces which are normal to the plane on which they act, acts parallel to the plane (see Figure 8.33).

$$\text{Shear Stress } \tau = \frac{F}{a^2}$$

$$\text{Shear Strain } y = \frac{\delta a}{a}$$

If Hooke's Law is obeyed,

$$\text{Shear Modulus } G = \frac{\tau}{ya} = \text{Modulus of Rigidity}$$

Amongst a number of loading systems which give rise to shear forces is the instance where parallel but opposite forces act through the centroids

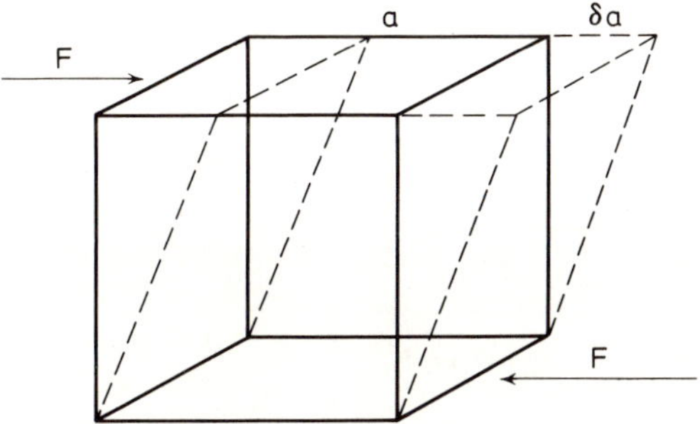

Figure 8.33. Material in shear

of sections that are spaced an infinitesimal distance apart. This would lead to pure shear but cannot be realised in practice; somewhere near an approximation comes from the *punch shear* test where a punch bears on a flat sheet of test material resting on a die; the nearer the internal diameter of the die and the external diameter of the punch, the closer the approximation to pure shear. This type has found wide popularity as a standard test for plastics.

Methods 305A and 305B of B.S. 2782:1970, shown in Figure 8.34, for respectively moulding material in the form of a disc 25·27±0·13 mm (0·995±0·005 in) diameter by 1·59±0·13 mm (0·0625±0·005 in) thick and sheet material in the form of a rectangular bar not less than 32 mm ($1\frac{1}{4}$ in) long by 6·35±0·25 mm (0·250±0·010 in) wide by the thickness of the sheet (up to 6·35 mm), both use the same punch and die assembly. The die is screwed home against the specimen in the holster.

The test is carried out with the tool mounted in a testing machine operated as for compression tests and must be completed between 15 and 45 s after first applying load.

For Method 305A, where the disc specimen is used:

$$\text{Shear Strength} = \frac{W}{\pi d T} = \frac{W}{1{\cdot}57T}$$

where W = force at fracture
πdT = shearing surface if
d = diameter of punch
T = thickness of specimen

In Method 305B, only two arcs of the circumference of the punch tool are causing shear; thus:

$$\text{Shear Strength} = \frac{W}{2{\cdot}096BT}$$

where W and T have the same meanings as before and
B = width of specimen

The units are as for tensile strength, for example (see Section 8.4.4).

A.S.T.M. D.732–46 allows the use of specimens of any thickness between 0·005 and 0·500 in; the specimen is drilled centrally to take a guide pin on the punch which it is suggested should be of 1 in diameter. (Torsional tests are one method of measuring shear forces, but these are not often used under static conditions (see Chapter 15 for use of torsional tests in dynamic applications); in fact torsional tests per se are almost non-existent as standard methods for examining plastics.)

8.7 BENDING PROPERTIES AND MEASUREMENT

8.7.1 Some Terms Explained

The property of most interest in this section is known by a number of different titles, some logical and some trivial, and it might help to set these down right at the beginning. Probably the best term is:

Cross-breaking strength which is used in B.S. 2782:1970 (Methods 304A, B, C and D); this standard acknowledges that the property is also known as *Flexural strength*, a term used in A.S.T.M. D.790–66. To avoid confusion,

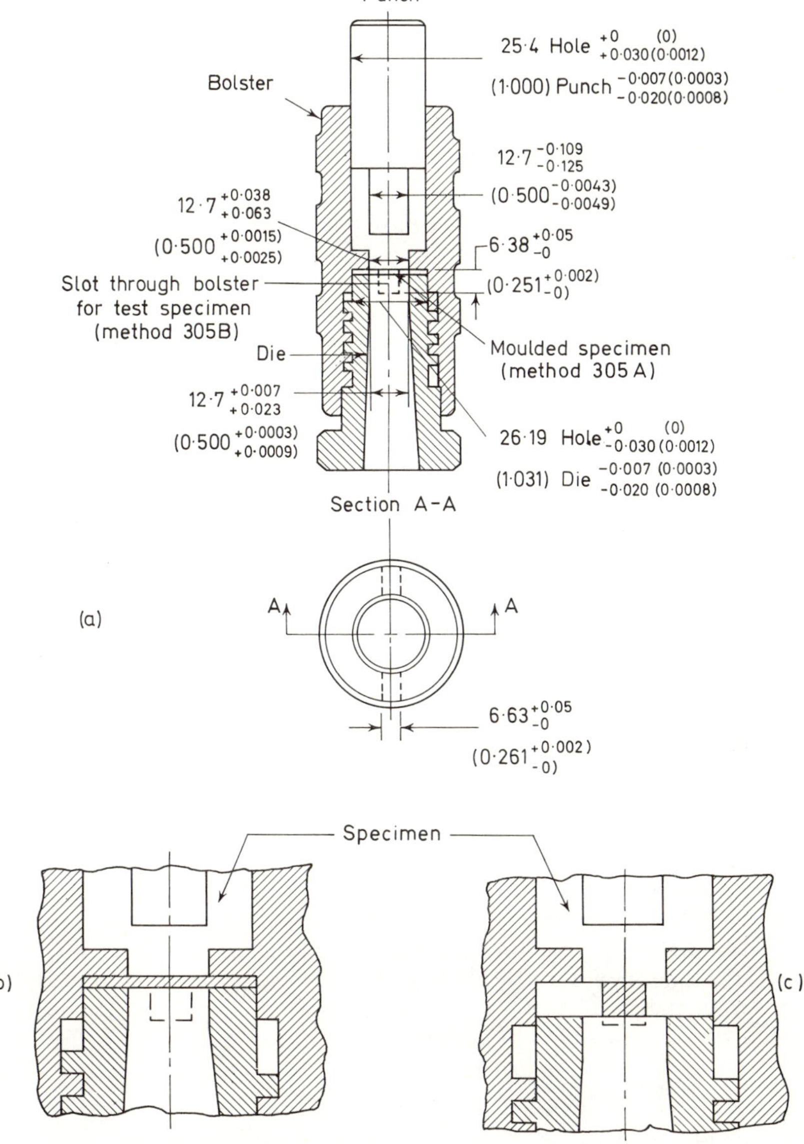

Figure 8.34. B.S. 2782 punch shear test assembly. (a) Dimensions in millimetres with inch equivalents in parentheses. (b) Enlarged view showing position of specimen used in Method 305A. (c) Enlarged view showing position of specimen used in Method 305B

no other terms should be used for this property (as defined below); however, to clarify the situation, it must be admitted that the following exist (or have existed):

Modulus of rupture (in A.S.T.M. D.790—it is no 'modulus' in the usually accepted meaning of the word)
Fibre stress (as maximum fibre stress, this has relevance)
Bending strength
Transverse strength
Coefficient of bending strength

8.7.2 Behaviour of Materials in Bending

When a rectangular beam is bent a continuous change occurs from maximum tensile stress on one surface through the thickness to maximum compressive stress on the other.

In Figure 8.35 (for a homogeneous isotropic material in pure bending), the top surface is in tension and the under surface in an equal compression;

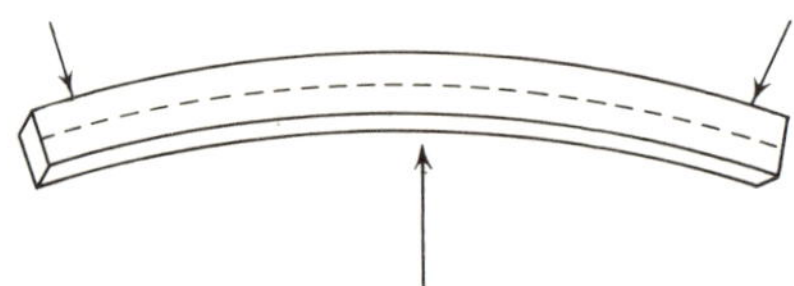

Figure 8.35. Material in bending

if the tensile and compressive moduli are equal, at the midpoint of thickness the stress is zero, where tension diminishes to zero before compression starts building up, and the dotted line represents the *neutral axis*.

It is beyond the scope of this book to develop the formulae for bending; suffice to state that pure bending, i.e. the combination of only tensile and compressive forces, is never achieved and there is always some transverse shear component.

The bending formulae given below have been derived assuming pure bending, which is assisted by having a span (length of loaded section) of beam large by comparison with the thickness so that bending approaches a true arc of a circle. For three point bending of plastics, as in the diagram above, the 'span/depth' ratio should be about 16:1, but this cannot be taken as invariably satisfactory; the appropriate figure for span/depth ratio depends on the characteristics of each individual material.

THREE-POINT BENDING

This is the most popular type of bending as far as standard tests are concerned.

For a rectangular beam (supported at mid point)

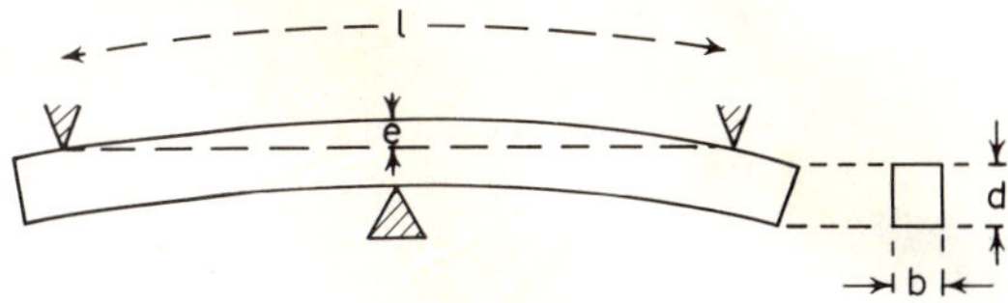

Figure 8.36. Three-point bending

Maximum Fibre Stress (A.S.T.M. D.790–66)

$$S = \frac{3Pl}{2bd^2}$$

where P = force at mid point
This occurs, as has already been stated, at the upper (tensile) surface.

At fracture:

$$\text{Cross-breaking Strength (flexural strength)} = \frac{3Wl}{2bd^2}$$

where W = Force at fracture

For large spans (greater than 16:1), according to A.S.T.M. D.790–66, 'Flexural Properties of Plastics':

$$\text{Cross-breaking Strength} \approx \frac{3Wl}{2bd^2}\left[1+6\left(\frac{e}{W}\right)^2-4\left(\frac{d}{W}\right)\left(\frac{e}{W}\right)\right]$$

[There are other formulae proposed for more accurate calculation of cross-breaking strength.]

Flexural yield strength and flexural offset yield strength are defined in an analogous manner to their tensile counterparts (Section 8.4.1), but here they are of value in ascribing a bending resistance to materials that do not actually break in the test and therefore do not give a cross-breaking strength. A.S.T.M. D.790 suggests that the latter property should not in fact be quoted for any material which fractures at more than 5% outer fibre strain (see below) in the test.

Maximum Strain (A.S.T.M. D.790)

$$r = \frac{6ed}{l^2}$$

which is the strain in the top surface of the beam above.

Thus, for Hookean behaviour

$$\text{'Modulus of Elasticity'} = \frac{\text{Stress}}{\text{Strain}} = \frac{Pl^3}{4bd^3e}$$

(Modulus of Bending)
(Bending Modulus)

(Flexural Modulus)
(Modulus in Flexure)
(Elastic Modulus in Bend)

A.S.T.M. D.790 defines a 'tangent modulus of elasticity' taken from the slope of the tangent drawn to the steepest initial straight line portion of the load deformation curve. Likewise a 'secant modulus of elasticity' is similarly defined in relation to a given point on the curve.

For a circular rod

$$\text{Cross-breaking strength} = \frac{8Wl}{\pi d^3}$$

$$\text{Modulus of elasticity} = \frac{4Pl^3}{3\pi d^4}\left(\frac{1}{e}\right)$$

Where W, l, P and e have the same significance as before and d = diameter of rod.

FOUR-POINT BENDING

For a rectangular beam (see Figure 8.37)

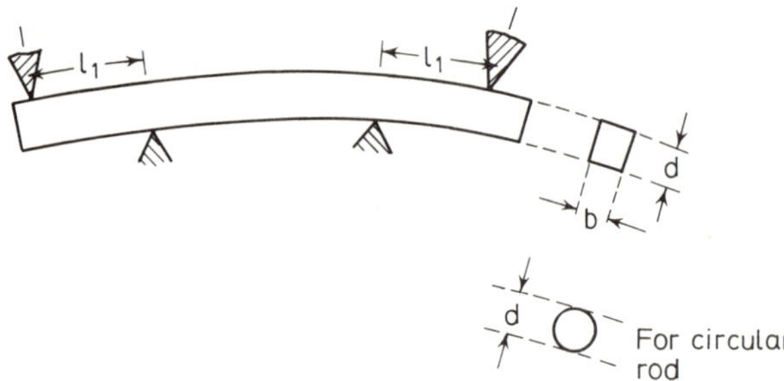

Figure 8.37. Four-point bending

$$\text{Cross-breaking strength} = \frac{6W_x l_1}{bd^2}$$

For a circular rod:

$$\text{Cross-breaking strength} = \frac{32W_x l_1}{\pi d^3}$$

where W_x = force on each bearing point, *i.e. half total force.*

CANTILEVER BENDING

For a rectangular beam (see Figure 8.38)

$$\text{Cross-breaking strength} = \frac{6Wl}{bd^2}$$

For a circular rod:

$$\text{Cross-breaking strength} = \frac{32Wl}{\pi d^3}$$

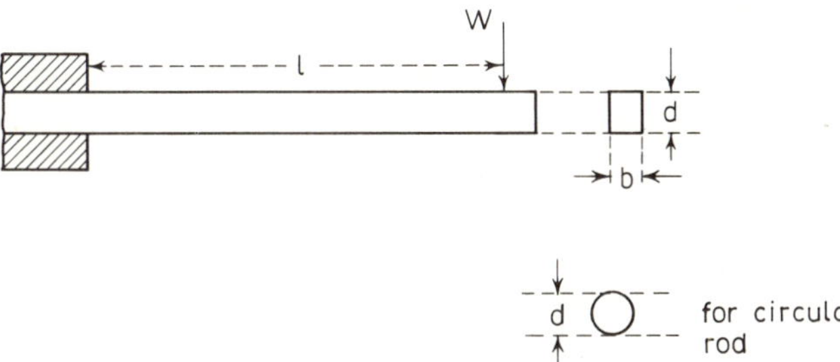

Figure 8.38. Cantilever bending

8.7.3 Some Details of Specific Tests

Bending tests may be carried out in tensile or compression test machines (see Figure 8.39).

In addition to the usual variables of specimen size and shape, test speed and temperature, there must be added the radius of curvature of the bearing rods, which must not be too sharp to cause fracture. Cross-breaking strength tests are popular because the test specimens are simple and therefore easy to prepare and the loads to be measured are relatively low; also gripping problems are eliminated and deflection data more easily obtained. Theoretically, in pure bending, cross-breaking strength and bending modulus are equal to tensile strength and Young's Modulus respectively, but because bending is not pure and because many materials tested are not homogenous and isotropic, these equalities are often very approximate at best.

Again, three-point bending is generally preferred in standard tests, but there are many advocates of four-point tests because in the latter the stress is equal, under practical conditions, over the whole of the span between the inner two supports (Section 8.7.2), in contrast to the local maximum stress which occurs opposite the centre support in three-point bending (Section 8.7.2) which is essentially non-pure; clearly errors may result if

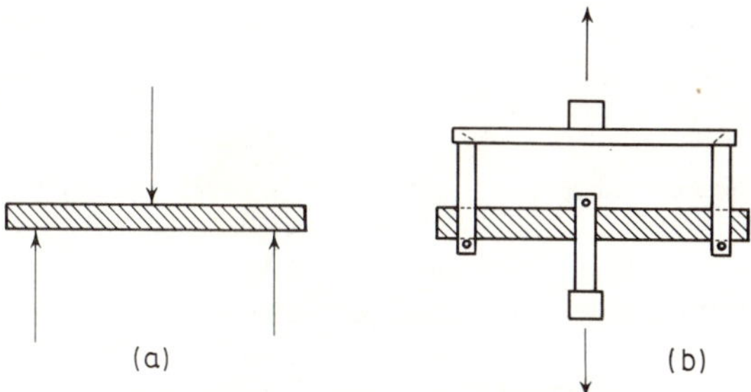

Figure 8.39. Bending tests in different types of testing machine. (a) Compression machine, (b) tensile machine

the material in the region of the centre support is not representative of the whole.

In B.S. 2782:1970 there are five variants of the standard cross-breaking strength test, Methods 304A, B, C, D and E, the last mentioned being essentially that of ISO/DR.1674 except for a relaxation on one tolerance of dimension in the B.S. 2782 method. All involve three-point loading (see Table 8.2).

For Methods 304A, B, C and D, load is increased steadily, so that fracture occurs in 15 to 45 s, on loading blocks of contact radii of approximately 1·6 mm ($\frac{1}{16}$ in) or 3·2 mm ($\frac{1}{8}$ in) and length not less than 25 mm (1 in). For Method 304E, the outer supports have bearing radii 4·8–5·2 mm for specimens up to 3 mm in thickness, and 1·8–2·2 mm above 3 mm. The radius of the loading nose is 4·9–5·1 mm.

Cross-breaking strength is calculated according to Section 8.7.2 above and expressed as MN/m^2, kgf/cm^2 or lbf/in^2.

Method 302D of the same British Standard is equivalent to ISO/R.178 and describes a test for *elastic modulus in bend*. For moulding material, two

Table 8.2 VARIANTS OF THE STANDARD CROSS-BREAKING STRENGTH TEST OF B.S. 2782: 1970

Test ref.	*Material*	*No. of test specimens*	*Size of test specimen* mm (in)	*Span*
304A	Thermosetting moulding material	3	$\geqslant 114l \times 12{\cdot}7b \times 9{\cdot}53d$ (4·5 × 0·50 × 0·375)	101·6 mm (4 in) (i.e. $\approx 10 \times d$)
304B	Laminated sheet (thermosetting)	5*	12·7b (0·50) × thickness of sheet [not above 9·5 ($\frac{3}{8}$)] × length 24–30 times thickness.	16 × d
304C	Casting and laminating resin systems	5*	12·7b (0·50) × 3·2 or 6·4d (0·125 or 0·25) × length $\geqslant$ 24 times thickness.	16 × d
304D	Laminated sheet (thermosetting) with coarse filler	5*	25b (1·00) × thickness of sheet [not above 9·5 ($\frac{3}{8}$)] × length 24–30 times thickness.	16 × d
304E	ISO/DR.1674 Method	5*	Length $\geqslant$ 80 mm, width 10 ± 0·5 mm and thickness 4 ± 0·2 mm for moulding materials and casting and laminating resins; for sheet, length $\geqslant$ 20 × thickness, and width 10–25 mm or 20–50 mm for materials with very coarse filler.	15–17 × d

*As Table 8.1 (Section 8.4.4)

specimens in the form of rectangular bars at least 95 mm long by 10 ± 0·5 mm wide by 4 ± 0·2 mm thick are used; for sheet, the length of each of the two specimens is not less than 24 times the thickness (1 mm–10 mm) and the width 10–25 mm. Again directional effects should be studied if they are likely to be present.

The outer supports have bearing radii 1·5–3·5 mm and span L equal to 16 × thickness of specimens. The load is applied centrally through a nose of radius 5 mm, at a rate so that the centre of the specimen is deflected

$$\frac{d}{2}\ \text{mm/min}\ \pm 50\%$$

where d = thickness of specimen

A load deflection curve is drawn and the elastic modulus in bend calculated from the initial (straight) part of the curve:

$$\text{Elastic modulus in bend} = \frac{100WL^3}{4bd^3D}$$

Where W = value of load selected from the straight part of the curve
D = deflection corresponding to the load W
L = distance between outer supports
b = width of specimen
d = thickness of specimen.

A *cantilever type of modulus test* is contained in Method 309A of B.S. 2782: 1970; however, large deflections are produced for easy reading on a scale and the test can be used for *comparative purposes only*. The specimen is 69·9 × 25·4 × 1·52 mm (2·75 × 1·00 × 0·060 in) with a hole drilled near one end for locating the load carrier, as shown in Figure 8.40.

The test is carried out in duplicate. The specimen is set up and located in the apparatus as shown in the figure and after 5 ± 1 min the vertical height of the free end of the loading arm (total mass 30 ± 0·5 g) is read to the nearest millimetre. Then a 50 g weight is added carefully and after 5 min ± 15 s, the vertical height of the loading arm is read again. The difference between the two readings is taken as the deflection under load in bend, but if the specimen is not precisely 1·5 mm (0·060 in) a correction factor is applied to the reading of deformation obtained.

The cantilever type of test is readily set up and carried out; a number of non-standard variants exist and, if a true straight-line relationship between load and deflection is obtained, elastic modulus in bend may be calculated from the formula:

$$\text{Elastic modulus in bend} = \frac{4WL^3}{bd^3D}$$

where all terms have the same significance as before with L being the total distance between the end support and the point of application of load W, at which deflection D is recorded.

The standardised methods used in the U.S.A. are laid down in A.S.T.M. D.790–66, to which reference has already been made in Sections 8.7.1 and 8.7.2 in referring to terms used. The A.S.T.M. lays down all the necessary variables for three point tests, giving two different procedures for materials which fracture with very little deflection and for those which bend considerably during test. A load–deflection curve is plotted and correction is made to recorded deflection for indentation of the specimen by the loading nose; two alternatives of the latter are provided to facilitate control of specimen indentation.

A.S.T.M. D.747–63, 'Stiffness of Plastics by Means of a Cantilever Beam', is *somewhat* similar to Method 309A of B.S. 2782 described above. Basically it also is a comparative method, but does not enjoy the same advantage as the B.S. technique of extreme simplicity of apparatus.

Three-point bending is also specified in DIN.53452 where the practice is to use rather low span:depth ratios of about 10:1. Amongst the different forms of test specimens allowed is one for the Dynstat apparatus[27] which will be found mentioned again in Section 8.10 below; it uses specimens 15 × 10 × 1·5–4·5 mm. The apparatus is described in Reference 28 and its particular merit is versatility and the use of small specimens. The apparatus is shown assembled for cross-breaking tests in Figure 8.41.

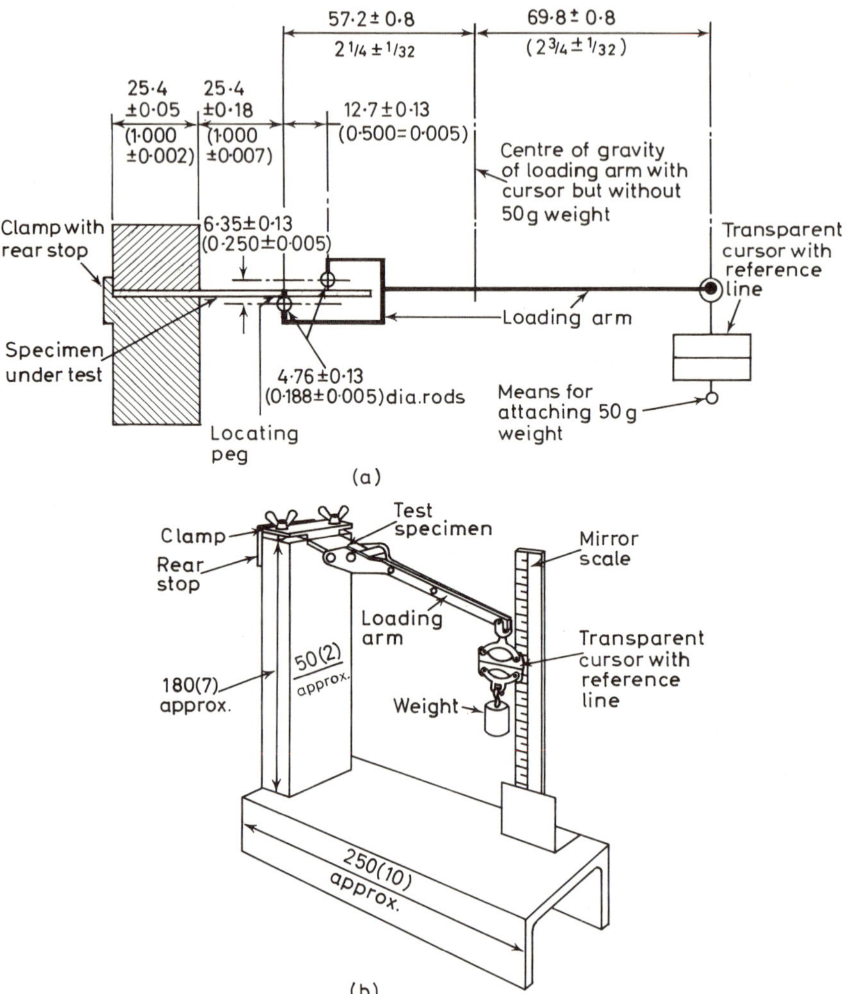

Figure 8.40. Test of deflection under load in bend (B.S. 2782). Dimensions in millimetres with inch equivalents in parentheses. (a) Essential features of test. Note: mass of loading arm with any attachments for measuring deflection to be 30 ± 0·5 g. (b) Suitable form of apparatus

The aluminium pendulum A, which can be fitted with interchangeable weights B_1, B_2, etc., so that the bending moment can be altered, is mounted in ball bearings at the centre of dial C which carries a removable clamp D. The dial, and hence the clamp, can be rotated with respect to the fixed dial E by turning a crank handle which operates a worm wheel F. The

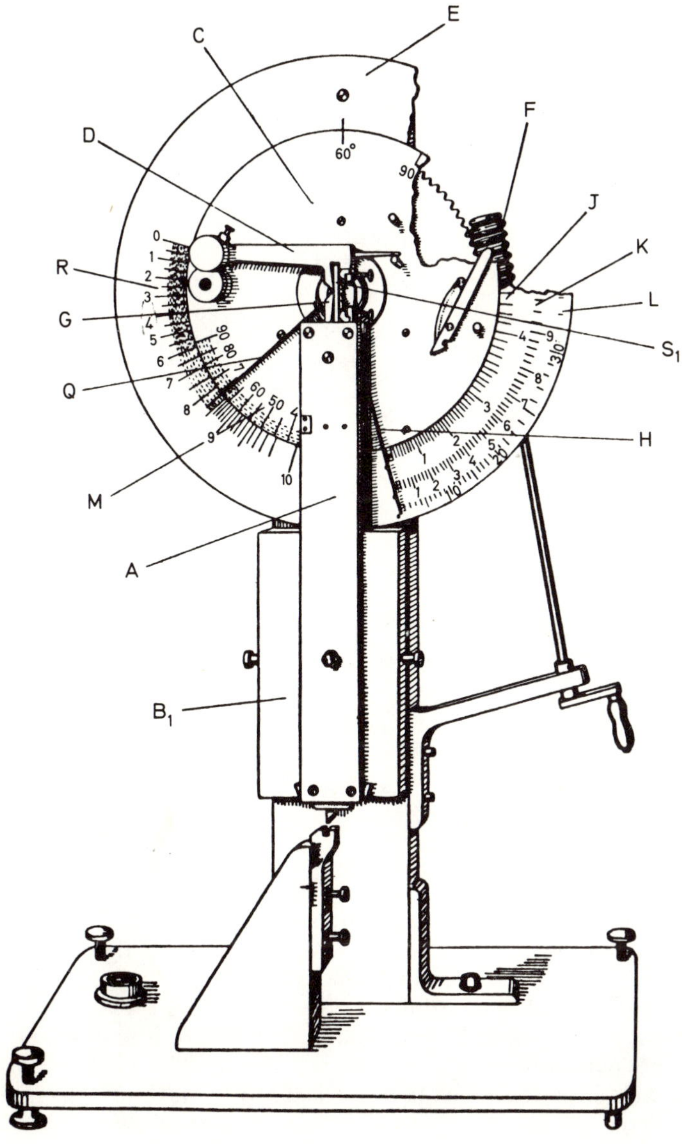

Figure 8.41. Dynstat apparatus for bending tests (B.S. 1330)

specimen S is mounted in clamps D and G attached respectively to dial C and pendulum A. The clamps apply three-point bending to specimen S_1 as shown in Figure 8.42.

Dial C is rotated by worm wheel F which causes clamp D to rotate counter clockwise so that pendulum A, appropriately weighted for the specimen under test, applies a bending moment to the specimen and rises

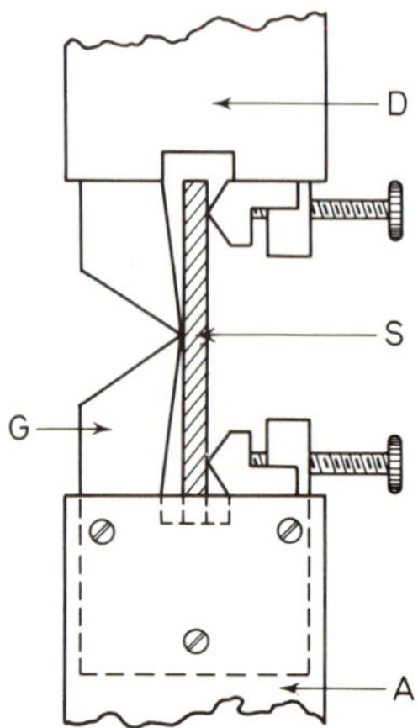

Figure 8.42. Mounting of specimen in Dynstat apparatus (B.S. 1330)

until the specimen breaks. The amount by which the pendulum has risen is indicated by the idle pointer H on the scales J, K and L which are calibrated directly for three alternative weights on the pendulum.

$$\text{Cross-breaking strength} = \frac{6M}{bd^2}$$

M = bending moment (read off scale J, K or L)
b = width of specimen
d = thickness of specimen.

Flexural tests on plastics are discussed by Loveless[29] and stress and strain formulae generally are described in detail by Roark[30]. Heap and Norman[31] have written a monograph on the subject. All the tests described above are designed to measure the bending properties of rigid and semi-rigid plastics materials. The rigidity or stiffness of flexible plastics is covered to a certain extent in the tests described in Chapter 11 for investigating the effect of lowering of temperature on such materials (Section 11.4.1). A test for ambient temperature application has been worked out by Stuart[32]; this relies solely on the shape taken up by a strip of flexible material when bent in a loop and placed on a horizontal surface.

$$\text{The 'bending length', } C = \left(\frac{B}{w}\right)^{\frac{1}{3}}$$

where B is the flexural rigidity of the strip and w is its weight per unit length. Stuart has established that $C = 1{\cdot}03\ \Upsilon_m$ where Υ_m is the height of the loop.

8.8 BEARING STRENGTH AND MEASUREMENT

Plastics materials, particularly laminated in sheet form, may be bolted in assembly and when thus joined will be subjected to stress around the bolt holes when the assembly is put to use. In this context, this particular stress is called the *bearing* stress and A.S.T.M. D.953–54 defines bearing strength as 'the bearing stress at the point on the stress–strain curve where the tangent is equal to the bearing stress divided by 4% of the bearing hole diameter'. The bearing area is the diameter of the hole multiplied by the thickness of the specimen. This area divided into the bearing load gives the bearing stress.

Two special procedures are described for determining bearing strength, one for tension testing and one for compression testing, for as stated in A.S.T.M. D.953: 'While it is known that higher strength materials will generally give higher bearing strengths, there is no satisfactory method by which bearing strength may be estimated from tensile or compressive properties of the material'. (No mention is made of shear properties!)

As already implied, a load–deflection curve is plotted and to aid this the speed of testing is stipulated to be not more than 0·05 in/min. Five specimens

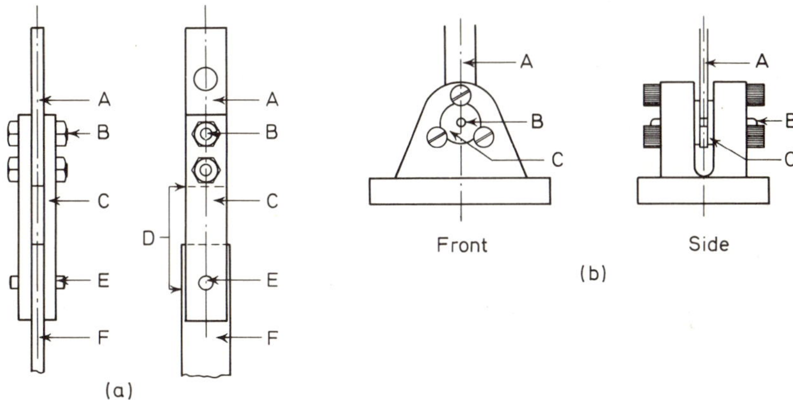

Figure 8.43. (a) Tension loading assembly (A.S.T.M. D.953). A. Hardened spacer plate. B. ¼ in steel bolts in reamed holes. C. Hardened side plate. D. Extensometer span. E. Hardened steel pin in reamed hole. F. Test specimen. (b) Compression loading assembly (A.S.T.M. D.953). A. Test specimen. B. Hardened steel pin. C. Hardened thrust bushing.

are used, or ten if directional effects are suspected. A technique involving the use of a template is described to obtain the 4% deflection figure uninfluenced by zero load errors, i.e. taking up the slack in the testing machine, test specimen assembly, etc. Figure 8.43 shows the A.S.T.M. assemblies.

8.9 HIGH SPEED PROPERTIES AND TESTS

8.9.1 Why Speed Affects the Strength of Plastics

Mention has frequently been made above of the sensitivity of the mechanical properties of plastics materials to change in rate of loading (and rate of extension). The more common usages of plastics as structural materials in any sense involve prolonged steady, pulsing or oscillating forces; this subject is dealt with in Chapters 14 and 15. However, the spectrum may be spread in the opposite direction, that is very fast build up of forces may be encountered—shock loading—in (modestly) the sudden taking up of slack in a polypropylene rope hawser or (more emphatically) the component part of a high speed aircraft or missile. High speed or shock tests in use comprise those for conventional mechanical properties, particularly tensile tests suitably speeded up, or the more 'ad hoc', those for impact and tensile impact strength. These two latter will be dealt with in the two sections following this, which is devoted to mainly high speed tensile tests. It must be admitted that whilst they are certainly 'short term' in nature, they are not routine quality control as are most of the tests that have gone before and some that follow in this chapter. This time scale justifies their inclusion here, but they go some way to meeting the general criticism of standard tests, that of not simulating user conditions.

The reasons why plastics materials as a class, and soft thermoplastics especially, should be time sensitive have received considerable attention. Various hypotheses have been put forward and the reader is referred to various standard works on the physics and properties of polymers and plastics for details[33–36]. The last mentioned, by Frith and Tuckett, is particularly well written from the point of view of gaining a simple non-mathematical idea of the subject as far as amorphous polymers are concerned. One explanation put forward is that (total) deformation D is the sum of the deformations from three mechanisms:

1. Bond bending of the individual atom to atom (usually C–C) links in the backbone chain of the polymer—'ordinary elasticity':OE
2. Uncoiling of the individual chains—'high elasticity':HE
3. Irreversible slipping of the chains past each other—'viscous flow':VISC

where

$$D = D_{OE} + D_{HE} + D_{VISC}$$

D_{OE} is assumed to be instantaneous, i.e. time independent, and its practical significance in terms of the interpretation of actual stress–strain behaviour is a little doubtful. It is usually small by comparison with D_{HE}. the other elastic or reversible component, which is very time dependent. D_{VISC} is both time dependent and irreversible, but need not be considered further here for the purposes of a brief mention of the importance of time scale in short term mechanical testing. D_{HE}, high elasticity deformation, has been related to uncoiling of the polymer chains and to this has been ascribed an 'orientation time' τ on the plausible basis that the rate of uncoiling should be proportional to the fraction of chains which have not responded to the (massive) deformation. On this assumption, that each chain uncoils independently of its neighbours—a unimolecular process—the fraction of the ultimate deformation achieved in time t will be

$$(1 - e^{-t/\tau})$$

and τ is the time taken for the deformation to reach $(1 - {}^{1}/e)$ of its final value. (τ is also sometimes known as 'relaxation time'.) It will not be unexpected to learn that this is a gross over-simplification of even a graphical representation such as that described, but it will suffice in the present instance to correlate deformation behaviour with the polymer chain and its characteristics. Since the chains, and particularly their attached groupings, vary from polymer type to polymer type, it may readily be envisaged that each has its own particular time response depending on the nature of the backbone links and the type and size of the side chains, quite apart from the external influences of plasticisers in particular.

1. If $t \leqslant \tau_m$, where τ_m is the average orientation time (to allow for deviation from the ideal behaviour considered above):

$$(1 - e^{-t/\tau_m}) \rightarrow t/\tau_m$$

and here D_{HE} will be very small, leaving only D_{OE} as a significant contributor to deformation D.

2. But if t/τ_m is large, $(1 - e^{-t/\tau_m}) \rightarrow 1$ and D_{HE} assumes significance in that it approaches its ultimate value in the time t available.

At ordinary temperatures and conventional testing speeds, polystyrene and polymethyl methacrylate for example are instances of alternative (1) whilst polyethylene and plasticised PVC behave as (2). However, if t is reduced to a thousandth or less of its normal value, by high speed testing, polyethylene and plasticised PVC will tend to behaviour (1) also. This trend is borne out by the results of various workers quoted below. A classic case of high sensitivity to speed is the so-called 'bouncing putty' which will bounce if dropped or 'snap' when stretched quickly, but flows if left as a lump on a flat surface.

Vincent[37], in one of an interesting series of articles entitled 'Strength of Plastics' which appeared between October 1961 and January 1964 in the U.K. journal *Plastics* (now merged with *British Plastics*), makes the point that it is important to distinguish between yield strength and brittle strength and criticises some of the data reported on plastics for failure to record which type of failure has occurred. Yield strength has always been found to increase with increase of straining rate, but the trend of brittle strength is less certain owing to its extreme dependence on quality of specimen preparation, amongst other factors. However, work cited indicates that it is not so speed dependent as is yield strength. Vincent points out:

1. If the yield strength is greater than the brittle strength, the material will break in brittle fashion.
2. But if the brittle strength is greater than the yield strength, the material will break in tough (ductile) fashion.

Thus it is obvious that for any plastics material, if the testing speed is low enough the yield strength will be less than the brittle strength and the material will therefore be tough. As the speed is increased the yield strength will overtake the brittle strength and the transition to brittle behaviour will occur. In terms of usability as a mechanical material, the type of failure is of utmost importance[38]; it is often identified by the physical appearance of the break—if there is a permanent deformation the break is tough, i.e. the yield strength has been evaluated, but this can be difficult to assess practically. A better method is to study the shape of the stress–strain curve, (Figure 8.3).

8.9.2 Standard Methods

It is believed that there is only one official standard covering this subject, A.S.T.M. D.2289–69, 'Tensile Properties of Plastics at High Speeds'. This standard spans the gap between conventional speeds of testing (A.S.T.M. D.638—see Section 8.4) and those near the speed of sound where stress wave propagation effects may become significant, that is the distribution of load along the specimen may not be uniform. Little detail is given except to allow grip separation measurement instead of actual strain and to require a control of testing speed of ±20% of that specified. A cathode ray oscilloscope may be run off the load cell and the grip displacement made the X axis by taking the output of a suitable transducer; the time scale may be estimated by allowing an audio oscillator to modulate the Z axis of a single beam oscilloscope. A camera is mounted on the latter if a record of the test is required. Suggested speeds of separation of the grips are 2·5, 25 and 250 m/min.

8.9.3 Other Techniques

There is, however, a fairly copious bibliography on the subject. Early reviews[39, 40] of speed effects give little of the actual techniques; likewise

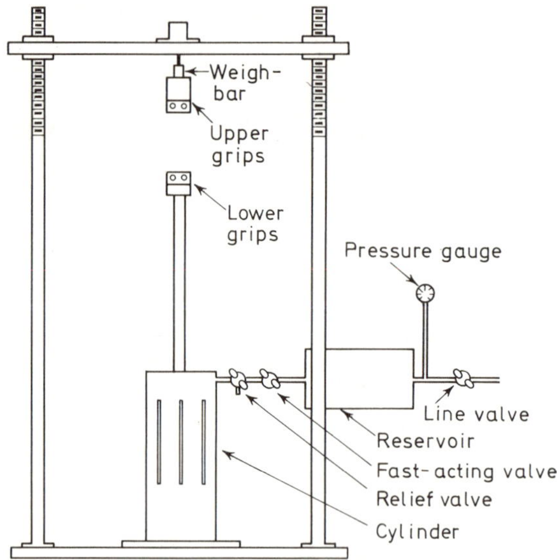

Figure 8.44. High speed tension testing machine (Strella et alia)

Hall[41] analyses published data in studying the transition from ductile to brittle behaviour. The subject is discussed generally by Supnik[42] whilst Goldfein[43] describes a mathematical procedure for obtaining high speed tensile data from figures obtained at conventional speeds, but low temperatures.

Strella et alia[44] follow the ideas of Dorsey, McGarry and Dietz[45] in using the rapid release of compressed gas to provide a high rate of loading. Strella and co-workers' apparatus is sketched in Figure 8.44.

A high pressure gas cylinder is attached to (the r.h.s. of) the storage reservoir via a needle valve. This reservoir is connected to a single stroke double acting air cylinder through a fast opening valve. The cylinder (4 in diameter) is 12 in long and is fitted with a piston and a connecting rod which is attached to the lower grip of an otherwise conventional testing

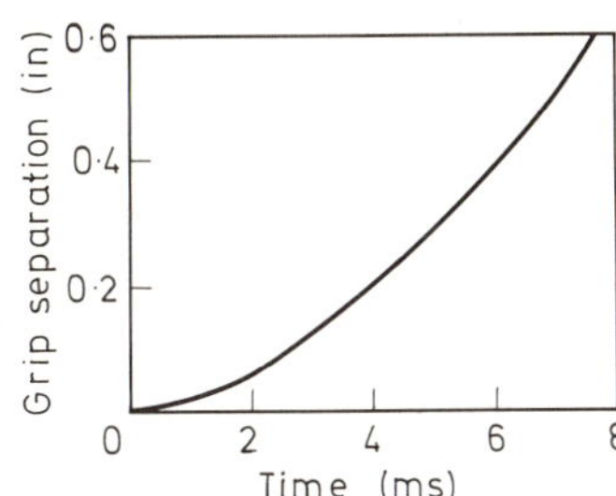

Figure 8.45. Speed variation in Strella et alia apparatus

machine design of the 'electronic' type. Six slots, 4·5 in long and $\frac{1}{8}$ in wide, cut into the cylinder wall as shown, serve to 'leak' away gas when the piston has travelled $2\frac{3}{8}$ in and effectively to stop the test.

Gas at high pressure is admitted into the reservoir and is then rapidly released into the cylinder by means of the fast opening valve. The gas forces the cylinder to move extremely fast and thus loads the specimen mounted between grips in a matter of milliseconds. Most materials fracture within the $2\frac{3}{8}$ in of piston travel, but highly extensible materials are also caused to fracture as the momentum of the piston causes it to carry on until it 'bounces' off the air cushion held by the (lower) closed end of the cylinder. Obviously the controlling factor is the gas pressure; Strella and his co-workers state that in their apparatus pressures of 1700–2000 lb/in^2 cause fracture in about 6–10 ms whilst those of about 150 lb/in^2 increase the failure time to 30–50 ms. To increase the time still further, a needle valve may be inserted between the fast opening valve and the cylinder. The speed of the grips is not quite steady and Figure 8.45 shows grip separation versus time, obtained from an oscilloscope trace.

The authors describe some results obtained with high density polyethylene (Figure 8.46) where the effect of increasing the rate of loading on modulus, strength and elongation is seen.

Modifications of this apparatus are described in Reference 46, improving the techniques for measuring the separation of the grips and for activating the gas release valve and tidying up the recording circuit. In this short paper Strella gives some results of tensile tests on polystyrene at conventional (slow) and high speeds (Figure 8.47).

Using a similar pneumatically operated machine, Ely[47] examined polymethyl methacrylate and cellulose acetate butyrate; he found a linear relationship for these materials between tensile strengths and log (fracture time) or log (strain rate), the former a negative correlation and the latter positive. For polymethyl methacrylate, the strength increased from 7000 lb/in^2 to 16 500 lb/in^2 as the straining rate was raised from 0·001 in/in/min

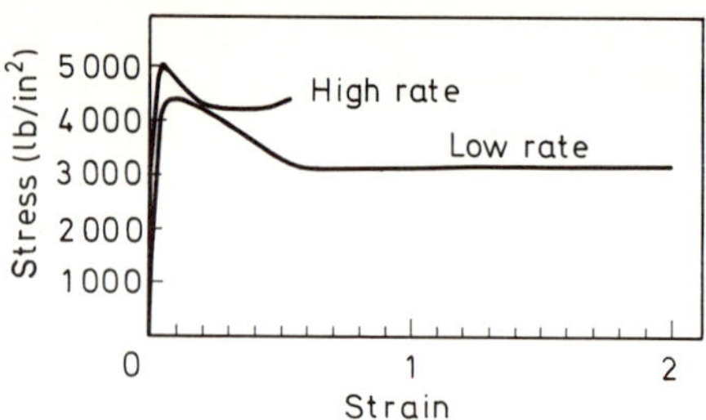

Figure 8.46. Effect of speed on tensile behaviour of high density polyethylene (Strella et alia)

to 1000 in/in/min. Ely also examined low density polyethylene and ethyl cellulose using this type of machine[48].

This type of apparatus is available commercially[49] and its use, with more detail of construction and data obtained, will be found in References 50 and 51. Further variants of the pneumatic method will be found in References 52 and 53, the latter presenting a wide range of data including from high speed tests on thermosets. Other data from this type of test method are described in References 54, 55 and 56.

Barnes, Hawkins and Davis[57] have developed a technique of high speed testing, based on the pneumatic principle, where the test can be interrupted

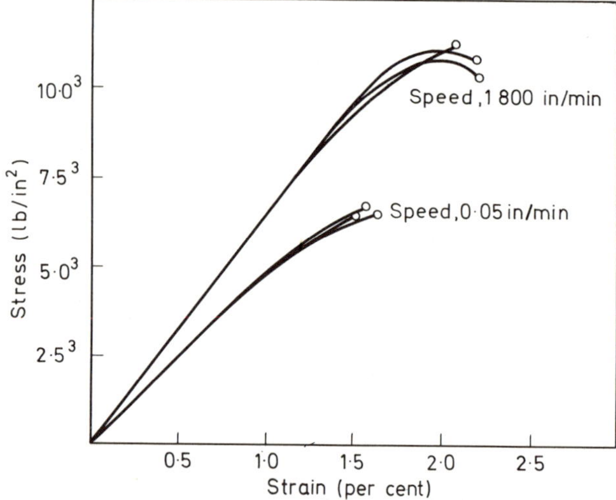

Figure 8.47. Effect of speed on tensile behaviour of polystyrene (Strella)

before the specimen fractures. The idea is to be able to test the specimens, after their high speed straining, at conventional low rates to ascertain the residual tensile properties after, say, a high strain rate forming technique.

Testing speeds of 100 000 in/min and more have been achieved by ballistic means, using 'smokeless powder' to drive the piston of a tensile or compression testing machine[58]. A low cost ballistically operated machine, operating at strain rates of 20 000 in/in/min and over, is described by D'Amato[59] in reporting some high speed work on polyethylenes.

Flywheel machines have been described, particularly for testing film test specimens at high speeds; here the principle is that the wheel is rotated at the required peripheral speed and has sufficient kinetic energy so that the percentage lost in fracturing the specimen is negligible and the strain rate remains essentially constant throughout the test. In the apparatus described by Patterson and Miller[60], a retractable fork permits the engaging of the flywheel to the specimen assembly when required, and the force is measured by strain gauge. A similar apparatus is described in detail by Amborski and Mecca[61], for speeds of 50–10 000 in/min (see Figure 8.48).

A 1 hp motor (A), rotating at 1750 rev/min, is coupled directly to the flywheel B, of 11·5 in diameter and weighing 69 lb, balanced both statically and dynamically. C is a 1 hp variable speed transmission of output 0–130 rev/min which controls the speed of the impellant wheel D, 25 in diameter with a lead tyre weighing approximately 39 lb. E is a 96 tooth steel spur gear engaging a 24 tooth fibre gear attached to a 9 in diameter polymethyl methacrylate disc. This latter has 90 holes precision drilled on a circumference which is $\frac{1}{2}$ in from the periphery. The motion of the holes is detected by a light source on one side and a photo-electric cell on the other of the disc and the cell is coupled to an oscilloscope to give the movement of the lower specimen grip of the machine.

On the periphery of the wheel D is a spring-loaded steel fork (F) which is retained by ratchet and does not normally project beyond the periphery of the lead tyre. When the test speed, preset on C, is attained the fork is released to engage cross bar G at one end of $\frac{1}{2}$ in wide steel tape H (adjustable in length to suit the specimen under test). The other end of the tape is

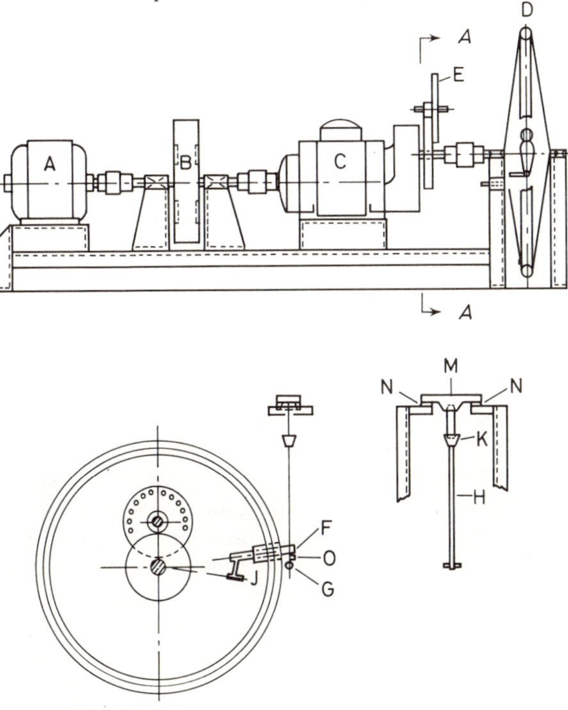

Figure 8.48. Flywheel machine for high speed testing (Amborski and Mecca[61])

connected to the lower specimen grip K holding the lower end of the specimen. The upper end of the latter is clamped in a low mass grip M which rests on a pair of piezoelectric crystals N which are used to record the load. When the fork F engages the crossbar G it wraps the steel tape round the wheel D until the specimen breaks.

Pendulum machines are most often used for tensile impact or flexural impact tests, which are essentially similar to the tests just described. Their use for high speed tensile testing is described by Grimminger and Jacobshagen[62], but detailed description will be found in Section 8.11 below.

Finally, in this by no means exhaustive review of literature on high speed testing, mention must be made of five volumes devoted specifically to the subject[63] and in which a number of the above-mentioned articles will be found reproduced; these references, however, also cover the fields of impact testing and stress relaxation.

8.10 IMPACT STRENGTH, TOUGHNESS AND SHOCK RESISTANCE AND THEIR MEASUREMENT

8.10.1 Introduction

It is ironical that in contrast to the scarcely standardised subject of (scientific) high speed mechanical testing in general, and for tensile, flexural, compressive and shear properties in particular, the decidedly ad hoc field of impact strength and testing has received considerable attention in official standards, materials data sheets and the literature at large; yet the result of an impact test is basically no more than one point on the general curve of studying strength properties as a function of speed of testing. It is only the attention impact tests have received in their own right, stemming from habits inherited from metals technology, that justifies their separation from the previous section of this chapter. The one advantage they offer is a ready measure of the actual energy required to break an (arbitrary) test piece, which information can only be calculated from stress–strain diagrams in, say, tensile or flexural tests with some effort.

'Toughness' is a concept most people can appreciate and a broadly accepted mathematical definition is the work done in breaking a specimen, moulding, article, etc; as such it may obviously be derived from the load–extension graph by computing the area under the curve (e.g. see Figure 8.4) since this will integrate all the units of force × distance (i.e. extension over which force operated) to give the total work, i.e. the toughness. As calculated from high speed tests such data may be quite useful for prophesying the behaviour of plane face articles, but the area under a conventional tensile or bending test curve is of very limited value because in practice we are mainly interested in toughness under conditions of rapid deformation, that is where an article is dropped or has something dropped on it—hence the concept of impact resistance or 'strength' and the introduction of impact tests, or shock resistance and its assessment. The subject is vast and at least one whole book has been given over to it[64] with a whole chapter (II) devoted to plastics and rubber. The reader is also referred to

References 65 and 66 for *general* descriptions of conventional impact testing.

Standard impact tests for plastics divide themselves into (a) those which use instruments where a pendulum of known energy strikes a specimen of defined size and shape and (b) those where weights or other impactors are allowed to fall freely through known heights on to specimens and the impact strength computed from the minimum combination of height and weight required to cause fracture. Tests of type (a) are further subdivided into cantilever (Izod) and supported beam (Charpy) variants, using specimens of flat plane face or more often containing a moulded or machined notch in order to assess sensitivity to weakening by notches. Data from tests of type (a), and particularly the Izod version, are open to much criticism; they and the other important standard tests will be described first, after which the practical significance or otherwise of the results is discussed briefly.

8.10.2 Pendulum Impact Tests

IZOD TESTS

The basic principle of this type of test is to allow a pendulum of known mass to fall through a known height, strike a standard specimen at the lowest point of its swing and record the height to which the pendulum continues its swing. If the striking edge of the pendulum is sited to coincide with the centre of percussion of the pendulum, the bearings of the pendulum are frictionless and there is no loss of energy to windage, then the product of the mass of the pendulum and the difference between the fall distance and the height it reaches after impacting the test specimen is the impact 'strength'* of the latter (plus the energy imparted in throwing off the broken piece—the 'toss factor'). The result is usually expressed as ft lbf or kgf cm and, since there is by and large no simple or reliable method of reducing the values to a property figure independent of the specimen cross-section and of the distance between the specimen support(s) and the point of striking (see below), the test values must be referred to the original specimen only; sometimes they are reduced to unit width across the specimen. The test may be carried out on plain rectangular bars—in which it is essentially a fast cross-breaking strength measurement—but most often a carefully defined notch is moulded or machined into the face to be struck. The reason for this is that impact tests are often regarded as a means of assessing the resistance of a material to shock where notches or 'stress raisers' generally are present; the ratio of the impact strengths unnotched/notched is sometimes regarded as a measure of the *notch sensitivity* of a material.

Method 306A of B.S. 2782:1970 uses a specimen 63·6 mm (2·50 in) long by 12·7 mm (0·50 in) thick with a notch in the centre of the width face (see Figure 8.49). For moulded specimens the standard shape is as follows.

*It will be seen that in fact whatever property is determined by this test it is certainly *not* a '*strength*' as so often it is called—even in official standards.

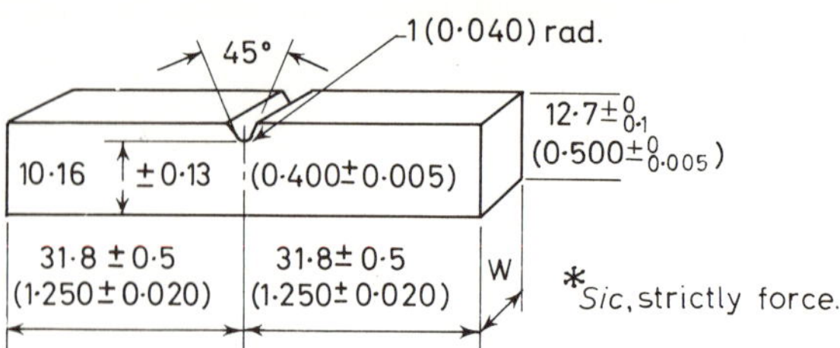

Figure 8.49. Izod impact strength specimen—moulded (B.S. 2782). The edges of the ends of the specimen may be radiused to 2·4 ($\frac{3}{32}$). Dimensions in millimetres with inch equivalents in parentheses

Three variants are specified:

'*A*'. *W* = 12·7 mm plus nil minus 0·3 mm (0·500 in plus nil minus 0·010 in) (The specimen may be machined but the notched face and notch must have moulded surfaces)

'*B*'. As '*A*', but the notch is machined

'*C*'. *W* = 6·35 mm plus nil minus 0·3 mm (0·250 in plus nil minus 0·010 in); the notch must be machined.

Further details are available from the relevant material specification. '*A*' is particularly used for thermosetting materials (compression moulded) and '*C*' for thermoplastics (injection moulded from the end).

The test is also used for laminated products. If the thickness of the sheet exceeds 6·35 mm (0·250 in) a specimen of length 63·5±0·5 mm (2·500±

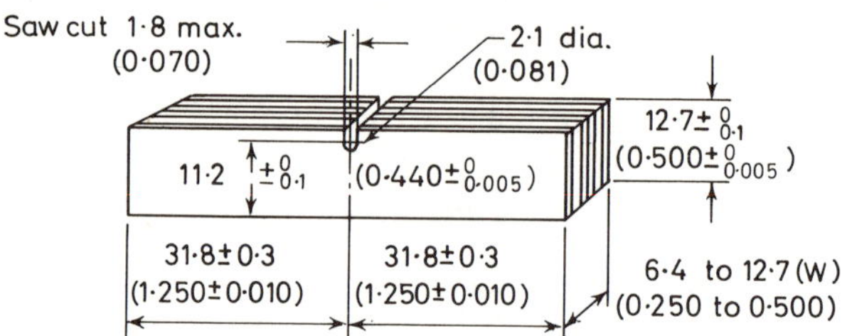

Figure 8.50. Izod impact strength specimen—machined laminate (B.S. 2782). Dimensions in millimetres with inch equivalents in parentheses

0·020 in) and 'width'* 12·7 mm plus nil minus 0·1 mm (0·500 in plus nil minus 0·005 in) is cut out; when the sheet thickness is over 12·7 mm (0·500 in) the excess is machined to 12·7 mm plus nil minus 0·1 mm from one face only. The rectangular specimen thus produced is notched *across the edge of the laminations* by drilling as shown in Figure 8.50 and opening up the drill hole with a saw cut.

As an alternative, a specimen may be machined as in Figure 8.49 with a milling cutter. If the thickness of the laminate is between 2·54 and 6·35 mm (0·100 and 0·250 in) the specimen may be built up to 12·7 mm (0·500 in)

*Not the width of the specimen as tested!

as nearly as possible, but no more, by placing together two or more specimens machined and notched as before.

The width of machined specimens (across the notch) is measured to 0·03 mm (0·001 in); for moulded specimens the widths are taken as 12·7 mm or 6·35 mm as appropriate.

In the B.S. Izod test the pendulum striking edge has a velocity of 2440 ± 30 mm (8·0 ± 0·1 ft) per second at the moment of impact, which is achieved by a free fall of 305 mm (1 ft). A typical commercial machine is shown in Figure 8.51.

The height to which the pendulum rises after impact is recorded by an idler pointer on a scale which reads directly in energy absorbed. Details of

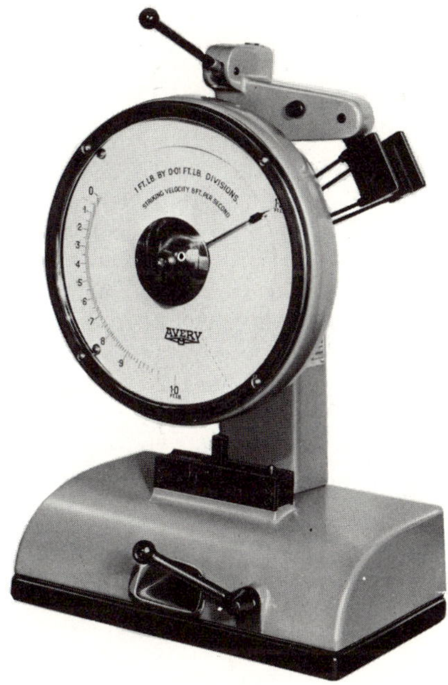

Figure 8.51. The Avery izod impact testing machine no. 6702

mounting the specimen in the vice of the machine are given in Figure 8.52, together with the position of the pendulum striker at the moment of impact.

Not surprisingly, to obtain reasonable accuracy the energy of the pendulum should not be too large in relation to the impact strength of the specimen. B.S. 2782, Method 306A requirements are set out in Table 8.3.

Regrettably, to cover the whole range this usually means three different machines, as interchangeable pendulum/striker assemblies are not generally supplied for fitting to one chassis. 11·5 J (8·5 ft lbf) will not cope with materials such as glass cloth laminates, but although larger machines are available (steels may require over 136 J (100 ft lbf) for their standard specimens) the test is unsuitable for such products as clean breaking is not usually achieved and results are biased by the ragged edge of the torn specimens causing a drag on the pendulum as it follows through.

B.S. 2782 requires the results to be calculated to give the energy absorbed per 12·7 mm (0·500 in) width of specimen.

A.S.T.M. D.256–56, Method A (and Method C) follow the same basic principles as above but, regrettably, with two very important differences

Table 8.3 B.S. 2782. IZOD IMPACT TEST. RELATIONSHIP BETWEEN IMPACT STRENGTH AND ENERGY OF PENDULUM

Energy absorbed				*Energy of blow*		
Greater than	*Up to and including*					
J	J	kgf cm	ft lbf	J	kgf cm	ft lbf
—	1·15	11·8	0·85	1·36	13·8	1·0
1·15	3·4	34·6	2·5	4·07	41·5	3·0
3·4	11·5	118	8·5	13·6	138	10

which make the results of the B.S. and A.S.T.M. tests quite irreconcilable (see below)—though it is the latter which is essentially similar to the ISO/R.180 method. These essential differences are in the root radius of the notch (0·010 in instead of 0·040 in) and, less importantly, because of the relatively small difference, the speed of the pendulum at impact (approx. 11 ft/s instead of 8 ft/s, the former being equivalent to a free fall of 2 ft).

A.S.T.M. D.256–56, Method A, goes into the details of the test machine a little more deeply than the B.S., making reference for instance to the effect of friction losses (which may be checked by allowing the pendulum to swing without a specimen in position) and the influence of the clamping pressure of the vice on certain materials. Apart from the notch, which must be machined in all cases, there are some differences in dimension tolerances and moulding instructions compared with B.S. 2782 and, for sheet of less than $\frac{1}{16}$ in thickness, plied specimens should be bonded together with a suitable adhesive to prevent buckling or twisting during test (B.S. 2782,

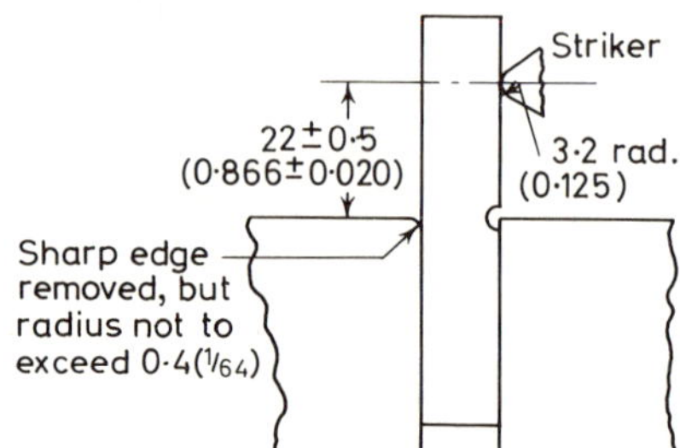

Figure 8.52. Specimen mounting for B.S. Izod test (B.S. 2782). Dimensions in millimetres with inch equivalents in parentheses

Method 306A, does not contemplate sheet below $\frac{1}{10}$ in). Fairly detailed advice is given on machining specimens.

The results are expressed as ft lb/in of notch by dividing the actual test values by the width of the specimens, though it is pointed out in a note (No. 5) that with some materials impact energy absorbed is *not* proportional to width of specimen! (see Section 8.10.4).

Method C of A.S.T.M. D.256–56 is essentially the same as Method A, but is recommended for materials of impact strength less than 0·5 ft lb/in of notch. In this region the 'toss factor' may be a large fraction of the impact strength as measured above and thus a technique is given for eliminating it, to yield an 'estimated net Izod impact strength'. The 'toss factor' is measured by replacing the free broken piece of the specimen on the still-clamped broken portion and striking again with the pendulum. The new reading is obtained by releasing the pendulum from a height corresponding to that to which it rose after breaking the test specimen, so as to impart to the broken piece of the specimen a velocity approximately the same as it had when first broken off. The 'toss factor' is calculated from the difference in reading between the new value thus obtained and the free-swing reading obtained from this height and again is expressed in ft lb/in width of notch.

CHARPY TESTS

Charpy impact tests are similar to Izod measurements in that they employ a pendulum of known energy striking a rectangular specimen; there are, however, a number of important differences of detail. The Charpy method has long been favoured on the Continent, is now standardised in I.S.O. and has consequently been taken up by the B.S.I. The A.S.T.M. variant

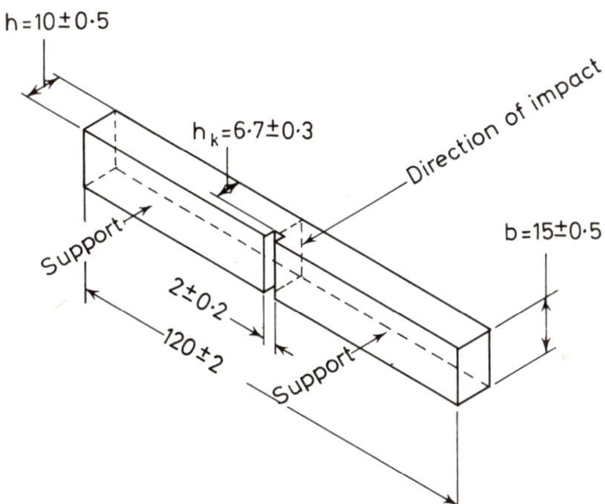

Figure 8.53. Charpy impact resistance specimen—moulded (B.S. 2782). Dimensions are in millimetres.

is not, however, identical with them. Charpy tests may be carried out on both notched and unnotched specimens and is more usually employed than the Izod method for assessing notch sensitivity from the ratio of the two values thus obtained.

In B.S. 2782, Methods 306D and 306E (which are the same in experimental details as those of the relevant parts of ISO/R.179) are for 'Charpy impact resistance' of unnotched and notched specimens respectively; being of continental origin it has been standardised directly in metric units and no British equivalents are given. The specimen is a rectangular bar of length 120±2 mm, width 15±0·5 mm and thickness 10±0·5 mm; in the case of Method 306E a notch is milled across one face (Figure 8.53) and the radius of the corners of the base of the notch must not exceed 0·2 mm.

Similar provisions are made for testing sheet materials as in Method 306A (see above) and the test is again carried out in quintuplicate, with the usual heed to directional effects. Likewise, a range of impact resistance scales is required according to the shock resistance of the material under examination. Readings should be between 10 and 80% of the scale in use, and accuracies are quoted for 0·5, 1·0, 4·0 and 15 J (5, 10, 40 or 150 kgf cm); the first three scales are used with an impact velocity of 2·9±0·5 m/s and the last with one of 3·8±0·5 m/s. The mode of supporting the test specimen is as shown in Figure 8.54.

Before test, the width and thickness of the specimen are measured to the nearest 0·1 mm; for sheet of less than 10 mm thickness tested flatwise, the supports of the machine are moved forward, by the amount the thickness of the specimen is less than 10 mm, with packing pieces inserted behind the supports.

The impact resistance of unnotched specimens (Method 306D) is calculated from the energy absorbed when the pendulum breaks the mounted specimen divided by the area of cross section of the specimen. In the notched method, 306E, the divisor is the area behind the notch (bh_k in Figure 8.53).

DIN.53453 follows a similar pattern, but also includes a smaller specimen 50 mm × 6 mm × 4 mm and another small scale test based on the Dynstat apparatus, which has already been mentioned in Section 8.7.3; the apparatus is specified in DIN.51222 and DIN.51230.

The Dynstat impact test uses a specimen 15±1 mm in length by width

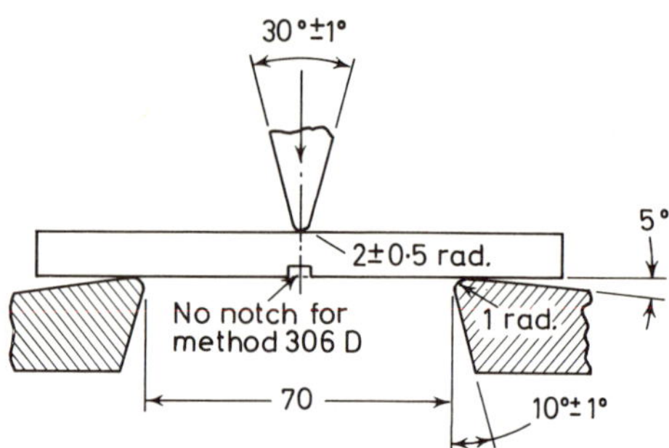

Figure 8.54. Specimen mounting for B.S. Charpy test (B.S. 2782). Dimensions in millimetres

10±0·5 mm and thickness 1·5–4·5 mm. The specimen may or may not be notched. Set up for impact resistance measurements, the Dynstat equipment operates as follows[28] (see Figure 8.55) the constructional details being as described before.

The specimen S_2 is supported in clamp N cantilever-fashion as in an Izod test. Clamp N can be raised or lowered such that the striking distance, that is the distance above the edge of the clamp at which the striker hits the specimen, can be varied between 2 and 9 mm, but DIN.53453 specifies this distance as 7 mm. The pendulum is held in clip T and rotation of dial C makes it possible to alter the angle through which the pendulum is

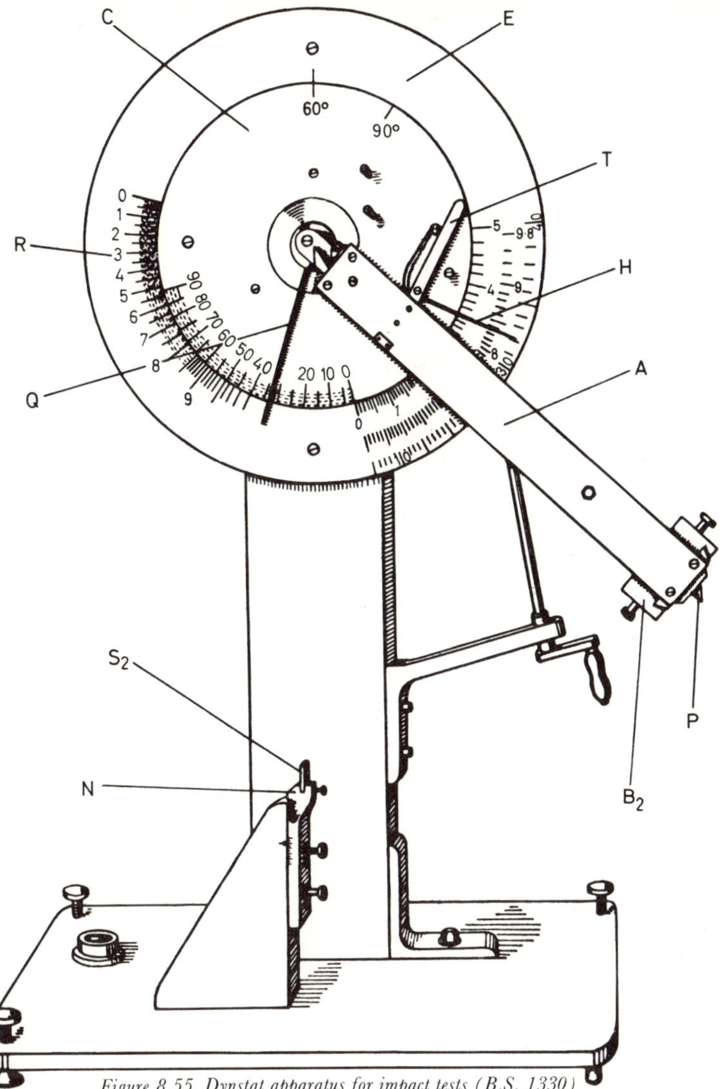

Figure 8.55. Dynstat apparatus for impact tests (B.S. 1330)

raised and held—again the DIN. standardises on one figure, 90°. The pendulum is 25 cm long, thus giving an impact velocity of 222 cm/s from 90°. When released the pendulum swings and hits the specimen with its tip P. Idler pointer Q indicates on scale R, calibrated in kg cm, the excess energy left in the pendulum which, subtracted from the starting energy, gives the impact strength of the specimen. DIN.53453 requires the value to be quoted after division of the cross-sectional area, or that behind the notch, as relevant. The automation of this equipment has been described by Hofmeier and Röver[67].

The Charpy test in A.S.T.M. D.256–56 (Method B) follows the same principles as the B.S. ISO and DIN methods but differs in specimen dimensions, notch shape and in impact velocity (it is also specified in British units).

Thus the specimen is 5 in long by $\frac{1}{2}$ in wide (or less as before) by 0·500 ± 0·006 in thick and the notch is precisely as in the A.S.T.M. Izod specimen. The impact velocity of the A.S.T.M. Charpy test, like its Izod counterpart, is approximately 11 ft per second.

A Charpy machine, very useful for comparative purposes but not related to a standard method, is manufactured by Tensometer Limited[68]; it uses quite small specimens, $1\frac{3}{4}$ in long by $\frac{5}{16}$ in diameter, and offers 7 ranges with interchangeable taps of $\frac{1}{32}$, $\frac{1}{16}$, $\frac{1}{8}$, $\frac{1}{4}$, $\frac{1}{2}$, 1 and 2 lb.

8.10.3 Falling Weight Tests

Methods 306B and 306C of B.S. 2782 describe two variants of an essentially simple impact strength test which assesses the resistance to shock of a thin flat disc of material by finding how resistant is such a specimen to fracture when impacted by allowing a 'ball' to fall centrally on its surface. The height of fall of the 'ball', and the mass thereof, may be altered and the minimum value of the product of fall height and mass which causes fracture is the impact strength of the material. The concept is simple, but the standardisation is not, and a little thought will soon indicate an important reason why. Thus, if the same disc is continually impacted by increasing impact blows, how is one to know how much 'damage' (not actually causing fracture) has been sustained before the impact value causing fracture is finally reached? If such a progressive damage is to be tolerated, how may it be standardised?

In practice, the B.S. method tests each specimen only *once* and the impact resistance is computed from the general spectrum of behaviour of 'go-not-go' fracture against regularly varied impact values; the method is therefore somewhat profligate in its use of test specimens.

The specimen is a disc 57–64 mm (2·25–2·5 in) in diameter or 57–64 mm (2·25–2·5 in) square. For moulded or extruded material the specimen is 1·52 ± 0·05 mm (0·060 ± 0·002 in) thick, for sheet the thickness thereof. At least twenty such specimens are required and no specimen is used more than once.

The apparatus may be of the form shown in Figure 8.56 (suitable for Method 306B).

The specimen support is a hollow cylinder of internal diameter 50·80 ± 0·05 mm (2·000 ± 0·002 in), external diameter not less than 57·2 mm (2·25 in) and length not less than 25·4 mm (1 in). The axis of the cylinder coincides with line of fall of the striker and a soft shock-absorbing disc is placed inside the cylinder to rest on its base. The specimen may be clamped on to the support.

The striker has a hardened hemispherical striking surface, 12·7 ± 0·05 mm (0·500 ± 0·002 in) diameter (the 'ball') and is fitted with a carriage to take weights so that a specified series of increments of energy may be obtained if the striker is released from a height of 610 ± 2 mm (24 ± 0·1 in) (Method 306B) or 305 ± 1 mm (12 ± 0·05 in) (Method 306C) above the upper surface of the specimen. The striker is allowed to fall with or without guides; in the former instance the fall must be substantially without friction. The striker

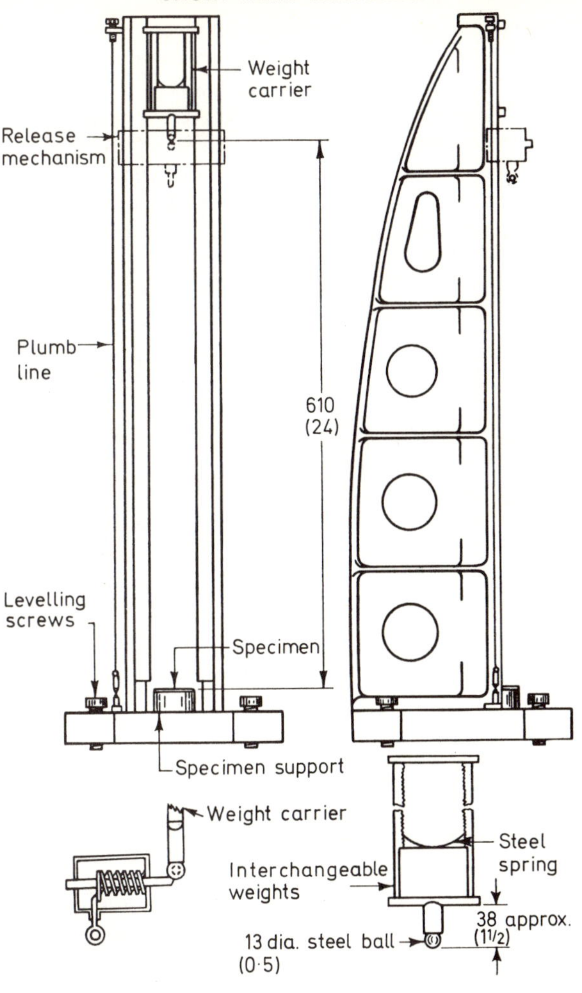

Figure 8.56. Falling weight impact strength apparatus (B.S. 2782). Dimensions in millimetres with inch equivalents in parentheses

can conveniently be supported electromagnetically and released by switching off the current.

To carry out the test a 'trial run' is undertaken first. The striker is loaded such that the product of weight and fall height is equal to the expected impact strength. If, on release, the specimen does not fracture, or cracks on one surface only, the result is recorded as 'unbroken' but if the specimen breaks or cracks or tears through from one surface to the other, the result is recorded as 'broken'. Thereafter:

1. If the result was 'broken', a second specimen is tested with an impact energy less than the first by a specified amount S. If this second specimen is 'broken' a third is tested with an impact energy S* less than that applied to the second—and so on until a specimen does not break.

2. If the first result was 'unbroken', a second specimen is tested with an impact energy greater by amount S obtained as before. As in 1. this is repeated, this time until a specimen does break.

The 'test run' is now carried out, using the remaining test specimens, the energy of the blow applied to any specimen being T more or less than that on the previous specimen, respectively according to whether the latter remained unbroken or was broken, with T being specified in the B.S. Method. The test run thus uses a maximum of 18 specimens (since at least two of the twenty must be required for the 'trial run'), but must not use less than twelve.

The impact strength is calculated as follows:

$$\text{Impact strength} = \frac{1}{21-m}\{Y_{m+1}+Y_{m+2}\ldots, Y_{21}\}$$

where

m = number of blows in trial run
Y_{m+1} = impact energy of first blow of the testing run
Y_{m+2} = impact energy of second blow of the testing run
Y_{21} = impact energy of 20th blow, decreased or increased by T according to whether the specimen broke or did not break.

[see also Chapter 17, Section 17.4.2].

8.10.4 Consideration of Impact Strength Data

Of the two classes of standard tests for impact strength just described, the pendulum method is the traditional and more widely used method. From the brief discussion in Section 8.9.1 of the effect of rate of strain and of other test variables, on mechanical properties of plastics, it may well be imagined that this type of test does not yield what might be termed 'fundamental data'; furthermore, accepting the arbitrary selection of variables which constitutes the standard A.S.T.M. Izod test for instance, the values obtained are still of doubtful significance, a point well illustrated by the correction made for 'toss factor' in Method C of A.S.T.M. D.256–56 (see above), which refinement is not included in any other of the standard tests of a similar nature.

Although, as has already been seen, the various plastics materials may show an infinite range of 'strength' values under a given set of test conditions, very broadly the mode of failure is generally tough or brittle. Which of these is exhibited depends as stated before on whether the material yields prior to crack initiation (tough failure) or a crack forms at a stress lower than that of the yield stress (brittle failure). Since brittle fracture is if possible to be avoided in service, the ostensible aim of impact testing is to investigate those conditions under which a material will fail in a brittle manner; the purpose of introducing the notch in the pendulum tests is to concentrate the stress and increase the straining rate at the root of the notch, thus inducing brittle behaviour in a wide range of materials at normal testing temperatures and simply obtained speeds of testing. Vincent[69] however, takes notched impact testing to task and makes, inter alia, the following points:

1. The test may be too severe—i.e. causes brittle behaviour unrealistically and penalises a satisfactory material.
2. It may give a result dependent more on crack propagation resistance than ability to resist crack initiation.
3. It chooses an arbitrary combination of speed, specimen dimensions, notch profile and temperature which may give quite misleading results even on a comparative basis.
4. It is an inaccurate test, very dependent on precisely how the test is carried out (see below).
5. Not surprisingly, in view of 1–4, the test conditions are probably unrelated to service conditions.

Telfair and Nason[70] have analysed the factors comprising 'impact strength' as measured by the Izod test as follows:

1. Energy to initiate fracture of specimen.
2. Energy to propagate fracture across specimen.
3. Energy to deform the specimen plastically.
4. Energy to throw the broken end of the test specimen.
5. Energy lost through vibration of the apparatus and its base, and through friction.

1 is probably the value of real interest though the addition of 2 and 3 have obvious practical significance. 4 is the 'toss factor' referred to in Method C of A.S.T.M. D.256–56 and the significance of the correction in certain cases is well illustrated in a general discussion on the Izod Impact test in Reference 35: with low impact strength materials the 'toss factor' can be the major proportion of the measured value. 5 requires a rigid design of pendulum arm in particular and a massive foundation.

Reference 35 also mentions further factors which may influence results and lead to errors, illustrating each with results from practical investigations:

1. Variation of clamping pressure in vice.
2. Failure to strike the specimen squarely.
3. When notches are machined, the state of the cutter and actual cutting technique used.

The effect of such variables as temperature and notch radius on the impact strength of polyolefins in particular has been reported by Horsley and Morris[71]. They[72] have also examined the influence of processing variables on falling weight impact strength. The B.S. 2782 technique of reducing impact strength to unit width of specimen is open to criticism as a result, e.g. of the work of Wolstenholme et alia[73] who showed that impact strength per unit width is *not* invariably independent of specimen width tested. A wide range of thermoplastics was examined at specimen notch widths between $\frac{1}{8}$ in and $\frac{1}{2}$ in and many showed marked differences in 'reduced' impact strength even from widths between $\frac{1}{4}$ in and $\frac{1}{2}$ in. (It has been stated that with polycarbonate, as little as 0·01 in difference in the width of specimens can alter the mode of failure from tough to brittle—see also Reference 74.)

B.S.I. Committee PLC/36, 'Plastics Design Data', referred to in Chapter 2, is known, at the time of writing, to be near finalising a document* recommending a practice for presenting impact data—the term 'impact strength' is specifically avoided. It is anticipated that it will suggest that data be

*To form Recommendation 1.2 of B.S. 4618.

derived from unnotched specimens and from notched specimens with various notch tip radii, at various temperatures; generally the effect of altering testing speed can be inferred from the effect of temperature, though the former may have to be examined specifically in certain cases. Impact energies should be calculated by dividing energy absorbed in the test by the area of cross section of specimen fractured (i.e. less notch) and the

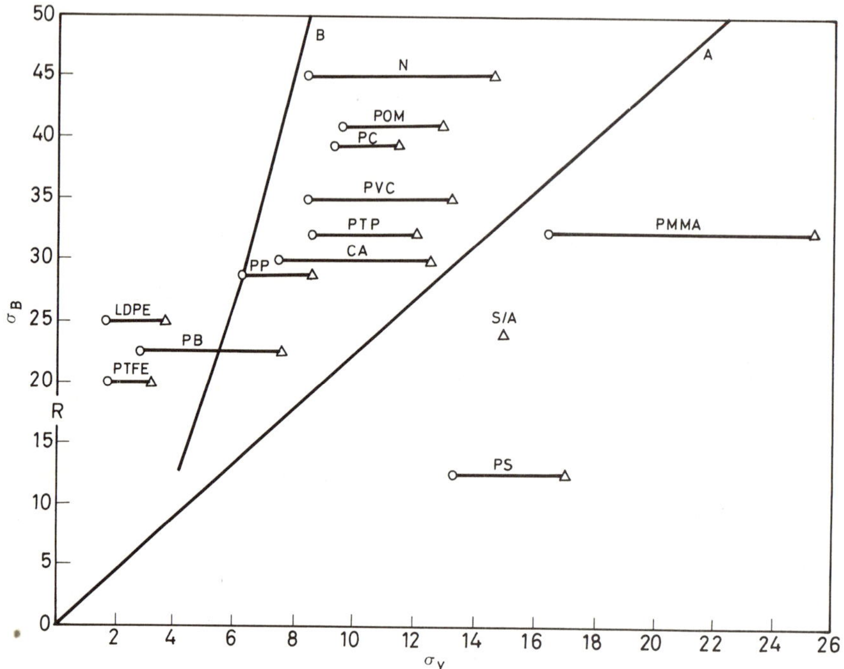

Figure 8.57. Brittle strength–yield stress diagram (Vincent[75])

impact energies for a given specimen plotted against temperature and for a given temperature against notch depth and/or notch tip radius. Data should also be compiled to demonstrate the effect of processing conditions. Drop impact tests, on sheet and other finished articles, are recommended and discussed in general terms; the importance of observing the mode of failure—brittle, intermediate or ductile—is stressed throughout.

It will be obvious that many of the criticisms applied to the pendulum type of impact test are not applicable to the falling weight method. It will be equally obvious, however, that this is basically a different test—applicable to (unnotched) sheet, possibly quite close to certain specific use conditions and essentially a fast bending strength test or impact flexural test (see Section 8.11 below). Reference 71 presents data from this test for thermoplastics, especially polyolefins, examining the effects of sheet thickness, processing conditions, and temperature on results.

As a means of assessing whether a material is likely to behave in brittle or ductile fashion, Vincent [75] presents an interesting technique based on relatively simple testing methods. The answer to the question, as has already been seen, reduces to whether brittle strength is greater or less than yield stress, i.e. σ_B/σ_Y is less than or not much greater than unity if the material is to behave in brittle fashion. Vincent suggests that σ_Y be obtained from a tensile test using a straining rate near 50% per minute; the temperature of test is +20°C, but the data given include figures at −20°C (which is not so readily achieved). As a convenient means of obtaining σ_B, a flexural test is carried out at about −180°C which is obtained quite simply by carrying out the test quickly after dipping the specimen in liquid nitrogen or liquid air. The values of σ_Y and σ_B so obtained are entered on Figure 8.57.

The line A divides brittle and tough materials, respectively right and left thereof. The slope is not unity, however, but around $2\frac{1}{4}$ though subject to substantial error either way. There are good reasons for this in the fact that the value of σ_B has been obtained for convenience in flexure and not in tension and has also been obtained at about −180°C and not +20°C. The line B, which is stated to be even more approximate, divides brittle and tough materials when notched—clearly an arbitrary division depending on notch dimensions in particular. In the typical values plotted the alphabetical abbreviations will generally be self-evident; the circles give σ_Y at +20°C and the triangles at −20°C. (The original diagram is without units.)

At the time of writing, a Plastics Institute monograph on impact tests and service performance is in course of preparation—to parallel that on flexural testing by Heap and Norman[31].

8.11 TENSILE IMPACT AND FLEXURAL IMPACT STRENGTH MEASUREMENT

The pendulum impact test is a flexural impact strength test in reality, especially in the variety which uses an unnotched rectangular specimen. However, we have seen the disadvantages of this test (for many of those listed for the notched test apply equally to the unnotched variant) and a more reliable quality control type of shock loading test is required; high speed tests such as given in Section 8.9 may be suitable especially if graphical data are produced from which energy to break may be computed, but the equipment is usually complex and costly. For this reason in particular the *tensile-impact* test has been evolved and standardised as A.S.T.M. D.1822–68.

Reference 35 describes three types of equipment. One is essentially a fast tensile test of the above type. The second gives straining rates of 20–80 ft/s;

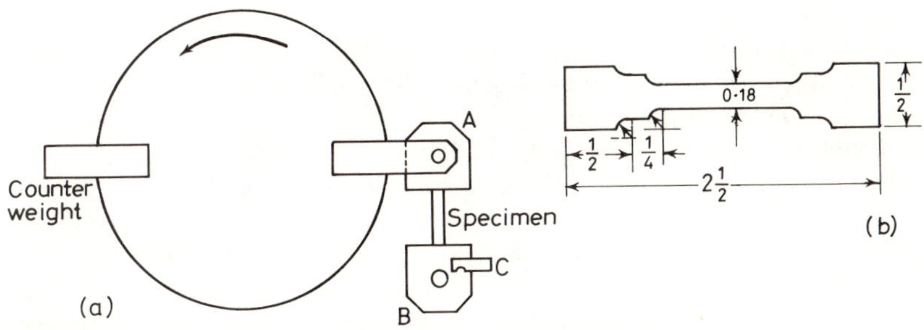

Figure 8.58. Essentials of (a) high speed tensile impact machine[35]; (b) test specimen. Dimensions in inches

the tensile test piece is mounted between two chucks as shown in Figure 8.58 on a massive flywheel.

The grip B is 'free' on the end of the specimen but C is a completely separate rigidly mounted arrester chuck. The flywheel is connected to a drive mechanism which gives the required angular velocity, at which point the drive is disconnected. The flywheel continues to rotate at essentially constant velocity, but as quickly as possible the chuck C is moved into position (to left) to halt chuck B so that the specimen is broken by tensile impact. From the change in angular velocity of the wheel the energy absorbed by the specimen is calculated, the moment of inertia of the flywheel being such that the change in angular velocity is small and thus the test is carried out at virtually constant speed.

The third type described is based on a pendulum machine, which is the essential feature of the one standardised method A.S.T.M. D.1822–68. The

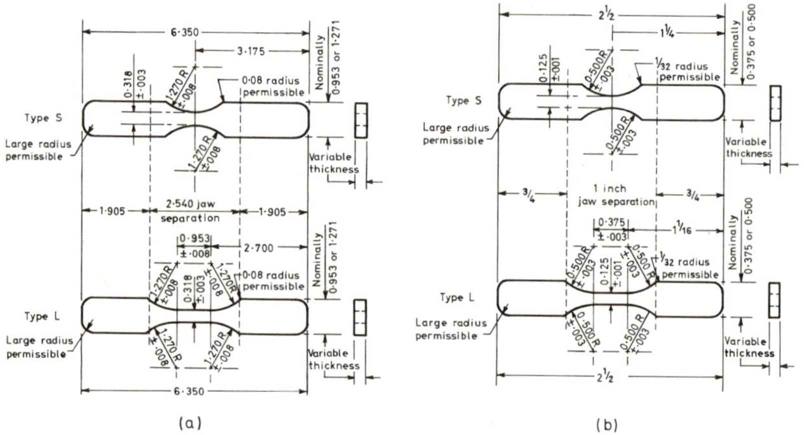

Figure 8.59. Tensile impact specimens (A.S.T.M. D.1822). (a) Dimensions in centimetres. (b) Dimensions in inches.

specimen (two types are allowed) is shown in Figure 8.59 and is fixed at one end to the 'head' of the pendulum, the chuck being so designed as to pass unhindered through the arrester of the machine which is situated at the pendulum vertical position where the vice of the normal impact test machine would be (see Figure 8.60). The other end of the specimen is gripped by a 'crosshead clamp' which is firmly arrested at the vertical position of the pendulum.

As in the A.S.T.M. Izod test—D.256—the velocity at fracture is 344·4 cm/s (11·3 ft/s), achieved by a fall height of 60·96 cm (2 ft). The need for rigidity and careful location of the centre of percussion are as before (Section 8.10.2).

At least five, and preferably ten, specimens are tested and the results calculated on a unit cross-sectional area basis after applying correction factors for friction and windage and the 'bounce' of the crosshead. The method is described in rather greater detail by Westover and Warner[76] who give typical test results obtained in the evolution of the method. Richter[77] has examined the problems of tensile impact testing with these different forms of test piece on a range of thermoplastic materials. Therberge and Hall[78] found a high degree of correlation between tensile impact values and unnotched 'impact strengths' for glass filled SAN, polystyrene, polycarbonate, polyethylene, polypropylene, PVC, polyurethane, polysulphone, polyester and nylons −6, −6·6 and −6·10; not surprisingly, the correlation did not exist for notched Izod impact strengths.

The DIN.53448 tensile impact is also based on a pendulum machine, as described in DIN.51222 (see Section 8.10.2).

Vincent[79] criticises the pendulum type of tensile impact test on such grounds as limited range and inaccuracy and advocates a falling weight method. This is in many ways similar to the methods described in Section 8.10.3 for assessing the impact resistance of sheet; the tensile specimen is hung vertically and securely gripped at its upper end by a fixed chuck of such a size and shape that the falling weight passes unhindered down past the fixed chuck but is firmly arrested by the larger chuck which simply grips the lower end of the specimen. Weights and heights may be varied

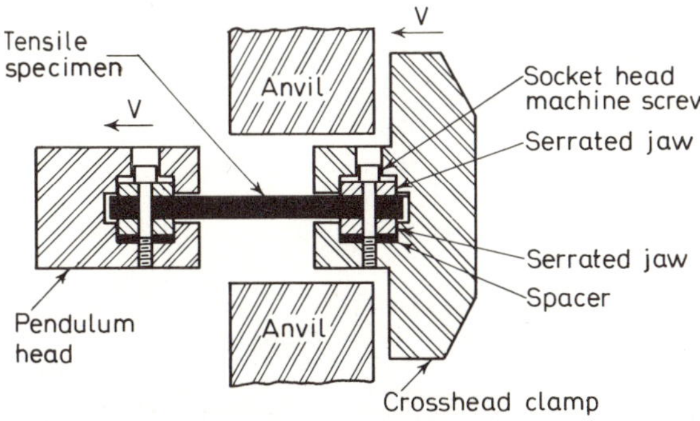

Figure 8.60. Specimens in tensile impact testing machine—schematic (A.S.T.M. D.1822)

to alter the energy of the weight at impact and the tensile impact value of the specimen may be calculated from a series of trials and tests as described before. Within convenient limits of height, the effect of speed on test values may be easily investigated and indeed one of Vincent's main claims of advantage for the technique is that of simplicity (and thus cheapness). A more sophisticated method, based on the same principles, involves the use of a hollow rectangular test piece, aptly described as the 'Racetrack'[80].

Donnelly and Ralston[81] describe a falling weight tensile impact tester of rather greater elaboration. Heights of fall between 6 in and $3\frac{1}{2}$ ft can be investigated and the stress–strain behaviour is followed from a load cell/transducer/oscilloscope set up (see Figure 8.61).

Vincent[82], in a further article in the series already referred to, describes the adaption of the falling weight technique to flexural impact strength tests, as being preferable to pendulum tests, which are basically the same as the B.S. 2782 method (Section 8.10.3) except that the specimen is a bar

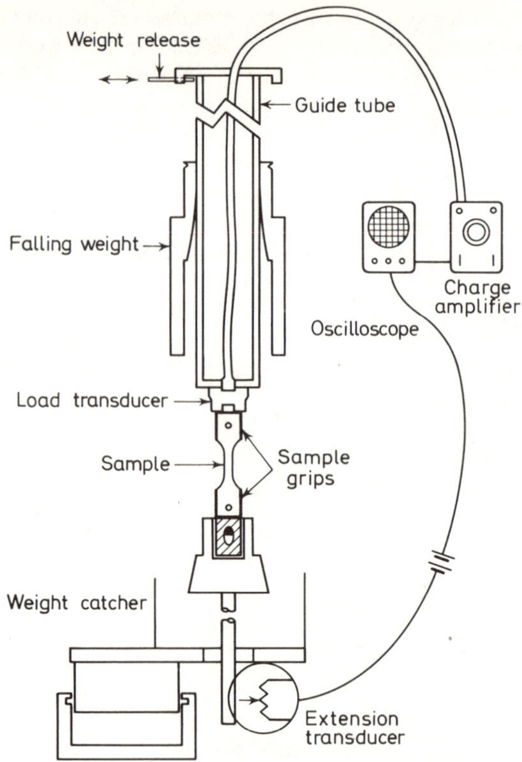

Figure 8.61. Falling weight tensile impact tester (Donnelly and Ralston)

resting on two outer supports and struck centrally by the falling weight on the opposite face.

8.12 LIMITATIONS OF STANDARD MECHANICAL TESTS

The limitations of the data obtained by the standard tests for tensile, bending, compression, shear and impact strength have already been mentioned throughout this chapter and in dealing with the general subject of short term mechanical properties indications have been given of how the spectrum of data may be extended in the direction of shorter time scale to deal with applications involving shock loadings. The measurement of strength data above and below normal ambient temperature is described in Chapter 11, but many—probably most—applications of plastics of a structural nature involve loading for prolonged periods under either constant or varying loads. Under constant stress, plastics materials continue to strain over weeks or years depending on the material in question, the temperature and the level of stress; this is generally known as 'creep'. Under constant strain the stress decays over much the same order of time scale; this is 'stress–relaxation'.

These subjects are dealt with in Chapter 14 and the behaviour under varying stress in Chapter 15. Useful discussions of the type of data required for design purposes are to be found in References 83 and 84.

REFERENCES

1. 'Measurement of Load by Elastic Devices', *Notes on Applied Science No. 21,* National Physical Laboratory, Her Majesty's Stationery Office (London) (1961)
2. CARSWELL, T. S. and NASON, H. K., 'Effect of Environmental Conditions on Mechanical Properties of Organic Plastics', *Mod. Plast.,* **21,** No. 10, 121–6, 158, 160 (June 1944)
3. NIELSEN, L. E., *Mechanical Properties of Polymers,* Chapter 5, Reinhold Publishing Corporation, New York (1962)
4. RITCHIE, P. D., (ed.) *Physics of Plastics, Plastics Institute Monograph,* Chapter 2, Iliffe Books Ltd., London (1965)
5. ALFREY, T. JR., *Mechanical Behaviour of High Polymers,* Chapter F, Interscience Publishers Inc., New York (1948)
6. OGORKIEWICZ, R. M., CULVER, L. E. and BOWYER, M. P., 'Deformation of Thermoplastics under Different Types of Tensile Loading', *S.P.E. J.,* **25,** No. 3, 43 (March 1969)
7. BATSON, R. G. and HYDE, J. H., 'Mechanical Testing', *Testing of Materials of Construction,* Volume 1, Chapman and Hall Ltd., London (1941)
8. LAMBLE, J. H., 'Some Points for Consideration in the Design of Testing Machines', *Bull. mech. Engng Educ.,* **2,** 93–109 (1963)
9. PAYNE, A. R. and SMITH, J. F., 'Dynamometer for Tensile Testing of High Polymers', *J. Sci. Instrum.,* **33,** 432 (November 1966)
10. EAGLES, A. E., PAYNE, A. R. and SMITH, J. F., 'Comparison of Electronic and Pendulum Dynamometers for Tensile Testing', Part 1, *Research Memorandum No. R.403,* Research Association of British Rubber Manufacturers (now Rubber and Plastics Research Association of Great Britain) Shawbury (1955)
11. B.S. 1610: 1964, *Methods for the Load Verification of Testing Machines,* British Standards Institution, London
12. A.S.T.M. E4–64, *Verification of Testing Machines,* American Society for Testing and Materials, Philadelphia
13. 'A Simple Way to Calibrate Tensile Testers', *Rubb. Age, Lond.,* **99,** No. 4, 72 (April 1967)
14. MORLEY, A., *Strength of Materials,* 10th edn, Longmans, Green & Co., London (1952)
15. KINLOCH, C. D. and WATERS, N. E., 'Extensometer for Semi-Rigid Materials', *J. Sci. Instrum.,* **37,** No. 3, 93 (March 1960)
16. FARLIE, E. D., HAWKES, J. E. and WATERS, N. E., 'A Simple Extensometer for Dumb-bell Tensile Test-pieces', *Rubb. J.,* **132,** No. 20, 648 (May 18th 1957)
17. EAGLES, A. E. and PAYNE, A. R., 'A Simple Extensometer for Tensile Testing of Polymers', *Rubb. Plast. Age,* **38,** No. 169, 811 (September 1957)
18. MEARDON, J. I., 'The Automatic Recording Extensometer Type K301 for the Hounsfield RTM', *Research Report No. 142,* Rubber and Plastics Research Association of Great Britain, Shawbury (1965)
19. BOUCHE, R. B. and TATE, D. R., 'Autographic Stress–Strain Recorders', *Bull. Amer. Soc. Test. Mat.,* **No. 228,** 33 (February 1958)
20. HARRIS, W. D., 'Measurement of Tensile Properties of Polymers', *Testing of Polymers* (Edited by W. E. Brown), Volume 4, Interscience Publishers, New York, 399 (1969)
21. PANOV, P. 'Effect of Absolute Specimen Size on the Tensile Strength of Plastics', *Plaste u. Kautsch.,* **14,** No. 7, 491 (July 1967)
22. HOLLAND, J. R., 'Self-Aligning Grip System and Tensile Specimens for Instron Machines,' *J. Sci. Instrum.,* **44,** No. 5, 389 (May 1967)
23. KOO, G. P. and ANDREWS, R. D., 'Mechanical Behavior of Polytetrafluoroethylene around the Room-Temperature First-Order Transitions', *Polymer Engng Sci.,* **9,** No. 4, 268 (July 1969)
24. HEAP, R. D., 'Order Statistics: The Estimation of Tensile Strength', *R.A.P.R.A. Bulletin,* Rubber and Plastics Research Association, Shawbury, 146 (Sept.–Oct. 1964) See also: *I.R.I.Trans.,* **41,** No. 3, T.127 (1965)
25. WILLIAMS, J. G. and FORD, H., 'Stress–Strain Relationships for Some Unreinforced Plastics', *J. mech. Engng Sci.,* **6,** No. 4, 405 (December 1964)
26. WILLIAMS, J. G., 'Plane–Strain Compression Testing of Polymers', *Trans. J. Plast. Inst.,* **35,** No. 117, 505 (June 1967)

27. Karl Frank G.m.b.H. 'Testing Equipment for Metals, Plastics, Rubber', Mannheim-Rheinau
28. B.S. 1330: 1946, *Interim Report on Suggested Methods of Testing Finished Mouldings (Plastics)*, British Standards Institution, London (Now withdrawn)
29. LOVELESS, H. S., 'Flexural Tests', *Testing of Polymers,* Volume 2 (Edited by J. V. Schmitz), Interscience Publishers, New York (1966)
30. ROARK, E. J., *Formulas* (sic) *for Stress and Strain,* McGraw-Hill Book Co. Inc., New York (1954)
31. HEAP, R. D. and NORMAN, R. H., *Flexural Testing of Plastics,* The Plastics Institute, London (1969)
32. STUART, I. M., 'A Loop Test for Bending Length and Rigidity', *Brit. J. appl. Phys.,* **17,** No. 9, 1215 (September 1966)
33. ALFREY, T. Jr., *Mechanical Behavior of High Polymers in High Polymers,* Volume VI, Chapter F, Section III, Interscience Publishers Inc., New York (1948)
34. GORDON, M., *High Polymers—Structure and Physical Properties,* 2nd edn, Chapter 3, Iliffe Books Ltd., London (1963)
35. RITCHIE, P. D., *Physics of Plastics,* Chapter 3, Iliffe Books Ltd., London (1965)
36. FRITH, ELIZABETH M. and TUCKETT, R. F., *Linear Polymers,* Chapter 8, Longmans, Green & Co., London (1951)
37. VINCENT, P. I., 'Strength of Plastics, Part 3, The Effect of Testing Speed', *Plastics,* **27,** No. 291, 115–7 (January 1962)
38. VINCENT, P. I., 'Testing for Brittle Facture', *Trans. Plast. Inst.,* **30,** No. 87, 157 (June 1962)
39. DIETZ, A. G. H., GAILUS, W. J. and YURENKA, S., 'Effect of Speed of Test upon Strength Properties of Plastics', *A.S.T.M. Publication, Symposium on Speed of Testing,* 32, American Society for Testing and Materials, Philadelphia (1948)
40. DIETZ, A. G. H. and MCGARRY, F. J., 'The Effects of Speed on the Mechanical Testing of Plastics', *Special Technical Publication No. 185, Symposium on Speed of Testing of Non-Metallic Materials, 30,* American Society for Testing and Materials, Philadelphia (1956)
41. HALL, WARBURTON, H., 'Mechanical Properties of Plastics at High Speeds of Stressing', *Int. Symp. on Plast. Test. Stand.,* Special Technical Publication No. 247, 137, American Society for Testing and Materials, Philadelphia (1959)
42. SUPNIK, R. H., 'Rate Sensitivity: Its Measurement and Significance', *Mater. Res. Stand.,* **2,** No. 6, 498 (June 1962)
43. GOLDFEIN, S., 'Computing High-Rate Tensile Strength from Static Strength Data', *Mod. Plast.,* **41,** No. 12, 149 (August 1964)
44. STRELLA, S., SIGLER, H., CHMURA, M. and HOLMAN, B., 'A High-Speed Tension Testing Machine', *Bull. Amer. Soc. Test. Mat.* **No. 228,** 50 (February 1958)
45. DORSEY, J., MCGARRY, F. J. and DIETZ, A. G. H., 'High Speed Tension Testing Machine for Plastics', *Bull. Amer. Soc. Test. Mat.* **No. 211,** 34 (January 1956)
46. STRELLA, S., 'An Improved High-Speed Tension Tester', *Bull. Amer. Soc. Test. Mat.* **No. 236,** 59 (February 1959)
47. ELY, R. E., 'High-Speed Tensile Tests of Thermoplastics', *Plast. Technol.,* **3,** No. 11, 900 (November 1957)
48. ELY, R. E., 'Tensile Properties of Three Thermoplastics over Six Decades of Rate', United States Department of Commerce, Office of Technical Services, *Report PB.151573* (31 October 1957)
49. 'Plastechon High-Speed Tester', *Rubb. Age, Lond.,* **85,** No. 1, 140 (April 1959)
50. SANDEK, L., 'High-Speed Testing', *Plast. Technol.,* **8,** No. 2, 26 (February 1962)
51. SILBERBERG, M. and SUPNIK, R. H., 'High-Speed Testing . . . A new Dimension in Polymer Evaluation', *S.P.E. Trans.,* **2,** No. 2, 140 (April 1962)
52. HEIMERL, G. L. and MANNING, C. R. Jr., 'A High-Speed Pneumatic Tension Testing Machine', *Mater. Res. Stand.,* **2,** No. 4, 270 (April 1962)
53. MCABEE, E. and CHMURA, M., 'High Rate Tensile Properties of Plastics', *S.P.E. J.,* **19,** No. 1, 83 (January 1963)
54. STRELLA, S. and GILMAN, L., 'High-Loading-Rate Tensile Properties of Thermoplastics', *Mod. Plast.,* **34,** No. 8, 158 (April 1957)
55. ELY, R. E., 'Some Tensile Properties of Cellulose Acetate and Butyrate', *Plast. Technol.,* **5,** No. 6, 36 (June 1959)
56. HOLT, D. L., 'The Modulus and Yield Stress of Glassy Poly(Methylmethacrylate) at Strain Rates up to 10^3 inch/inch/second', *J. appl. Polymer Sci.,* **12,** No. 7, 1653 (July 1968)

57. BARNES, C. B., HAWKINS, E. L. and DAVIS, M. V., 'A Method for Interrupting a High Strain Rate Tension Test before Specimen Fracture', *Mater. Res. Stand.*, **6**, No. 11, 560 (November 1966)
58. JONES, J. W., 'Tensile Testing of Elastomers at Ultra-High Strain Rates', *J. appl. Polymer Sci.*, **4**, No. 12, 284 (Nov.–Dec. 1960)
59. D'AMATO, D. A., 'Adaption of a Ballistic Tool for Obtaining Strain Rates Above 20 000 Inch/Inch/Minute', *J. appl. Polymer Sci.*, **8**, No. 1, 197 (Jan.–Feb. 1964)
60. PATTERSON, G. D. Jr. and MILLER, W. H. Jr., 'New High-Speed Tester for Polymer Films', *J. appl. Polymer Sci.*, **4**, No. 12, 291 (Nov.–Dec. 1960)
61. AMBORSKI, L. E. and MECCA, T. D., 'A Study of Polymer Film Brittleness', *J. appl. Polymer Sci.*, **4**, No. 12, 332 (Nov.–Dec. 1960)
62. GRIMMINGER, H. and JACOBSHAGEN, E., 'High Speed Tension Tester for Film, Recording the Force-Deformation Curve', *Kunststoffe*, **52**, No. 5, 254 (May 1962)
63. DIETZ, A. G. H. and EIRICH, F. R., *High Speed Testing*, Interscience Publishers: John Wiley and Sons Inc., New York, Volume I (1960), Volume II (1961), Volume III (1962), Volume IV (1964), Volume V (1965)
64. SPÄTH, W., *Impact Testing of Materials* (Revised and adapted by M. Rosner), Thames and Hudson, London (1961)
65. DAVIS, H. E., TROXELL, G. E. and WISKOCIL, G. T., *The Testing and Inspection of Engineering Materials*, 3rd edn, Chapter 8, McGraw-Hill Book Company, New York (1964)
66. FENNER, A. J., *Mechanical Testing of Materials*, Chapter 7, Philosophical Library Inc., New York (1965)
67. HOFMEIER, H. and ROVER, M., 'Automatic Equipment for the Impact Bending Test According to DIN 53453', *Kunststoffe*, **57**, No. 11, 877 (November 1967)
68. The Hounsfield Impact Testing Machine, Tensometer Ltd., 81 Morland Road, Croydon CR9 6HG, Surrey
69. VINCENT, P. I., 'Strength of Plastics, 5, The Effect of Notches', *Plastics, Lond.*, **27**, No. 294, 116 (April 1962)
70. TELFAIR, D. and NASON, H. K., 'Impact Testing of Plastics. I. Energy Considerations', *Mod. Plast.*, **20**, No. 11, 85 (July 1943)
71. HORSLEY, R. A. and MORRIS, A. C., *Shell Polyolefins Engineering Design Data*, Shell International Chemical Company Ltd. and Shell Chemicals U.K. Ltd., London (1966)
72. HORSLEY, R. A. and MORRIS, A. C., 'Impact Tests—Guide to Thermoplastics Performance', *Plastics, Lond.*, **31**, No. 350, 1551 (December 1966)
73. WOLSTENHOLME, W. E., PREGUN, S. E. and STARK, C. F., 'Factors Influencing Izod Impact Properties of Thermoplastics Measured with the Autographic Impact Tester', *High Speed Testing*, Volume IV, Interscience Publishers: John Wiley & Sons Inc. New York, 19 (1964)
74. SHOULBERG, R. H. and GOUZA, J. J., 'Impact Tests, Their Correlation and Significance', *S.P.E. J.* **23**, No. 12, 32 (December 1967)
75. VINCENT, P. I., 'Strength of Plastics 6. Conclusion—Yield Stress and Brittle Strength', *Plastics, Lond.*, **29**, No. 315, 79 (January 1964)
76. WESTOVER, R. F. and WARNER, W. C., 'Tensile Impact Test for Plastics', *Mater. Res. Stand.*, **1**, No. 11, 867 (November 1961)
77. RICHTER, K., 'Problems Associated with Tensile-Impact Tests', *Plaste u. Kautsch.*, **14**, No. 1, 23 (January 1967)
78. THERBERGE, J. E. and HALL, N. T., 'Impact Behaviour of Glass-Fortified Thermoplastics', *Mod. Plast.*, **46**, No. 7, 114 (July 1969)
79. VINCENT, P. I., 'Strength of Plastics 6. Tensile Impact Tests', *Plastics, Lond.*, **27**, No. 295, 133 (May 1962)
80. DANNIS, M. L. and WATTING, R. E., 'Stress–Strain Properties of Plastics at High Strain Rates', *Rev. Sci. Instrum.*, **37**, No. 12, 1716 (December 1966)
81. DONNELLY, P. I. and RALSTON, R. H., 'Comparison of Impact Properties of Six Materials in Four Types of Tensile Machines', *High Speed Testing*, Volume V, 71, Interscience Publishers: John Wiley & Sons Inc., New York (1965)
82. VINCENT, P. I., 'Strength of Plastics, 7, Flexural Impact Tests', *Plastics, Lond.*, **27**, No. 296, 136 (June 1962)
83. TURNER, S., 'The Mechanical Properties of Plastics Relevant to Design', *Trans. J. Plast. Inst.*, **33**, No. 106, 95 (August 1965)
84. RATCLIFFE, W. F. and TURNER, S., 'Engineering Design: Data Required for Plastics Materials', *Trans. J. Plast. Inst.*, **34**, No. 111, 137 (June 1966)

9

Mechanical Properties of Surfaces

9.1 GENERAL COMMENT

The subjects to be considered in this chapter are hardness, softness and resistance to indentation generally, coefficient of friction, abrasion, mar and wear resistance and scratch hardness. As a whole they have little in common except that they are all characteristics of the *surface* or the outer layer of the material. They range from the reasonably well understood subject of hardness—as a function of one or more fundamental mechanical properties (Chapter 8)—to the essentially empirical subject of abrasion resistance, which is only just being analysed into its essential parameters in order to design test methods simulating use conditions on a scientific rather than an ad hoc basis.

Methods for measuring these surface properties are not generally well standardised and, probably as a direct result, many different techniques are available, which themselves are not very adequately documented collectively. Except for friction there is no very clear dividing line between the properties considered in this chapter and hence 'spill over' of one section into another is unavoidable.

9.2 HARDNESS, SOFTNESS AND RESISTANCE TO INDENTATION

9.2.1 Introduction

'Hardness' is not a fundamental property—indeed the very interpretation of the word is decidedly subjective; the most widely accepted concept is probably that of resistance to indentation, but others widely held include *scratch resistance* and rebound resilience. The last-mentioned is considered in Chapter 15 as a dynamic property and scratch resistance later in this chapter.

Resistance to indentation, which is invariably measured before fracture occurs, is readily envisaged as some function of rigidity or modulus and like this property, measurement of hardness is subject to all the effects of temperature, time and other test variables mentioned in Chapter 8. In common with the traditional methods used for metals, hardness measurement of plastics usually takes the form of forcing a standard indentor—often a hardened steel ball—under known load into a flat surface of the material under examination and measuring the resultant degree of penetration. The viscoelastic nature of plastics introduces two complications for, in addition to the dependence of depth of penetration on the time of application of the load, the results of the metals-type test are calculated from the diameter of the indentation (or some other characteristic parameter if the indentor is not sphere-ended) and so the values may well depend on how speedily the diameter can be measured after removal of the loaded indentor. A fairly successful way of overcoming this problem in one type of test is to interpose a piece of carbon paper between the ball-indentor and the test surface (carbon side to the latter) and measure the diameter of the carbon impression left after removal of the load.

There are a few attempts in the literature to relate indentation of plastics, but more often elastomers, with modulus. Detailed consideration of the theories is unnecessary here and the reader is referred to References 1–5 for instance, the last-mentioned being also a useful if now somewhat out-of-date review of methods for hardness testing of elastomers. Suffice to state here that what with the already-mentioned effects of temperature and time and the nonlinear response of strain to stress, we have in addition the fact that, at best, indentation is a complex function of properties including modulus (if such a single value property exists for a plastics material!), force, and indentor profile. It therefore follows that hardness values according to one method cannot generally be compared with those derived from another, though some ad hoc correlations have been published.

General reviews of hardness testing methods will be found in References 6 and 7 and for plastics specifically in References 8–13.

9.2.2 National Standard Tests

UNITED KINGDOM

There is only one method for resistance to indentation in B.S. 2782:1970, 'Methods of Testing Plastics'; this is Method 307A for *softness number* of flexible PVC extrusion compound and it uses basically the same apparatus as that of B.S. 903:Part A7:1957 'Methods of Testing Vulcanised Rubber. Determination of Hardness'* in that the same indentor and major and minor loads are used. However, whereas the latter expresses the results on an arbitrary scale from 0 to 100 'degrees', with highest values for hardest rubbers,[2–4] the B.S. 2782 method gives (softness) values directly related to measured indentation.

The test is carried out at 23 ± 1°C with an indentor in the form of a steel ball of diameter 2·38 mm ($\frac{3}{32}$ in). The test specimen is a sheet of thickness 10·2 ± 0·6 mm (0·400 ± 0·025 in) and must be maintained at the test temperature for seven days before measurement. This is because the properties

*The revised document, B.S. 903: Part A26: 1969, 'Methods of Testing Vulcanised Rubber. Determination of Hardness', has slightly but not significantly altered the picture. I.S.O. methods and SI units have been incorporated; Normal Method N is the relevant one of the five methods now described.

of plasticised PVC compounds take several days to equilibriate after gelling[14]. The specimen is laid on a rigid, horizontal flat base. A plunger terminating in the indentor already described is pressed under a load of 294 mN (30 gf) on to the specimen for 5 s; the position of test shall not be less than 13 mm (0·5 in) from the nearest edge. The vertical position of the plunger relative to, say, the base, is read to 0·01 mm and then an additional load of 5·25 N (535 gf) applied for 30 s, with gentle vibration (e.g. from an electric buzzer) to overcome any friction. At the end of the 30 s the position of the plunger is read again; the softness number is the difference between the two readings, in units of 0·01 mm.

Two readings are taken on each face of the specimen, from tests not less than 12·7 mm ($\frac{1}{2}$ in) apart. The result is expressed as the mean of the four readings so obtained.

This somewhat anonymous description taken from the B.S. gives little idea of what the apparatus looks like 'in the flesh'; a typical commercial instrument is shown in Figure 9.1.

Some idea of the importance of control of thickness of test specimen may be gauged from the work of Waters[15] using the B.S. 903: Part A7: 1957* test on rubber.

UNITED STATES

One A.S.T.M. standardised procedure for determining hardness of plastics is based on the two *Shore Durometers* A and D, and described in A.S.T.M. D.2240–68, 'Indentation Hardness of Rubber and Plastics by Means of a Durometer'. Type A is for soft and Type D for hard materials (see also ISO/R.868).

The indentors of the two types of instrument are shown in Figure 9.2.

The force on the indentor is applied by a calibrated spring, formulae being given relating force to the hardness of the material under test for each type of indentor, and the hardness reading taken from an arbitrary scale (on a dial gauge) reading zero when the indentor meets no resistance (extension $2{\cdot}50 \pm 0{\cdot}4$ mm) and 100 when the extension is zero, i.e. when the presser foot (shown hatched in Figure 9.2) and indentor rest on a glass plate.

Specimen thickness and temperature are generally as the B.S. 2782 method already described, but readings are taken within 1 second of placing the presser foot in contact with the specimen. One form of the apparatus is similar to that illustrated in Figure 9.6 below.

An apparatus is described for calibrating the spring, but nevertheless the method emphasises that the technique is an empirical one and generally suitable only for quality control purposes.

Barcol hardness is frequently quoted in the literature, but it is necessary immediately to point out that there are three so-called Barcol Impressors[16]:

GYZJ 934.1 for soft metals and some of the harder plastics
GYZJ 935 for very soft metals and softer plastics
GYZJ 936 for extremely soft materials such as leather

Of these, only the first is standardised, as D.2583–67, 'Indentation Hardness of Plastics by Means of a Barcol Impressor'.

*Now out of date (see above).

A diagram of the general style of the impressors is given in Figure 9.3.

The indentor is a hardened steel truncated cone, having an angle of 26° with a flat top of 0·157 mm (0·0062 in) diameter. The indicating device has 100 divisions, each representing a depth of 0·0076 mm (0·0003 in) penetration—the higher the reading the harder the material. Test specimens

Figure 9.1. B.S. 2782. Method 307A. Dead load softness tester (Courtesy H. W. Wallace and Co. Ltd.)*

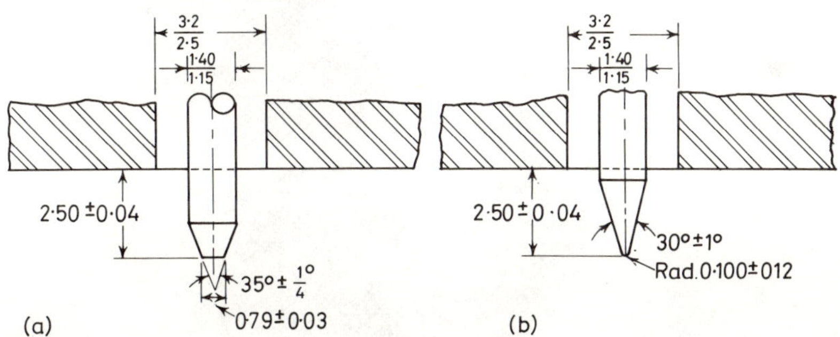

Figure 9.2. Shore durometer indentors. (a) Type A. (b) Type D. (A.S.T.M. D.2240). Dimensions are in millimetres

must be at least 1·5 mm ($\frac{1}{16}$ in) thick and large enough in area to ensure a minimum distance of 3 mm ($\frac{1}{8}$ in) in any direction from the point of measurement to the edge of the specimen.

When testing, the housing of the indentor is applied quickly by sufficient

*172 St. James's Road, Croydon, CR9 2HR.

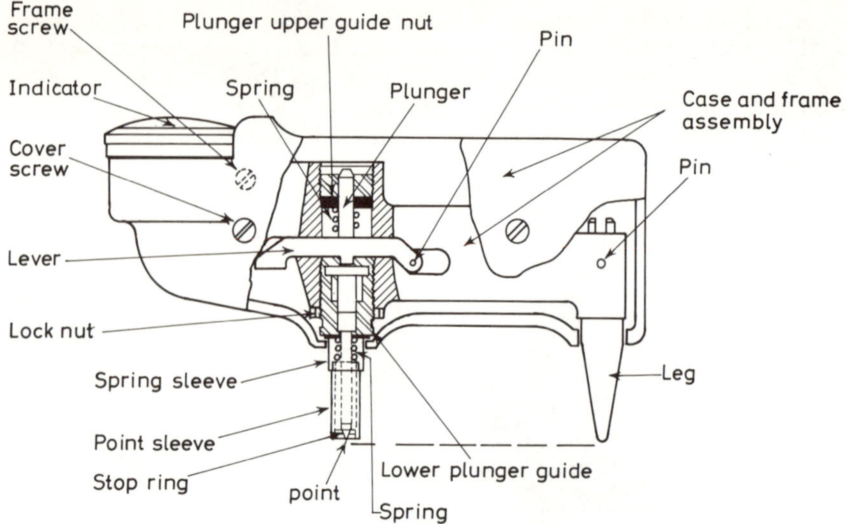

Figure 9.3. Barcol impressor (A.S.T.M. D.2583)

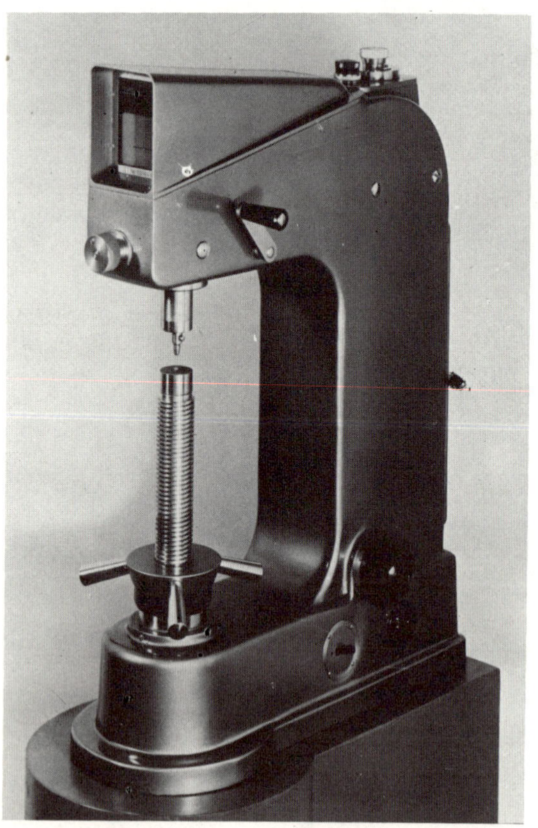

Figure 9.4. Galileo Rockwell and Brinell hardness tester type A200 (Courtesy Metallurgical Services Laboratories Ltd.)

hand pressure to ensure firm contact with the test specimen and the highest dial reading is noted; the number of readings necessary ranges upwards from five for homogeneous materials.

Another A.S.T.M. standard procedure makes use of the well-known *Rockwell hardness tester* and is described in A.S.T.M. D.785–65; a typical commercial apparatus is shown in Figure 9.4 which is also designed to carry out determinations of Brinell hardness.

Five scales are included in A.S.T.M. D.785 (see Table 9.1).

Table 9.1 ROCKWELL SCALES (A.S.T.M. D.785)

Rockwell hardness scale	*Minor load* kg	*Major load* kg	*Indentor diameter* in	cm
R	10	60	0·5000 ± 0·0001	1·27000 ± 0·00025
L	10	60	0·2500 ± 0·0001	0·63500 ± 0·00025
M	10	100	0·2500 ± 0·0001	0·63500 ± 0·00025
E	10	100	0·1250 ± 0·0001	0·31750 ± 0·00025
K	10	150	0·1250 ± 0·0001	0·31750 ± 0·00025

The scales overlap to a certain degree—the test increases in severity down the table—but correlation of scales is stated not to be desirable.

The test specimen (sheet 0·6 cm thick) rests on a flat anvil of at least 5 cm diameter. Two procedures are described.

In Procedure A, essentially the minor load is applied for 10 s and then the major one for 15 s; the hardness reading is taken off the scale 15 s after the major load has been removed, but with the minor loading still operating. In Procedure B, which unlike A is limited to scale R only, the essential principle is that indentation is recorded 15 s after application of the major load, but with the latter still applied. The test values are quoted as α (alpha) Rockwell Hardness Numbers and are obtained by subtracting the indentation reading from 150.

Further information on the Rockwell test as applied to plastics will be found in References 9 and 10, the latter including three scales not included in the above table but applicable to plastics.

GERMANY

DIN.53456 employs a tester with a spherical indentor of 5 mm diameter; the specimen is 4 mm thick and a general instruction about avoiding edge effects is included. A preliminary load of 1 kp is applied carefully and the dial gauge measuring indentation is set to zero. The testing load is then applied within 2 s, but again 'without shock'; this load may be 5 ('A'), 13·5 ('B'), 36·5 ('C'), or 98 ('D') kp according to which gives a (net) indentation between 0·13 and 0·35 mm. The indentation h is in fact measured 10 and 60 s after the testing load F has been applied and is read to within 0·005 mm. 10 tests are carried out.

$$\text{Ball-indentation hardness } H = \frac{1}{\pi D} \cdot \frac{F}{h} = \frac{0{\cdot}064F}{h}$$

where D = ball diameter (5 mm)
F = testing load, kp
h = indentation depth, mm

and H is annotated with the measuring delay time, and the test load, e.g. H_{A60}.

DIN.53505 covers the measurements of *Shore hardness* (A, B and C) of rubber.

9.2.3 Some Other Methods

There are many other arbitrary hardness scales available for rubber and plastics and many which, fortunately, over the years have fallen into disuse. No attempt has been made to produce a comprehensive survey of all these methods and attention is directed below to only the more important and the more recent.

The *Brinell* method already mentioned uses a spherical indentor and the hardness number is calculated from the formula:

$$HB = \frac{2F}{\pi D^2 \left\{ 1 - \left[1 - \left(\frac{d}{D} \right)^2 \right]^{\frac{1}{2}} \right\}}$$

where F = Load, kg
D = Diameter of indentor, mm
d = Diameter of impression produced by indentor during period of 15 s, mm

The indentors used are normally 1, 2, 5 or 10 mm diameter and loads are selected to give F/D^2 ratios of 30, 10, 5 or 1; the method was originally designed for metals and therefore F/D^2 ratios of 5 or 1 are usually needed for plastics materials. The *HB* value is quoted with the F/D^2 ratio employed in the test.

The fundamental objection to the test for plastics materials is their recovery after removal of the load, and to some extent the indefinition of the indentation during test; the use of carbon paper (Section 9.2.1) affords some means of overcoming the former problem. Hounsfield[17] suggests the use of a Tensometer (Chapter 8) to record the depth of penetration of the ball during test, but points out that the general principle is unsound as load is not proportional to depth of penetration; he prefers the use of a *paraboloid* indentor where the cross-section is proportional to the distance from the apex, i.e. the depth of penetration. He shows plots of load against depth of penetration for 'Perspex' and (low density) polyethylene which are very nearly linear. The apparatus depicted in Figure 9.4 has conversion scales relating diameter of indentation to depth of penetration of some of its ball indentors, so that Brinell hardness can be obtained essentially whilst the load is still applied and the problem of recovery is thus overcome.

Brinell testing of metals is covered by B.S. 240: Part 1: 1962 and A.S.T.M. E.10–66, the details of which differ in some degree to the statements made above.

The *(Vickers) diamond pyramid* test is another standard technique for metals which has been borrowed for hard plastics, from time to time. Its use for metals is described in B.S. 427: Part 1: 1961 and A.S.T.M. E.92–67. The test employs a right diamond pyramid on a square base, with an apex angle of 136° between opposite facets, the origin of which was to simulate one indentor in the Brinell range. The test is conducted in similar manner and the test values computed as follows:

$$HV = \frac{2F \sin(\theta/2)}{d^2}$$

where F = applied load, kg
d = mean diagonal width, mm
θ = apex angle of the pyramid = 136°

i.e.

$$HV = \frac{1{\cdot}844F}{d^2}$$

A 5 kg load is normally used for plastics.

Campbell[18] has conducted an investigation of the variables encountered

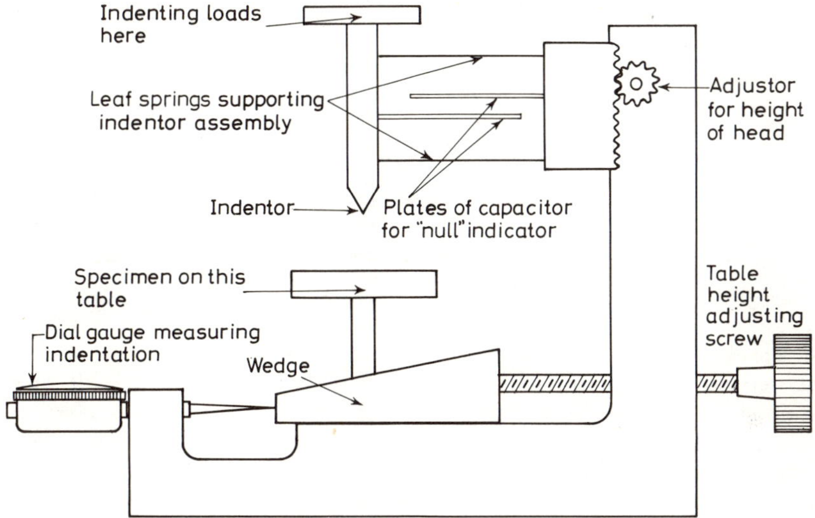

Figure 9.5. Diagrammatic view of Wallace micro-hardness tester

in the Vickers test and included acrylics in his assessment of time effects, etc.

The Vickers diamond principle is used in the *Wallace micro-hardness tester*[19, 20] The apparatus is shown diagrammatically in Figure 9.5.

The loads employed are very small (a fraction of a gram up to 3·5 kg) and the indentations correspondingly so; the test figures necessarily relate only to the surface of the test material, but the test is virtually non-destructive. The table on which the specimen is mounted is raised by the adjusting screw so that the indentor is returned to the exact position it occupied before the major load was applied. The indentor assembly is supported by leaf springs as shown, so that friction is eliminated, and as the principle is to return the indentor to the 'null' position, the leaf springs do not affect the result.

The 'null' position used to be identified by means of headphones into which two oscillatory circuits were fed, beating at nearly the same frequency. One of the circuits contained the capacitor shown and matters were so arranged that the frequencies were identical at the 'null' position, so that the latter would be readily identified by ear. More recently an electronic 'null indicator' has been substituted. The wedge system and a suitable dial gauge permit indentation to be measured to 0·00001 in, clearly illustrating just how small the total indentation is in the test.

The same apparatus, but with a ball indentor $0{\cdot}395 \pm 0{\cdot}005$ mm diameter, a contact force of $8{\cdot}3 \pm 0{\cdot}5$ mN and an indenting force increment of $145 \pm 0{\cdot}5$ mN, is used for the micro-hardness testing of rubber (B.S. 903: Part A26: 1969, Method M). The small size of the specimens required allows hardness measurements of small O-rings and irregularly-shaped articles.

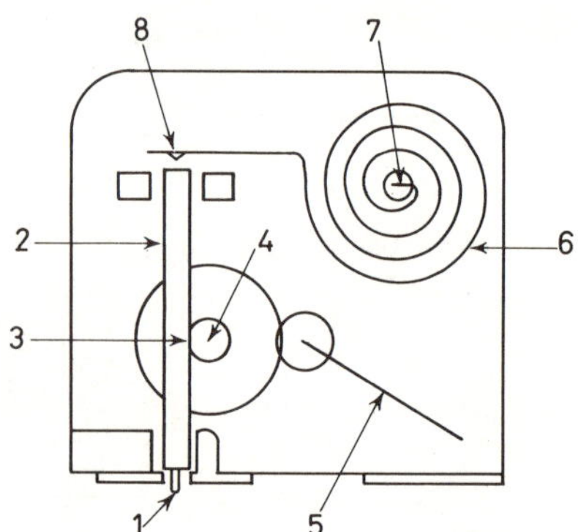

Figure 9.6. Mechanism of Wallace pocket hardness meter (B.S. 2719)

The results are read directly in 'International Rubber Hardness Degrees' but depending on a number of variables may not compare with those obtained from the already-mentioned Normal Method, N, of B.S. 903: Part A26: 1969. The micro-test is best used as a method in its own right.

Likewise use of a suitable dial gauge will give a 'micro-test' suitable for softness numbers of PVC, but the agreement with B.S. 2782, Method 307A (see above) is not good and this variation is *not* standardised.

The microtest for rubber has been investigated, as to its suitability for I.S.O. standardisation, by Morris and Holloway[21] using a variation of a design by S. Oberto[22].

Before leaving the closely allied subject of rubber hardness, mention must be made of B.S. 2719: 1956, 'Pocket Type Rubber Hardness Meters (Methods of Use and Calibration)'. One of the instruments described is the Wallace pocket meter, the mechanism of which is shown in Figure 9.6 to illustrate one method of using springs for this purpose.

'The indentor 1 is fixed to a shaft 2, having a rack 3 rotating a pinion 4, which transmits movement through a train of gears with hairspring to the spindle carrying the pointer 5. Upward movement of the indentor and shaft is resisted by the clock type spring 6, fixed at its inner end on the split spigot 7, and having a "pip" on its free end 8 bearing on the top of the shaft'.

Spigot 7 is part of a circular plate on the back of the instrument, rotatable by means of a key to loosen or tighten the spring.

Knoop Hardness Number is obtained by using a Knoop indentor on the commercial instrument known as the 'Tukon' tester[23] which is also capable of carrying out the Vickers (136°) diamond test. The Knoop indentor is again a diamond, but with the ratio of long and short diagonals approximately 7:1 and, as a result, the depth of indentation is only about $\frac{1}{30}$ th of its length. Recovery problems are less[9] and only light loads need be applied, rendering the liability to shatter of brittle materials less than in the 136° test.

The *T.N.O. hardness tester*[24] employs an indentor of a sapphire polished into the form of a Vickers diamond pyramid and uses a capacitive technique for determining depth of penetration.

The *Sward* hardness test has been suggested for evaluating plastics materials non-destructively, its original development being for the examination of the hardness of paint films[25].

The indentation testing of flooring materials is considered in detail by Gavan and Wein[26].

The *Scleroscope* is sometimes described as giving hardness values, but it is essentially a rebound (resilience) test. Likewise *Moh's* test and *pencil hardness* are really for scratch hardness and are accordingly described in Section 9.3. As an introduction thereto may be mentioned the article by Boor et alia[27] which compares results from a number of the above hardness tests (several Rockwell scales, Knoop, Barcol etc.) with those from scratch hardness (Bierbaum, pencil (Kohinoor) and Moh) and abrasion (A.S.T.M. Mar and Taber), examining a wide range of plastics materials. It is concluded, not surprisingly, that they mostly measure different parameters and therefore cannot be expected to give results which compare—even in placing materials in order of merit. Less comprehensive comparisons will be found in References 8 and 9.

From time to time hot penetration tests have been suggested for checking the degree of cure of thermoset plastics (e.g. see References 28 and 29); the relationships, if any, are essentially empirical and subject to the influence of many compositional variables.

9.3 SCRATCH HARDNESS

Almost as old as using the index finger or the thumb-nail to assess hardness is *Moh's scale,* whereby the 'hardness' of a surface is measured by finding which of a defined range of minerals, of increasing hardness, causes the test surface to be damaged. The test is not recommended and scarcely used now for plastics or any other materials: if details are required, reference may be made to the out-of-date B.S. 1344:1947, 'Method of Testing Vitreous Enamel Finishes' (page 12). Nine materials are used: Talc, Selenite, Calcite, Fluor-spar, Apatite, Felspar, Quartz, Topaz and Corundum, and the powders used should be between 60 and 72 mesh (B.S. 410).

Very slightly more scientific is the pencil hardness test[30] which makes use, in a similar way, of the increasing hardness of draughtsmen's pencils, through from 6B (softest) to 9H (hardest).

This test is also examined by Woodruff[31] in its application to the assessment of the hardness of dried paint films. Two techniques are described: (a) 'disbonding' or chipping out the film, and (b) 'indentation' of the film, in each case ascertaining the grade of pencil which just causes damage. An apparatus, for applying the pencil at a fixed angle, under a fixed load and at a standardised speed, is described.

Probably the most widely-known test in this category is that due to *Bierbaum*[32] which at one time appeared in the A.S.T.M. standards as A.S.T.M. D.1526–58T, but has since been discontinued, possibly due to lack of reproducibility. The basic principle of the test is that a diamond is drawn, under a vertical load, steadily over a horizontal test surface:

$$\text{Bierbaum Scratch Hardness} = \frac{\text{Load (kg)}}{w^2}$$

where w = measured width of scratch, mm.

In the A.S.T.M. method a diamond indentor taken from the corner of a cube was used (Figure 9.7) and drawn across the test surface at 0·25 to 0·30 mm/s under a load of 3 g ± 10 mg.

The width of scratch produced was measured by means of a microscope. Haward[33] deals with the subject at some length and in two articles[34, 35]

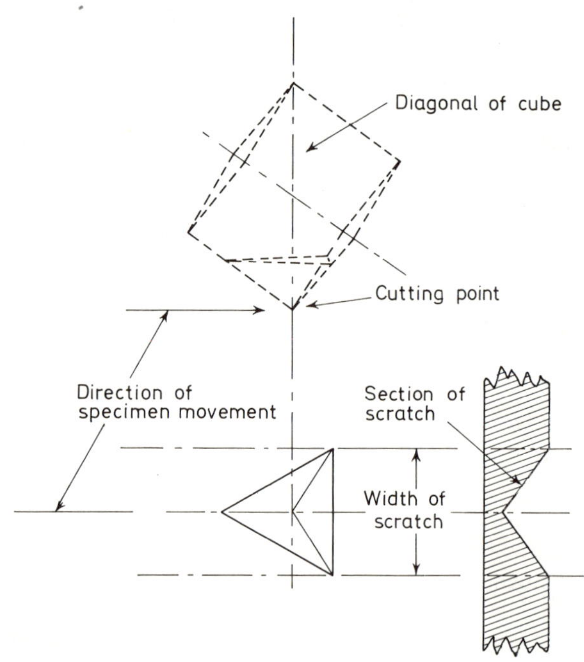

Figure 9.7. Scratch indentor (A.S.T.M. D.1526-58T)

Bernhardt first traces a history of scratch hardness tests and secondly considers tests suitable for plastics.

Bierbaum scratch hardness should not be confused with the scratch hardness test used for paints, where a hardened steel needle is drawn across the surface and the extent of damage noted (see, for instance, Part E2,

'Scratch Resistance Test', of B.S. 3900:1966, 'Methods of Test for Paint', based on Method No. 14 of Ministry of Defence specification DEF.1053). Complete breakthrough to the metal substrate may be identified by completion of an electrical circuit between the substrate and needle, lighting up an indicator lamp, for instance.

9.4 ABRASION AND MAR RESISTANCE

9.4.1 General Comments

Abrasion resistance, or resistance to surface damage caused by wear or rubbing, is of obvious importance in many important uses of plastics—flooring, shoe soling, bearings, gear wheels, to name a few. Many attempts have been made to develop laboratory tests to measure this property, but the efforts have been mainly ad hoc and isolated in approach. As a result there are many tests available, and very many more are used in the rubber industry where the bulk use of elastomers in car tyres has given the necessary impetus. Not surprisingly, the various tests do not give comparable results—abrasion is a complex mixture of mechanically degradative effects of which perhaps one or at most a few may be simulated in any given test—and probably no test in existence permits the prediction of wear life. Perhaps less obviously, the various tests assign different orders of merit to a given range of materials.

A review of early test methods for plastics is given by Haward[33] and a list of references to many standardised and other methods is provided by Lever and Rhys[36]; Scott[37] reviews the methods available for rubber and Haldenwanger[38] those for plastics, itemising some 44 techniques. Gouza[12] also describes a wide range of tests for wear of plastics, as does Gavan[39] in a later volume of the same series.

Some idea of the complexity of establishing a standard test suitable for an important application of plastics will be gained by a study of the work carried out at the Building Research Station over the last few years. The factors influencing the rate of deterioration of a flooring material are numerous. The following are some of the variables involved: force applied, nature of force, presence and nature of any dirt present, humidity and temperature.

It is not difficult to imagine that rates of wear will be quite different according to whether, for instance, the surface suffers only straight line walking or is at a point where the passer by turns on his feet round a corner; whether it is in the innermost parts of a large building and invariably dry or near the door to a street and frequently wet: whether it is kept chemically clean or remains permanently covered with a sharp gritty grime. It is further not difficult to envisage that materials as widely different as rigid vinyl, rubber, linoleum and wood will react each in a unique manner to the influences of these variables, so that no one 'order of merit' will hold for all possible user conditions, still less will comparative quantitative data be obtained. It is a fact that there are enough 'standard' abrasion methods, employing various test conditions, to be able to choose at least one to give any selected flooring the most advantageous rating. The more reliable methods, e.g. the Taber Abraser (see below), may be employed safely only to compare similar materials, which effectively rules out assessing say, rigid vinyls against flexible vinyls of the true 'plastics family' even if the borderline 'Hypalon' is ignored.

At the present moment there is no reliable substitute in floor wear testing for the well organised user trial and the only acceleration feasible is obtainable by using areas which encounter heavy traffic, as for example underground station booking halls. Nevertheless, the Building Research Station has made a careful study of the forces which bear on a floor[40] with the aim of gaining a thorough insight into what physical processes are involved so that a meaningful laboratory test may be evolved. At the time of writing a prototype is under evaluation which promises to be successful if somewhat costly!

Comparisons of the data obtained from various abrasion tests applied to flooring materials will be found in References 41–43, all of which confirm the above comments on the value of any single test only too well. In passing, it is of interest to note that at the Open Days of the Building Research Station held in September 1966, some data were presented indicating a significant correlation between the logarithm of the 'work done', obtained from a fairly slow tensile test, and wear loss of the same material in practical trials on flooring. 'Work done' was computed from the area under the load–extension curve and the idea followed an article by Ratner[44]. Two thermoplastics tiles, two linoleums, four flexible PVC sheets, a vinyl asbestos sheeting and rubber were compared.

9.4.2 Standard Tests

Abrasion tests do not feature very strongly in U.K. Standards, though the well-known Taber method appears in one form in B.S. 3794: 1964 'Decorative Laminated Plastics Sheet'; the basic apparatus is described in Reference 45 and, in its latest version, from Reference 46. B.S. 3794 employs the method on a comparative basis against standard reference sheet material* and the result of 'wear resistance' is calculated from the relative amounts of rubbing required to produce the same degree of visual wear. (For further details, see below).

A form of abrasion test is contained, for instance, in B.S. 1763: 1967, 'Thin PVC Sheeting (Flexible, Unsupported)', by referring to Method 310B of B.S. 2782: 1970, 'Methods of Testing Plastics', which is for determining *print adhesion*. The apparatus is as shown in Figure 9.8.

The abrading member A is a brass peg the centre line of which is at right angles to glass plate B over which the test surface is clamped by grips C and D. The brass peg exerts a force of 9N (907 gf; 2 lbf) on the test piece, through a piece of carefully specified bleached cotton fabric secured round the peg. The rate of reciprocation is 15 ± 2 cycles (each of two strokes) per minute.

Test pieces 229 mm (9 in) × 51 mm (2 in) are used and first washed with

*Obtainable from the British Standards Institution.

a soap solution, rinsed, dried and conditioned. The number of cycles to cause visible damage of the print is recorded.

A.S.T.M. D.1044–56 covers measurements of abrasion of transparent plastics using the *Taber Abraser* with 'Calibrase' CS-10F abrading wheels. The basic principle is the rotation, at a prescribed speed, of test discs 4 in diameter, or 4 in squares, under the abrading wheels freely resting thereon with loads acting on them of 250, 500 or 1000 g. To obtain abrasion the wheels are misaligned. Abrasion resistance is measured by the amount of transmitted light which is diffused by the abraded track, using the A.S.T.M.

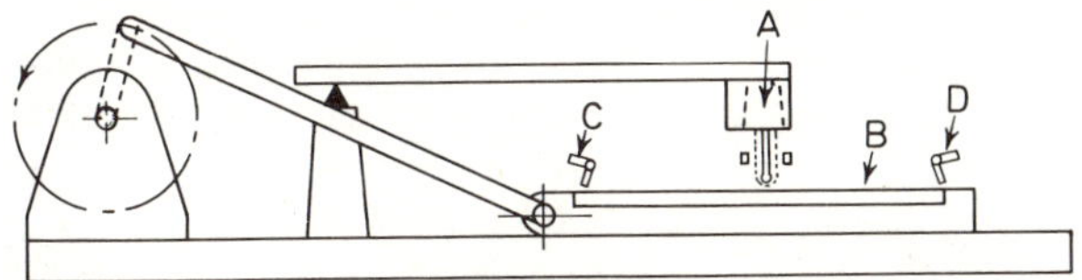

Figure 9.8. Print adhesion test (B.S. 2782)

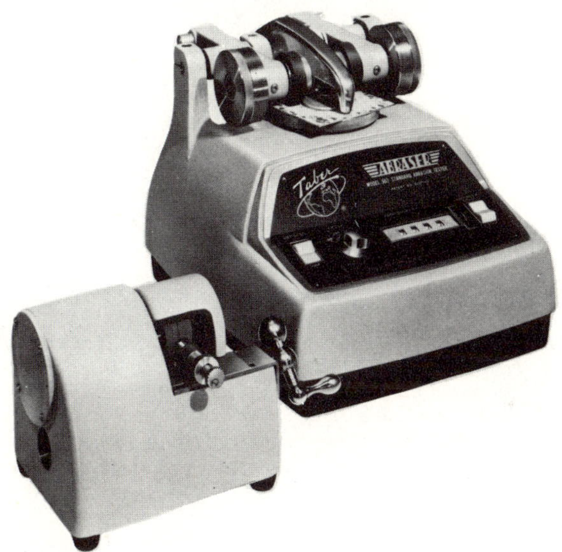

Figure 9.9. Model 503 Taber abraser (Courtesy sole U.K. agents: Funditor Ltd.)

D.1003 'Test for Haze and Luminous Transmittance of Transparent Plastics' (Chapter 16).

The more conventional use of the Taber Abraser is by determining the weight loss of the test disc or square induced by the abrading wheel under a specified load and as a result of a specified number of revolutions—with a particular type of the four variants of resilient wheels ('Calibrase') or four variants of hard wheels ('Calibrade'). This form of test is included in Test LPS-106 of the National Electrical Manufacturers' Association standard for decorative laminated plastics. Great care must be exercised in the use of the wheels, for example they must be kept clean at all times—an air jet usually blows off abraded material during the test—and the flexible wheels have a very limited life. A study of the reliability of the Taber method has been made by Hill and Nick[47] who conclude that it is best to take the mean of a considerable number of test runs; they also correlate results with the hardness of the (flexible) test wheels.

Two methods of testing the abrasion resistance of plastics are included in A.S.T.M. D.1242–56. Method A uses loose abrasive applied under defined conditions to the test surface and results are expressed as volume loss (from weight loss and specific gravity). Method B uses a 'Bonded Abrasive Abrading Machine' (see Figure 9.10).

Essentially the machine is comprised of two independent units. On the right hand side is a vertical unit which carries the test specimens (on P for

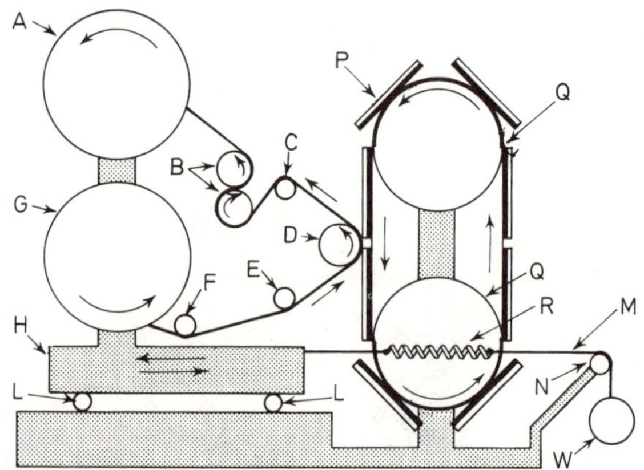

Figure 9.10. Bonded abrasive abrading machine (A.S.T.M.D.1242). A. Take-up drum. B. Constant speed driven rolls. C. Abrasive tape. D. Steel contact roll. E. Slotted guide roll. F. Slotted guide roll. G. Abrasive tape supply drum. H. Carriage. L. Roller bearings. M. Spring cable. N. Pulley. P. Specimen mounting plates. Q. Constant speed driven pulleys with continuous link belt. R. Spring. W. Dead weight

example) to meet the abrasive tape C which rotates in the opposite direction and is kept in contact by being mounted on a carriage H on roller bearings L and tensioned by weight W through spring R. Results are again given in volume loss.

The resistance of glossy plastics to abrasion may be measured by loss of gloss caused by the abrasive action of impacting carborundum grit; such is the principle of A.S.T.M. D.673–44 'Mar Resistance of Plastics'. Abrasive grit is allowed to fall on to the test specimen resting in groove E (Figure 9.11) from a hopper B rotated to produce a controlled rate of feed. The grit falls down tube A on to the specimen and then into receptacle F.

The gloss before and after abrasion is measured with a gloss meter (see Figure 9.12).

The photoelectric receptor is set at 45° and the galvanometer reading taken (I_1); the reading at 60° is then taken (I_2)

$$\text{Gloss, \%} = 100 \; \frac{I_1 - I_2}{I_2}$$

Wiinikainen[48] has described a 'round robin' of scratch resistance and abrasion tests on transparent plastics materials where the latter property, evaluated by increase in haze, looks promising as the basis of a new standard test.

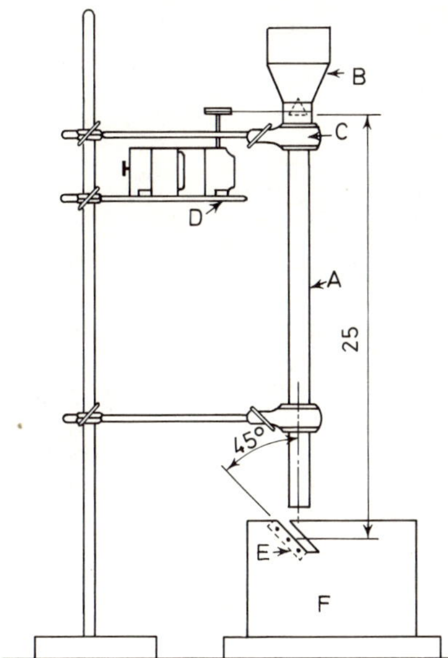

Figure 9.11. Mar resistance abrader (A.S.T.M. D.673). Dimensions in inches

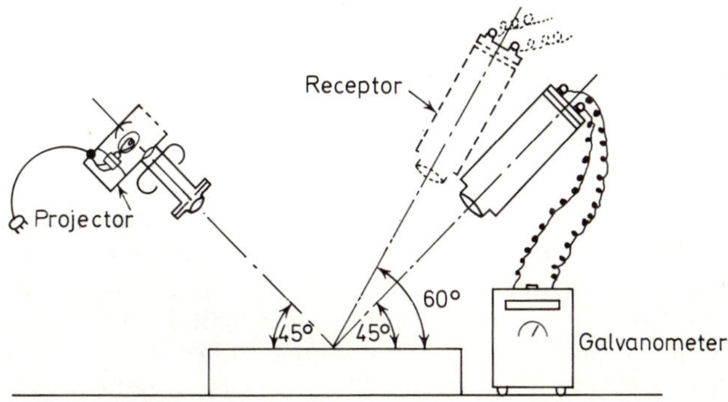

Figure 9.12. Mar resistance gloss meter (A.S.T.M. D.673)

(Further details of gloss and haze measurements in Chapter 16).

Not specifically related to plastics, but highly relevant to the comments made in Section 9.4.1 above, is DIN.51954, describing laboratory tests for 'organic floor coverings'. Two methods are described: (a) a wear test using a Böhme-type abrading disc as covered in DIN.52108 with evaluation based on thickness loss, and (b) a complex test involving cycles of moistening, u.v. radiation, etc. in a so-called Stuttgart wear tester.*

9.5 FRICTION

9.5.1 Introduction

If two surfaces are in contact, W the normal force acting between them and F_s the minimum force required to initiate sliding

$$\frac{F_S}{W} = \mu_S$$

where μ_S is the *static* coefficient of friction.

If F_D is the minimum force required to maintain steady sliding

$$\frac{F_D}{W} = \mu_D$$

where μ_D is the dynamic or kinetic coefficient. (Generally $\mu_S > \mu_D$.)

In the case of a body sliding over a horizontal surface, W is the weight of the body.

The friction of plastics and its relationship to other properties have been discussed by Hulse[49] and Nielsen[50] in general detail. The reader interested in the development of the theory of plastics friction is referred particularly to References 51–54.

9.5.2 Standard Tests

There is a dearth of standard methods of test for friction of plastics. There is one each in the B.S. and A.S.T.M. series, both for film.

Method 311A of B.S. 2782: 1970 is designed to measure both the coefficients of static and dynamic friction of polyethylene film over itself, according to the principles mentioned above. A sled, 150 ± 1 mm long by 100 ± 1 mm wide, carries on its underside a 3 mm thick layer of carefully specified latex foam rubber; the total weight W is such as to apply a pressure of 490 ± 10 N/m^2 ($5{\cdot}0 \pm 0{\cdot}1$ gf/cm^2) to the area on which the foam rests. A specimen of the film, 100 ± 1 mm wide, is attached by adhesive tape to the sled so as to cover the whole of the length of the foam rubber under the sled. This assembly rests on a second piece of the test film held by means of a peripheral vacuum clamp on a flat bed, of dimensions sufficient to allow the sled to travel at least 40 cm and of width sufficient to essentially avoid any effects due to the clamp (55×20 cm is a suitable size for the bed).

The sled assembly is drawn over the vacuum clamped film at a rate of 80 ± 8 cm/min by a driving mechanism into which is built a strain gauge or other suitable device for measuring (a) load A necessary to initiate sliding and (b) mean load B necessary to maintain smooth motion.

*This DIN standard is available in English and the interested reader is recommended to study the complex technique first-hand!

$$\text{Coefficient of static friction} = \frac{A}{W}$$

$$\text{Coefficient of kinetic friction} = \frac{B}{W}$$

The test is conducted at 23 ± 1°C at a relative humidity of 50 ± 5%.

A.S.T.M. D.1894–63 is for plastics film generally, but the principle of the test is precisely as given above; somewhat more variation of the apparatus

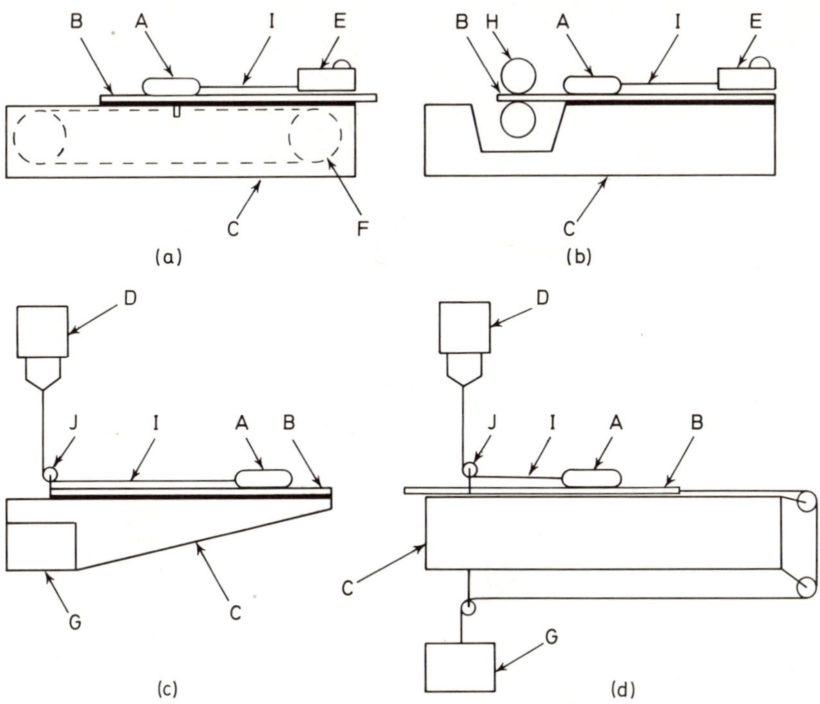

Figure 9.13. Apparatus for measurement of coefficient of friction of film (A.S.T.M. D.1894). A. Sled. B. Plane. C. Supporting base. D. Strain gauge. E. Spring gauge. F. Constant speed chain drive. G. Constant speed tensile tester crosshead. H. Constant speed drive rolls. I. Nylon monofilament. J. Low friction pulley

is allowed (Figure 9.13), for not only is the moving sled/horizontal plane set up standardised, but also one combining stationary sled and moving plane.

9.5.3 Other Tests

In contrast to the paucity of standardised methods of measurement of friction of plastics, there are a number of more refined methods described in the literature; it is not possible to give here more than a brief outline of some of them.

An apparatus which has been commercialised[55] is described by Gough[56].

Basically it consists of a test platform BB′ (Figure 9.14) supported at its four corners by four identical levers. Each lever is connected by a pivot to the corner of the platform and is itself carried by a spindle C, which latter is connected to a fixed main framework. Each lever also carries a counterweight D at its upper end, the mass centre of D being on the line BC extended. Distance CC′ on the fixed framework equals BB′ on the platform so that CC′ and BB′ are always parallel and the four levers are always parallel

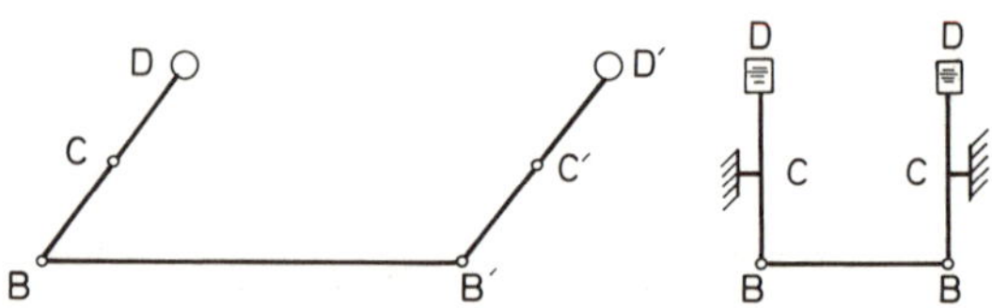

Figure 9.14. Essential features of Dunlop direct reading friction meter (Gough)

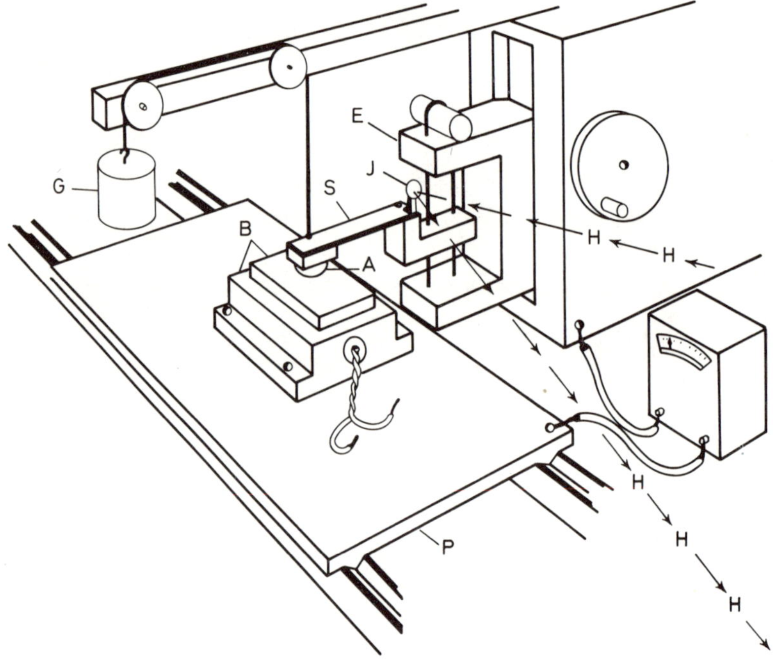

Figure 9.15. Friction apparatus derived from surface grinder (Bowers et alia) (Reproduction by permission of Modern Plastics Magazine, McGraw-Hill, Inc.)

whatever their inclination. Suitable counterweights are placed at D so that the mechanism is in neutral equilibrium.

One of the arms CD moves over a scale to indicate directly the coefficient of friction when the test surface is drawn over platform BB′ by hand, for the coefficient is in fact the tangent of the angle between the position taken up by the lever arms and the vertical.

A surface grinder has been adapted by Bowers et alia[57] who present frictional data of a number of thermoplastics against a variety of surfaces. Figure 9.15 shows the basic features.

'The lower test specimen B is screwed to the table P, which is hydraulically driven at a uniform velocity. To apply the load, the upper ball shaped specimen A attached to the frame E is lowered with the weight G in the position shown until it just makes contact with the plate shaped specimen of platen B. Upon removing the weight, the restoring force in the flat steel spring S causes the two specimens to be pressed together with a force equal to the weight G. The light beam H is reflected from the mirror J attached to the bifilar suspension, and is focused on a modified oscilloscope type magazine camera located 675 cm from the mirror. A tangential force of 150 g will deflect the light beam 1 cm on the camera'. D. I. James[58] describes a somewhat similar method based on a lathe, with the frictional force being balanced against a transducer/proof ring dynamometer.

Frictional 'drag' fed to the secondary winding of a differential transformer via a leaf spring has been used by Lauer and Friel[59] in a machine they

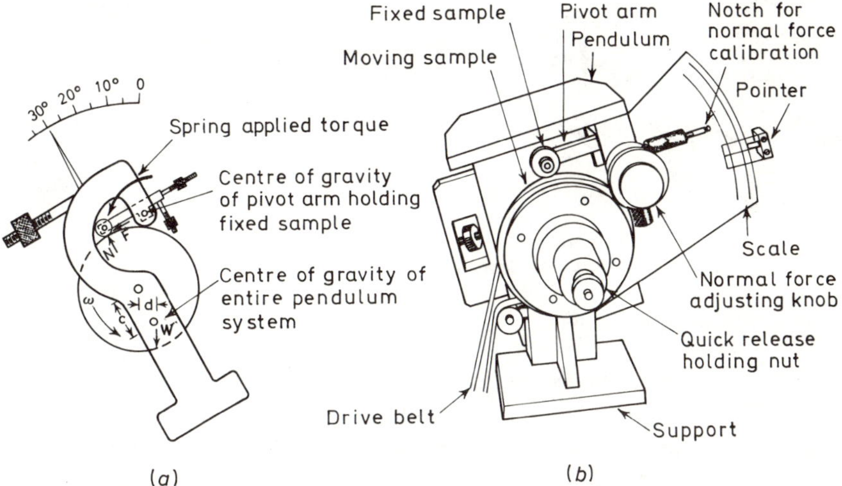

Figure 9.16. Variable speed frictionometer (Westover and Vroom). (a) Principle of operation, (b) support and pendulum assembly (Courtesy The Society of Plastics Engineers, Inc.)

claim has sensitivity over a wide range of values. Marcucci[60] gives data for thermosets and thermoplastics in an article which includes a description of an apparatus designed to investigate the effects of extended periods of contact, high speeds and loadings.

The effect of velocity on the coefficients of friction of a number of thermoplastics has been studied by Westover and Vroom[61] who described 'a variable speed frictionometer', which is commercially available[62, 63]. The essential features of the instrument are given in the two diagrams of Figure 9.16.

A 10 cm diameter drum is rotated under controlled conditions and its surface forms one of the friction members. A tachometer measures the linear velocity of the surface—speeds up to 500 cm/s second are stated to be available. The fixed sample is attached, through a counter balanced pivot arm, to a pendulum which is free to rotate about the same axis as the drum. The pivot point of the pivot arm lies on a tangent to the drum at the point of friction contact 'so that the friction force has no feedback effect upon the normal force which is applied by a spiral spring encased by the adjusting knob. The centre of gravity of the pivot arm system is located at the pivot point so that the pendulum attitude will have no effect upon the normal force'.

Readings are taken directly off a scale calibrated from the equation

$$\mu = \frac{(\text{Max. pendulum moment}) \sin \theta}{5N}$$

where θ = angular pendulum displacement and the maximum pendulum moment and normal force N are fixed.

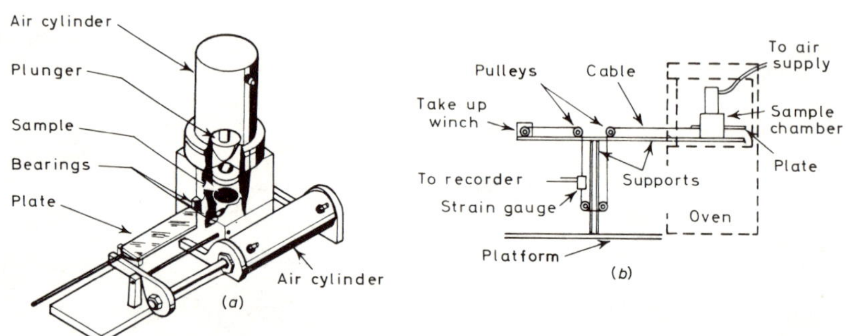

Figure 9.17. Apparatus for measuring friction (Friehe). (a) General assembly, (b) strain gauge/drive mechanism (Courtesy The Society of Plastics Engineers, Inc.)

Finally, an apparatus suitable for operation in an oven, to obtain frictional data over a temperature range is used by Friehe[64] who also gives data for ethylene polymers and copolymers.

The mode of operation will be obvious from Figure 9.17, and all that needs comment is the neat method of applying normal force, by air pressure.

9.5.4 Data

As the data on frictional properties of plastics are scattered far and wide over the technical press, a few selected references (65–74 inclusive) are given for the reader interested in obtaining specific values—these are in addition to those already mentioned, of course. Some articles contain in addition, test methods and/or correlations with other plastics properties. The position has, however, been neatly summarised in Reference 75.

REFERENCES

1. NIELSEN, L. E., *Mechanical Properties of Polymers,* Reinhold Publishing Corporation, New York, 220 (1962)
2. RITCHIE, P. D. (ed.) *Physics of Plastics,* Iliffe Books Ltd., London, 185 (1965)
3. SCOTT, J. R., *Physical Testing of Rubbers,* Chapter 4, Maclaren & Sons Ltd., London (1965)
4. GENT, A. N., 'On the Relation between Indentation Hardness and Young's Modulus', *I.R.I. Trans.,* **34,** No. 2, 46 (April 1958)

5. SODEN, A. L., *A Practical Manual of Rubber Hardness Testing*, Maclaren & Sons Ltd., London (1951)
6. DAVIES, H. E., TROXALL, G. E. and WISKOCIL, C. T., *The Testing and Inspection of Engineering Materials,* 3rd edn, Chapter 7, McGraw-Hill Book Company, New York (1964)
7. FENNER, A. J., *Mechanical Testing of Materials,* Chapter 8, Philosophical Library Inc., New York (1965)
8. MAXWELL, BRYCE, 'Hardness Testing of Plastics', *Mod. Plast.,* **32,** No. 9, 125 (May 1955)
9. LYSAGHT, V. E., 'How to Make and Interpret Hardness Tests on Plastics', *Mater. & Meth.,* **27,** 84 (May 1948)
10. BOOR, L., 'Hardness, Abrasion and Wear Resistance Testing of Plastics', *Bull. Amer. Soc. Test. Mat.* **No. 244,** 43 (February 1960)
11. HALDENWANGER, H., 'The Determination of the Indentation Hardness of Plastics', *Kunststoffe,* **51,** No. 2, 82 (February 1961)
12. GOUZA, J. J., 'Methods of Test for Hardness and Wear of Plastics', *Testing of Polymers',* Volume 2, Chapter 7, (Edited by J. V. Schmitz), Interscience Publishers, New York (1966)
13. LIVINGSTON, D. I., 'Indentation Hardness Testing', *Testing of Polymers,* Volume 3 (Edited by J. V. Schmitz and W. E. Brown), Interscience Publishers, New York (1967)
14. BESSANT, K. H. C., DILKE, M. G., HOLLIS, C. E. and MILLANE, J. J., 'Physical Evaluation of Small Samples of Plastics', *J. appl. Chem.,* **2,** No. 9, 501 (September 1952)
15. WATERS, N. E., 'Variation of Indentation Hardness with Sample Thickness', *I.R.I.J.,* **1,** No. 1, 51 (Jan.–Feb. 1967)
16. Barber Colman Company, Rockford, Illinois, U.S.A.
17. HOUNSFIELD, L. H., *Commercial Testing, Part II, Plastics,* Tensometer Ltd., 81 Morland Road, Croydon CR9 6 HG, Surrey
18. CAMPBELL, R. F., 'Dependence of Diamond Pyramid Hardness Number on Experimental Variables', *Mater. Res. Stand.,* **7,** No. 10, 443 (October 1967)
19. 'Hardness Tester for Plastics', *Brit. Plast.,* **28,** No. 3, 110 (March 1955)
20. BENNETT, F. N. B. and HAYES, R., 'Measurement of the Degree of Cure of Polyester Resins by the Wallace Micro Indentation Tester', *Plastics, Lond.,* **20,** No. 217, 282 (August 1955)
21. MORRIS, R. E. and HOLLOWAY, J. M., 'Evaluation of a Microhardness Tester', *Bull. Amer. Soc. Test. Mat.* No. 222, 45 (May 1957)
22. OBERTO, S., 'New Instruments for Measuring Hardness', *Rubb. Chem. Technol.,* **28,** No. 4, 1054 (Oct.–Dec. 1955)
23. LYSAGHT, V. E., 'The Knoop Indenter as Applied to Testing Nonmetallic Materials Ranging from Plastics to Diamond', *Bull. Amer. Soc. Test. Mat.* **No. 138,** 39 (January 1946)
24. GRODZINSKI, P., 'Hardness Testing of Plastics', *Plastics, Lond.,* **18,** No. 194, 312 (September 1953)
25. ROBERTS, J. and STEEL, M. A., 'The Sward Hardness Number and Mechanical Properties of Plastics', *J. appl. Polymer Sci.,* **10,** No. 9, 1343 (September 1966)
26. GAVAN, F. M. and WEIN, J. T. Jr., 'Indentation and Compression Testing of Floor Coverings', *Testing of Polymers,* Volume 1, Chapter 12, (Edited by J. V. Schmitz), Interscience Publishers, New York (1965)
27. BOOR, L., RYAN, J. D., MARKS, M. E. and BARTOE, W. F., 'Hardness and Abrasion Resistance of Plastics', *Proc. Amer. Soc. Test. Mater.,* 17, 1017 (1947)
28. BENNITT, J. H. and AVENALL, C. E., 'Indentation Testing of Phenolic Mouldings by Brinell and Hot-Needle Methods', *Chemy Ind.,* No. 39, 936 (September 27, 1952)
29. MCSHEEHY, W. H., 'Quality Control of the Cure-State of Encapsulated Electronic Modules', *Plast. Technol.,* **10,** No. 11, 44 (November 1964)
30. SMITH, W. T., 'Standardisation of the Pencil Hardness Test', *Off. Dig. Fed. Paint Varn. Prod. Cl.,* **28,** No. 374, 232 (March 1956)
31. WOODRUFF, H. C., 'Experimental Resurvey of the Pencil Hardness Test for the Evaluation of the Hardness of Dry Films', *J. Paint Technol.,* **38,** No. 502, 691 (November 1966)
32. BIERBAUM, C. H., 'The Microcharacter, Its Application in the Study of the Hardness of Case-Hardened, Nitrided and Chrome Plated Surfaces', *Trans. Amer. Soc. Steel Treat.,* **18,** 1009 (1930)
33. HAWARD, R. N., *The Strength of Plastics and Glass,* Chapter 5, Cleaver-Hume Press Ltd., London (1949)
34. BERNHARDT, E. C., 'A History of Hardness Tests Based on Scratch Resistance Measurements', *Bull. Amer. Soc. Test. Mat.* **No. 157,** 49 (March 1949)
35. BERNHARDT, E. C., 'Scratch Resistance of Plastics', *Mod. Plast.,* **26,** No. 2, 123 (October 1948)

36. LEVER, A. E. and RHYS, J., *The Properties and Testing of Plastics Materials,* 3rd edn, Temple Press Books, London, 101 (1968)
37. SCOTT, J. R., *Physical Testing of Rubbers,* Maclaren & Sons Ltd., London, 120 (1965)
38. HALDENWANGER, H., 'On the Determination of the Abrasion Properties of Plastics Materials, Part 1, Abrasion Testing', *Kunststoffe Rundschau,* **12,** No. 1, 1 (January 1965)
39. GAVAN, F. M., 'Wear Testing', *Testing of Polymers,* Volume 3 (Edited by J. V. Schmitz and W. E. Brown), Interscience Publishers, New York (1967)
40. HARPER, F. C., WARLOW, W. J. and CLARKE, B. L., *The Forces Applied to the Floor by the Foot in Walking, I, Walking on a Level Surface,* H.M.S.O., London (1961)
41. HARPER, F. C., 'The Abrasion Resistance of Flooring Materials. A Review of Methods of Testing', *Wear,* **4,** No. 6, 461 (Nov.–Dec. 1961)
42. 'Performance of Abrasion Machines for Flooring Materials', Report of an Investigation by the International Study Committee for Wear Tests of Flooring Materials, *Wear,* **4,** No. 6, 479 (Nov.–Dec. 1961)
43. 'Note on the Performance of Abrasion Machines for Flooring Materials', *Wear,* **1,** No. 3, 302 (May–June 1964)
44. RATNER, S. B., 'Connection Between Wear Resistance of Plastics and Other Mechanical Properties', *Sovetsk. Plast.,* **7,** 37 (1964)
45. 'Abrasion Testing Set' (Taber Model 140), *Plastics, Lond.,* **21,** No. 229, 277 (October 1956)
46. Funditor Ltd., South Way, Wembley, Middlesex HA9 OHE, and Taber Instrument Corporation, 111 Goundry Street, North Tonawanda, N.Y., U.S.A.
47. HILL, H. E. and NICK, D. P., 'A Study of the Reliability of Taber Abrasion Results', *J. Paint Technol.,* **38,** No. 494, 123 (March 1966)
48. WIINIKAINEN, R. A., 'Scratch and Abrasion Testing of Transparent Plastics', *Mater. Res. Stand.,* **9,** No. 12, 17 (December 1969)
49. RITCHIE, P. D., (ed.) *Physics of Plastics,* Chapter 3, 188, Iliffe Books Ltd., London (1965)
50. NIELSEN, L. E., *Mechanical Properties of Polymers,* Chapter 9, Reinhold Publishing Corporation, New York, 222 (1962)
51. SHOOTER, K. V. and TABOR, D., 'The Frictional Properties of Plastics', *Proc. phys. Soc. Lond.,* **65B,** 661 (1952)
52. PASCOE, M. W. and TABOR, D., 'The Friction and Deformation of Polymers', *Proc. Roy. Soc.,* **A.235,** 210 (1956)
53. HUFFINGTON, J. D., 'Theory of Friction of Polymers', *Res. Appl. Ind.,* **12,** 443 (Oct.–Nov. 1959)
54. BARTENEV, G. M. and ELKIN, A. I., 'The Nature of Friction of Polymers in High Elastic and Glasslike States', *J. Polymer Sci.,* Part C, No. 16, *Polymer Symposia—International Symposium on Macromolecular Chemistry,* Part 3, Prague, 1965, 1673, Interscience Publishers, New York (1967)
55. 'Dunlop Friction Meter', *Rubb. Age Lond.,* **80,** No. 1, 140 (October 1956)
56. GOUGH, V. E., 'A Simple Direct-Reading Friction Meter', *J. Sci. Instrum.,* **30,** 345 (October 1953)
57. BOWERS, R. C., CLINTON, W. C. and ZISMAN, W. A., 'Frictional Properties of Plastics', *Mod. Plast.,* **31,** No. 6, 131 (February 1954)
58. JAMES, D. I., 'Measurement of Friction between Rubber-Like Polymers and Steel', *J. Sci. Instrum.,* **38,** No. 7, 294 (July 1961)
59. LAUER, J. L. and FRIEL, P. J., 'Device for Measuring Friction', *Rev. Sci. Instrum.,* **28,** No. 4, 294 (April 1957)
60. MARCUCCI, M. A., 'Friction and Abrasion Characteristics of Plastics Materials', *S.P.E. J.,* **14,** No. 2, 30 (February 1958)
61. WESTOVER, R. F. and VROOM, W. I., 'A Variable Speed Frictionometer', *S.P.E. J.,* **19,** No. 10, 1093 (October 1963)
62. 'Frictionometer', Model CS–131, Custom Scientific Instruments Inc., 541 Devon Street, Kearny, N. J., U.S.A.
63. BULTMAN, H. J., 'Dynamic Friction Measured Fast, Accurately', *Mater. Engng,* **66,** No. 3, 86 (September 1967)
64. FRIEHE, C. A., 'Static Coefficients of Friction for Polyethylenes', *Polymer Engng Sci.,* **6,** No. 2, 135 (April 1966)
65. THOMPSON, J. B., TURREL, G. C. and SANDT, B. W., 'The Sliding Friction of Teflon', *Plastics, Lond.,* **20,** No. 215, 213 (June 1955)
66. REES, B. L., 'Static Friction of Bulk Polymers over a Temperature Range', *Research, Lond.,* **10,** No. 8, 331 (August 1957)

67. JAMES, D. I., NORMAN, R. H. and PAYNE, A. R., 'The Relation Between Coefficient of Friction and Dynamic Properties of Polyvinyl Chloride', *Physical Properties of Polymers,* S.C.I. Monograph No. 5, Society of Chemical Industry, London, 233 (1959)

68. BUECHE, M. A. and FLOM, D. G., 'Surface Friction and Dynamic Mechanical Properties of Polymers', *Wear,* **2,** No. 2, 168 (February 1959)

69. FLOM, D. G., 'Rolling Friction of Polymeric Materials. II Thermoplastics', *J. appl. Phys.,* **32,** No. 8, 1426 (August 1961)

70. MCLAREN, K. G. and TABOR, D., 'Viscoelastic Properties and the Friction of Solids', *Nature, Lond.,* **197,** No. 4870, 856 (March 2, 1963)

71. BAUMGÄRTEL, K. and RICHTER, E. F., 'Speed-Dependency of Friction Coefficient and Abrasion Behaviour of some Thermoplasts', *Kunststoffe,* **53,** No. 10, 747 (October 1963)

72. BOWERS, R. C. and ZISMAN, W. A., 'Frictional Properties of Polyethylenes and Perfluorocarbon Polymers', *Mod. Plast.,* **41,** No. 4, 139 (December 1963)

73. STEIJN, R. P., 'Friction and Wear of Plastics', *Metals Engng Quart.,* **7,** No. 2, 9 (May 1967)

74. BARTENEV, G. M. and ELKIN, A. I., 'The Friction of Polymers at the Initial Stage of Sliding in Various Temperature Ranges', *Wear,* **11,** No. 6, 393 (June 1968)

75. 'The Frictional Properties of Thermoplastics', I.C.I., *Plastics, Lond.,* **26,** No. 281, 117 (March 1961) See also: Technical Service Note G.111, Imperial Chemical Industries Ltd., Plastics Division, Welwyn Garden City, Herts.

10

Electrical Properties

10.1 THE FUNCTION OF PLASTICS AS ELECTRICAL INSULATORS AND SOME DEFINITIONS

10.1.1 Introduction

The primary function of electrical insulation is to separate metallic conductors at different potentials. A simple example such as two mounted terminals can usefully serve as an illustration. The insulation must maintain the separation under a wide variety of environmental hazards which include humidity, temperature, vibration, radiation, the presence of gases, moisture and other contamination, and be sufficiently strong physically to withstand the mechanical forces which may be exerted by the conductors and sufficiently strong electrically to withstand the applied voltage and resulting current flow, which may itself give rise to secondary effects, such as temperature rise, mechanical stress and electrochemical action.

The ability of an insulator to resist the passage of direct current is not necessarily synonymous with its ability to resist alternating currents, and whereas in the former case the resistivity of the material is of prime importance, in the latter the power factor and permittivity are more relevant since they determine the power loss occurring in alternating fields, which loss may vary enormously with applied frequency. The three properties referred to may be loosely described as weak field phenomena since they are mainly relevant under conditions of low electric stress, where they are substantially independent of the stress and unchanged by its application. More importantly, these properties are specific and characteristic of the material, and from the measurement viewpoint are largely independent of the shape and geometry of the test piece employed.

10.1.2 Definition of Resistivity[1]

The *volume resistivity* of a material is simply defined as the electrical resistance between opposite faces of a unit cube, whereas the *surface resistivity* may be defined as the resistance between surface mounted electrodes of unit width and unit spacing (volume currents are assumed to be minimal by appropriate electrode design and measuring technique). In passing, it may be noted that *insulation resistance* is the resistance measured between any two particular electrodes mounted on or in an insulator, by inference the arrangement being such that neither the volume nor surface currents are known nor can they be calculated, so that the property in general is a complex function of both surface and volume resistivities and the specific geometry of the system.

10.1.3 Definition of Power Factor* and Permittivity[2]

The properties of *power factor* and *permittivity,* loosely coupled as the *dielectric properties* (dielectric is synonymous with insulation), are not strictly dependent upon one another, although closely related. The power factor is due to the translation or rotation of dipoles and the movement of charged ions in an alternating field, and this field causes the polarisability or charge displacement of such dipoles and ions and also atoms, contributing to the permittivity. The power factor is that fraction of the electrical energy applied to a material which is dissipated as heat and is therefore non-recoverable electrically. The *relative permittivity* (formerly *dielectric constant*), may be defined practically as the ratio of the charge which can be stored by a sheet of insulating material filling the space between two metallic plates to that which can be stored when the material is replaced by air (strictly vacuum), all other things being equal. The adjective 'relative' is omitted henceforth but is understood to be implicit.

In the majority of applications it is important to restrict the current in the insulation; thus under d.c. conditions it is evident a high resistivity material will permit the use of a smaller amount of insulation to achieve a given resistance path, and similar economy will result under a.c. conditions by employing a low power factor material. It should be noted here that, although the power factor is a fraction representing the ratio of energy lost to energy applied, the actual power dissipated in a material placed in an alternating field is proportional to the product of power factor and permittivity and frequency, all else being constant; therefore a low permittivity is also essential where the power available is small as in many telecommunication applications, particularly at the highest frequencies.

10.1.4 Importance of High Permittivity Materials

The major field of application for dielectric materials where a high permittivity is the chief factor is in capacitors. Here the emphasis is continually towards reduction in physical size and since the capacitance of a given electrode/dielectric system is proportional to the permittivity of the material (capacitance being defined as the ratio of charge to applied voltage), the importance of a high permittivity need hardly be stressed. As far as solid plastics materials as capacitor dielectrics are concerned, the available range of permittivity is approximately 2–20, so that the possibilities of size reduction by choice of permittivity alone are limited, and these possibilities are reduced even further when power factor is considered, since virtually all the

*See p. 224 for definition of the closely related term loss tangent (dissipation factor in the USA and on the Continent), and IEC Publication 250 (Reference 2) for precise definitions.

low power factor materials have permittivities at the lower end of the permittivity range. Capacitance is also inversely proportional to the dielectric thickness and so the only real direction of advance at present is in the use of thinner films; plastics foils of thickness 1 μm or less have been developed and used to a limited extent, limited since the problems of handling and manipulating such material are severe, and some work is in progress on the vapour deposition of polymeric dielectric films in vacuo.

10.1.5 Application of High and Low Loss Materials

It has already been mentioned that in many insulation applications power losses are important (and undesirable) and low values of both power factor and permittivity are desirable. Since capacitance is proportional to permittivity and in most cases the capacitance of a given conductor/insulator system is an unwanted effect which must be tolerated, the capacitance must be kept low by using low permittivity material. In screened cables, for example, of small diameter and hence low thickness of dielectric, it is fairly common practice to use dielectrics of cellular plastics materials to reduce the capacitance, and such materials may have permittivities approaching 1, the value for air. The problem of unwanted capacitance becomes more acute with increasing frequency since for practical design reasons circuit values of capacitance decrease.

One application where high dielectric losses may be preferred is in the radio frequency heating field, where, for example, synthetic resin adhesives used in furniture and cabinets may be cured by the application of strong fields. The use of water soluble adhesives such as those based on urea–formaldehyde eases the problem since the permittivity of water is approximately 80 and it has a high power factor also. Another rather more specialised application is in energy absorbent materials for use at microwave frequencies.

10.1.6 Definition of Electric Strength

So far, we have discussed only weak field phenomena and those properties which are fairly well defined and constant for given environmental conditions, and the tests themselves do not cause any irreversible alteration to the insulation. Under the influence of relatively strong fields, however, plastics materials may be permanently altered, either in form or character, and under continual stress degradation may occur and ultimately lead to failure. The main properties which will be covered under this heading are those of electric strength, tracking and arc resistance.

Electric strength is defined either in terms of the field strength (voltage divided by thickness) required to produce failure of the insulation or the voltage stress which a material can just survive without failure, under given experimental conditions. The difficulties associated with the definition and the measurement of the property arise because of the importance of a large number of variables which affect the result, and because of the number of modes in which breakdown can occur, all or some of which may contribute to failure in a given situation. Breakdown is a disruptive phenomenon and as such is very dependent upon the macroscopic, and even microscopic, structure of the insulation. It may occur through dielectric heating causing an excessive temperature rise, and with many materials power factor increases with temperature, resulting in further heating and ultimately thermal 'runaway'; the voltage stress may be sufficiently high to accelerate electrons through the material, causing an electron avalanche; gaseous voids inside the material may ionise, causing bombardment of the material and subsequent degradation; the stress may even be high enough to break the chemical bonds in the polymer structure and cause decomposition into less resistant materials, and so on.

Breakdown is a 'weak point' phenomenon and as such it may be instructively compared with, say, a tensile strength failure on a fabric specimen where one thread invariably parts first, thus initiating rupture; similarly breakdown commences where the stress is highest and one of the major difficulties in electric strength tests lies in producing a uniform field in the test specimen. Irregularities in electrode surfaces, and in the specimen itself due to thickness variation or inhomogeneity, difficulty in avoiding spurious discharges in the surrounding media which may cause degradation, and the basic fact that field strength near an electrode edge increases inversely as the electrode curvature radius—all these factors make precise measurements impossible and comparisons between materials valid only when experimental conditions are very strictly controlled. Under the most carefully controlled conditions, when all spurious discharges are eliminated and all secondary mechanisms of breakdown are avoided, it is possible to achieve a consistent value for a material, known as the *intrinsic electric strength*, which is independent of thickness, and rather surprisingly, the range of values obtained for many homogeneous plastics materials is small and of the order of 0·5–1 MV/mm (12 500–25 000 V/mil; 1 mil ≡ 0·001 in). Such measurements are of great interest to the theoretician, but in practice industrial measurements are usually made under relatively crude conditions, in order to simulate more closely practical situations, and breakdown can occur by a number of modes, sometimes operating simultaneously. Since breakdown is a time-dependent phenomenon, and the rate of voltage application is analogous with rate of loading in the tensile example given earlier, it is important to standardise on the time of voltage application and even the rate of initial voltage increase, and so-called industrial electrical strength tests are careful to specify such details. The results of such tests are very thickness dependent in contrast to intrinsic strength tests, and the values obtained in general rarely approach even one-tenth of the intrinsic values.

10.1.7 Tracking and Arc Resistance

The remaining high-stress properties which are of practical importance are those of *tracking* and *arc resistance* and discharge resistance, which may be considered together since they all involve disruptive tests concerned essentially with surface failure of insulation as distinct from the breakdown through the volume of the material in the electric strength tests. It is beyond the scope of this handbook to consider these properties in detail, as the phenomena are complex, involving as they do the state and nature of surfaces and the influence of extraneous contaminants. Tracking may be defined generally as the failure of an insulator caused by the application of an electric field to its surface, resulting in the formation of a carbonised conducting path. Since the presence of carbon in the insulator is a prerequisite for this type of failure, it might be expected that plastics in general would be prone to tracking, but this does not follow and some of the richest carbon containing polymers such as polystyrene are non-tracking, although other adverse effects such as erosion occur in this particular case. Tracking is influenced considerably by the presence of surface films of moisture and dirt, etc., and the present test methods invariably involve the introduction of conducting media such as salt solutions on to the surface in the form of droplets or by dipping, in order to produce an acceleration of the natural processes. Although tracking may occur in a variety of practical situations, two common extremes being the failure of domestic light switches and car distributor caps, the problem is more usually associated with areas of high atmospheric pollution and those near the coast and on board ships, where salt spray is a natural hazard.

In arc resistance tests, an arc produced between two surface mounted electrodes is allowed to play along the surface of the insulation so that the predominant mode of failure is due to thermal degradation. However, the failure is accelerated by the surface carbonisation that occurs resulting in current actually flowing in the surface layers of the material. This property is important in high current switchgear and circuit breakers where arcs occur on making and breaking contacts.

In the vicinity of high voltage conductors ionisation of the air occurs due to the high local stresses, and insulation subject to prolonged exposure to the consequent discharges that are present may fail through degradation or erosion. Discharge resistance is therefore of practical importance in high voltage bushings and insulators, which now often utilise epoxide resins and other thermosets in addition to the more common porcelain or ceramic materials.

Since appreciable amounts of plastics materials are used as insulation for both low and high voltage applications outdoors, the problems of selection and improvement of existing materials and the development of new ones are complex and the testing procedures rigorous, as failure under electrical stress may involve any one or all of the various modes so far discussed.

10.1.8 Miscellaneous Tests

The properties outlined above are considered to be the prime ones and in all applications of plastics involving electrical insulation, some knowledge of one or more of these properties is required to ensure the proper and efficient use of the appropriate material. There are, however, many other properties, and test methods therefor, which have greater or lesser relevance in a given application and some mention of these is appropriate here. Examples are the electrolytic corrosion resistance of electrical insulating tapes with respect to copper and other wires, the compatibility of various encapsulating resins with components and conductors, the solvent extractability tests which give a measure of the ionic activity of insulation and the so-called 'antistatic' tests which assess the ability of insulating materials to build up static charge. This last consequence of the generally excellent insulating properties of plastics materials may have disastrous consequences, for example if a high static charge is built up between the rubber treads of a hospital trolley and a PVC floor and the charge sparks to earth in an atmosphere heavily laden with ether.

10.1.9 General References

The following general books which deal with the electrical properties of polymers will be found useful (with specific authors in brackets).
Physics of Plastics edited by Ritchie[3] [Parkman]
Plastics Materials by Brydson[4]
Physical Testing of Rubbers by J. R. Scott[5]
Testing of Polymers Vol. 1 edited by Schmitz [Sharbaugh[6], Scott[7], Tucker[8], Warfield[9], Dakin[10], and Brunton[11], contributing chapters on different aspects of electrical tests].

Fairly elementary descriptions of dielectrics and dielectric processes mostly with special reference to polymeric materials are given by Moullin[12], Swiss and Dakin[13], Hoffman[14], Vail[15], Sharbaugh[16] and Devins and Sharbaugh[17]. These references are interesting, not least for the variety of approaches to the subject adopted by the authors.

Broader descriptions of electrical insulation properties and test methods considered mainly from the application viewpoint are given by Mason[18], Stark[19] and Baker[20].

10.2 RESISTIVITY MEASUREMENTS

10.2.1 Range of Values

The volume resistivity of common plastics covers a span of some ten decades between, very approximately, 10^{10} and 10^{20} Ωcm, and with the conventional size of specimen in sheet form, the corresponding resistances to be measured lie between 10^8 Ω and 10^{18} Ω. At the upper extremity of this range, above approximately 10^{17} Ωcm, lie the non-polar plastics, PTFE, polyethylene, polystyrene, etc., and measurements on these materials are considered sufficiently difficult (and the results perhaps of rather academic interest only) to be excluded from our present terms of reference. The reader is referred to References 21 and 22 for further details.

10.2.2 Measurement of Resistance

Only the measurement of resistance up to approximately 10^{15} Ω, which is the upper limit of commercially available resistance meters, will be considered, the normal lower limit of such instruments being of the order of 10^6 Ω. As surface resistance values for plastics are in general of the same order or lower than the volume resistance for standard specimens, what follows applies to both properties. The usual measurement technique is best illustrated by considering the simple case of a two terminal resistance, for example a resistor, before turning to the additional complications involved with measurements on insulation. In general, the unknown resistance is connected in series with a standard resistance, whose magnitude is usually several orders smaller, and a large potential applied across the combination. The potential is large to provide the necessary sensitivity and is commonly of the order of 100–500 V d.c. The small voltage developed across the standard is then measured by a suitable meter, which is usually either a valve voltmeter or an electrometer circuit, whose input resistance is sufficiently high so as not to shunt significantly the value of the standard. The unknown resistance is then given with sufficient accuracy by the ratio of test voltage to voltage across standard multiplied by the standard resistance, since at best the accuracy is limited to a few percent.

In the commercial megohmmeter, one of a number of standard resistances in decade steps can be selected by switching, the potential is provided internally, and the scale is suitably calibrated in terms of the unknown resistance so that the instrument is direct reading at some specified voltage. Well known instruments of this type available in the U.K. include those made by Electronic Instruments[23], Pye[24], British Physical Laboratories[25], Hewlett Packard[26] and General Radio[27]. Hitchcox[28] has written extensively on high resistance measurements and the use of electronic techniques for extending the upper limits of measurement.

10.2.3 Surface and Volume Resistivity

When measurements on insulating materials are considered it is usual in specification tests to use a sheet specimen in the form of a disc or a square. A two terminal specimen consisting of such a disc with a suitable electrode applied to each surface would be simple to measure, but the current passing through the body of the material would be augmented by current passing over the surface and round the edge of the specimen, so that it would be impossible to separate the volume resistance from the total measured resistance. To obviate this difficulty a third electrode, called a guard ring, is placed concentrically around one of the existing electrodes, which is reduced in size and separated from the guard electrode. By modified circuitry the desired volume resistance can now be measured direct, and further, by a simple interchange of connections the surface resistance can be measured also, using the same specimen. The electrodes and circuit arrangements are shown in Figures 10.1–10.3, the symbols R_v, R_s and R_g representing the volume, surface and guard resistances respectively, and R the standard resistor in the measuring instrument.

In connecting the specimen to the measuring instrument attention should be paid to the following details. It is good practice to use non-microphonic coaxial cable for the centre electrode connection since the relevant terminal on the measuring instrument (usually marked *H*) is sensitive to 'noise' pick-up. *H* stands for 'high' referring not to potential, since this terminal is usually very nearly at earth potential, but to the impedance to earth that is to say the resistance to earth of any component connected to this terminal must be high to avoid shunting the standard resistor. In the circuits of Figures 10.2 and 10.3 the standard resistor is shunted respectively by the surface resistance in one case and the volume resistance in the other. The values of these two components will in general be approximately of the same order as the resistance being measured, which as mentioned earlier is many times that of the standard, and consequently the resulting

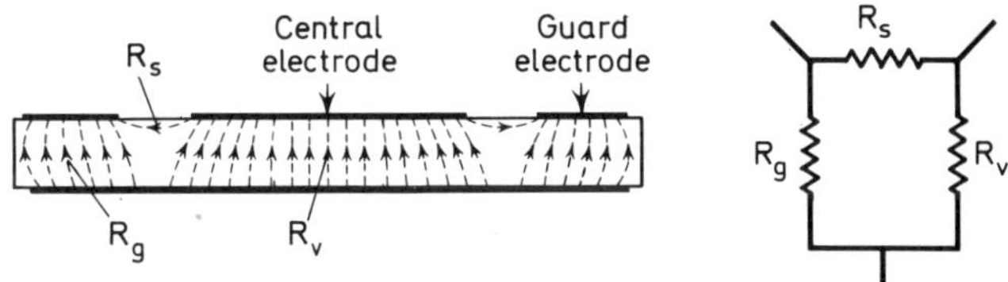

Figure 10.1. Electrode arrangement and equivalent circuit in resistivity measurement

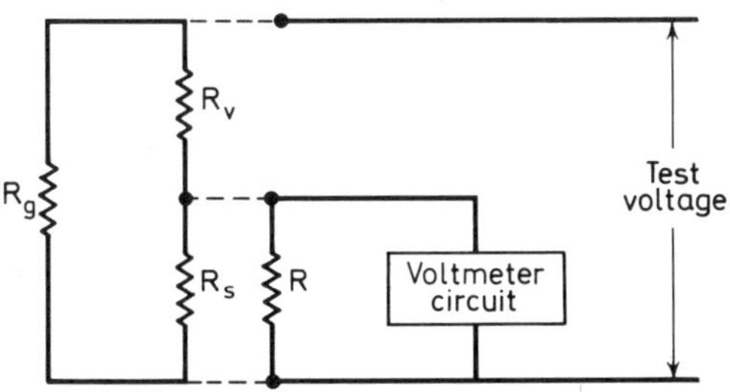

Figure 10.2. Volume resistivity measurement

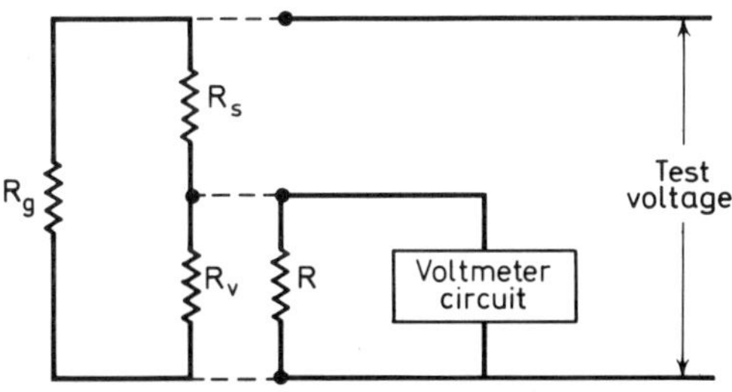

Figure 10.3. Surface resistivity measurement

error is negligibly small. The guard resistance in both cases is a shunt across the supply voltage and therefore has no effect on the measurement. Ordinary screened cable can produce quite large amounts of noise when flexed, but non-microphonic types are available in which the noise level is considerably reduced by the insertion of a semiconducting layer between the dielectric and the screening braid.

It is essential at the higher end of the resistance range to have adequate screening of the specimen and electrode system, otherwise the proximity of the operator may cause errors, or stray fields cause instability problems.

In the authors' experience, the measurement of high resistivity values is beset with difficulties. To avoid uncertainty about the functioning of the measuring apparatus it is essential to have a set of high resistances to hand, ranging from 10^6 to $10^{13}\,\Omega$. Such resistances are now available with accuracies to 2% or better, even for the higher values, but care should be taken with the selection, since the voltage coefficient is extremely large. This is normally negative, so that a nominally $10^{13}\,\Omega$ component can have a value 5 to 10 times lower, at 500 V, than the manufacturers' reported value, usually measured at a few volts only. France[29] gives details of measurement of high value resistors.

Since all insulators are dielectrics, practical specimens will have an associated capacitance and this can cause difficulties, more particularly at high values of resistance. The capacitance has to 'charge up' after the test voltage is applied and the time constant of the circuit, which is a function of the product of capacitance and circuit resistance, may be large. For non-polar materials it may be of the order of many hours before steady conduction occurs and a true resistivity value is obtained. Additionally, however, any change in applied voltage during the measurement has a considerable effect on the charging process and it can be shown that to measure a resistance of $10^{17}\,\Omega$ in parallel with 100 pF at 100 V with an accuracy of 10%, requires that the test voltage should not change faster than 1 µV/s, a rate which few power supplies at present are capable of achieving. Errors can often be caused by residual charges on the test specimen, and it is necessary to eliminate these by short-circuiting the electrodes together before test and checking that any residual charge is negligible before applying the test voltage. Munick[30] has investigated transient currents in plastics due to residual charge effects.

Since the charge on the associated capacitance increases exponentially and the charging current likewise decreases exponentially from a high initial value to an eventual constant value, short time measurements give an apparently low value of resistance. To standardise, in reporting values of resistivity it is customary to adopt the value attained one minute after the initial application of voltage, and strictly such values should be reported as apparent resistivity (one minute value) though this is seldom done. These values for polar materials will generally be within an order or two of the true infinite time resistivity. Reddish[31] has shown the errors that can occur with plasticised PVC compounds, where the 'one-minute' value is predominantly due to polarisation current and is useless as a criterion, even for comparing such compounds.

Resistivity measurements are seldom capable of very great precision and in the $\log_{10}$ units which are increasingly being used to avoid the large multiplying factors involved, one decimal place is usually adequate, a volume resistivity of $2 \times 10^{14}\,\Omega$cm being reported as 14·3 in $\log_{10}$ (Ωcm) units. As a general guide to the accuracy expected from industrial or routine measurements, two materials would not be considered very significantly different unless their resistivities differed by at least an order. Duplicate measurements on different specimens of the same material are regarded as essential, since considerable variations can occur due to the influences and hazards of manufacturing technique such as moulding parameters, mixing processes and so on.

10.2.4 Electrodes

Turning now to the electrodes themselves, there are many types available, but among the most widely used are the following: mercury, graphite, conducting paints, metallic foils applied with a thin layer of petroleum jelly, vacuum deposited metallic coatings and conducting rubber sheets. Most of these, mercury excepted, require solid metal backing plates to contact the electrodes proper, but whichever type is used, intimate contact with the specimen is essential and the choice will be dictated by environmental conditions. Mercury is probably simplest in application, needing only containing rings, the specimen being floated on a pool of the liquid in a dish, its weight being supported by submerged knife edges, but its toxicity is a hazard and for this reason it is unsuitable for elevated temperature work.

Colloidal graphite suspensions painted on to the test specimen are often successful, but are not very suitable with some materials, particularly flexible ones, due to poor adhesion since the coatings tend to lift. Conducting paints, often based on a suspension of metallic silver in a suitable resin or resin–solvent system, have the virtue of a high conductivity, but care should be taken that the solvent does not adversely affect the specimen, and residual solvent can produce odd results. All the brushed-on coatings require care in ensuring that the guard ring gap is not encroached upon.

Metallic foils such as tin or lead of a few thousandths of an inch thickness, applied with a thin layer of petroleum jelly or silicone grease, and rolled very hard with a metal roller, are easily removed and therefore useful where other tests may be required to be made on the same specimen. Their effectiveness depends on the thickness of grease being extremely small so that, despite its fairly high resistivity, the actual volume resistance is relatively low. This limitation makes foils generally unsuitable for the lower resistivity materials and also, since it is difficult to avoid contamination of the guard gap, they are not recommended for surface resistivity measurement. Gold leaf does not suffer from this limitation since it may be applied without grease. Conducting rubber sheets have been used for rapid comparative measurements on materials, but fairly high contact pressure is required and some conducting rubbers tend to become less conducting with age; their lack of resistance to elevated temperatures is also a limitation.

The dimensions of electrodes are not critical, but the B.S. 2782* sizes for a 100 mm test disc, approximately 3 mm thick, are a 50 mm inner electrode diameter, a 10 mm gap and a 5 mm wide guard ring. The lower (since normally the guard and guarded electrodes are uppermost) electrode diameter should be at least equal to that of the outer edge of the guard ring, i.e. 80 mm. The electrode area for volume resistivity measurements is taken as the area of the inner electrode and the resistivity is given by $AR_v/t\,\Omega$cm, where A is the area (cm^2), t the thickness (cm), and R_v the measured volume resistance (ohm) between centre and lower electrodes. The surface resistivity

*B.S. 2782: 1970, 'Methods of Testing Plastics'.

is given by

$$\frac{2\pi R_s}{\log_e \frac{D}{d}}$$

where D and d are the diameters corresponding to the outside and inside of the guard gap respectively, and R_s is the surface resistance measured between the two upper electrodes. Note the absence of length dimensions in the surface resistivity expression, which is thus stated in ohms. The units of volume resistivity are often misquoted as Ω/cm^3 or Ω/cc which are dimensionally incorrect and misleading.

It will be noted that in the calculation of volume resistivity the area of the central electrode is used rather than some effective area based on the mid-diameter of the guard gap. The resulting error is small for the electrodes described (*c.* 20%), compared with other errors generally outside the control of the operator, and in this country it is disregarded. In the U.S.A. (see A.S.T.M. D.257–66 'D.C. Resistance or Conductance of Insulating

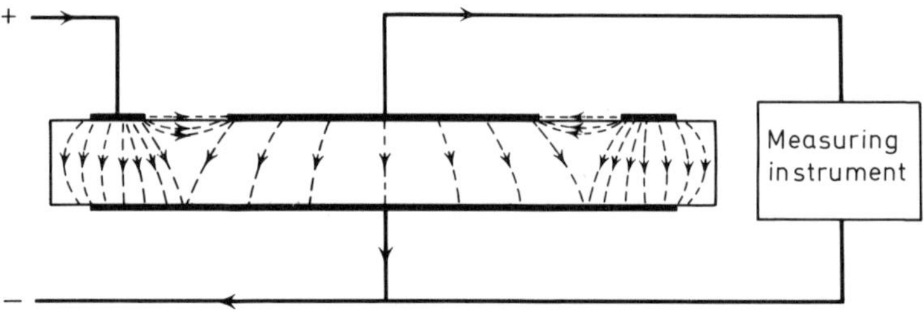

Figure 10.4. Current distribution in surface resistivity measurement

Materials') not only is the mid-diameter used, but an additional fringing error is also applied, based on the dimensions of the electrodes and the thickness of the specimen, where a high precision is considered necessary!

Resistivity measurements are very susceptible to environmental conditions such as humidity and temperature, surface resistivity changing by several orders over the humidity range 20–100% r.h. depending on the nature of the plastics material under test (e.g. nylon), and it is well to consider exactly what is being measured. Figure 10.4 shows the disposition of the electrodes and the idealised lines of current flowing during a typical surface resistance measurement, these latter being somewhat exaggerated to illustrate the point.

The current passing into the outer electrode is seen to have two components, one which passes through the volume of the specimen to the lower electrode, and which therefore bypasses the measuring circuit, and one which passes directly to the central electrode and thence to the measuring circuit. This latter component corresponds to the surface resistance as defined by the particular electrode system, but the important feature of it is that it is not *precisely* defined. Part of the measured current indeed passes over the surface of the specimen, but some part of it must inevitably pass through the layers of insulation adjacent to the surface so that surface resistivity is really a complex function involving volume resistivity also. This point need not be stressed unduly, since the surface resistivity is only of practical importance when the surface current, as distinct from the volume currents near the surface, dominates. In the rare cases when it does not, then the material is characterised by the volume resistivity only, which is in general more easily measured, and less susceptible to environmental changes. Electrodes have been designed by Kao[32] which truly enable surface resistance to be measured, but they are not suitable for general use since they involve rather special moulded or machined specimens.

10.2.5 Effect of Environment

Not only is surface resistivity particularly dependent upon ambient humidity, but in many cases it responds very quickly to humidity changes, as may be observed by breathing on the surface during a measurement. The importance of ambient conditions upon the measurement of resistivity is underlined by the requirements of most specifications, which generally call for tolerances of $\pm 5\%$ in relative humidity and ± 2 degC for the preconditioning period, usually of a minimum 18 h duration. In view of what has been said about the speed of response to humidity changes, some official specifications quite inadequately specify the conditions during measurement, which should be as rigorous as those of the pre-conditioning cycle where surface resistivity is concerned. Sensitivity to temperature change is possibly more marked for volume resistivity than surface resistivity, but it is advisable always to consider the possible effects of humidity and temperature variations on both properties.

10.2.6 Measurements at Low and High Temperatures

Resistivity is a temperature dependent property and for non-metallic materials it invariably decreases with increasing temperature, unlike the behaviour of many metals. The decrease is always relatively large and of the order of several decades, going from ambient to, say, 100°C. It is important in measurements at any temperature to ensure that the specimen temperature is maintained very constant whilst the test voltage is applied. Because of the temperature dependence of resistivity, fluctuations in temperature produce changes in measured current which cannot be distinguished from the normal time dependent charging current, and this may cause errors greatly in excess of those due to the temperature dependence itself.

It is usual to plot log resistivity against the reciprocal of absolute temperature $1/T$, in common with many other phenomena, and frequently the result is a straight line for the long-time steady values. A precaution in determinations at elevated temperatures is to measure with increasing temperature and to ensure that the specimen is fully discharged and short circuited before changing temperature. It is possible to produce a residual charge on a specimen by cooling with the voltage applied, which charge may only disappear after a very long time interval.

It is always advisable finally to cool and retest, to ensure that initial and final measurements at room temperature agree, any difference suggesting a change in cure, for example, or some other non-reversible temperature effect.

At sub-ambient temperatures an additional hazard is caused by the possible condensation of moisture on insulators, and this can be avoided by using an evacuated system or one which can be mounted in a vacuum chamber placed in a low temperature bath. Alternatively the electrodes can be swept continuously with dry nitrogen.

10.2.7 Insulation Resistance

The comments made so far apply with equal force to the measurement of insulation resistance, which is a complex function of both surface and volume resistivity and of the shape of the test piece and disposition of electrodes. In this respect, it is useless to quote a value for this property without carefully defining the foregoing parameters and in practice the utility of the test is confined to comparative measurements on a few materials of similar type, for example synthetic resin bonded paper, and other laminates. In such cases the test is commonly used to assess the effect of moisture penetration.

10.2.8 Specification Tests

In the U.K., B.S. 2782: 1970, 'Methods of Testing Plastics', gives methods of test for both volume and surface resistivity, respectively Methods 202 and 203, based on the same electrode system which may, however, vary widely in size. The central electrode is 50–150 mm diameter, the gap 10 mm wide and the guard ring 5 mm wide, whilst the lower electrode has to be of the same diameter at least as the outside of the guard ring. Only mercury, foil and graphite are permitted as electrode materials and in the case of the latter, two thick brass backing plates are required.

Figure 10.5 shows a typical mercury system based on the smallest permitted dimensions. Method 202B allows the guard ring to be dispensed with when the surface resistance is not required and is sufficiently high to be neglected. The measurement is made at 500 V d.c. after one minute's electrification and the resistivities calculated in the normal way:

$$VR = \frac{AR_v}{t}\ \Omega\ \text{cm} \quad \text{and} \quad SR = \frac{2\pi R_s}{\log_e \dfrac{D}{d}}\ \Omega, \text{ where the dimensions are in cm.}$$

The results of measurements on duplicate specimens are reported as $\log_{10}$ units. (The S.I. units for volume resistivity are Ωm, so that 10^{14} Ωcm becomes 10^{12} Ωm; the magnitude of surface resistivity is the same in both systems of course.)

Details of specimen sizes, preparation, preconditioning, and conditions of test are not given in B.S. 2782, but are laid down in the appropriate material specification. Generally, specimens are 100 mm minimum diameter and of 3 mm thickness, and preconditioning is for a minimum period of 18 h at 20 ± 5°C, 65 ± 5% r.h. For rubber (B.S. 903) the surface resistivity has to be determined in the conditioning atmosphere of 20 ± 2°C, 65% r.h., but more commonly the surface resistivity test is required after 24 h immersion in distilled water at 23°C (B.S. 1322: 1956, 'Aminoplastic Moulding Materials', for example).

Volume resistivity measurements are seldom required to be made at elevated temperatures, but B.S. 3815: 1964, 'Epoxide Resin Casting Systems for Electrical Applications', is an exception and a test at 90°C is involved. Here colloidal silver or graphite electrodes are stipulated.

The blanket A.S.T.M. specification covering resistivity measurements is D.257–66, 'D.C. Resistance or Conductance of Insulating Materials'. The general methods are similar to B.S. 2782 practice, but rather more scope is allowed. For instance, the electrode materials encompass conducting silver paint, sprayed or evaporated metal, foils, mercury, graphite and conducting rubber. Graphite is only permitted where a relatively dry preconditioning treatment is involved. Electrode sizes are not stipulated, nor even the shape, since round, square or rectangular types are permitted. In the case of surface

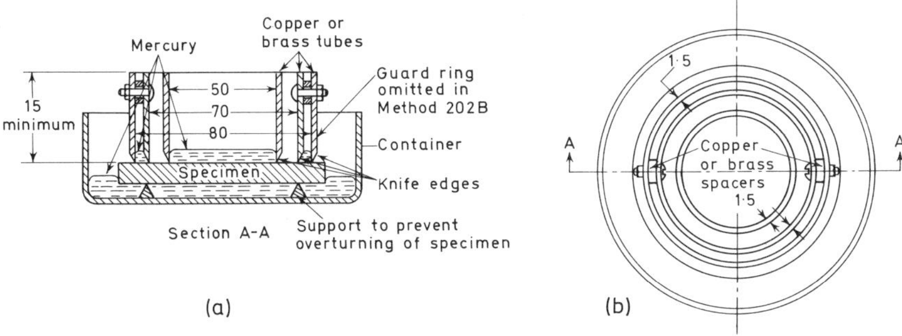

Figure 10.5. Mercury resistivity electrodes (B.S. 2782). Dimensions in millimetres

resistivity, the gap width between guard ring and centre electrode is made approximately equal to twice the specimen thickness. The test voltage is usually 500 V applied for one minute as in the British test. The expressions for the resistivity are as in B.S. 2782, except that the effective area for volume measurements is that corresponding to the mid-diameter of the gap. For electrodes equivalent to the smallest British size, the A.S.T.M. calculations would give a volume resistivity about 40% greater than the B.S. expressions. This may sound serious, but it serves to put the accuracy of resistivity measurements into some perspective. Duplicate measurements are normal.

The A.S.T.M. specification gives a great deal of additional information about the techniques and hazards of measurement and also covers methods for tubing, flexible tapes and films.

Continental methods are covered by IEC 93, 'Recommended Methods of Test for Volume and Surface Resistivities of Electrical Insulating Materials,'[1] with which the B.S. methods broadly agree, and DIN.53482, 'Testing of Electrical Insulating Materials; Methods for Determining Electrical Resistance Value'. The IEC document is much more informative than the British one and gives a very comprehensive survey of sources of error and precautions needed. The latest version of DIN.53482 (Jan. 1967) has been brought substantially in line with IEC 93 and IEC 167.

Four insulation resistance tests, all for laminated materials, are given in B.S. 2782, Methods 204A, B, C, and D. The first two cover sheet and tube (or rod) materials respectively using long-established test specimen assemblies, viz. for sheet a square with three sets of 2 BA nuts, bolts and washers arranged in an equilateral triangle at 31·8 mm ($1\frac{1}{4}$ in) centres, and for rod or tube a 25 mm (1 in) length with plane parallel ends clamped between flat metal plates under a pressure of 35 kN/m^2 (350 gf/cm^2; 5 lbf/in^2)—see

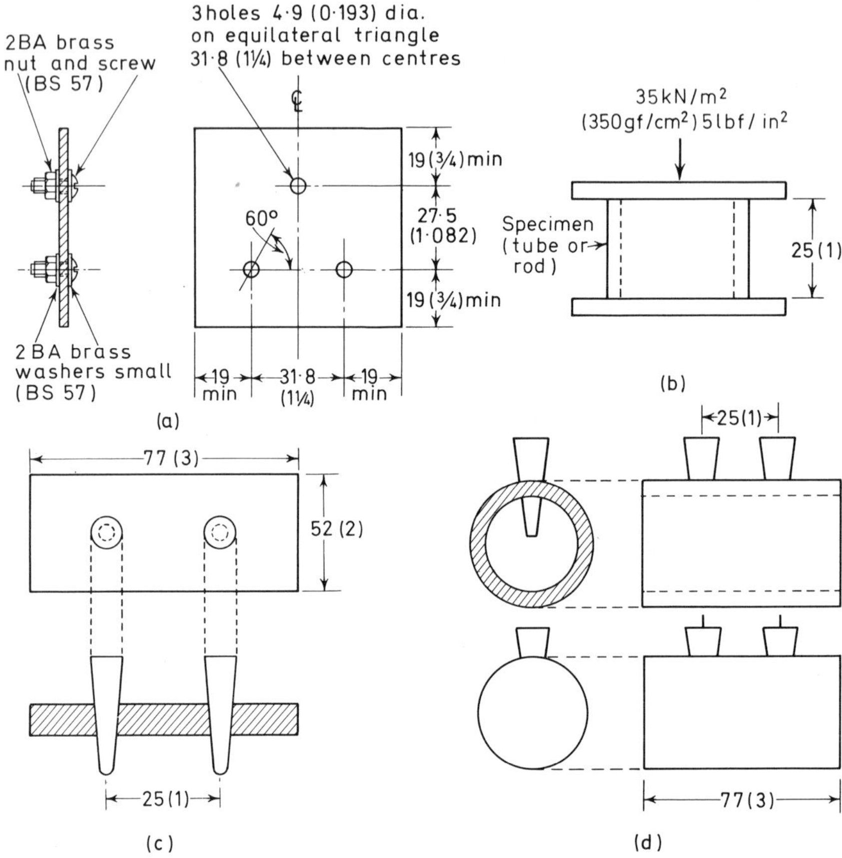

Figure 10.6. Insulation resistance electrodes and specimens (B.S. 2782). (a) Method 204A, (b) 204B, (c) 204C, (d) 204D. Dimensions in millimetres with inch equivalents in parentheses

Figure 10.6. Triplicate specimens are used in each case and these are pre-dried at 50°C for 24 h followed by desiccator cooling and 24 h immersion in distilled water at 25°C (20°C in some British Standards). After immersion the specimens are quickly wiped dry with a cloth and tested within five minutes of removal.

Methods 204C and D are based on IEC 167 'Methods of Test for the Determination of the Insulation Resistance of Solid Insulating Materials'[33], and respectively relate to sheet and tube (or rod). The sheet test employs a 77 by 52 mm specimen and taper pin electrodes are mounted normal to the plane of the sheet 25 mm apart in holes of approximately 5 mm mean diameter. The tube and rod test uses a specimen 77 mm long with the taper pins 25 mm apart mounted along the length. The pre-treatment is as for the Methods 204A and B tests except that the water temperature is 23°C, and also the electrodes are inserted after and not before the conditioning. The measurement of resistance is made at 15–40°C in an ambient humidity of less than 75% r.h. on single specimens, after one minute at 500 V.

Figure 10.6 shows the arrangement of the four B.S. systems.

The A.S.T.M. D.257 insulation resistance methods are somewhat similar to the B.S. ones. Thus for sheet material, nuts, bolts and washers are used spaced 32 mm ($1\frac{1}{4}$ in) apart in a line, or sometimes in a circle (e.g. A.S.T.M. D.700–68, 'Phenolic Moulding Materials') or alternatively taper pins with a similar geometry to Method 204C are specified, and this latter arrangement is also used for tube or rod as in Method 204D. The preconditioning treatment is 96 h at 35°C, 90% r.h. and measurements are made in situ, usually at 500 V after one minute.

For the measurement of adhesive tapes both A.S.T.M. and B.S. methods

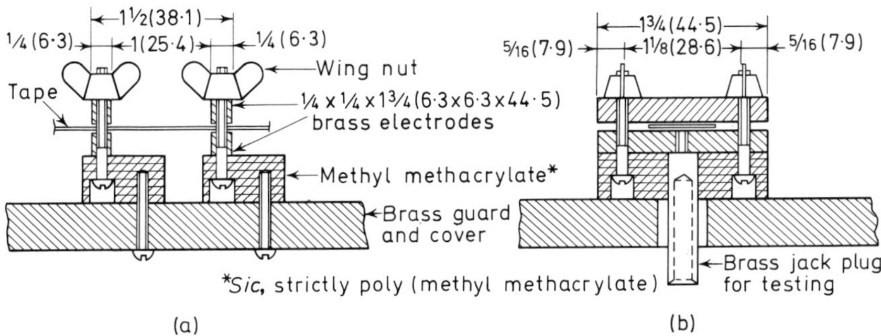

Figure 10.7. Insulation resistance electrodes for adhesive tapes (B.S. 3924). Dimensions in inches with millimetre equivalents in parentheses

use a similar electrode arrangement, consisting of pairs of brass bars of $\frac{1}{4}$ in square section tightly clamped on the specimen at a spacing of 1 in.

In A.S.T.M. D.1000–68, 'Testing Pressure-sensitive Adhesive Coated Tapes used for Electrical Insulation', the assembled specimen is preconditioned at 23°C, 96% r.h. for 18 h and the resistance between pairs of electrodes measured at 115 V after 15 s. The B.S. 3924: 1965, 'Pressure Sensitive Adhesive Tapes for Electrical Purposes', test is generally very similar, as shown in Figure 10.7. Five separate measurements are obtained per specimen.

Insulation resistance tests on PVC tubing are covered by A.S.T.M. D.876–65, 'Testing Non-rigid Vinyl Chloride Polymer Tubing', which requires that a U-shaped specimen filled with a dilute salt solution is immersed in a salt water bath and the resistance measured between a wire inserted through the length of the specimen and the external water. This method is similar to both A.S.T.M. and B.S. methods for measuring the insulation or sheath of cables (cf. B.S. 6004: 1969, 'PVC-Insulated Cables (non-armoured) and Flexible Cords for Electric Power and Lighting').

No British Standard specifically covers the measurement of insulation resistance or resistivity of films, but A.S.T.M. D.2305–68, 'Testing Polymeric Films used for Electrical Insulation', covers both properties. The insulation resistance test involves the brass bars of A.S.T.M. D.257, the method being otherwise as for adhesive tape (A.S.T.M. D.1000). The volume resistivity test is made using 1 in diameter conductive silver electrodes painted on each side of a test specimen and no guard ring is required for films under 0·005 in thickness. The test voltage is 100 V d.c. applied for one minute.

The expression and presentation of resistivity data is one of the subjects under review by Committee PLC 36 of the B.S.I., which has the remit to consider the presentation of design data for plastics over the whole field of properties. Recommendations 2.3 and 2.4, covering volume resistivity and surface resistivity respectively, are expected to be published soon, as parts of B.S. 4618, 'Recommendations for the Presentation of Plastics Design Data.'

10.3 POWER FACTOR AND PERMITTIVITY MEASUREMENTS

10.3.1 Introduction

The measurement of power factor and permittivity, over the wide band of frequencies in which these properties are important, is rendered difficult by the limitations of existing test equipment and the necessarily large variety of methods required to cover this range. Insulating materials are used as dielectrics at commercial frequencies between approximately 50 Hz* and 10^{11} Hz and the variety of test methods used reflects the many differing practical techniques necessary to generate and distribute power at different parts of this frequency spectrum. The range up to 10^8 Hz, it is true, can be covered by only two types of measuring equipment, but between approximately 10^9 and 10^{11} Hz a different set of equipment is generally required for each specific frequency of interest, since the waveguides (conductors) used at these frequencies are of particular dimensions directly related to the wavelength. The two kinds of techniques related to these two frequency ranges (above and below 10^8 Hz) are quite different and are sometimes loosely differentiated by the terms 'lumped circuit' and 'distributed circuit'. In the former case the circuitry involves the use of the three basic electrical components: inductors, capacitors and resistors; a passive circuit (one containing no sources of voltage), however complex, comprising these elements can be represented at a given frequency by an equivalent circuit having a single value inductance, capacitance and resistance. This notion of equivalence is a powerful aid to the solution of such complex circuits. At the higher frequencies, say 10^9 Hz and above, the three basic parameters can no longer be 'lumped' in this fashion and the waveguide behaves as though these parameters were distributed properties, so that solutions can only be found in terms of the electro-magnetic equations relating the associated electric and magnetic field distributions in and around the waveguide.

Dielectric measurements at u.l.f. (ultra low frequency) are of some interest in view of the light they throw upon the basic structure of some materials, notably glasses, but are outside the brief of this book and the

*Hz = Hertz = c/s.

reader is referred to references[31, 34, 35,] and [36] for details. Such measurements have been made down to 0·001 Hz and lower.

The power factor and permittivity of an insulator (or a capacitor) are so closely related that in practice many measuring circuits are designed to give both properties simultaneously, and as in any case the determination of one property can only be made with accuracy by adjustment of a circuit component which compensates for the presence of the second property, it is generally expedient to measure the two together.

10.3.2 Equivalent Circuits

The power factor of a material has been defined in Section 10.1.3 as that fraction of the electrical energy applied to a material which is dissipated as heat and is irrecoverable. Since it is seldom practical to measure these energies directly it is necessary perforce to seek some related property and

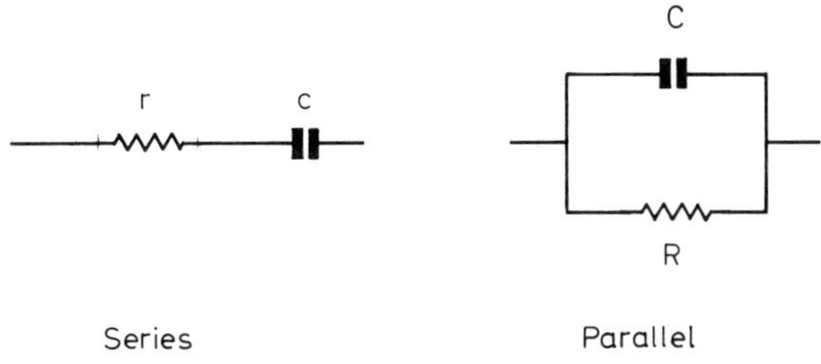

Figure 10.8. Equivalent circuits

some further definitions are needed. The use of 'lumped' circuits in measurements at frequencies up to 10^8 Hz, involving components of resistance, capacitance and inductance, has already been mentioned. The behaviour of a dielectric material, which may be very complex over this frequency range, can be simply represented at a *single* frequency by a lumped circuit comprising a pure capacitance and a pure resistance either in series or in parallel (see Figure 10.8).

Which representation is used in a particular instance is a question of convenience, but although, by definition, both circuits are exact equivalents (that is the impedance, or a.c. resistance, and the power factor are both identical), the two pairs of parameters r, R and c, C are not identical (although for most practical purposes the capacitances are very nearly equal). This latter point may be important in precision measurements and therefore the choice of the equivalent circuit is usually made on the basis of the measuring circuit employed, since in general the measuring circuit components are so arranged to give the appropriate capacitance and power factor in terms of either the series or parallel equivalents.

The condition that the series and parallel equivalents are identical at a given frequency is that their impedance (a.c. resistance) is the same with respect to both magnitude and phase. Using complex notation ($j^2 = -1$) the impedances of the series and parallel equivalents are respectively:

$$r-\frac{\mathrm{j}}{\omega c} \quad \text{and} \quad \left(\frac{R}{1+\omega^2C^2R^2}-\frac{\mathrm{j}\omega CR^2}{1+\omega^2C^2R^2}\right)$$

where $\omega = (2\,\pi \times \text{frequency})$ and the units of capacitance and resistance are respectively farads and ohms. Equating real and imaginary parts leads to the following relations:

$$r = R\frac{t^2}{1+t^2} \quad \text{and} \quad c = C(1+t^2)$$

where $t = \tan\delta = 1/\tan\theta$, θ being the phase angle of the circuits and δ the complementary angle ($\delta = 90-\theta$). The power dissipated in each circuit is $EI\cos\theta$ for a sine wave, where E and I are respectively r.m.s. values of voltage and current in the circuit, whereas the apparent power applied to the circuit is simply EI. The power factor by our original definition is the ratio of these two quantities, i.e. $\cos\theta$. Since for most insulators θ approaches 90°, it is more usual to use the complementary angle δ, called the loss angle, and the power factor is then equal to $\sin\delta$.

It is, however, common practice in dielectric work to use the tangent of the loss angle rather than the sine, since the equations relating the equivalent circuit parameters with $\tan\delta$ are much simpler. As a corollary to this the measuring circuits employed in bridge and resonant circuits invariably give simpler equations also. Thus the equations for the series and parallel circuits above are respectively, neglecting the sign,

$$\tan\delta = \omega cr \quad \text{and} \quad \tan\delta = \frac{1}{\omega CR}.$$

For values of $\tan\delta = 0{\cdot}1$ or less (in practice 0·1 would be regarded as a high power factor, 0·001 or less as low), the difference between the loss tangent and power factor is less than 1%, which is generally negligible, and the two earlier equations above reduce to:

$$r = R\tan^2\delta \text{ and } c = C \text{ (both within 1\%)}$$

To illustrate the significance of these equations, consider the example of a low loss specimen of polyethylene with a capacitance of 10 pF and loss tangent of 0·000 1. At $\omega = 10^4$ rad/s (frequency = 1600 Hz approximately), the series equivalents would be $r = 1000\ \Omega$ and $c = 10$ pF and the parallel equivalents $R = 10^{11}\ \Omega$ and $C = 10$ pF. At $\omega = 10^7$ rad/s ($f = 1{\cdot}6$ MHz approximately), the equivalents would be $r = 1\ \Omega$, $R = 10^8\Omega$, both capacitances again being 10 pF. The small value of r at the higher frequency also illustrates the importance of keeping any lead resistances in the measuring circuit small, since if they are included in the measurement the loss tangent will be augmented accordingly by an amount $\omega cr'$, r' being the extraneous resistance and c the measured capacitance. At the lower frequencies the reverse situation applies and it is important to ensure that no extraneous resistance in parallel with the specimen is included in the measurement unless it is negligibly high compared with the specimen effective parallel resistance. This latter resistance should not be confused with the d.c. resistance of the specimen or component and it cannot be measured by d.c. methods.

10.3.3 Two and Three Terminal Measurements

Before turning to a description of the types of measuring equipment and techniques employed, it is appropriate at this point to consider briefly the use of two and three terminal electrode systems which are covered more fully in Section 10.3.4. The choice is governed mainly by the type of measuring equipment available and by and large three terminal systems are more accurate but limited mainly to bridge measurements, since they become rather impracticable above about 1 MHz, although there is an increasing tendency for radio frequency bridges to be designed to accommodate them. Above 10 MHz two terminal systems are standard.

As in resistivity measurements, a sheet dielectric material with an electrode applied to each surface would be simple to measure, but the capacitance between the electrodes due to the specimen would be augmented by extraneous capacitance round the edges of the specimen either wholly or partly in air, which would be difficult to assess and allow for. A third electrode placed concentrically around one of the existing electrodes (which is diminished in area) and as close to it as possible without touching serves as a guard ring, and when used with appropriate circuitry reduces the uncertainty of the measured capacitance to an extremely small amount, since the effective area of the guarded electrode very nearly corresponds to that based on the mid-gap radius. Usually the requirement of the measuring circuit is that the guarded and guard electrodes are maintained at exactly the same potential and phase during measurement to ensure that the capacitance and associated loss between them do not contribute to the measured values, or alternatively the capacitances of the guard/guarded electrode and the guard/non-guarded electrode may with some circuits be connected so that they do not affect the balance or accuracy.

10.3.4 Measurements at 50 Hz–20 kHz

INTRODUCTION

Bridge circuits are invariably employed in the audio frequency range (50 Hz–20 kHz) and beyond up to 100 kHz, but the upper limit is probably of the order of 10 MHz, although bridges suitable for the latter frequency have to be of rather special construction. The basic bridge circuit (shown in Figure 10.9) consists of four arms, and the simplest case involving four resistances is that of the Wheatstone bridge for measuring resistance. The circuit is completed by the addition of a voltage source across one diagonal of the bridge and a suitable voltage detector across the other. One terminal of the bridge is invariably earthed.

At least one of the arms must be variable and the bridge is balanced when the detector indicates zero voltage; the relation $P/R = Q/S$ then holds. Thus one unknown component can be measured in terms of three known ones. For a.c. measurements at least one other arm must contain a reactive component, that is one having inductance (self or mutual) or capacitance,

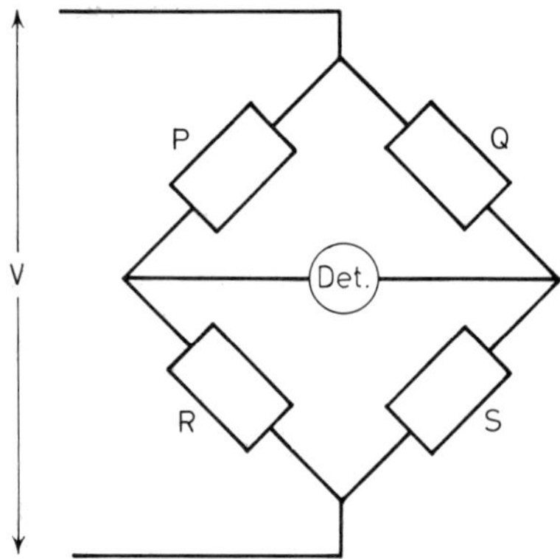

Figure 10.9. General bridge circuit

and two examples will be considered in some detail since most present day determinations are based on them or their variants. These are the Schering bridge and the transformer ratio-arm bridge.

SCHERING BRIDGE

The Schering bridge in its simplest form comprises two equal value resistance arms and two capacitance arms, one of which is a relatively loss-free variable standard capacitor and the other the unknown, i.e. the specimen. At balance the total capacitance in the unknown arm is equal to the total capacitance in the standard arm, whilst the loss balance is achieved by a variable capacitor across the resistance arm diagonally opposite the unknown arm, the loss tangent being a function of the capacitance and resistance of this arm and the unknown capacitance.

This arrangement is seldom used in practice, since the residual stray capacitances which are inevitably associated with each arm of the bridge are difficult to delineate and allow for. Instead, a substitution method is employed using a bridge containing two capacitive arms and two resistive arms, the bridge being initially balanced with the unknown disconnected and then re-balanced with the unknown connected across one capacitive arm containing a low loss calibrated standard. The diagonally opposite resistive arm has a calibrated loss balancing capacitor in parallel and the fourth resistive arm usually also contains an uncalibrated capacitor to facilitate initial balancing. The capacitance and loss are given by the changes in settings of the standard and the loss capacitor, providing that the remaining components are left untouched after initial balance. The virtue of this arrangement is that all the residual capacitances and losses in the bridge remain constant during the measurement and do not enter into the bridge equations.

Figure 10.10 shows a typical conjugate Schering bridge, C_1 being the low loss standard capacitor of known calibration, C_2 a low loss uncalibrated balancing capacitor, and C_3 and C_4 two capacitors of lesser quality, the latter suitably calibrated. With the unknown C_x (of loss tan δ_x) disconnected and assuming equal ratio-arms ($R_3 = R_4$) the initial balance gives $C_1 = C_2$ where these symbols denote the total capacitance in arms 1 and 2 including extraneous capacitance. Any difference in loss in arms 1 and 2 will be balanced out by the difference in settings of C_3 and C_4 (again including stray capacitance). Denoting the initial settings of C_1 and C_4 by single primes and the final settings with the unknown connected by double primes the balance equations give:

$$C_x = C'_1 - C'_1 \qquad \text{or } C_x = \Delta C_1$$

$$\tan \delta_x = \omega C'_1 R_4 \frac{(C'_4 - C''_4)}{(C'_1 - C''_1)} \quad \text{or} \quad \tan \delta_x = \omega \frac{C'_1}{10^{12}} R_4 \frac{\Delta C_4}{\Delta C_1}$$

where ω = $2\,\pi \times$ frequency
ΔC_1 = change in capacitance of C_1, pF
ΔC_4 = change in capacitance of C_4, pF
C'_1 = initial setting of C_1, pF
$R_4 = R_3$ = ratio-arm resistance, Ω

It should be noted that the residual losses in C_1 and C_2, although small, are finite, and they have been taken into account in the derivation of the above equations, but they disappear providing the loss in C_1 is due to a constant equivalent parallel resistance, which is usually the case for good quality capacitors.

In the two-terminal example given it is important to connect the unknown via a low capacitance switch as shown, where any alterations in the distribution of the stray capacitances from the specimen electrodes to earth

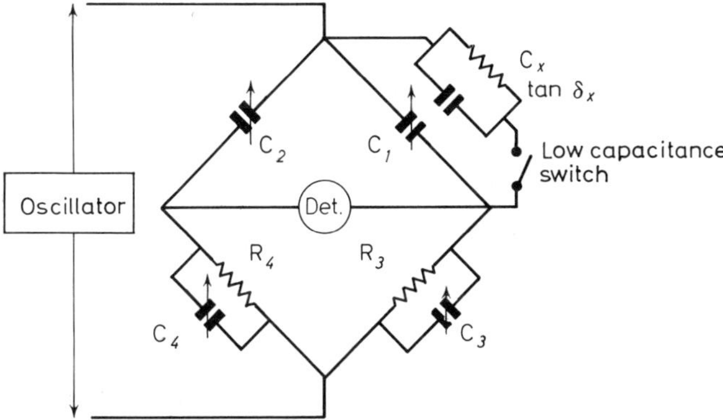

Figure 10.10. Schering bridge for two terminal measurements

between the two balances has a negligible effect. If the switch was in the upper or high voltage lead, a large stray capacitance to earth could appear across arm 4 on the initial balance, which would disappear on final balance and the resultant change in the capacitance of this arm would be reflected as a spurious contribution to the loss balance ΔC_4. The values of C_3 and C_4 are often of the order of 500–1000 pF, while C_1 and C_2 are usually 1000 pF or less. The values of R_3 and R_4 are typically 1000–10 000 Ω; R_3 and R_4

are sometimes made unequal, the ratio giving a multiplying factor by which both capacitance and loss ranges may be extended.

Three terminal electrode systems are accommodated on the Schering bridge by the provision of a further two arms similar to one half of the existing bridge, which form a Wagner earth circuit as shown in Figure 10.11. There are many forms of this arrangement, the one shown being fairly typical. The main capacitance in the electrode system to be measured is connected across C_1, with the larger electrode to D, the high voltage side of the bridge, and the smaller central electrode to B (the electrode system is

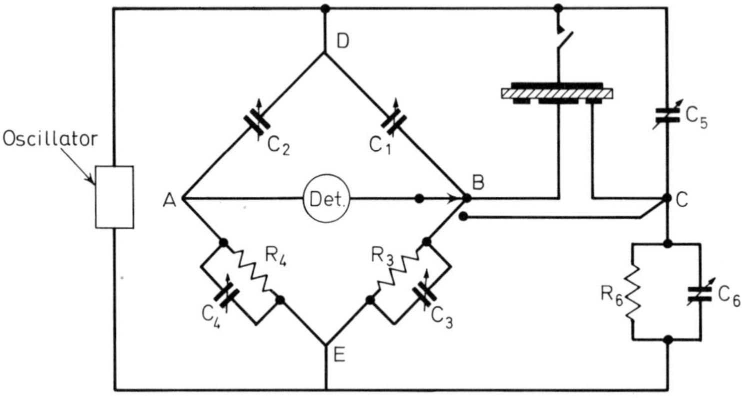

Figure 10.11. Schering bridge with Wagner Earth

drawn inverted for convenience). The guard electrode is connected to the junction of the auxiliary arms at C. The main feature of the Wagner circuit is that the detector can be switched either across AB, the bridge proper, or across AC, so that when the detector indicates a null in both positions the points A, B and C are at the same potential and phase and the guarding of the electrode system is perfect. The procedure in the substitution method is as follows:

1. With specimen main (high voltage) electrode disconnected, balance the main bridge using C_2 and C_3, C_1 and C_4 being meanwhile left at their zero settings (corresponding usually to maximum capacitance and minimum capacitance respectively).
2. Transfer detector to AC and balance auxiliary circuit components C_5 and C_6.
3. Alternate detector between AB, balancing C_2 and C_3, and AC, balancing C_5 and C_6 until both positions give a null simultaneously. Hereafter C_2 and C_3 must be left undisturbed.
4. Connect main electrode and re-balance main bridge with C_1 and C_4 only (detector at AB) followed by auxiliary bridge with C_5 and C_6 (detector at AC). Alternate between bridges until both are in balance simultaneously as before.

The loss tangent and capacitance are then derived as for the two terminal arrangement.

Hague's excellent book[37] on a.c. measurements, although somewhat dated, gives an extremely detailed account of Schering's bridge and many others.

TRANSFORMER RATIO-ARM BRIDGE

The other type of bridge circuit which will be considered in detail is the transformer ratio-arm bridge, which, although known for many years, has nevertheless not been widely used until recently, its development being somewhat delayed until advances in transformer design and materials enabled its virtues to be better exploited.

Figure 10.12 shows such a bridge in its simplest form, with the transformer secondary winding divided electrically exactly into two, forming two of the bridge arms, and the unknown and standard capacitances C_x and C_s respectively forming the other arms. The primary winding of the transformer is connected to an a.c. source and the detector completes the circuit.

For equal capacitors of equal loss tangent, at balance $C_s = C_x$. For measuring low loss capacitors it is usual to provide a series variable resistance in the standard arm and for higher power factors a variable resistance in parallel with the standard capacitor, the values of loss tangent and capacitance of the unknown being then given in terms of the appropriate equivalent circuits. For three terminal unknowns the transformer centre tap is normally earthed and becomes the guard point. The capacitances to guard from the remaining two terminals or electrodes of the unknown are thus either a shunt on the detector, or across one half of the transformer winding. The effect is either to decrease the sensitivity of the detector (negligible for capacitances of the order of a few hundred pF) or to impose an additional

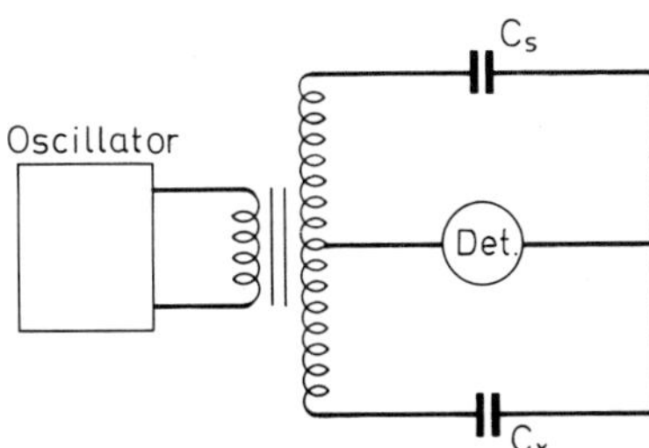

Figure 10.12. Simple transformer ratio-arm bridge

loading on the transformer winding. This latter effect is in general small, with modern constructional methods and the improved magnetic core materials now available.

The simplicity of the basic circuit conceals the versatility of the transformer ratio-arm bridge and its outstanding advantages are its range and accuracy. Figure 10.13 shows a modern transformer ratio-arm bridge and Figure 10.14 a simplified circuit diagram of the same bridge[38].

Here the single standard capacitor is replaced by a number of standards in decade steps, say 1 pF, 10 pF, 100 pF, etc., each one connected to a

Figure 10.13. Transformer ratio-arm bridge (Courtesy General Radio (U.K.) Ltd.)

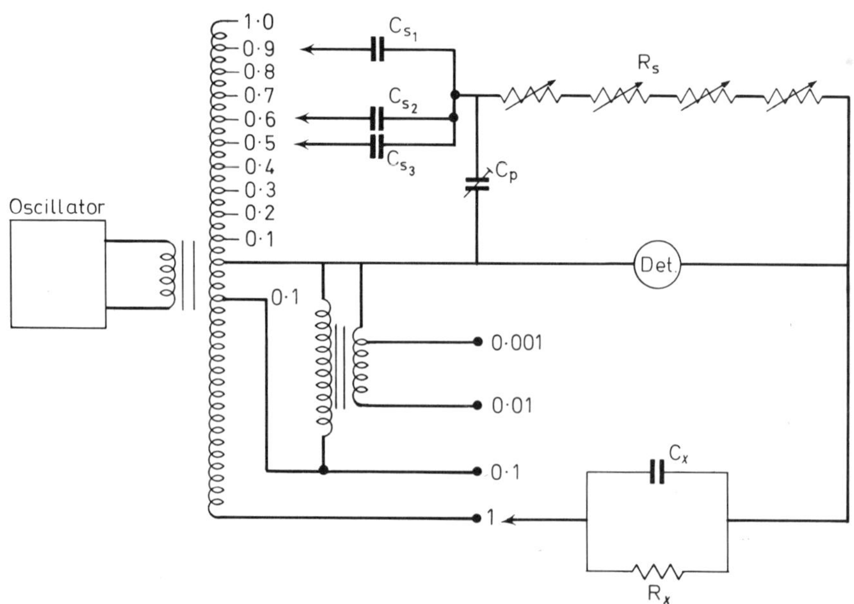

Figure 10.14. Simplified circuit of bridge in Figure 10.13

separate ten-way switch. The 'standards' side of the transformer secondary is tapped at 1/10, 2/10, 3/10 and so on, of the winding and the switches arranged so that any standard capacitor can be switched to any ratio. The total capacitance of the 'standard' arm is then given by the sum of the standards multiplied by their respective ratios. For the circuit shown, if $C_{s1} = 100$ pF, $C_{s2} = 10$ pF and $C_{s3} = 1$ pF, the effective capacitance is 96·5 pF, which is the value of C_x in this instance, assuming balance. It can be shown that the loss tangent of C_x would be given at balance by $\omega R_s (C_s + C_p)$, where C_p is a variable padding capacitance, and C_s is the total value of the standards in circuit, in this case 111 pF. By making the sum $C_s + C_p$ (which includes stray wiring capacitance also) equal to some multiple of 1592 pF (viz, numerically $10\,000/2\,\pi$) the resistance R_s, which is a series of decade units, can be made to correspond exactly, numerically, to the loss tangent at some convenient frequency, say 1000 Hz. In the example above, if $R_s = 15\,\Omega$ and $C_s + C_p = 1592$ pF, at 1000 Hz the loss tangent would be 0·000 15 so that the decade resistance is numerically equal to this value, allowing for the decimal point.

Thus, the positions of the capacitance switches give the unknown capacitance direct and the positions of the resistance switches give the loss tangent direct. The position of the decimal point is usually taken care of mechanically, this being linked with the range change switch. Ranges are altered by switching the appropriate capacitance standard from one decade switch to another so that not necessarily all standards are in use on a given range. To avoid the use of large value, low loss and hence expensive standards, e.g. 0·1 and 1 μF, the effective range of the existing standards can be increased by 10, 100 and 1000 times, for example, by means of an auxiliary transformer, shown in Figure 10.14, connected across the 0·1 tap of the 'unknown' half of the main transformer. This auxiliary transformer is tapped at $\frac{1}{10}$ and $\frac{1}{100}$ of its electrical length so that the unknown is effectively multiplied by the factors 0·1, 0·01 or 0·001.

The inherent accuracy of the bridge comes from the precision with which modern transformers can be wound and tapped, combined with the use of precision fixed value capacitors. In the best bridges the ratios can be made as accurately as 1 p.p.m. or better and this accuracy is invariable and cannot change with age. The use of a number of relatively cheap fixed precision standard capacitors in place of expensive variable or decade capacitors, required in conventional bridges, results in a considerable economy. The modern transformer bridge has a discrimination of 1 p.p.m. in both loss and capacitance and an accuracy of 0·01% in the latter variable. Not the least of its virtues are its self-guarding facility, simple balancing procedure and freedom from corrections. It can be used over the whole of the frequency range from 50 Hz to 10 kHz, although the loss tangent reading must be multiplied by the ratio (measurement frequency/calibration frequency). This implies that for a given loss tangent the actual bridge indication will be larger the lower the frequency. This advantage is somewhat diminished by the increasing difficulties of making a balance as mains frequency and its lower order harmonics are approached.

Automatic self-balancing bridges with digital read-outs based on transformer ratio-arms are now available[26, 27], the advantages of which hardly need extolling where rapid routine measurements are required. Manufacturers of transformer ratio-arm bridges include Hewlett-Packard[26], General Radio[27]. Wayne Kerr[39], Sullivan[40], and Marconi[41].

ELECTRODES

Electrode Materials

Consideration will now be given to the means by which contact is established between dielectric specimen and measuring instrument. The importance of understanding the principles and techniques involved cannot be overstated, since probably more difficulties and errors are associated with electrodes than with the measurement itself. Much of what has been said in Section 10.2.4 about electrodes in connection with resistivity measurements is applicable here also, with the recognition that the requirements are probably more stringent with dielectric measurements, since greater accuracy can be obtained and is usually desired.

In what follows essentially only sheet specimens are considered, but the techniques are equally applicable to tubes with appropriate changes in the permittivity formulae (see Reference 2). It is worth noting in passing that loss tangent or power factor is a property which is independent of specimen geometry, unlike capacitance.

All electrodes are designed either to ensure as intimate a contact as possible between specimen and measuring apparatus, or alternatively and rather paradoxically, to ensure a precisely defined air or liquid gap between specimen and electrodes, for which accurate allowance can be made. These diverse methods may broadly be referred to as contacting and non-contacting systems respectively. Amongst the former appear conducting paints, vacuum deposited metallic films, mercury, conducting rubbers, metal foils applied with grease, and colloidal graphite. Although each of these types may have some utility under particular circumstances, the most widely used is probably that involving metal foils, and possibly the best that utilising metal films deposited in vacuo. Conducting paints and graphite suspensions tend to give relatively high resistance films, which cause spuriously high loss values on low-loss materials, and, the solvent base may attack the material. Conducting rubbers likewise have relatively high resistance and are thus only suitable for 'lossy' materials, but have the virtue of being speedy and 'clean' to apply. Mercury is a toxic hazard and requires carefully designed clamps and considerable care in avoiding air bubbles. Vacuum deposition of metallic films such as gold or aluminium ensures intimate contact, and since the thickness may be controlled fairly accurately, the resulting film is very uniform; appropriate masks enable the electrode diameter to be defined accurately. Suitable equipment is available from the manufacturers of standard vacuum pumps.

Tin foils applied with silicone grease or petroleum jelly and rolled on very hard with a roller are convenient and simple. The foils are usually 0·03–0·05 mm thick and the rolling process produces an extremely thin film of grease, which must have a capacitance at least two orders greater than the specimen to reduce the measurement error to 1%. This is not difficult to achieve with low permittivity specimens 2 mm thick or so, but errors may become very large with high permittivity or thin specimens ($< 0{\cdot}5$ mm say) and foils are not generally suitable for measurements on films. Tin foils are not easy to centre accurately on the specimen and also tend to increase in area with repeated rolling. Gold foil does not require grease but is difficult to handle; on the other hand it does provide very intimate contact and can be used even on thin films. It should be emphasised that no one electrode material or technique is superior to all others and necessarily gives the 'correct' answer. The only procedure with a doubtful result is to substantiate it by at least one other completely independent method.

Two and Three Terminal Electrodes

With all contacting electrodes, metal backing plates are required of the same size as the electrodes proper, although their edges may be rounded to reduce local stress concentrations. For guarded (three terminal) electrode systems, the gap between guard and guarded electrodes should be kept fairly small if high accuracy is required, say below 0·5 mm, and if d_0 is the mean diameter mid-way in the gap, the permittivity is given by

$$\varepsilon = \frac{144tC}{d_0^2}$$

where d_0 = mean diameter, mm
t = specimen thickness, mm
C = capacitance, pF

For two terminal electrode systems, when the specimen is at least equal in diameter to the electrodes, the permittivity is given by

$$\varepsilon = \frac{144t(C-C_e)}{d^2}$$

where d = electrode diameter, mm
t = specimen thickness, mm
C = capacitance of specimen/electrodes, pF
C_e = edge capacitance of electrodes, pF

The edge correction may be obtained either from the Kirchhoff relation, as published in references 42 and 43, or by calibration with specimens of known permittivity.

Air Gap Electrodes

A difficulty with most contacting electrodes is that any pre-conditioning of the specimen must be carried out before their application, but this difficulty does not rise with non-contacting electrodes. With air gap systems, the electrodes are spaced apart and the specimen introduced leaving a deliberate gap. Two measuring techniques are then available, viz. fixed gap and fixed capacitance. In the first, the capacitance and loss tangent of the system is measured with the specimen in and then with the specimen removed, the electrode spacing being left untouched throughout. The values for the specimen are then given by

$$\varepsilon = \frac{t_s}{t_s - t_o\left(1 - \dfrac{C_o}{C_i}\right)}$$

$$\tan\delta = \tan\delta_i\left[1 + \varepsilon\,\frac{t_o - t_s}{t_s}\right] - \varepsilon\frac{t_o}{t_s}\tan\delta_o$$

where t_s = specimen thickness
t_o = electrode separation
C_i = capacitance, specimen in
C_o = capacitance, specimen out
$\tan \delta_i$ = loss tangent, specimen in
$\tan \delta_o$ = loss tangent, specimen out
ε = specimen permittivity
$\tan \delta$ = specimen loss tangent

For a properly designed electrode system, where $\tan \delta_o$ would normally be negligibly small compared with $\tan \delta_i$, the second term would vanish.

The units of capacitance and distance are arbitrary (but consistent) since only ratios are involved. The equations are rather cumbersome, but their virtue is that the calibration of the electrodes need not be known, since their area is not involved, and no electrode adjustment is required, although this limits the permissible specimen thickness as it is desirable for the sake of sensitivity to keep the gap small. As a guide, the gap should be of the order of 10–20% of the specimen thickness. Smaller gaps tend to cause difficulty due to dust particles and fibres, and with larger gaps the electric field tends to distort and cause errors.

With a fixed capacitance system, measurements are made of capacitance and loss tangent as before, with the specimen in, but the electrode spacing is adjustable and after removal of the specimen the spacing is reduced until the capacitance of the empty electrodes is the same as when the specimen was in place. The equations are then

$$\varepsilon = \frac{t_s}{t_s - t_x}$$

$$\tan \delta = (\tan \delta_i - \tan \delta_o) \frac{t_o}{t_s - t_x}$$

where t_s = specimen thickness
t_x = electrode movement between specimen in and out positions
t_o = electrode separation, specimen out
$\tan \delta_i$ = loss tangent, specimen in
$\tan \delta_o$ = loss tangent, specimen out
ε = specimen permittivity
$\tan \delta$ = specimen loss tangent

Figure 10.15 shows a micrometer electrode system with non-rotating anvil suitable for specimens of diameter 50 mm or greater. Such electrodes are available from Wayne-Kerr[39] and Yarsley Laboratories[44].

With suitable specimens, a precision of the order of 1 part in 1000 is achievable for permittivity, such precision being required for example in the control of polythene used in submarine cables. Lynch[45] gives useful information on the use and construction of micrometer air gap electrodes with a full analysis of electrode and specimen imperfections and B.S. 4542: 1970, 'Determination of Loss Tangent and Permittivity of Electrical Insulating Materials in Sheet Form', is largely based on his work. This standard is complementary to B.S. 2067 (see Section 10.3.5) and principally covers the frequency range 1–100 kHz.

In recent years liquid immersion electrodes have been developed, with which very precise measurements of permittivity can be made, and these are based on the fixed gap techniques, but with the gap filled with a suitable non-polar liquid. Principally developed through the submarine cable requirement referred to above, a high accuracy is achieved by using a fluid of closely similar permittivity to that of the test specimen so that only small changes in the electrode capacitance and loss occur when the specimen is inserted. If

Figure 10.15. Micrometer electrode system

there were no change at all in these properties on insertion then the specimen would have an identical permittivity and loss tangent to that of the liquid; otherwise the specimen properties are given by:

$$\varepsilon = \varepsilon_l \pm \frac{\Delta C}{C} \cdot \frac{t_o}{t_s}$$

$$\tan \delta = \tan \delta_l + (\tan \delta_i - \tan \delta_l) \frac{t_o}{t_s}$$

where t_o = electrode separation
t_s = specimen thickness
ΔC = change in capacitance of liquid filled electrodes when specimen inserted, pF
C = capacitance of electrodes, empty, pF
ε = permittivity of specimen
ε_l = permittivity of liquid
$\tan \delta$ = loss tangent of specimen
$\tan \delta_l$ = loss tangent of liquid
$\tan \delta_i$ = loss tangent of liquid filled electrodes with specimen inserted.

In the permittivity expression, the sign is positive if ΔC is an increase in capacitance of the system and vice versa. This method of measuring very precise permittivity is mainly limited to materials permittivities of which can be matched by readily available liquids. A.S.T.M. D.1531–62, 'Dielectric Constant and Dissipation Factor of Polyethylene by Liquid Displacement Procedure', gives details of the requirements for low density polyethythene measurement using either benzene or silicone fluid. The equations given apply only where the liquid and solid permittivities are within a few percent of each other.

10.3.5 Measurements at 10 kHz–100 MHz

INTRODUCTION

The frequency range 10 kHz–100 MHz corresponds approximately to that of radio transmissions, and telecommunication techniques based upon the tuned or resonant circuit are widely used for measurement purposes. The resonant circuit, comprising basically an inductance and capacitance in parallel, has the important property that it is frequency sensitive and at one particular frequency its sensitivity to relatively slight changes in either frequency or capacitance is enormously greater than at any other frequency. These two features are exploited in the Q-meter and in the susceptance variation methods of measuring dielectric properties (susceptance is the reciprocal of reactance), whereby the addition of a capacitance in the form of a dielectric test specimen to the circuit causes a change in resonant frequency, in the first case the change being compensated by frequency adjustment and in the second by capacitance adjustment. The Q-meter will not be dealt with since it is used rather less in the U.K. than the second method, although it is probably capable of as great an accuracy in its refined forms[46, 47].

HARTSHORN AND WARD APPARATUS

The susceptance variation method was originated by Hartshorn and Ward[48] in 1936 and is dealt with in B.S. 2067: 1953 'Determination of Power Factor and Permittivity of Insulating Materials (Hartshorn and Ward Method)', in some detail. Recent refinements by Barrie[49] enable it to be used to measure loss tangents of the order of 0·000 1 to a few μR (1μR = 0·000 001). The apparatus is shown in Figure 10.16.

In this method a resonant circuit comprising an inductance coil in parallel with a variable air capacitor, which also forms the specimen holder, is loosely coupled electromagnetically to a stable oscillator. A valve voltmeter with a square law relationship (that is to say one whose output is proportional to the square of its input voltage) is connected in parallel with the circuit and serves both to indicate resonance and to measure the degree to which the circuit is adjusted off resonance during a measurement. The voltmeter output is connected to a reflecting galvanometer with a 1 metre scale, to achieve the high sensitivity required. A second small capacitor is also connected in parallel with the main circuit and is used to produce a given degree of detuning in the circuit. The circuit is shown in Figure 10.17.

In practice a specimen is placed between the electrodes C and the oscillator

Figure 10.16. Hartshorn and Ward dielectric test set (H. Tinsley & Co. Ltd.)

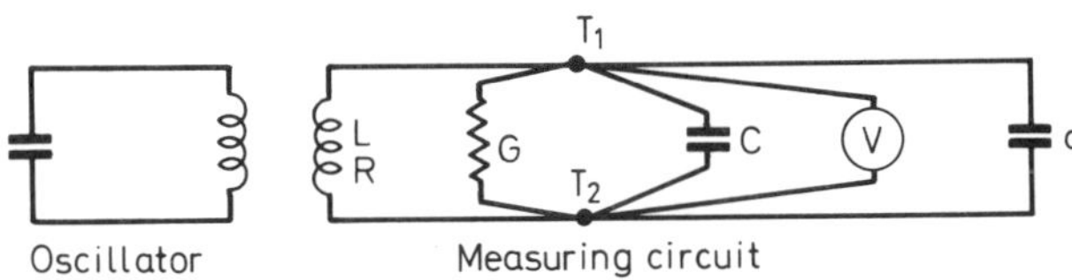

Figure 10.17. Basic circuit of Hartshorn and Ward apparatus

frequency adjusted to resonance as indicated by maximum galvanometer deflection θ. The circuit is then de-tuned using capacitor c to produce a deflection of $\theta/2$ first on one side of resonance by increasing the value of c and then on the other side of resonance by decreasing c, the total change of capacitance being denoted by Δc_i. Finally c is adjusted back to resonance. The specimen is now removed and resonance re-established by increasing the capacitance of C, after which the de-tuning to half deflection is repeated using c, the total change in its value being denoted by Δc_0 (less than Δc_i).

The true specimen capacitance C_s' is thus given by the change in setting of the main calibrated capacitor C after appropriate corrections for the specimen size (referred to later), and the loss tangent is given by the expression

$$\frac{\Delta c_i - \Delta c_0}{2C_s'}$$

Figure 10.18 illustrates the process of measuring the degree of de-tuning and shows the total circuit capacitance C (not required to be known) plotted

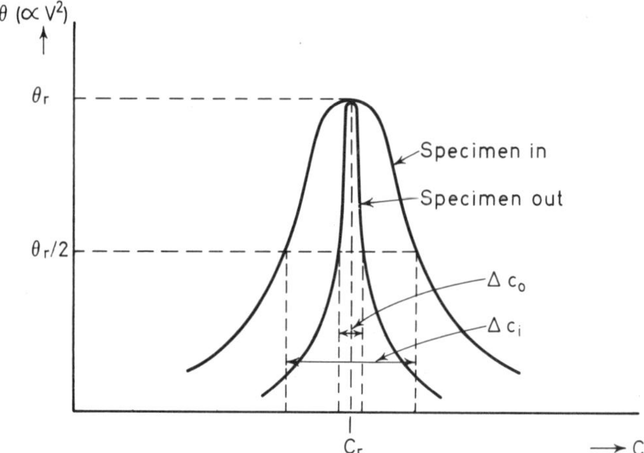

Figure 10.18. Resonance curves for Hartshorn and Ward apparatus with specimen 'in' and 'out'

against the galvanometer deflection at one frequency for both the specimen 'in' and the specimen 'out' conditions.

The deflections at resonance for both conditions are shown here equal for convenience, but this is not necessary, although for accuracy they should be kept as large as possible and this is achieved by moving the oscillator nearer or away as appropriate.

For low loss tangents, of the order of 0·000 2 or less, the difference $\Delta c_i - \Delta c_0$ becomes too small to measure accurately in this way (for example, with $C_s' = 20$ pF typically $\Delta c_i = 0{\cdot}65$ and $\Delta c_0 = 0{\cdot}64$ pF at 1 MHz for $\tan\delta = 0{\cdot}000\,25$) and an alternative method based on the ratio of deflections at resonance is used. It can be shown that

$$\frac{\Delta c_i}{\Delta c_0} = \frac{\theta_0}{\theta_i},$$

where θ_0, θ_i are the deflections at resonance corresponding to specimen 'out' and specimen 'in' respectively, *providing the oscillator is not disturbed during the measurements* (i.e. the coupling is not altered). By re-arranging,

$$\Delta c_i - \Delta c_0 = \left[\left(\frac{\theta_0}{\theta_i}\right)^{\frac{1}{2}} - 1\right] \Delta c_0$$

so that the loss equation becomes

$$\frac{\left[\left(\frac{\theta_0}{\theta_i}\right)^{\frac{1}{2}} - 1\right] \Delta c_0}{2C_s'}$$

and for small losses this approximates further to

$$\frac{\Delta\theta \Delta c_0}{4\theta_i C_s'} \quad \text{where } \Delta\theta = \theta_0 - \theta_i.$$

Since Δc_0 is a constant for a particular apparatus and frequency, by using a fixed value of θ_i a determination is reduced to a measurement of $\Delta\theta$ and C_s' only. For the example given above of a loss tangent of 0·000 25, $\Delta\theta$ would be 25 mm using a metre scale and θ_i = 800 mm so that measurements down to 0·000 01, corresponding to $\Delta\theta$ = 1 mm are quite feasible. Some care in ensuring stability of the deflections is essential at these low levels and Barrie's[49] paper is particularly helpful.

10.3.6 Measurements above 100 MHz

At frequencies above 10^8 Hz the number of methods available for dielectric measurements is very restricted and the two of principal interest are those employing standing wave methods and cavity resonator techniques.

In the standing wave method a waveguide is employed appropriate to the particular frequency range of interest and a dielectric specimen of rectangular or cylindrical section, depending on the type of waveguide, which must be a close fit in the guide, is inserted in one end. The length of the specimen required is a function of its permittivity and the wavelength corresponding to the test frequency and it may be necessary to adjust the length by machining after an initial measurement. The frequency of a generator coupled to the opposite end of the guide is adjusted to resonance, a suitable crystal probe inserted in a slot in the guide serving as a detector. The detector may be accurately positioned along the axis of the guide by means of a vernier slide so that the positions of the standing wave nodes can be evaluated. The standing wave pattern is studied with the specimen both in position and removed, and the permittivity and loss tangent are given by rather complicated equations, which yield a number of solutions, and some care is needed in obtaining the correct one.

Further details are given by Roberts and Von Hippel[50], Westphal[51] and Gevers[52].

With cavity resonators the specimen is normally a disc, which forms the end of a cavity, coupled to a waveguide. As with the standing wave methods a generator and crystal detector (in this case usually a square law type) are also coupled to the guide and the frequency adjusted to resonance. Measurements are now made of the bandwidth of the system, usually at half power points, by means of a fine frequency control (accurately calibrated) in a manner somewhat analogous to those using the Hartshorn and Ward apparatus. The solutions are again complex and tedious to solve. Gevers[52], Parry[53], Horner et al[54] and Brydon and Hepplestone[55] describe in detail various cavity resonators.

The two methods described are suitable for the frequency range from approximately 3×10^8 Hz to 3×10^{10} Hz.

10.3.7 Specification Tests

Power factor and permittivity tests are required by a number of British Standards for materials. In general, B.S. 2782: 1970 is invoked for both low and high frequency methods, and for the latter Method 207 cross-references back to B.S. 2067—the Hartshorn and Ward method.

The low frequency methods (Method 205 for 50 Hz, Method 206 for 800–1600 Hz) both specify a Schering bridge with Wagner earth as mandatory, but neither gives any details. The applied voltage is stated to be preferably 1000 V at 50 Hz, or less for thin or low electric strength specimens, and 100 V for 800–1600 Hz. Electrodes are mercury or metal foil only, although at 50 Hz plain brass discs applied with pressure are permitted, providing they give the same results. At 800–1600 Hz, power factor, but not permittivity, measurements are allowed to be made without a guard ring. No electrode sizes are given, but a guard ring width of 25 mm and a gap of approximately 1·5 mm are stipulated at both frequencies, the only limitation on central electrode area being that it should be sufficiently large to enable capacitance to be determined to $\pm 5\%$. For thin specimens the guard ring width is excessive and in practice this requirement means that the approximately 51 mm diameter Hartshorn and Ward specimen cannot be used for low frequency measurements.

Two specifications which give further details of the Schering bridge are B.S. 234: 1957, 'Loaded and Unloaded Ebonite for Electrical Purposes', and B.S. 903, 'Methods of Testing Vulcanised Rubber', Part C3: 1956, covering ebonite and vulcanised rubber respectively. The circuits and descriptions are identical and Figure 10.19 shows the arrangement for substitution measurements.

Measurements at 50 Hz and 800–1600 Hz are required by several other British Standards, but although in general a Schering bridge is mandatory, the methods are noteworthy for the wide variety of differences, often trivial, relating to permitted electrode materials and sizes which are allowed or specified, as the Table 10.1 indicates.

Although the Harshorn and Ward apparatus is mandatory in B.S. 2782, B.S. 903 allows any other apparatus to be used which can be shown to give the same results. Both B.S. 234 and B.S. 903 require only one specimen to be tested, whereas B.S. 2782 stipulates duplicate specimens.

The blanket A.S.T.M. specification covering dielectric measurements is D.150–68, 'a.c. Loss Characteristics and Dielectric Constant (Permittivity) of Solid Electrical Insulating Materials'. Some confusion has existed in the past between American and British terms used in connection with dielectrics. *Dielectric constant* is the preferred U.S. term for permittivity, but the latter term is probably more widely used in the U.K. Tan δ (or loss tangent in the U.K.) is called *dissipation factor* and given the symbol D in the U.S. The product permittivity × loss tangent, which is a measure of the energy loss in a material, is now called the *loss index* internationally, but formerly it was known in the U.S. as the *loss factor*.

Although extremely helpful and informative, A.S.T.M. D.150 is written in quite general terms and makes very few specific recommendations about methods or apparatus. For example, no electrode sizes are even suggested

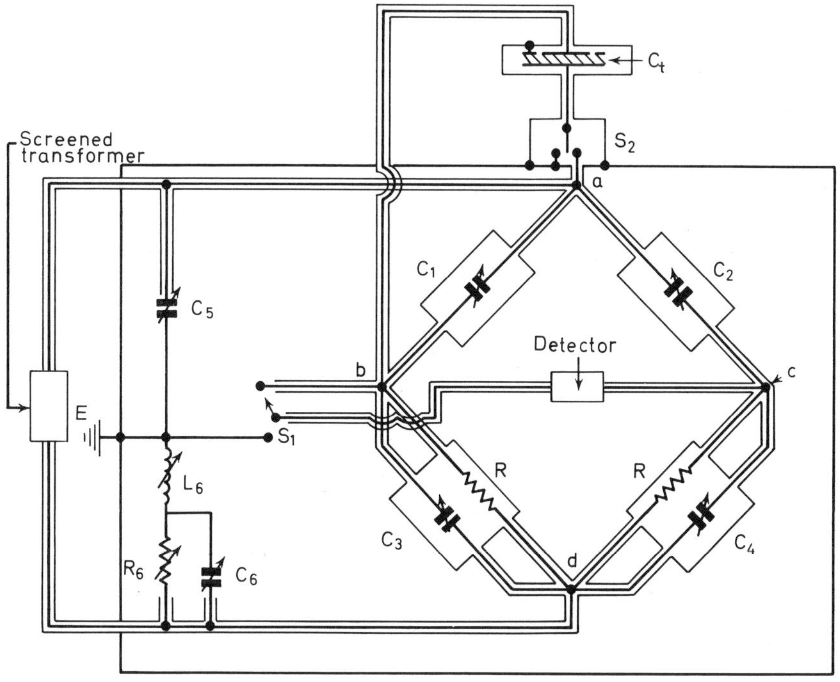

Figure 10.19. Schering bridge (B.S. 234 and B.S. 903)

and the only recommendations made are that the guard gap should be as small as possible and the guard width should be at least twice the specimen thickness. No particular type of apparatus is required to be used, but an appendix gives details of a number of bridge and other circuits which are suggested.

Some details are given of the liquid immersion technique, but this method is dealt with at some length in D.1531–62, 'Test for Dielectric Constant and Dissipation Factor of Polyethylene by Liquid Displacement Procedure'. The cell is a two terminal type with a double electrode arrangement requiring two specimens $2\frac{11}{16} \times 3\frac{15}{16} \times 0{\cdot}050$ in. The fluid is either very pure dry benzene or silicone oil of viscosity 1·0 cS, both of permittivity near 2·28, and only low density polyethylene is covered by the method. The frequency range covered is between 1000 Hz and 1 MHz and although no one type of

Table 10.1 BRITISH STANDARD POWER FACTOR AND PERMITTIVITY METHODS

Specification	Material	Bridge	Frequency (Hz)	Voltage	Electrode materials	Electrode sizes: Main, low voltage	Main, high voltage	Guard, width	Guard, gap	Comments
B.S. 2782: 1970	Plastics	Schering and Wagner earth	50	1 000 or less	Hg, foil or brass discs alone	any size		25 mm (1 in)	1·5 mm ($\frac{1}{16}$ in)	Guard ring may be omitted if no error caused
		Schering and Wagner earth	800–1 600	100	Hg, foil	any size		25 mm (1 in)	1·5 mm ($\frac{1}{16}$ in)	
B.S. 234: 1957	Ebonite	Schering and Wagner earth	800–1 600	100–150	Graphite, foil, vacuum deposited metal	50 mm	100 mm	23 mm	2 mm	
						150 mm	200 mm	46 mm	2 mm	
B.S. 903: 1956	Rubber	Exactly as for B.S. 234, except that mercury is also permitted								
B.S. 1137: 1966	Phenolic resin-bonded paper	Schering, earthing device not specified	50	500–1 000	Hg, foil or brass discs alone	6 in	7 in	$\frac{7}{16}$ in	$\frac{1}{16}$ in	Guard ring may be omitted if no error caused
			800–1 600	100	Hg, foil	6 in	7 in	$\frac{7}{16}$ in	$\frac{1}{16}$ in	
B.S. 2966: 1958	Phenolic resin-bonded cotton	Schering or any other suitable bridge	50	500–1 000	Foil (preferred), Hg	6 in	7 in	$\frac{7}{16}$ in	$\frac{1}{16}$ in	Guard ring may be omitted if no error caused
			1 000	100	Foil (preferred), Hg	6 in	7 in	$\frac{7}{16}$ in	$\frac{1}{16}$ in	
B.S. 3816: 1964	Cast epoxide material	Generally as B.S. 2782, except that for tests at 90°C electrodes should be silver or graphite dispersions								
B.S. 3815: 1964	Epoxide resin casting materials	As B.S. 3816								

apparatus is mandatory, equations are given for direct reading bridges and Q-meters. The lower limit of loss tangent which can be determined is 0·000 01. Figure 10.20 shows the cell.

A.S.T.M. D.1673–61, 'Tests for Dielectric Constant and Dissipation Factor of Expanded Cellular Plastics used for Electrical Insulation', although invoking A.S.T.M. D.150 for test methods, is interesting in that it takes account of the heterogeneity of such materials and their availability in large pieces by permitting specimens of up to 16 in square by 2 in thickness to be tested, albeit at a somewhat reduced frequency (<1 MHz).

A.S.T.M. D.1674–67, 'Testing Polymerisable Embedding Compounds Used for Electrical Insulation', is also relevant in that a simple two terminal

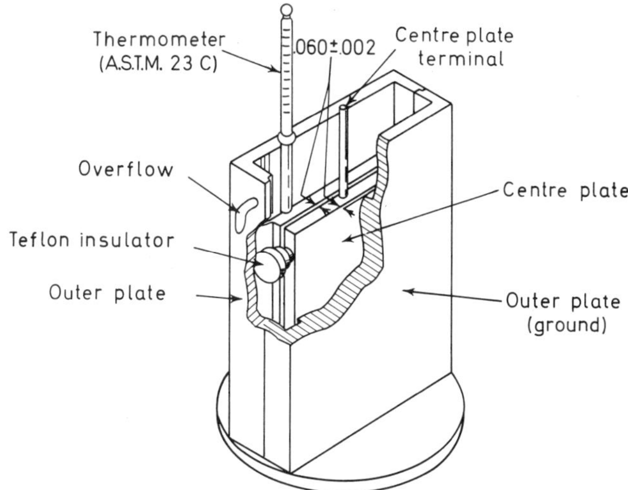

Figure 10.20. Liquid displacement cell (A.S.T.M. D.1531). Dimension in inches

electrode system suitable for elevated temperature measurements up to 300°C is described. Figure 10.21 shows the details.

The German standard for dielectric measurements is DIN.53483, 'Testing of Insulating Materials, Determination of Relative Permittivity and Loss Factor', covering frequencies between 15 Hz and 10 000 MHz. As with A.S.T.M. D.150 a good deal of scope is allowed and very few requirements are mandatory. For example, it is suggested that the specimen capacitance should be no smaller than 70 pF (for bridge measurements) and that specimens of dimensions 120 mm × 120 mm or greater are necessary, or at least desirable. Electrodes permitted are flat metal plates alone (which it is suggested are best suited for films), graphite and silver paints, sprayed zinc (for high temperature resistant materials only), vacuum deposited metal and fired on metallic salts (for glass and ceramic materials).

The use of flat plates alone is permitted both in A.S.T.M. D.150 and in B.S. 2782 (for 50 Hz only), but this is qualified by the requirement to use them only under very high pressure (possibly 100 kgf/cm^2 or more) or only if they can be shown to give the same results as accepted methods. DIN.53483 suggests pressures of the order 10–20 gf/cm^2, which the authors consider totally inadequate, particularly for films.

DIN.53483 is one of the few national specifications to give details of microwave methods. Up to 4000 MHz a coaxial line is used with tubular specimens and above this frequency cylindrical specimens are measured in a waveguide, using a standing-wave technique.

IEC 250[2] gives general recommendations for dielectric measurements and guidance on apparatus and techniques, as does ERA Report L/S 9[42].

Finally, an attempt to rationalise methods of test and the expression of design data for plastics is currently being undertaken by Committee PLC 36 of the B.S.I. Recommendations 2.1 and 2.2 of B.S. 4618:1970 cover permittivity and loss tangent respectively. These recommendations give valuable advice for selecting frequency and temperature intervals, etc., and

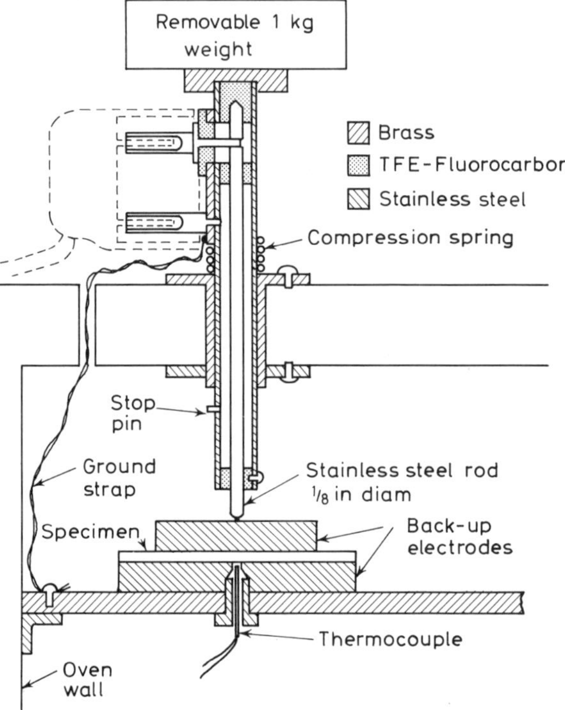

Figure 10.21. Elevated temperature apparatus (A.S.T.M. D.1674)

graphically displaying the results in a clear, standardised way. If these documents become widely accepted then the present confusion resulting from the use of different methods, frequencies, temperatures, stresses, etc., by manufacturers may yet disappear.

10.4 ELECTRIC STRENGTH MEASUREMENTS

10.4.1 Introduction

As already explained in Section 10.1.6, intrinsic values apart, the electric strength of insulation is dependent on so many variables that there is no body of good practice to rely upon, nor is there a 'correct' value to aim at, and consistency of results between one experimenter and another can only be achieved by mutual attention to the test variables. In other words, specifications are essential, but it should not be expected that two apparently fairly similar test methods will give similar results. The tests to be described commonly measure the 'industrial electric strength' under discharge conditions, and over a time period of the order of one minute.

10.4.2 United Kingdom Tests

ELECTRODES AND SPECIMEN FORMS

B.S. 2918: 1957, 'Electric Strength of Solid Insulating Materials at Power Frequencies', is the blanket document, but although published over 10 years

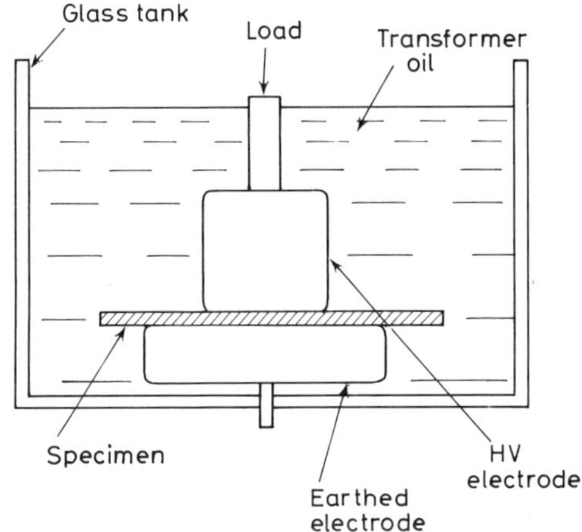

Figure 10.22. Electric strength electrodes for sheet materials (B.S. 2782)

ago, oddly it is not generally invoked by other, later, specifications containing electric strength requirements, and many of these specifications give the method of test in full detail. Specifications covering plastics materials (at least the later documents) invoke B.S. 2782: 1970, Methods 201.

The most frequently occurring test is that used for flat mouldings or sheet material, and the standard electrode system for measurements on 4 in (100 mm) discs or squares (usually $1\frac{1}{8}$ in ($3 \pm 0{\cdot}2$ mm) thick), consists of a pair of solid brass cylinders, the lower earthed one being 3 in (75 mm) diameter by 1 in (25 mm) thick and the upper high voltage electrode being 1·5 in (38 mm) diameter by 1·5 in (38 mm) in height. The arrangement is shown in Figure 10.22, and is completed by the addition of some means of applying loading to the top electrode. For most tests the electrodes are immersed in transformer oil (complying with B.S. 148: 1959, 'Insulating Oil for Transformers and Switchgear') and the temperature is either 20 ± 5°C or 90 ± 2°C for plastics materials.

The electrodes have radiused edges to reduce the local stress and until recently the radius was $\frac{1}{32}$ in (0·8 mm) maximum approximately, throughout British Standards. However, B.S. 2782 was recently amended to conform to metric units and, more particularly, to IEC practice [see the IEC 243 document covering standard methods of test for electric strength[56]] and the radius is now specified as 3 mm; also, the diameter of the smaller electrode is reduced to 25 mm.

Since, with the more recent specifications for plastics involving electric strength tests [e.g. B.S. 3784: 1964, 'Polytetrafluoroethylene Sheet', B.S. 3816: 1964, 'Cast Epoxide Resin . . .'], B.S. 2782 is invoked direct, it appears that the earlier material specifications may ultimately be brought into line with the latter document; therefore attention will be focused on B.S. 2782 methods to avoid confusion. Seven methods are detailed, numbered 201A–G inclusive. Methods 201A, B, C, and G employ the electrodes described above and respectively cover moulding material, flexible extrusion compound, sheet and casting/laminating resin systems. The methods differ only in respect of specimen preparation and specimen thickness. The latter is required to be $3 \pm 0{\cdot}2$ mm for mouldings and castings, whereas for sheet it is as supplied; for extrusion compounds it is also 3 mm, or as stipulated in the appropriate material specification. Method 201E also covers sheet material, but is intended principally for laminated materials whose electric strength may be much lower in the plane of the laminate than normal to it. The test is therefore conducted on specimens 100 ± 2 mm long by $25 \pm 0{\cdot}2$ mm wide placed edgewise between metal plates large enough to overlap all round. It is important that the specimen has edges truly machined and to assist the mechanical stability of the arrangement, two or three specimens may be used to support the electrodes, although of course fresh specimens are required for replicate tests.

Method 201D covers measurements on tubes in a direction normal to the wall. The outer electrode is a 25 mm wide band of sheet metal, and the inner a metal mandrel or a sheet metal split cylinder sprung into position and overlapping the outer electrode by at least 25 mm each end. For very large tubes (internal diameter > 100 mm) the internal electrode is a 38 mm metal disc flexible enough to 'conform to the curvature of the inner surface of the tube' and the outer electrode band is increased threefold in width.

Method 201F also covers tube (but this time for measurements parallel to the axis) both round and square, and also rod, and normally specimens $25 \pm 0{\cdot}2$ mm long with true ends are tested in a similar manner to laminated materials between metal plates.

Figure 10.23 shows typical electrode arrangements.

The other common size of B.S. electrode is that for narrow strips or tapes, consisting of vertical $\frac{1}{4}$ in coaxial brass rods radiused to $\frac{1}{32}$ in, the upper being of weight 50 g. These are mounted five pairs at a time in suitable insulating blocks so that the quintuplicate tests usually specified can be rapidly carried out. Essentially the test is for thin materials tested in air and consequently it is limited to voltages of about 10 kV or less. B.S. 3924: 1965, 'Pressure-sensitive Adhesive Tapes for Electrical Purposes', gives details and Figure 10.24 shows the electrode arrangement.

An unusually small electrode system is given in B.S. 3873: 1965, 'Polytetrafluoroethylene Moulded Basic Shapes'. As the title implies, a variety of specimen sizes may be involved and the electrodes are designed so that very small pieces can be tested. A test specimen $\frac{3}{8}$ in (9·5 mm) diameter by 0·03 in (0·76 mm) thickness is prepared by cutting from a moulding and it is important that the specimen has smooth square edges and plane parallel faces. The specimen is inserted in the centre of a 2 in long non-rigid electrical grade sleeving which must be of such a bore as to grip it tightly, and the assembly mounted in the electrodes shown in Figure 10.25, which are then immersed in transformer oil. The specimen is subjected to a proof voltage of either 300 V/mil or 600 V/mil (depending on grade) for one minute. Five specimens are tested.

In B.S. 903, 'Methods of Test for Vulcanised Rubber', Part C4: 1957, the electric strength test is generally similar to those of B.S. 2782, except that

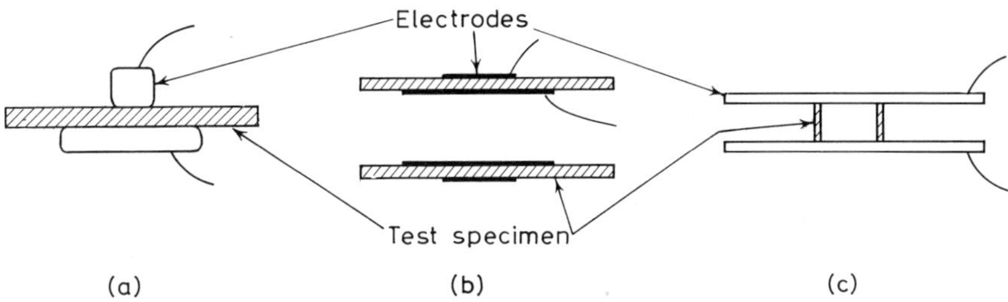

Figure 10.23. Electric strength electrode/specimen arrangements (B.S. 2782). (a) Methods 201A, B, C, and G, (b) 201D and (c) 201E

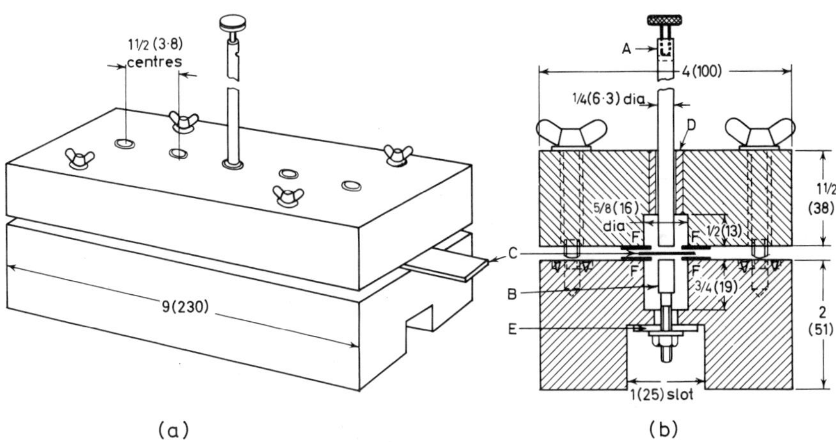

Figure 10.24. Electric strength electrodes for tapes (B.S. 3924). (a) General arrangement of apparatus, (b) section of apparatus through electrodes with top slightly raised. Dimensions in inches with millimetre equivalents in parentheses

the older cylindrical electrodes, 38 mm and 76 mm diameter with 0·8 mm radiused edges are used, and the specimen is only 1·25 mm thick (3·2 mm for ebonite). For the 20 s step-by-step method (see below), triplicate specimens plus the trial specimen are required. This test is given as an alternative to the 'one minute' electric strength which requires at least ten specimens. Additionally, a one minute proof test is also described, the details being as before, except that five specimens are tested at the required voltage level.

Other special kinds of electrodes may be required for finished products. For example, B.S. 2848: 1957, 'Flexible Insulating Sleeving for Electrical Purposes', which includes sleeving of PTFE, silicone rubber, PVC and polyethylene, gives an 'ad hoc' test in which a specimen filled with one or more copper wires is wound upon a metal mandrel and proof tested. Further details are given in Section 10.7.3.

VOLTAGE APPLICATION

For tests other than proof tests ('go-no-go' tests at a given voltage stress for a specified period), a trial specimen is usually tested initially by raising the

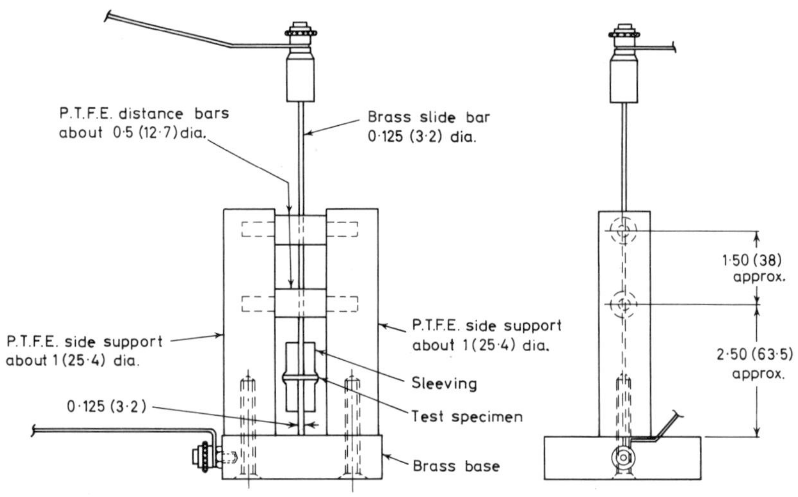

Figure 10.25. Electrodes for testing small PTFE mouldings (B.S. 3873). Dimensions in inches with millimetre equivalents in parentheses

voltage from zero at a smooth, uniform rate in order to produce a breakdown in 10–20 s. The starting voltage for the test proper is then taken as 40% of the trial specimen failure voltage and the test proceeds in defined voltage steps in the series 10, 11, 12 . . . 20, 22, 24 . . . 50, 55, 60 . . . 100 kV etc. or submultiples thereof, each step being maintained for 20 s.

The electric strength is defined as the maximum voltage withstood for 20 s in the case of the axial tests on 25 mm long tube and rod specimens, and the parallel-to-plane tests on 25 mm laminates, and the maximum voltage withstood divided by the specimen thickness for the remaining tests. (In the U.S. the electric strength (or rather 'dielectric strength') is defined as the actual breakdown voltage divided by thickness.) Two specimens are normally tested, apart from any 'trials', and most tests are conducted under transformer oil at room temperature. The thermosetting plastics more usually are tested at 90°C (see B.S. 771: 1959, 'Phenolic Moulding Materials', for example).

BREAKDOWN TESTER

Safety apart, there are a number of hazards associated with the tester itself, amongst which are errors caused by loading of the transformer due to excessive specimen leakage current or discharge currents, distortion of waveform by poor transformer design, insufficient power output, malfunctioning of circuit breakers, and so on. Since it is the peak value of voltage which determines breakdown, whereas many indicating meters are average value instruments (although r.m.s.—scaled) it is essential to check the meter readings against a peak measuring device, such as a sphere gap; B.S. 358: 1960, 'Method for the Measurement of Voltage with Sphere Gaps', is particularly helpful here and 5 cm spheres are appropriate for the range 10–50 kV, which is relevant to most of the requirements in plastics specifications.

10.4.3 United States Tests

The A.S.T.M. blanket test for electric strength is D.149–64, 'Dielectric Breakdown Voltage and Dielectric Strength of Electrical Insulating Materials at Commercial Power Frequencies'. Electrodes for sheet and film materials and rubber are 2 in diameter by 1 in height with the radius on the edges of $\frac{1}{4}$ in; for other solid materials the diameter is reduced to 1 in and the radius to $\frac{1}{8}$ in, the height being unchanged, but it is not very clear why the two sets of electrodes should be so close together in size, nor why and when one set is used in preference to the other. For tapes and films the electrodes are identical to the $\frac{1}{4}$ in British Standard ones.

Three types of voltage application are described, as follows:

1. *Short-time test*—voltage increased uniformly from zero to breakdown at four rates of rise between 100 and 3000 V/s. This test is usually applied to the trial specimen.
2. *Slow-rate-of-rise test*—50% of the short-time failure voltage applied initially and then increased smoothly to failure at a rate defined by the appropriate material specification.
3. *Step-by-step*—50% of the short-time failure voltage applied initially, then voltage increased in steps defined by the material specification.

Most of the tests require quintuplicate replication.

10.4.4 DIN Tests

DIN.53481, 'Testing of Electrical Insulating Materials: Determination of Breakdown Voltage and Electric Strength at Power Frequency', Parts 1 and 2, covers German practice. In Part 1 the principal electrodes for sheet materials are a sphere–plane disc system, the sizes varying from 10 mm/25 mm to 50 mm/100 mm, and the specimen thickness is 3 mm. The electrode system for tapes and films appears very similar to the B.S. and A.S.T.M. types.

Apart from the standard one minute electric strength determinations provision is also made for five, ten and thirty minute tests.

Part 2 appears to cover more rigorous techniques, approaching those used for intrinsic values, since specially contoured electrodes are specified with very large radius edges and also sphere–sphere electrodes are detailed.

10.5 TRACKING RESISTANCE AND ARC RESISTANCE MEASUREMENTS

10.5.1 Tracking Resistance

INTRODUCTION

Numerous tracking resistance tests have been developed in attempts to simulate the results of practical failures in the field and in general no completely satisfactory test has yet been devised which places materials in the same order of merit as experienced in the field. However, in Europe and in the U.K. test divergencies have been reduced until virtually a single test method is now common to both areas with minor, although not insignificant, differences.

BRITISH STANDARD TESTS

In the U.K. the accepted low voltage test is now incorporated in B.S. 3781: 1964, 'Method for Determining the Comparative Tracking Index of Solid Insulating Material'. This test involves the application of two chisel shaped electrodes (Figure 10.26) of brass, 4 mm apart, to the surface of a test specimen with a prescribed load (100 gf) at the tip and with a specified 50 Hz voltage between them. Drops of 0·1% NH_4Cl solution of size 20–25 mm^3 are allowed to fall at the rate of one every 30 ± 5 s, until tracking occurs or until 100 drops have fallen, whichever is the sooner, tracking being defined arbitrarily by the current in the circuit reaching a specified level.

The test is performed at a number of different voltage levels and a curve drawn of the number of drops to track at each level against test voltage. The comparative tracking index (c.t.i.) is defined by the asymptote to the curve, which is usually of roughly hyperbolic shape, or alternatively by the voltage at the 50 drop level, and is expressed as a pure number, not as a voltage.

Figure 10.27 shows the kind of curves that can be obtained with this test, although frequently they are not quite as smooth as indicated.

The test described is really only suitable for test voltages up to 500 V. Above this level flashovers tend to occur and the end point becomes increasingly difficult to define. The significance of the test voltage and its relationship with practical use voltages has not been established and the comparative tracking index obtained should not be regarded as a design level below which it is safe to operate. This is the reason why the index is expressed as a number, to dissociate it from an actual voltage and to emphasise its true importance as a simple guide for comparing materials.

Some care should be taken in applying the test. It is usually necessary to make measurements at about six different voltage levels to define a curve adequately and the scatter of results is high. This partly reflects the inhomogeneity of most filled plastics materials, but also it is an inherent feature

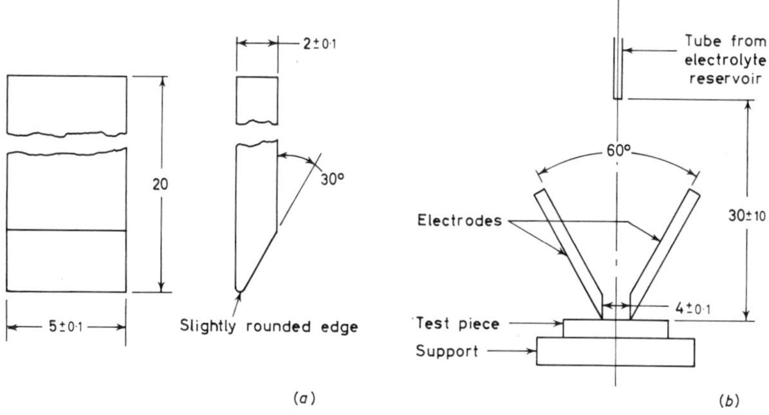

Figure 10.26. Comparative tracking index electrodes (B.S. 3781). (a) Electrode, (b) electrode arrangement. Dimensions in millimetres

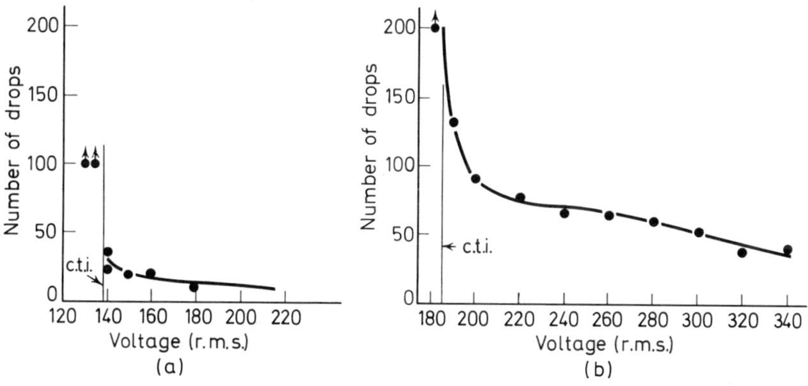

Figure 10.27. Example of curves for determining the comparative tracking index to B.S. 3781. (a) Curve obtained with a material having a c.t.i. coinciding with the 50 drop point, (b) curve obtained with a material where the 50 drop point cannot be used to establish the c.t.i.

of a test which endeavours to measure such a complex phenomenon as surface failure, dependent as it must be on the formation of a random path across the insulation surface. It is often possible to obtain the odd tracking failure at a voltage below the c.t.i. and there are cases on record of materials becoming non-tracking at significantly higher voltages than the c.t.i. Electrode materials other than brass are permitted providing it is reported and it is accepted that different results may be obtained.

Some materials do not track, but can fail by erosion, material between the electrodes volatilising to leave a crater. As in practice such behaviour would be undesirable, since terminals could become loosened, erosion is recognised in the test as a mode of failure (polymethyl methacrylate shows this behaviour for example).

One other tracking test appears in a British Standard, and that is for assessing aminoplastics in B.S. 1322: 1956, 'Aminoplastic Moulding Materials'. The test is crude and simple and rather spectacular. Two brass rod electrodes, 5 mm in diameter, are moulded into, or are otherwise made to fit tightly in holes drilled in a flat moulding, at a separation of $1\frac{1}{4}$ in, as shown in Figure 10.28. The electrodes are connected to a 200–250 V, 50 Hz supply via an 8 A fuse and a 10% NaCl solution poured on to the horizontal moulding. After the initial soaking, fresh solution is added at five minute intervals until 30 min has elapsed, or a track has occurred. In the absence of a track the moulding is washed repeatedly in water, finally wiped free of

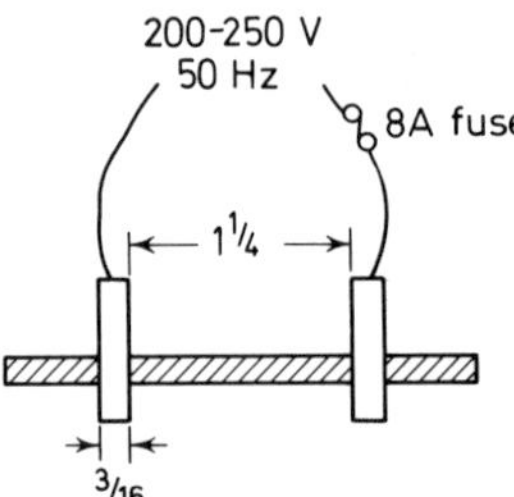

Figure 10.28. Tracking test for aminoplastics (B.S. 1322). Dimensions in inches

moisture and its insulation resistance measured between the terminals. The specification requirements are that the material shall not track and the insulation resistance should exceed 1 MΩ.

One snag with this test is that often erosion occurs around one or both electrodes, so loosening them, and rendering the insulation resistance test impracticable and rather meaningless.

A commercial version of the B.S. 3781 tracking tester is available from Yarsley Laboratories[44].

IEC TEST

The widely accepted European test from which the B.S. 3781 method derives is that of IEC Document 112, 'Recommended Method for Determining the Comparative Tracking Index of Solid Insulating Materials under Moist Conditions', and the essential difference with this test is that the c.t.i. is defined only in terms of the 50 drop voltage level, so that the two tests can give differing values with materials having a curve similar to Figure 10.27b, for example.

The IEC test also gives an alternative procedure whereby it is used as a proof test at a particular user voltage of interest. This is a dubious alternative, since a material successfully passing such a test may still fail in practice at the voltage concerned under other forms of contamination or after much longer time scales than arise in the test.

A.S.T.M. TESTS

The earliest U.S. method for tracking is given in A.S.T.M. D.2132–68, 'Test for Dust-and-Fog Tracking and Erosion Resistance of Electrical Insulating Materials'. The principle is that $2 \times \frac{1}{2}$ in brass or copper electrodes are placed flat upon the surface of a test piece, the assembly is coated with a synthetic dust and subsequently wetted by fine water spray in an

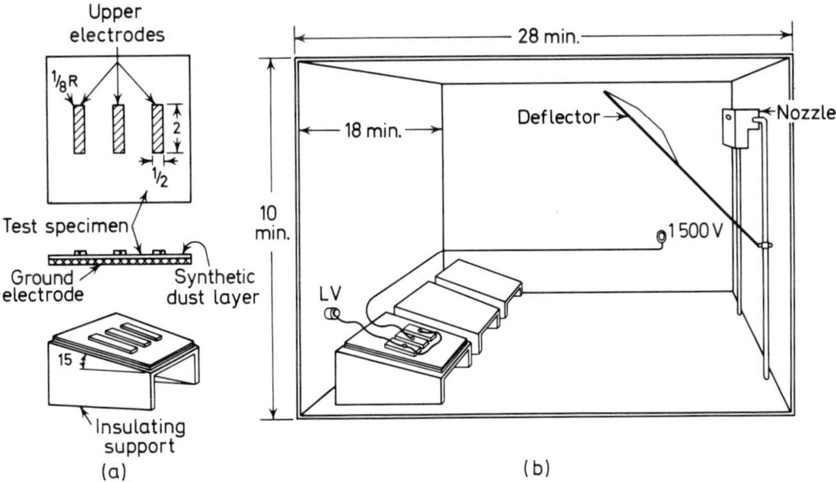

Figure 10.29. Dust-fog tracking and erosion test arrangement (A.S.T.M. D.2132). (a) Test arrangement of electrode system, (b) dust and fog test chamber, minimum recommended size. Dimensions in inches

appropriate chamber. A relatively high voltage of up to 1500 V applied to the electrodes induces tracking and/or erosion effects along the surface or through the thickness of the specimen and the test is continued until a permanent track ensues or the specimen is penetrated, the time to such failure being recorded in hours. Classification of tracking and erosion resistance is into several groups, a tracking resistant or erosion resistant material being defined arbitrarily as one withstanding respectively 100 h or 200 h without failure by the relevant mode. The specimen size is normally $5 \times 5 \times \frac{1}{16}$ in.

Although the test is simple in principle, the test details relating to the dust formulation, the fog deposition rate, the spray nozzle dimensions, the conductivity of the water used, etc., are carefully stipulated, as are the circuit details. Figure 10.29 shows the general arrangement.

Two other tracking methods are given in A.S.T.M. D.2302–69, 'Test for Differential Wet Tracking Resistance of Electrical Insulating Materials with Controlled Water-to-Metal Discharges', and A.S.T.M. D.2303–68, 'Test for Liquid Contaminant, Inclined–Plane Tracking and Erosion of Insulating Materials'.

The differential test is extremely rapid compared with the dust-fog and inclined plane tests; the test conditions are very severe and lead to failure in a matter of minutes. A test specimen with a small hole drilled in it is immersed at an angle in a trough of 1% NH_4Cl solution so that the hole just fills with liquid. A 500 V–3 kV voltage is applied in stages between the liquid and a square stainless steel electrode mounted on the dry emergent portion of the specimen, oriented so that one corner is close to the hole. Four stages of test conditions are imposed for different time periods, and the complete sequence of tests is repeated at increasing power levels. Although the test is supposed to correlate reasonably with the dust-fog test, a warning is given that comparisons are approximate and anomalies are to be expected.

In the inclined plane test a large flat specimen is inclined at 45° and electrolyte allowed to flow down its under surface, forming a bridge between two electrodes pressed against it. The electrolyte stream boils under the influence of the test voltage, which can lie between 1 and 10 kV, and the

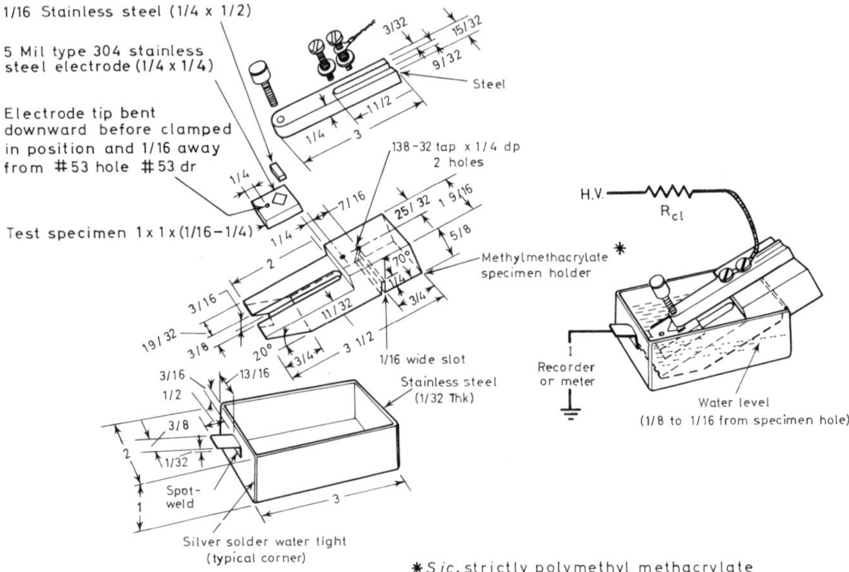

Figure 10.30. Differential wet tracking electrode assembly (A.S.T.M. D.2302). Dimensions in inches

resulting discontinuities occurring in the conducting path produce scintillation near the lower electrode. The test is stopped when a track has progressed up the specimen a distance of 1 in from the bottom electrode. The advantage of this test is that the conditions can be carefully controlled so that steady scintillation occurs, unlike the intermittent and somewhat uncontrolled scintillation which produces failure in 'drop' tests.

Both the last two tests are much more complicated to set up than the illustrations in Figures 10.30 and 10.31 might indicate.

DIN TEST

The tracking test of DIN.53480, 'Testing of Insulating Materials—Determination of Tracking Resistance at Operating Voltages below 1 kV', is

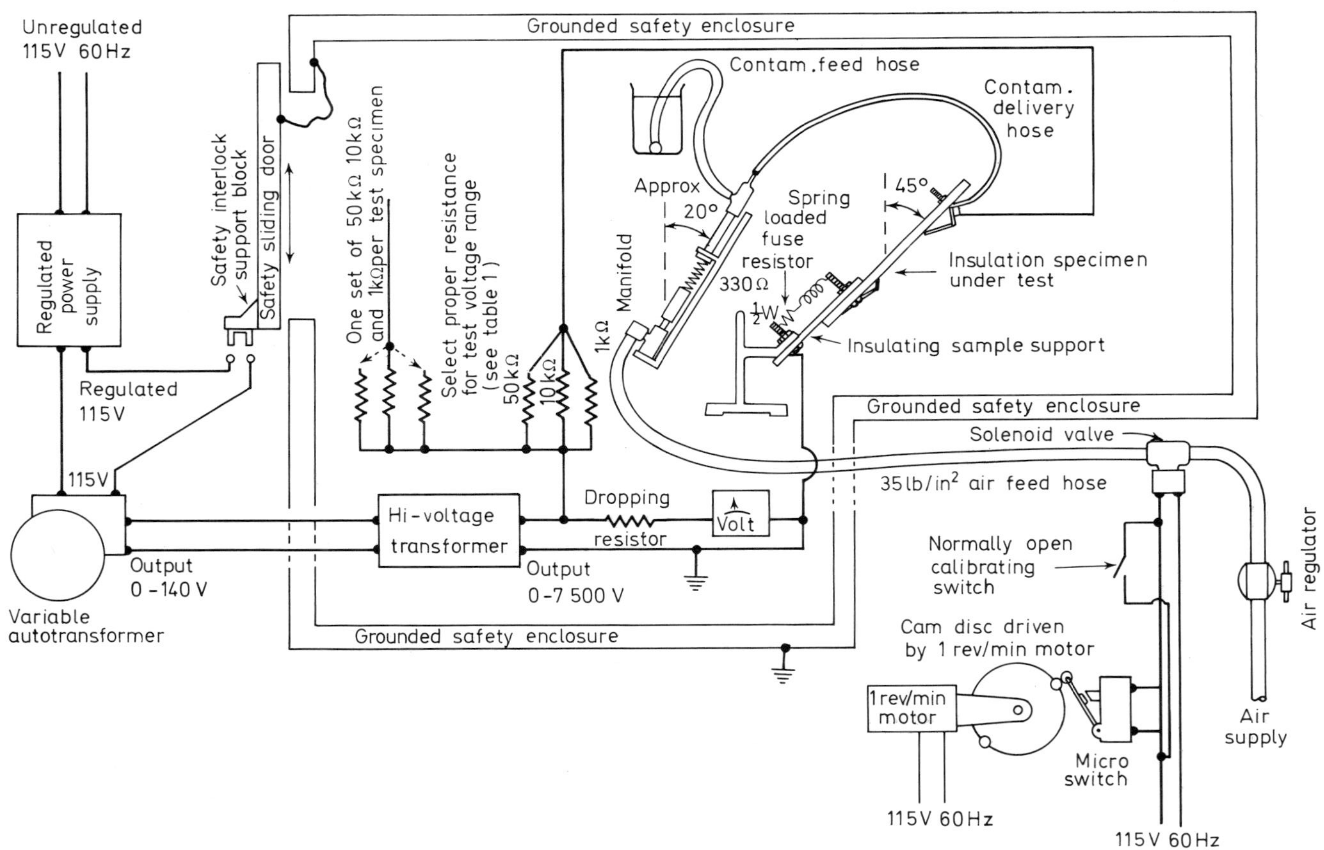

Figure 10.31. Inclined plane tracking apparatus (A.S.T.M. D.2303)

also very similar to the B.S. 3781 and IEC 112 methods. However, the electrodes are of platinum or at least platinum tipped brass, with a less acute angle at the chisel end, and the flat of the chisel is arranged to press on the specimen rather than the sharp edge, as in the other two methods. Two solutions are allowed, both containing the standard 0·1% NH_4Cl, but in one 0·5% of a specified wetting agent is incorporated.

The test is carried out at different voltage levels as usual, or alternatively it is used as a proof test at 380 V, the result being classified according to the number of drops required, or if the material withstood 100 drops, by the severity of the erosion produced.

The justification given for using platinum electrodes, in preference to copper or brass, is that they are more corrosion resistant. This argument ignores the fact that brass is widely used as a terminal material in electrical applications and therefore a test based on brass electrodes is likely to be more realistic. In case this appears to be a trivial point, it should be mentioned that the order in which the test classifies different materials may vary and even be reversed with differing electrode materials.

10.5.2 Arc Resistance

BRITISH STANDARD TESTS

Two arc resistance tests appear in B.S. 3497: 1967, 'Unimpregnated Asbestos Cement Boards (incombustible) for Electrical Purposes', but as these are not invoked by any specification involving plastics materials they will not be described in great detail. In the fuse wire arc test, a piece of fuse wire sandwiched between two specimens is ruptured by a 100 A d.c. current, the circuit being arranged so that immediately following rupture a 500 V peak 50 Hz voltage is applied to the fused ends of the wire. Fresh fuse wire is inserted in the same position each time and the test repeated, until either conduction occurs or 10 tests have been completed.

In the carbon arc test a 45 V, 10 A, 50 Hz arc is struck between two inclined carbon electrodes, which are mounted in a horizontal plane and press against the vertical specimen fixed to a hinged frame. A metal backing plate behind the specimen connected to an appropriate point in the circuit enables the penetration of the arc through the material to be detected. The arc resistance is the time in seconds from arc ignition to penetration.

A.S.T.M. TEST

The A.S.T.M. arc test is specified in D.495–61, 'Test for High Voltage, Low-current Arc Resistance of Solid Electrical Insulating Materials'. In this test, which is of a fairly complicated nature, a high voltage arc (15 kV) is struck between two surface electrodes placed on the specimen, the current initially being of a few milliamps. The arc is interrupted by an appropriate commutator in the circuit, the on/off intervals being carefully defined. As the test progresses the time sequence alters and the off period decreases until finally in the later stages the arc is continuous, but its severity is increased by increasing the arc current. The maximum time of the test if completed is 420 s and materials are classified by their time to failure.

The electrodes are either tungsten rods with chisel shaped ends mounted at 45° to the surface, or stainless steel strip electrodes placed corner to corner, in both cases the separation being 0·25 in: quintuplicate tests are a minimum requirement.

Needless to state, the test conditions and components used are rigorously detailed. Nevertheless, some ambiguity over the interpretation of the end-point, which is defined as the extinction of the arc (i.e. the current is entirely in the specimen surface), can occur, and some materials are not suitable for test because of melting, flaming, fouling of electrodes, and so on.

The apparatus is available in the U.K. from Yarsley Laboratories[44].

DIN TEST

The DIN. carbon arc test is described in DIN.53484, 'Testing of Insulating Materials: Determination of the Arc Resistance'. The test is very simple and involves the production of an arc between two 8 mm carbon rods inclined at 90° to each other and resting on the test surface. The electrodes are then separated mechanically at a maximum rate of 1 mm/s and the behaviour of the material categorised into one of six classes. The supply voltage is 220 V d.c. and the current is limited to approximately 10 A.

10.6 ANTISTATIC TESTS

10.6.1 Introduction

The problem of electrostatic charges, which can only occur on good insulators, is a serious one in several industries. Many situations involving rubbers and plastics, where 'static' can be built up to such an extent that a spark occurs, have an inherent fire hazard because of the presence of flammable solvents or explodable gas mixtures; mining, the petroleum industry and printing come easily to mind. The less well known hazard which exists in hospital operating theatres has already been referred to.

To a lesser extent the packaging industry has an aesthetic interest in the problem of dust and dirt pick-up on plastics which is aggravated by electrostatic attraction due to surface charges.

On the credit side, polymeric materials in fine particle form can be electrostatically (and very uniformly) coated on to an article by a spraying technique, which depends on the electrically insulating nature of the polymer, and outside the field of plastics altogether, the electrostatic precipitation of dust clouds is a vital process in many industries.

10.6.2 Direct Tests

Direct tests are those in which the electrostatic charge is measured directly by some appropriate high impedance instrument such as an electrometer or electrostatic voltmeter. Electrostatic field meters are described by Cross[57] and Shashoua[58]; Langdon[59] gives details of suitable enclosures and techniques for comparing fabricated articles and films. Commercial field strength meters are made by Davenport[60]. Figure 10.32 shows Langdon's apparatus for measuring the charge induced on moulded articles.

As far as the authors are aware, no specification tests exist for the direct measurement of surface charges on plastics materials, at the time of writing. However, methods of test for plastics films are currently being considered by sub-Committee PLC 17/2 of the B.S.I., and it is probable that a referee

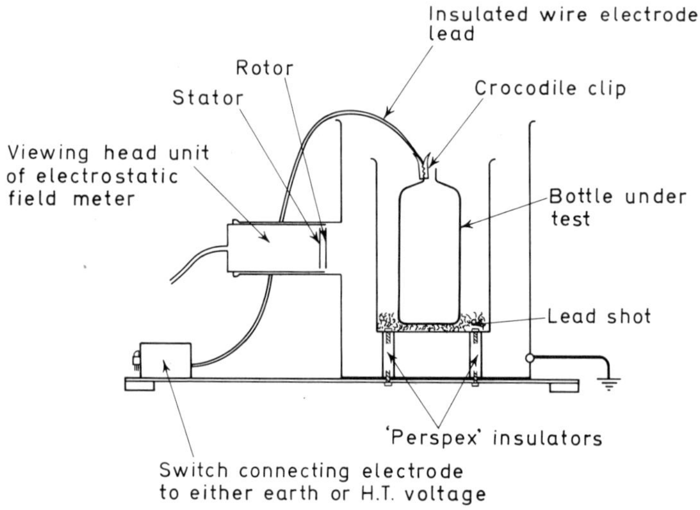

Figure 10.32. Antistatic testing of moulded articles (Langdon)

method based on the measurement of the time required for a charged specimen to lose half its charge (the half charge decay time) will appear in B.S. 2782 in due course.

10.6.3 Indirect Tests

There are two main aspects of electrostatic charge, vis à vis specific materials, first the amount of charge which is generated on a given material under a given set of conditions, and second the rate at which the charge is lost or dissipated. Little is known about the first aspect, although numerous attempts have been made to list materials in an order of decreasing susceptibility to charge, the so-called 'tribo-electric series', in which usually the non-polar plastics appear near the top. However, the position of each material may vary haphazardly with the test conditions and the experimenter alike, and the mechanism of charge generation is very little understood, although it is believed now to be entirely a contact separation phenomenon. Two RAPRA reviews by Morris[61] and Gale and Pacitti[62], cover the theory and practice of the subject in some detail and contain extensive bibliographies. The rate at which a given amount of charge, once deposited on an insulator, is dissipated, is a function of the resistivity of the material and both surface and volume resistances are involved here. Frequently, therefore, assessments of antistatic behaviour are made in terms of resistivity measurements and B.S. 2050: 1961, 'Electrical Resistance of Conductive and Anti-Static Products made from Flexible Polymeric Material', is typical. Under the conditions of the tests, which are mixtures of surface and/or volume resistance measurements, 'antistatic' is defined as a resistance of between $5 \times 10^4 \Omega$ and $10^8 \Omega$ and 'conducting' as a resistance of $< 5 \times 10^4 \Omega$. A conducting material is necessarily an antistatic one but the lower limit is essential where it is important to safeguard against the possibility of shock from mains operated electrical appliances, by the provision of adequate insulation from earth.

The test methods described are based on the use of small metal electrodes applied in conjunction with a suitable liquid wetting agent and are designed for use in the field on a wide variety of manufactured articles and products (e.g. anaesthetic masks, flooring for explosive areas, aircraft tyres and conveyor belts for mining). Rubber, both antistatic and conductive, is covered more specifically by B.S. 2044: 1953, 'Laboratory Tests for Resistivity of Conductive and Anti-Static Rubbers', which is intended for measurements on sheet material in the laboratory. Methods are given for both two- and four-terminal (potential drop) techniques, the latter using knife edge potential electrodes, and rather surprisingly an appendix includes constructional details of a valve electrometer circuit deemed essential for one of the methods.

10.7 MISCELLANEOUS TESTS

10.7.1 Introduction

Many electrical tests exist for specific materials or products and it would be impossible to cover them all in this chapter. Most of them involve in some way the application of one or other of the principles or tests already covered, although perhaps in a disguised form, and only one or two of these tests will be described here.

10.7.2 Tests for Adhesive Tapes

One of the more important requirements of adhesive electrical tapes is that the adhesive should not corrode copper, as this is a particular hazard in high humidity, high temperature atmospheres. In B.S. 3924: 1965, 'Pressure-sensitive Adhesive Tapes for Electrical Purposes', a test for electrolytic corrosion involves a length of tape wound adhesive-side-out on a large glass tube (see Figure 10.33) with two bare copper wires wound under pressure over the top of it and anchored. The wires are connected to a 120 V d.c. supply and the assembly subjected to a humidity cycling test for three days, after which the positive wire and a 'blank' sample have their breaking loads tested in a tensile testing machine. A corrosion liability factor (CLF) is then calculated as the percentage (loss in strength of the tested wire divided by the strength of the untested wire).

In practice, ten specimens are employed, half with wires in contact with the adhesive side, and half the non-adhesive side of the tape. All positive wires are tested and the standard deviation reported. Water soluble impurities in either the adhesive or the backing tend to be cationic and are

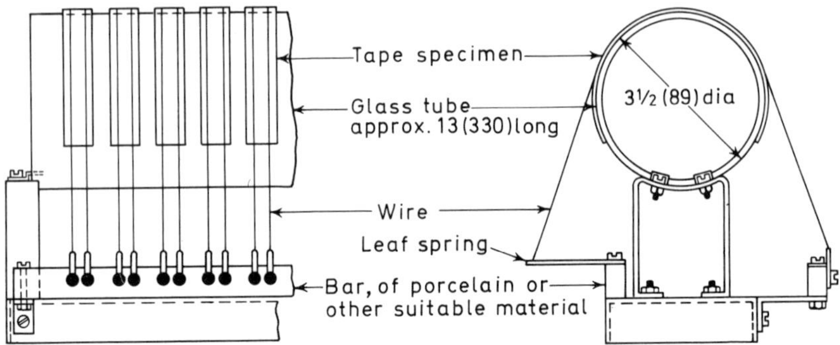

Figure 10.33. Electrolytic corrosion test apparatus (B.S. 3924). Dimensions in inches with millimetre equivalents in parentheses

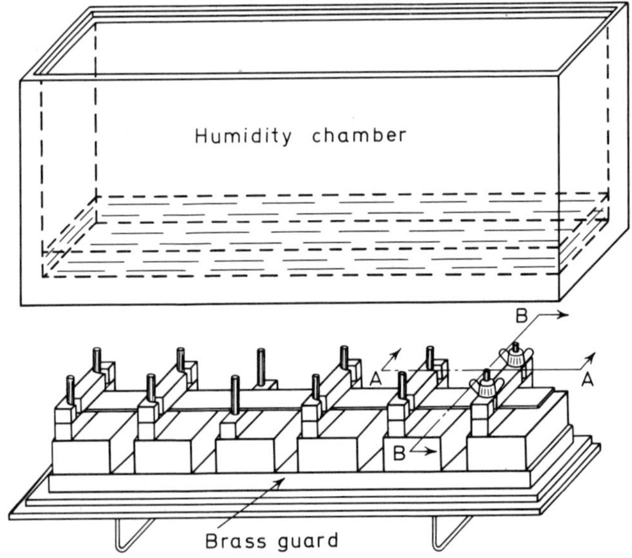

Figure 10.34. Insulation resistance electrodes for tapes (B.S. 3924)

therefore attracted to the positive wire which they attack and weaken. A very similar test formerly appeared in A.S.T.M. D.1000–62, 'Testing Pressure-sensitive Adhesive Coated Tapes used for Electrical Insulation', but has now been withdrawn. The property is now determined indirectly in the latest version (1968) on specimens mounted in the insulation resistance brass electrode system similar to that in Figure 10.34, which is taken from the corresponding B.S. document. The electrodes are conditioned at high humidity and any impurities which can form electrolytes cause a low insulation resistance to be obtained when the test voltage is applied between adjacent brass terminals. The B.S. test is very similar, but is termed 'insulation resistance'.

10.7.3 Tests for Insulating Sleeving

Tests on insulating sleeving are usually restricted to electric strength and insulation resistance. In B.S. 2848: 1957, 'Flexible Insulating Sleeving for Electrical Purposes', specimens for electric strength are filled with copper wire inserts (a single wire for the smaller bores) and then wrapped round a metal mandrel. A proof voltage is applied between mandrel and cores and the test is carried out either at normal temperature (with water immersion usually) or at maximum operating temperature in an oven. Insulation resistance is determined on specimens with a copper wire or tube insert, three foil electrodes being wound on top and connected so that end leakage is avoided. The test is conducted either at 75% r.h., 20°C, or after three cycles of accelerated damp heat conditioning. A.S.T.M. D.350–68, 'Testing Flexible Treated Sleeving Used for Electrical Insulation', gives only a breakdown voltage test on a specimen with foil outer and rod or tube inner electrodes.

10.7.4 Tests for Insulating Varnish

Insulating varnishes are tested in both B.S. and A.S.T.M. documents for electric strength only. In B.S. 2778: 1956, 'Organic Baking Impregnating Varnishes for Electrical Purposes', suitable specimens are prepared by dip-coating Kraft capacitor paper, whilst in A.S.T.M. D.115–55, 'Testing Varnishes Used for Electrical Insulation', the coating is on copper sheet. The B.S. test uses the standard $1\frac{1}{2}$ in on 3 in electrodes, whereas a single $\frac{1}{4}$ in rod is used in the A.S.T.M. test.

REFERENCES

1. 'Recommended methods of test for volume and surface resistivities of electrical insulating materials', *I.E.C. Publication 93,* International Electrotechnical Commission, Geneva (1958)
2. 'Recommended methods for the determination of the permittivity and dielectric dissipation factor of electrical insulating materials at power, audio and radio frequencies including metre wavelengths', *I.E.C. Publication 250,* International Electrotechnical Commission, Geneva (1969)
3. PARKMAN, N., 'Electrical Properties of High Polymers', *Physics of Plastics,* Chapter 6 (Edited by P. D. Ritchie), Iliffe Books Ltd., London (1965)
4. BRYDSON, J. A., 'Relation of Structure to Electrical and Optical Properties', *Plastics Materials,* Chapter 6, Iliffe Books Ltd., London (1966)
5. SCOTT, J. R., 'Electrical Tests', *Physical Testing of Rubbers,* Chapter 9, Maclaren & Sons Ltd., London (1965)

6. SHARBAUGH, A. H., 'Introduction to Electrical Property Tests', *Testing of Polymers,* Chapter 5, (Edited by J. V. Schmitz), Volume 1, Interscience Publishers, New York (1965)
7. SCOTT, A. H., 'D.C. Dielectric Conductance and Conductivity Measurements', *Testing of Polymers,* Chapter 6, (Edited by J. V. Schmitz), Volume 1, Interscience Publishers, New York (1965)
8. TUCKER, R. W., 'Dielectric Constant and Loss Measurements', *Testing of Polymers,* Chapter 7, (Edited by J. V. Schmitz), Volume 1, Interscience Publishers, New York (1965)
9. WARFIELD, R. W., 'Characterisation of Polymers by Electrical Resistivity Techniques', Chapter 8, *Testing of Polymers* (Edited by J. V. Schmitz), Volume 1, Interscience Publishers, New York (1965)
10. DAKIN, T. W., 'High Voltage Electrical Testing of Polymers', *Testing of Polymers,* Chapter 9, (Edited by J. V. Schmitz), Volume 1, Interscience Publishers, New York (1965)
11. BRUNTUN, J. H., 'Cavitation Erosion Testing of Polymers', *Testing of Polymers,* Chapter 10, (Edited by J. V. Schmitz), Volume 1, Interscience Publishers, New York (1965)
12. MOULLIN, E. B., 'An Elementary Description of Some Molecular Concepts of the Structure of Dielectrics', *J. Instn elect. Engrs,* Part 1, **91,** No. 48, 448 (December 1944)
13. SWISS, J., 'Fundamentals of Electrical Insulation, I, Chemistry of Insulation', DAKIN, T. W., 'II, Physics of Insulation', *Westingho. Engr,* **14,** No. 3, 114 (May 1954)
14. HOFFMANN, J. D., 'The Mechanical and Electrical Properties of Polymers: An Elementary Molecular Approach', *Trans. Brit. Instn Radio Engrs.,* **C.P. 4,** 42 (June 1957)
15. VAIL, C. B., 'Molecular Behaviour of Composite Electrical Insulation', *Electro-Technol.,* **69,** 82 (February 1962)
16. SHARBAUGH, A. H., 'Dielectric Constant and Loss—Their Significance in Insulation Selection', *Electro-Technol.,* **69,** 146 (February 1962)
17. DEVINS, J. C. and SHARBAUGH, A. H., 'The Fundamental Nature of Electrical Breakdown', *Electro-Technol.,* **68,** 104 (February 1961)
18. MASON, J. H., 'Testing Electrical Insulation for High-voltage Application', *Trans. Plast. Inst., Lond.,* **30,** No. 87, 171 (June 1962)
19. STARK, K. H., 'Testing and Evaluation of Plastics for use as Electrical Insulation', *Trans. Plast. Inst., Lond.,* **30,** No. 87, 184 (June 1962)
20. BAKER, W. P., *Electrical Insulation Measurements,* George Newnes Ltd., London (1965)
21. DORCAS, D. S. and SCOTT, R. N., 'Instrumentation for Measuring the D.C. Conductivity of Very High Resistivity Materials', *Rev. Sci. Instrum.,* **35,** No. 9, 1175 (September 1964)
22. SAZHIN, B. I. and SKURIKHINA, V. S., 'Method of Measuring Resistivities of the Order of 10^{17} to 10^{19} ohm cm.', *High Molec. Comp.,* **10,** 1535 (1960) (Russian)
23. Electronic Instruments Ltd., Richmond, Surrey.
24. W. G. Pye & Co. Ltd., Cambridge.
25. British Physical Laboratories Ltd., Radlett, Herts.
26. Hewlett–Packard Ltd., Slough, Bucks.
27. General Radio (U.K.) Ltd., Bourne End, Bucks.
28. HITCHCOX, G., 'Extending the Limits of Resistance Measurement using Electronic Techniques', *J. Brit. Instn Radio Engrs,* **16,** 299 (June 1956)
29. FRANCE, G., 'High Value Resistors and their Measurement', *Electron. Engng,* **29,** No. 6, 30 (June 1957)
30. MUNICK, R. J., 'Transient Electric Currents from Plastic Insulators', *J. appl. Phys.,* **27,** No. 10, 1114 (October 1956)
31. REDDISH, W., 'Conduction and Polarisation Processes in Plasticised Polyvinyl Chloride Compounds', *The Physical Properties of Polymers, Monograph No. 5,* Society of Chemical Industry, London, 138 (1959)
32. KAO, K. C., 'A New Electrode System for the Measurement of Surface Resistance', *J. Sci. Instrum.,* **39,** No. 5, 208 (May 1962)
33. 'Methods of Test for the Determination of the Insulation Resistance of Solid Insulating Materials', *I.E.C. Publication 167,* International Electrotechnical Commission, Geneva (1964)
34. VINCE, P. M., 'An Apparatus for the Measurement of the Permittivity and Loss Tangent of Glasses at Audio and Sub-audio Frequencies', *I.E.E. Conf. Dielect. Insul. Mater., Lond.* (1964)
35. SCHEIBER, D. J., 'An Ultra-Low-Frequency Bridge for Dielectric Measurements', *J. Res. nat. Bur. Stand.,* **65c,** 23 (1961)
36. BARNEY, W. M., 'Low Frequency Dielectric Investigations of Glasses', 1961 *Ann. Conf. Elect. Insul.,* **NAS-NRC 973,** 59
37. HAGUE, B., *Alternating Current Bridge Methods,* 5th edn, Pitman & Sons Ltd., London (1943)

38. The General Radio Experimenter, 36, Nos. 8/9 (Aug.–Sept. 1962). General Radio (U.K.) Ltd., Bourne End, Bucks.
39. Wayne Kerr Co. Ltd., New Malden, Surrey
40. H. W. Sullivan & Co. Ltd., Orpington, Kent.
41. Marconi Instruments Ltd., St. Albans, Herts.
42. 'Methods of Testing Permittivity and Loss Tangent of Dielectric Materials', *Technical Report L/S9,* Electrical Research Association, Leatherhead (1958)
43. HARTSHORN, L., *Radio Frequency Measurements by Bridge and Resonance Methods,* Chapman & Hall Ltd., London (1941)
44. Yarsley Laboratories, Chessington, Surrey.
45. LYNCH, A. C., 'Measurement of the Dielectric Properties of Low-loss Materials', *Proc. Instn elect. Engrs.,* **112,** No. 2, 426 (February 1965)
46. WESTON, D., 'Power Factor and Dielectric Constant of Polythene Compounds', *Plastics, Lond.,* **27,** No. 302, 105 (December 1962)
47. HAZEN, T., 'Some Special Electrical Test Methods for Telecommunication Cable Insulation', *Conf. Plast. Telecomm. Cables,* Plastics Institute (London) (April 1967)
48. HARTSHORN, L. and WARD, W. H., 'The Measurement of the Permittivity and Power Factor of Dielectrics at Frequencies from 10^4 to 10^8 cycles per second', *J. Instn elect. Engrs.,* **79,** 597 (1936)
49. BARRIE, I. T., 'Measurement of Very Low Dielectric Losses at Radio Frequencies', *Proc. Instn elect. Engrs.,* **112,** No. 2, 408 (February 1965)
50. ROBERTS, S. and HIPPEL A. VON, 'A New Method for Measuring Dielectric Constant and Loss in the Range of Centimetre Waves', *J. appl. Phys.,* **17,** 610 (1946)
51. WESTPHAL, W. B., 'Distributed Circuits', Section II A 2, *Dielectric Materials and Applications,* The Technology Press of MIT and John Wiley & Sons Inc., New York (1954)
52. GEVERS, M., 'Measurement of Dielectric Properties of Solids at Microwave Frequencies', *Precision Electrical Measurements—Proc. N.P.L. Symp.,* H.M.S.O., London (1955)
53. PARRY, J. V. L., 'The Measurement of Permittivity and Power Factor of Dielectrics at Frequencies from 300–600 Mc/s', *Proc. Instn elect. Engrs.,* **98,** Part III, 303 (1951)
54. HORNER, F., TAYLOR, T. A., DUNSMUIR, R., LAMB, J. and JACKSON, W., 'Resonance Methods of Dielectric Measurement at Centimetre Wavelengths', *J. Instn elect. Engrs.,* **93,** (III), No. 21, 53 (January 1946)
55. BRYDON, G. M. and HEPPLESTONE, D. J., 'The Microwave Measurement of Dielectric Constant and Tan δ over the Temperature Range 20°C–700°C', *Instn elect. Engrs. Conf. Dielect. Insul. Mater., Lond.* (April 1964)
56. 'Recommended Methods of Test for Electric Strength of Solid Insulating Materials at Power Frequencies', *I.E.C. Publication 243,* International Electrotechnical Commission, Geneva (1967)
57. CROSS, A. S., 'Two Electrostatic Field-Meters', Static Electrification, *Brit. J. appl. Phys.* Supplement No. 2 (1953)
58. SHASHOUA, V. E., 'Static Electricity in Polymers, I, Theory and Measurement', *J. Polymer Sci.,* **33,** No. 126, 65 (1958)
59. LANGDON, S. J., 'Assessment of Antistatic Properties of Plastics Materials', *Plastics, Lond.,* **29,** No. 322, 43 (August 1964)
60. Davenport (London) Ltd., Tewin Road, Welwyn Garden City, Herts.
61. MORRIS, W. T., 'Static Electrification of Polymers', *Technical Review 47,* Rubber & Plastics Research Association of Great Britain, Shawbury (1969)
62. GALE, G. M. and PACITTI, J., 'Anti-Static Agents: A Critical Review of the Literature', *Technical Review 43,* Rubber & Plastics Research Association of Great Britain, Shawbury (1968)

11

Tests for Temperature Dependence of Properties and Thermal Endurance

11.1 INTRODUCTION

Plastics materials as a whole, and thermoplastics in particular, are sensitive to temperature; whilst the ductile or flexible ones may demonstrate shortcomings in becoming brittle or rigid at some low temperature, far more serious as a rule is the softening and even degradation that may occur when the temperature is raised to only moderate levels, that is 'moderate' in the eyes of the designer, engineer or artisan more familiar with traditional materials such as metal, wood and stone.

Much effort has been expended in designing tests to measure the upper and lower temperatures limits of performance of plastics, for most of them do not demonstrate sharp melting points or *first order* transitions, but on the contrary change their physical condition over a range of temperature, the extent and position of which is dependent on a host of variables such as physical form, processing cycle, moisture content, time scale of measurement and physical parameter of interest. For those few plastics materials which show well defined *melting points*, suitable methods of test have already been described in Section 6.2 of Chapter 6.

Standardised procedures for assessing the softening characteristics in particular of plastics are invariably ad hoc in nature, selecting some mechanical property—usually bending stiffness or hardness—which is characterised indirectly by a measurement of deflection and the end point taken when some arbitrarily chosen value of deflection is reached. Rate of heating, and level of stress are equally arbitrarily chosen. These tests are described in the early sections of this chapter, but it will be realised at the outset that such tests yield figures *which have no absolute significance;* they are only intended as quality control methods and the data they yield should not be regarded as anything else. In the ultimate, the only satisfactory way of determining the upper or lower limits of usage temperature is to identify the property or properties of significance in the intended use (they may be electrical rather than mechanical) and then to make appropriate fundamental measurements to follow the deterioration of value with change in temperature. Some hints on this subject are given in the following sections.

Prior to all this, however, what might be termed a more fundamental parameter relating polymer behaviour to temperature is discussed; this involves the assessment of the change in flexibility or mobility of the polymer chains with temperature. The determination of *second order transition point* or *glass transition temperature* has also the merit of requiring only a small amount of sample.

Finally, completing the treatment of *reversible* effects of temperature, methods for measuring coefficients of linear and volumetric thermal expansion are described.

Most plastics materials, being based on a carbon–carbon chain backbone or containing a high content of this bond and of carbon–hydrogen bonding, have very limited resistance to actual degradation by heat. Depending on the temperature, the nature of the polymer and the presence or absence of oxygen, so initially breakdown or crosslinking (of thermoplastics) may result; however ultimately and invariably breakdown occurs into a variety of products ranging from carbon through to virtually pure monomer—again depending on the conditions, but particularly the nature of the polymer. The last sections of this chapter are therefore devoted to methods of test and investigational methods for the *irreversible* effects of temperature.

11.2 GLASS TRANSITION TEMPERATURE (T_g)

11.2.1 Brief Description of T_g

The Glass Transition Temperature, T_g, has been well described in a number of references[1–7]—many of which include typical values and techniques of determination—and it is only necessary to include some brief notes here.

A phase transition such as that from solid to liquid is termed a *first-order transition;* if such transition is manifested, as it is with small molecules but rarely with polymers, primary thermodynamic functions such as specific heat and specific volume show a discontinuous change with temperature at the transition point. A *second order transition* has been defined as one where the plot of a primary function against temperature remains continuous, but abruptly changes in slope. The *glass transition* is such a second-order transition.

It has already been seen in Chapter 8 how the deformation behaviour of a polymer is dependent on the time scale of the experiment, and how the concept of orientation time τ_m arises. This is a function of the mobility of the polymer chains; heating the polymers will increase their energy and mobility and reduce τ_m and cooling will have the opposite effects. Thus determination of the glass transition temperature, where the polymer changes from 'rubbery' to 'glassy', or vice versa, via a study of suitable thermodynamic function, cannot in fact yield some fundamental parameter, independent of arbitrary choice of experimental conditions, since the transition point is that temperature where the time scale of the experiment equates with the orientation time. A 'fast' method will equate with a lower τ_m than a 'slow' method and, since rise in temperature lowers τ_m, the 'fast' method will give a higher value for transition point than the 'slow' one. Brydson[7] quotes values for a polyoxacyclobutane polymer which ranged between 7°C and 32°C for respectively dilatometric and electrical (property) methods, where the frequencies (time scales) were correspondingly 10^{-2} Hz and 1000 Hz. Indeed, if the time scale is extended far enough the transition point can be made to disappear altogether (see Frith and Tuckett[3], p. 307). Notwithstanding these criticisms, determination of T_g is often an extremely useful means of characterising one aspect of the thermal behaviour of a polymer; the data can be quite reasonably reproducible if the same method is followed carefully and, as already stated, the technique has the great merit of requiring very little material. Furthermore, it should not be beyond the wit of the experimenter to select his technique—or more correctly the time scale thereof—to suit the type of property transition in which he is interested.

11.2.2 Methods for Obtaining T_g

There are no standardised methods (though in one official document a dilatometric method is recommended[8]), but a number of different techniques have been evolved, following the variation of various properties with temperature. Basically any derivative of the free energy may be followed, to assess T_g, or a property or its temperature coefficient may be used which is related to one such derivative or a combination of such derivatives. The important criterion is that the degree of change at T_g is sufficiently marked to yield an unambiguous answer; according to the nature of the polymer under examination, so the most suitable method or methods must be selected.

To treat the subject of the measurement of T_g of polymers comprehensively would take a disproportionate space in a book on general testing of plastics; little more therefore is attempted below than to list some of the more important and/or interesting techniques, give a very brief description and draw attention to some useful references.

METHODS BASED ON DIMENSIONAL CHANGE

For many polymers measurement of thermal expansion against temperature will provide a simple and inexpensive means of determining T_g. Either linear or cubical expansion may be used and, for the latter for example, a plot of this general type will be obtained ideally (see Figure 11.1).

The standardised techniques for determination of coefficients of expansion (see below Section 11.6) may be used, though they may not be the most experimentally convenient. A study of various dilatometric measurements was made some years ago by Boyer and Spencer[9] who quoted transition temperatures and coefficients of cubical expansion above and below them (straight line portions to left and to right of T_g in Figure 11.1). Just how sharp the transition point can be will be understood when it is observed that Boyer and Spencer found ratios of the v/t slopes of up to 2 to 1. Practical details of volume dilatometry are given by Millane and Mclaren[10] and a 'microdilatometer' is described by Meares[11]. The use of a density gradient column for the study of transitions is advocated by Gordon and Macnab[12] who applied it to polystyrene.

The examination of film samples may be achieved simply and rapidly—providing one has an Instron (see Chapter 8) or similar 'electronic' tester to hand—according to Rothstein and Spechler[13] who coupled a differential transformer to a table model Instron to measure linear displacement of films surrounded by a suitable oven within the working area of the machine; coefficients of linear expansion and transition temperatures are thus obtained.

It will be obvious that an interferometer should provide a convenient means of measuring small linear displacements, such as result from thermal expansion and contraction, and Wood et alia[14], for example, have applied this technique to the study of soft rubber. Likewise strain gauges provide another sensitive method of following linear dimensional change and their use in this direction has been tried successfully by Eisenberg and Rovira[15]. Heller and Lyman[16] used a differential pressure transducer to measure T_g, the transducer detecting changes in pressure between two containers, one containing the test sample and the other air for reference.

OPTICAL METHODS

In suitable cases a plot of refractive index (Chapter 16) against temperature will yield a plot of the type shown in Figure 11.2.

This technique has been described particularly by Wiley and co-workers in a series of papers[17]. Details of insulation for sub-ambient temperature

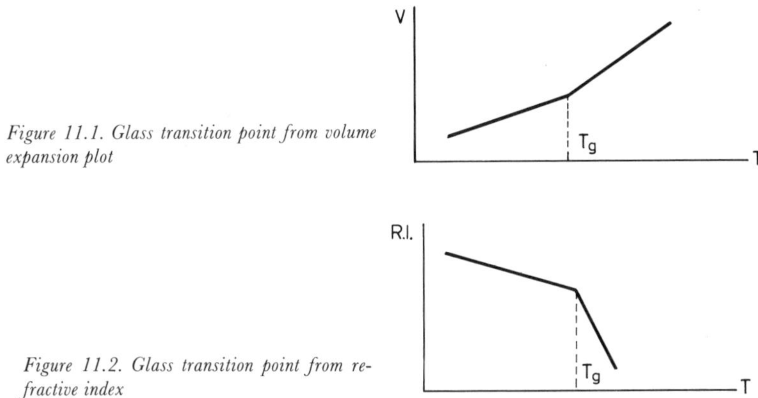

Figure 11.1. Glass transition point from volume expansion plot

Figure 11.2. Glass transition point from refractive index

measurements are given amongst other points of practical importance. Intimate contact between polymer sample and prisms is essential and may be achieved, for instance, by thermal softening of the material followed by pressing between the prisms.

Sheldon[18] states that T_g may be obtained by observing a sample, on the hot stage of a polarising microscope, for the appearance and disappearance of birefringence.

STATIC AND DYNAMIC MECHANICAL METHODS

Although 'static' mechanical methods of examining the temperature behaviour of polymers belong mainly to the ad hoc and softening point methods described in Section 11.3 below, one somewhat refined technique will be mentioned in the more illustrious company of T_g measurements. This is the so-called Du Pont *thermomechanical analyser*, the use of which is described in Garrett[19] in work on ethylene–propylene rubbers. It is an accessory to the Du Pont *differential thermal analyser* (the use of D.T.A. equipment is touched on below in this section and Section 11.8.2) and is basically a very sensitive penetrometer provided with accurate temperature programming equipment. Because surface hardness, as measured by penetrometer, is not a bulk property such as volume (as used in dilatometry), temperature changes used in the measurement need only affect the surface layers of the specimen and thus the technique affords a rapid method, for Garrett states that rates of heating and cooling of about 5 deg C/min may be employed successfully.

The subject of dynamic mechanical properties is considered in Chapter 15, and therefore the use of these measurements for obtaining transition temperatures is given with the absolute minimum of description. Nielsen and co-workers[20], for instance, describe the use of a torsion pendulum for examination of plasticised PVC compositions. The dissipation factor, or damping, plotted against temperature shows a pronounced peak. Reference 21 will also be found useful in presenting T_g data from dynamic mechanical methods in comparison with data from nuclear magnetic resonance (see below) for a number of polymers. The use of a vibrating reed for measuring dynamic mechanical properties, in a study of transition points and their meaning and significance, is described in Willbourn[22].

Gordon introduces the idea of measuring T_g by a ball rebound in Chapter 3 of his monograph (see Reference 2), where a peak is observed at the transition point in the plot of temperature against energy absorption of the ball bouncing on the polymer sample. The apparatus is described in detail in Reference 23.

Transition points from peaks in the hysteresis loss–temperature curves may be obtained by use of the 'rolling ball loss spectrometer' described by Cheetham[24]; a ball rolling over a perfectly elastic surface will move unhindered since the indentation caused at any point by the ball feeds back the stored energy to the ball as it moves on. The higher the hysteresis of the material, the more the movement of the ball is retarded. Cheetham's apparatus employs this principle in automatic fashion.

ELECTRICAL METHODS

The measurement of the electrical properties of polymers and plastics has been considered in Chapter 10. Inasmuch as their dielectric properties in particular are a function of the polar character of the interatomic bonds in the molecules and that, although these polar bonds may be on the side chains the movement of the latter must be influenced by the mobility of the backbone chain, it should obviously be feasible to measure transition points from the variation of certain electrical properties with temperature (see, for example, Reference 3).

Fuoss[25] has examined a number of polymers and plotted *loss factor* against temperature at frequencies of 60 Hz, 1000 Hz and 10 000 Hz. Characteristic peaks were obtained in most cases; further data is reported by Mead and Fuoss[26]. Nielsen et alia[20] give data which compare transition temperatures from dielectric measurements with those from the dynamic methods already mentioned. Reddish[27] has carried out similar investigations with PVC and chlorinated PVC.

Warfield[28] claims that second order transition points can be obtained from plots of volume resistivity against temperature, the dependence being somewhat haphazard apparently, as the data presented—for variously cured epoxides, a filled epoxide and a polyester—sometimes show a greater slope above the transition point and sometimes a lower one.

THERMAL METHODS

Whilst conventional methods such as plotting specific heat, thermal diffusivity or thermal conductivity against temperature, e.g. as considered by Wunderlich[29], Steere[30], Griskey[31] and Manley[32], have been used to identify T_g, probably the most used methods currently are those based on *differential thermal analysis*. This technique, as a general method for studying the thermal behaviour of polymers, is considered in Section 11.8.2 below.

A glass transition causes a shift in the thermogram base line, this being the plot of temperature difference (between heated polymer sample and

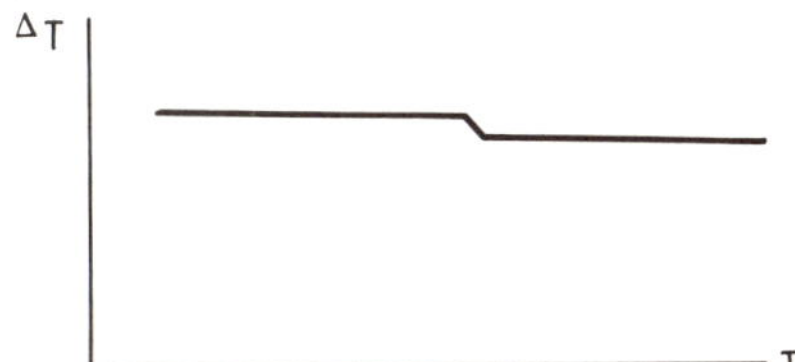

Figure 11.3. T_g from differential thermal analysis

similarly heated inert reference material) against temperature; a plot of the type of Figure 11.3 results if T_g can be obtained, i.e. if the technique is suitable for the polymer under evaluation, that is if there is a sharp enough change in heat capacity at the transition point.

The reader is referred to page 393 et seq of Reference 33 and Chapter 2 of Reference 34 for detailed expositions of the subject. Four papers[35] will be found useful for practical details.

NUCLEAR MAGNETIC RESONANCE

A description of nuclear magnetic resonance (N.M.R.) is beyond the scope of this book. The reader is referred to Chapter 8 in Reference 33 for a general description of the phenomenon and to References 21 and 36 for its application to measurement of transition temperatures.

11.3 THERMAL YIELD TESTS (SOFTENING POINT, ETC.)

The results of the ad hoc tests, which are referred to in Section 11.1 and which have found wide favour as standardised methods, flourish under a number of 'trivial' names, particularly: *softening point, plastic yield, heat distortion point, deformation under load* and *temperature of deflection under load.* They measure the temperature response of the material under test to a variety of stress conditions, especially in bending or under a penetrative load. There are two methods of obtaining an end point; the most popular is to select some arbitrary stress condition and raise the temperature until the strain has reached some arbitrary level—essentially a measure of the temperature at which the modulus deteriorates to some prescribed level (but without reference to the value of the modulus at normal ambient conditions!). The other principal variant is to select the test temperature and measure the results in terms of the degree of yield resulting, at this temperature, from standardised conditions of stress.

The size of the specimen requirements of the various types of test are quite different; for instance a penetration type test requires but a small area and thickness whereas a cantilever bending method may need a specimen of substantial cross-section. Since obviously the temperature of the surrounding heat exchange medium (water bath, air or whatever) must reflect accurately the temperature of at least the stressed area of the specimen, the rates of temperature rise of the heat exchange media may have to be quite different in the different methods—and they are!

Whilst not wishing to detract from the value of softening point determinations and the like as *quality control tests*, it cannot be stressed too strongly

Table 11.1 'SOFTENING POINTS' (°C) BY VARIOUS STANDARD METHODS

Material	*Cantilever B.S. 2782 M: 102C*	*D.T.L. (A.S.T.M.D.648)* 264 lb/in²	66 lb/in²	*Vicat B.S. 2782 M: 102D*
Polystyrene	95	90	97	98
Toughened Polystyrene	84	72	85	86
A.B.S.	94	84	96	95
Polymethyl Methacrylate	95	80	97	90
Cellulose Acetate	76	64	77	72
Rigid PVC	78	70	82	82
Polyethylene				
(L–D)	Too flexible	Too flexible	45	85
(M–D)	90	35	69	105
(H–D)	115	45	75	125
Polypropylene	145	60	140	150
Nylon 66	180	75	183	185
Acetal	170	120	165	175

just how ad hoc and arbitrary they are, and how their various results cannot be compared. Tordella et alia[37], Stephenson and Willbourn[38], Sherr[39] and Horsley[40] all quote results which emphasise this point and Table 11.1 from the last-mentioned reference serves to emphasise it.

Not only will the wide differences between results in each column be noted, but also the inconsistent differences and, in some cases, the changed orders of merit!

11.3.1 Cantilever Techniques

Method 102A of B.S. 2782: 1970 is for determining the 'Plastic Yield of Moulding Material' and requires two specimens each 200 ± 1 mm long by a square cross-section of side 15 mm + 0, − 0·300 mm. A notch is moulded or machined into one long face of each specimen, the root of the notch being 5 mm from one end. The notched surface is one of those perpendicular to the direction of moulding pressure.

The test is carried out in an oven capable of being maintained throughout the test within 2 degC of the specified test temperature; this latter is normally quoted in the relevant product specification but if not it shall be one of 55°C, 70°C, 100°C, 140°C, or 180°C, the temperature being selected so that the result ('plastic yield') does not exceed 6 mm.

The test set up is shown in Figure 11.4.

The clamped specimen, without weight W but with stirrup plus attachment (combined weight not exceeding 20 g), is placed in the oven at the test temperature; 15 min afterwards the height of the stirrup is taken, with respect to some suitable datum point, to the nearest 0·1 mm and then the weight W is attached to the stirrup (W + stirrup + attachment = 450 g).

After a further 6 h ± 10 mins the height of the stirrup with respect to the same datum is again taken. The difference between the two height figures, in mm, is the plastic yield; the results for the two specimens are averaged.

Using a similar principle, but a different test specimen, Part D2 of B.S. 903 describes a test for ebonite. Part D.1, however, employs the same type of test specimen and, by using a rising temperature, determines the 'yield temperature' at which the deflection becomes 2 mm.

Method 102B of B.S. 2782 describes a somewhat similar test to that of Method 102A, but it is termed 'Deformation in Bend Under Load at Elevated Temperature of Laminated Sheet'. Again two specimens are used, each approximately 200 mm (8 in) long and 25·4 ± 0·5 mm (1·00 ± 0·02 in) wide; the thickness is that of the sheet unless this exceeds 12·7 mm ($\frac{1}{2}$ in) in which case *one* face shall be machined to yield a thickness of 12·7 ± 0·5 mm (0·50 ± 0·02 in). A shallow notch is machined in one original face across the width, approximately 3 mm ($\frac{1}{8}$ in) from one end.

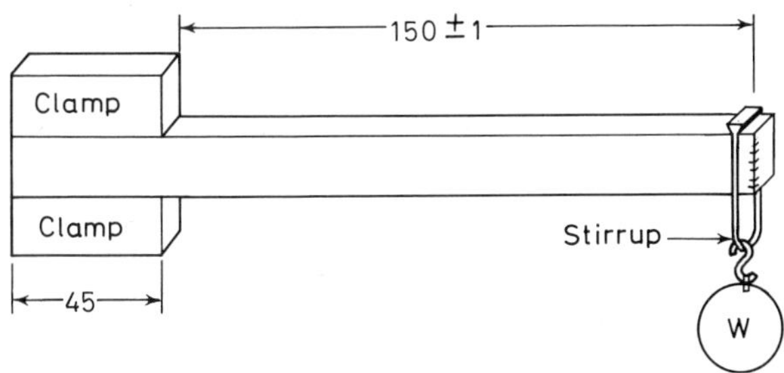

Figure 11.4. Plastic yield test (B.S. 2782). Dimensions in millimetres

The test is carried out at 90 ± 2°C, but otherwise generally as described above. The distance between the edge of the clamp and the notch root is

152 mm (6 in). The weight added shall be such as to give a maximum surface stress in bend in the specimen of 6·9 MN/m^2 (70 kgf/cm^2; 1000 lbf/in^2)

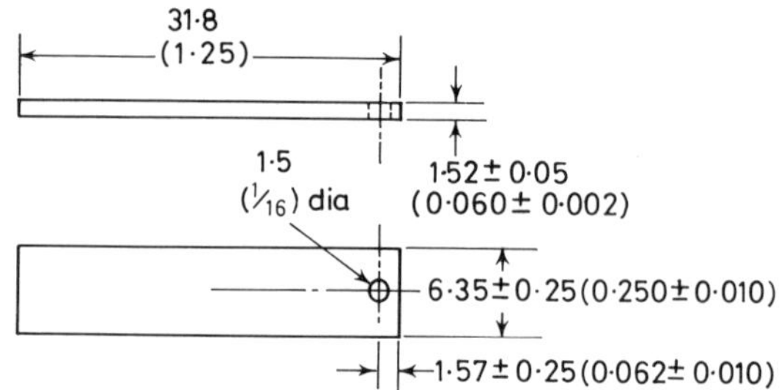

Figure 11.5. Softening point (bending test) specimen (B.S.2782). Dimensions in millimetres with inch equivalents in parentheses

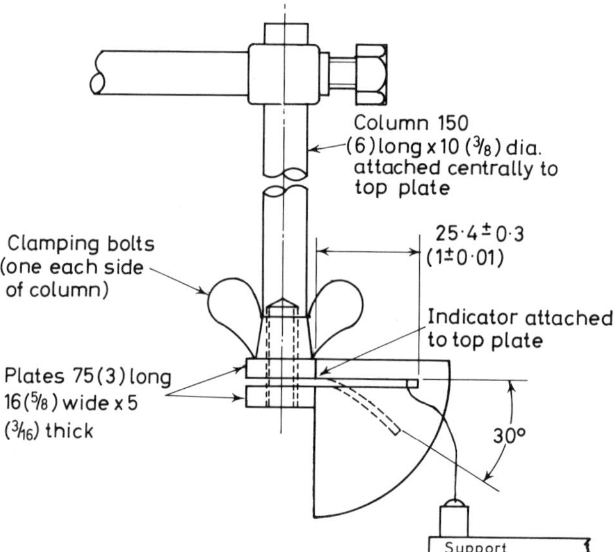

Figure 11.6. Test assembly for softening point (bending test) (B.S. 2782). Dimensions in millimetres with inch equivalents in parentheses

within 2%, calculated from the formula:

$$W = \frac{700\ BT^2}{6L} = 0{\cdot}77\ BT^2$$

where W = Requisite mass, g
B = Width of specimen, mm
T = Thickness of specimen, mm
and L = Effective length, of cantilever (152 mm)

The second stirrup height reading is taken after 60 min in this test, and the result is expressed in inches.

Finally, in cantilever methods, there is Method 102C of B.S. 2782 for 'Softening Point of Thermoplastic Moulding Material (Bending Test)'. The specimen is shown in Figure 11.5.

The specimen is mounted horizontally in a clamp, as shown in Figure 11.6. Initially the 20 g weight is supported so that there is no load on the specimen. The assembly is immersed in a bath of appropriate liquid, at a temperature 25–30 degC below the expected softening point. (The 'appropriate liquid' must be without effect on the material under test: liquid paraffin is suitable for cellulose acetate and glycerol for polystyrene, toughened polystyrene and rigid PVC.) The support is then removed from under the load and the temperature of the bath raised at $1{\cdot}0 \pm 0{\cdot}2$ degC/min with adequate stirring.

The temperature of the bath is noted at which the upper edge of the free end of the specimen coincides with a 30° graduation marked on the quadrant plate.

The mean of the temperatures obtained from two specimens is taken as the softening point.

11.3.2 Three Point Bending Techniques

The method of A.S.T.M. D.648 is historically known as yielding the '*Heat Distortion Temperature*'. A.S.T.M. D.648–56 describes the test result as 'Deflection Temperature of Plastics Under Load'. Directly derived therefrom, ISO/R.75 is called '*Determination of Temperature of Deflection Under Load*', as in the closely related Methods 102G and 102H of B.S. 2782. DIN.53461 is similarly related to the I.S.O. Method. As the various national and international standards are so closely related, only the B.S. method will be described.

To accommodate materials of different initial (room temperature) stiffness, two variants of the method are available, one applying a maximum surface stress of 1·81 MN/m^2 (18·5 kgf/cm^2; 264 lbf/in^2), which is Method 102G, and the other one of 0·45 MN/m^2 (4·6 kgf/cm^2; 66 lbf/in^2) which is Method 102H. Test specimens are as follows:

Moulding Material and Extrusion Compounds—Rectangular bar of minimum length 110 mm, width 9·8–12·8 mm and thickness 3·0–4·2 mm, moulded as appropriate to the material in question with the direction of pressure perpendicular to the largest faces of the specimen.

Sheet—Length and width as above, but with thickness that of the sheet (which shall be 3–13 mm).

Casting and Laminating Resin Systems—Length and width as above but 'thickness 3–7 mm'.

The apparatus used is shown in Figure 11.7.

The parallel cylindrical metal blocks which form the outer supports for the specimen are 100 ± 2 mm apart; these and the loading block, which rests centrally across the width of the specimen, all have contacting radii of $3 \pm 0{\cdot}2$ mm. All vertical members of the supports should have the same coefficient of expansion so that expansion of the apparatus cancels out and no error is thus introduced into the reading of the deflection of the specimen through uncompensated expansion in the apparatus; even so, it is as well to check this by carrying out a 'control' run beforehand using a test specimen of very low expension coefficient, for example Invar steel or borosilicate glass.

The heating bath into which the above assembly is immersed, is filled with a liquid which it has been shown is without effect on the material under test (liquid paraffin, transformer oil, glycerol and silicone fluid are suggested for trial). The bath must be capable of being heated so that the temperature can be raised at a rate of 2 degC/min, with the temperature at any time being not more than 1 degC from the value corresponding to this rate.

To undertake the test, the cross-sectional dimensions of the specimen are measured to the nearest 0·02 mm and then it is placed in the apparatus so that the direction of application of load is parallel to the plane of the largest faces. In this position, the vertical height is designated depth d and horizontal

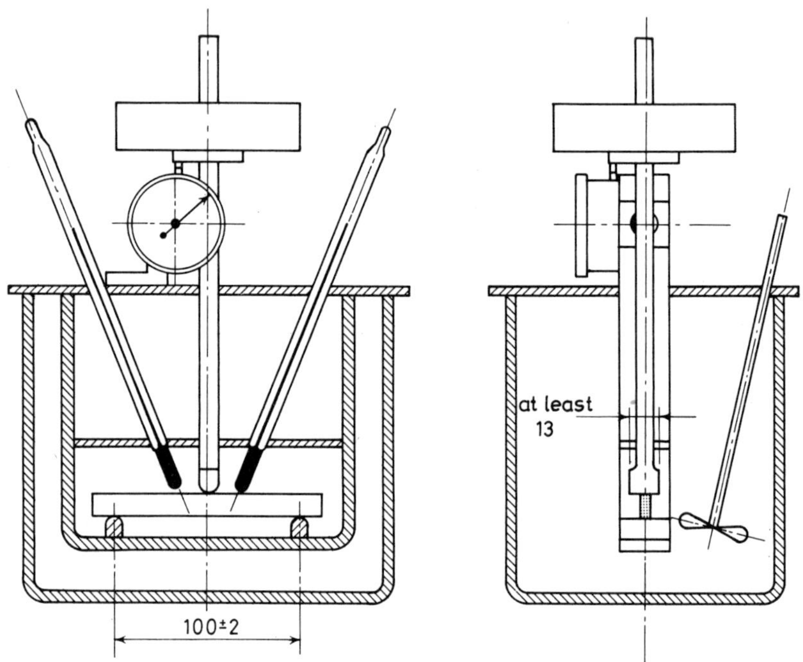

Figure 11.7. Apparatus for determination of temperature of deflection under load (B.S. 2782). Dimensions in millimetres

width from one large vertical face to the other is designated breadth b. The specimen is loaded under a bending stress of 1·81 MN/m^2 (18·5 kgf/cm^2) or 0·45 MN/m^2 (4·6 kgf/cm^2) (as appropriate) by applying a load as follows (making due allowance for any force exerted by the spring of the dial gauge):

$$\text{Load in kg} = \frac{Sbd^2}{150L}$$

where S = 18·5 for Method 102G and 4·6 for Method 102H
b = 'breadth', mm
d = 'depth', mm
and L = distance between supports in mm

The initial temperature of the bath is generally 20–23°C, but if it can be proved that the use of a higher initial temperature is without effect on the test result, then this may be used, providing it is not less than 30 degC below the expected temperature of deflection under load. 5 min after the load has been applied the deflection measuring device is set to zero (or its reading recorded) and the temperature of the bath raised as described above. The temperature in °C is noted when the deflection increase is reached which corresponds to that specified in a table relating deflections to depths *d* (0·33 mm deflection for 9·8 to 9·9 mm falling linearly to 0·25 mm for 12·8 mm).

The mean of the results on the two specimens is taken.

The higher alternative applied stress is too low for high modulus materials, particularly glass fibre reinforced resins, and as a result this test does not yield deflection temperatures low enough to be measured conveniently, say below 200°C. This is not only a question of experimental convenience, however, as it is most probable that at some temperature within the normal test range, such materials will have lost a significant proportion of that initial (high) value which was the very reason for their selection. In an endeavour to overcome this drawback, a British Standard product specification (B.S. 3534: Part 2: 1964, 'Epoxide Resin Systems for Glass Fibre Reinforced Plastics. Part 2. Pre-Impregnating Systems') has extended the test by applying a bending stress of 360 kgf/cm^2 on a specimen approximately 150 mm long and 12·5 mm wide by the prepared thickness of 3·2 ± 0·1 mm. The span is at least 50 mm.

Deflection is recorded against temperature to a value of the latter at least 30 degC above that at which the 'critical deformation' is achieved; best straight lines are drawn through the two sections of the curve (below and above the 'critical' point) and the point of intersection is designated the temperature of deflection. One specimen is used for a trial run and the test result is taken as the average of the figures from the second and third.

The extension of the technique is taken one stage further by B.S. 4045: 1966, 'Epoxide Resin Pre-Impregnated Glass Fibre Fabrics' by applying a stress equivalent to 10% of the (measured) ultimate cross-breaking strength. The specimen is as in B.S. 3534: Part 2: 1964 and the technique of obtaining the deformation point is precisely as before. This two fold test is somewhat inelegant and the choice of 10% of the normal temperature crossbreaking strength rather than some fraction of the bending modulus is decidedly surprising.

11.3.3 Four Point Bending Techniques

The *Martens'* test is described in DIN.53458 and 53462 and is an ingenious way of converting four point bending into magnified vertical movement of the end of an arm (Figure 11.8).

The apparatus is carefully specified and values of weight *G* given for each of the three different sizes of specimen permitted. The rate of heating is 50 ± 1 degC/h and the Martens' temperature taken when the deflection at the end of the lever reaches 6 mm. The values of *G* are selected to give the same maximum surface stress to each of the three alternative forms of test piece, but the maximum strain is necessarily different. The ratio of the two (stress–strain) is stiffness which is a function of temperature and therefore test results differ according to which of the standard test pieces has been used.

Gohl[41] discusses the limitations of the Martens' test and concludes that it is unsuitable for materials of modulus less than 50 000 kgf/cm^2. Wallhauser and Christmann found a reasonable correlation between Martens'

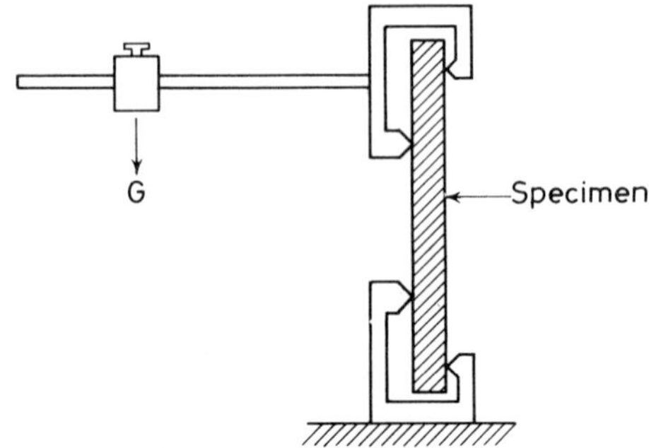

Figure 11.8. Martens' test (DIN.53462)

test results and those of the ISO/R.75 method when used on phenol-formaldehyde moulding materials[42].

11.3.4 Tensile Techniques

For sheets and films of thickness from 0·025–1·5 mm (0·001–0·060 in) and of modulus greater than 700 kg/cm^2 (10 000 lb/in^2) at 23°C, a test for *tensile heat distortion temperature* is described in A.S.T.M. D.1637–61. It measures the temperature at which thermoplastic sheeting begins to deform appreciably (either to stretch or shrink) under a small tensile force.

Two forms of apparatus are shown in Figure 11.9.

The oven or heating bath must be capable of being raised in temperature at a constant rate of $2 \pm 0{\cdot}2$ degC/min. The weight(s) required must be appropriate to apply a stress of 3·52 kg/cm^2 (50 lb/in^2) to the specimen.

The test specimens are strips 0·6–2·5 cm (0·25–1·0 in) wide and 5–18 cm (2–7 in) long. The initial distance between the grips is between 2·5 and 12·5 cm (1 and 5 in). Materials of less than 0·075 mm in thickness and capable of being readily cemented together (or otherwise joined) are forced into a loop, with an overlap of not more than 0·6 cm, and tested in this manner.

After the width and thickness have been carefully measured, the specimen is preconditioned and then mounted in the apparatus with the appropriate weights attached to the moving grip to give the specified stress of 3·52 kg/cm^2. The assembly is placed in the oven or bath held at a temperature approximately 20 deg/C below that at which the specimen is expected to begin to deform. The temperature is then raised at $2 \pm 0{\cdot}2$ degC/min and extension recorded against temperature until either the specimen has elongated 50% or the temperature has reached 250°C. It is suggested that a temperature correction (which should not exceed 3 degC) be applied to allow for any temperature lag between heating medium and specimen, which correction can be determined once and for all for each apparatus by attaching a thermocouple to a test specimen and comparing the heating curve for the thermocouple with that from the thermometer in the oven or bath.

The tensile heat distortion temperature is determined from a plot of percentage change in length against temperature and is the temperature at which a 2% shrinkage or a 2% extension occurs.

In the product specification A.S.T.M. D.1430–65T 'Polychlorotrifluoroethylene (PCTFE) Plastics' a test is included for '*zero strength time*'

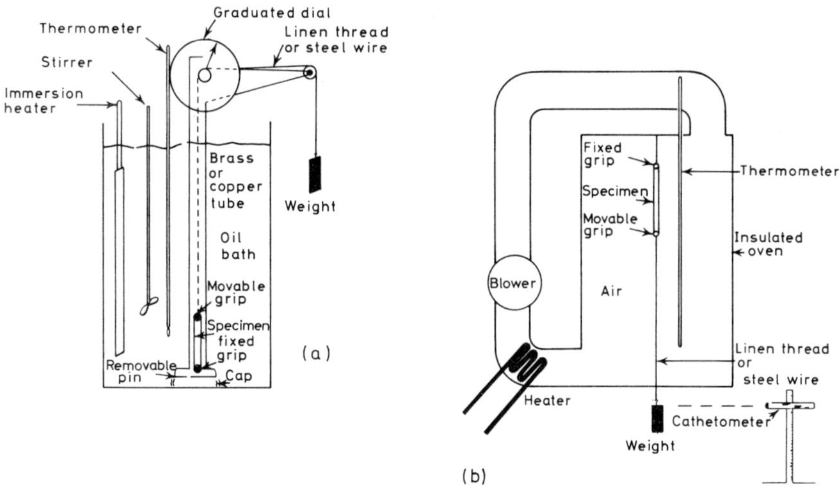

Figure 11.9. Tensile heat distortion apparatus (A.S.T.M. D.1637) (a) Heat bath method, (b) air oven method

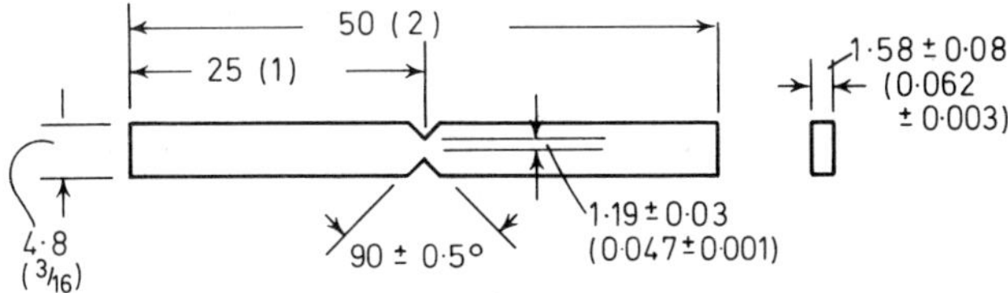

Figure 11.10. Z.S.T. test specimen (A.S.T.M. D.1430). Dimensions in millimetres with inch equivalents in parentheses

(Z.S.T.) which it is stated is useful for control of molecular weight of this type of polymer. The specimen is a notched strip (Figure 11.10) cut and punched out of sheet 1·58 ± 0·08 mm (0·062 ± 0·003 in) thick preformed and moulded from the powder in specified form.

One end of the specimen is hung from a specimen holder and a weight of 7·5 ± 0·1 g is attached to the other. Two such weighted specimens are inserted one each into two cylindrical holes in a brass 'thermostat' at 250° ± 1°C and timers started. The time is noted when each specimen breaks, denoted by the weight dropping through the bottom of the 'furnace' (Figure 11.11)

The average of two readings in seconds is taken as the ZST; if the two values differ by more than 10% the test is repeated.

Sandiford and Buckingham[43] describe the tensile heat distortion test in some detail, which they term the 'T.D.T.' test (tensile deformation versus

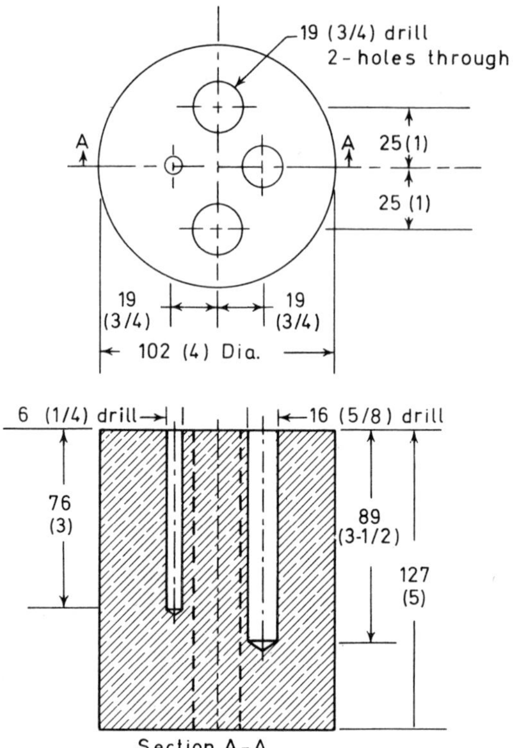

Figure 11.11. Z.S.T. thermostat (A.S.T.M. D.1430). Dimensions in millimetres with inch equivalents in parentheses

temperature), give details of single specimen and six specimen apparatus and present some typical values for thermoplastics films.

11.3.5 Compressive Techniques

Two test methods are contained in A.S.T.M. D.621–64 'Deformation of Plastics Under Load'. Method A is for rigid plastics and is designed to assess the ability of such materials to withstand compressive load, e.g. when held by bolts, without yielding and loosening the assembly; Method B is intended to determine the ability of non-rigid plastics to return to their original shape after having been deformed.

Method A

The essentials of the apparatus are that it shall be capable of exerting constant forces of 250 lb, 500 lb, and 1000 lb, all $\pm 10\%$, and that it is equipped with a dial gauge measuring the relative movement of the faces of the anvils to 0·001 in or less. A suggested apparatus is shown in Figure 11.12.

It will be realised that many compression testing machines (Chapter 8), if fitted with temperature controlled ovens, are adaptable to meet the specification requirements—if they can be spared for the duration of this test!

Test specimens are $\frac{1}{2}$ in cubes, either solid or composite, machined down or built up as necessary. Surfaces must be plane and parallel.

After preconditioning, if necessary, the specimen is placed between the anvils and the load applied without shock. After ten seconds the dial reading is taken and again after 24 h at the test temperature of 23°C, 50°C or 70°C,

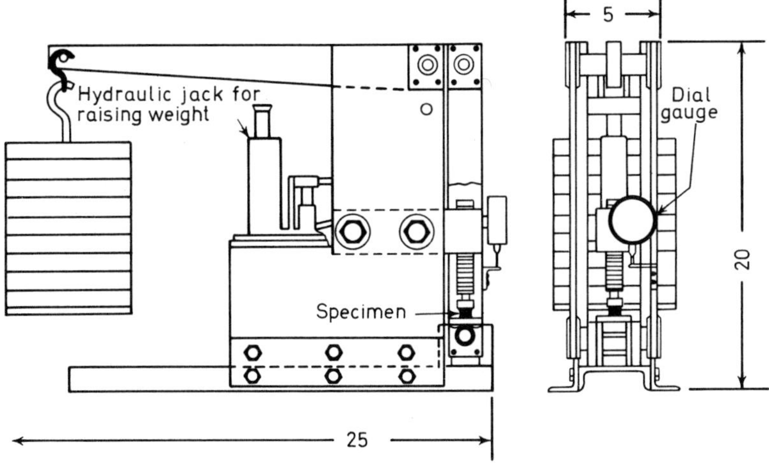

Figure 11.12. Deformation testing machine (A.S.T.M. D.621). Dimensions in inches

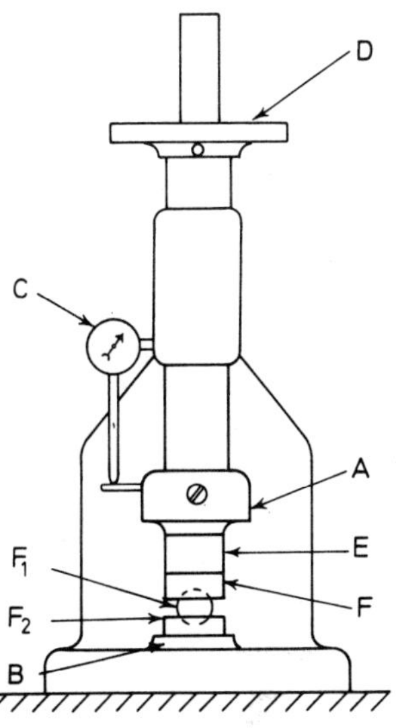

Figure 11.13. Low pressure deformation tester (A.S.T.M. D.621). A. Movable anvil. B. stationary plate. C. thickness indicator. D. weight platform. E. test specimen. F. lower anvil mounted on ball support F_1 and loose plate F_2

all ± 1 degC. The specimen is removed, its thickness measured and the original height (B) calculated from this plus the change in dial reading (A).

Deformation, % $= A/B \times 100$

Method B

This is basically the same as Method A but the stress is 100 lb/in$^2 \pm 1$% and the period of test only 3 h. One apparatus suitable for this is given in Figure 11.13.

The test specimens are 1·129 in diameter by $\frac{1}{2}$ in thickness and they are suitable also if the apparatus of Figure 11.12 is used without weights. After preconditioning, the procedure of Method A is followed with the exception given above and a difference in the dial reading sequence. The zero point is obtained at the beginning of the test by determining the dial reading with the anvils together under full load and the thickness of the specimen (H_0) measured by micrometer, with low pressure ratchet attachment, to the nearest 0·002 in.

After determining the thickness of the specimen at the end of three hours (H_1), it is removed from the apparatus and left in the test chamber for one hour. Finally the specimen is removed from the chamber, kept at room temperature for $\frac{1}{2}$ h and then its thickness redetermined by micrometer (H_2).

$$\text{Deformation, \%} = \frac{H_0 - H_1}{H_0} \times 100$$

$$\text{Recovery, \%} = \frac{H_2 - H_1}{H_0 - H_1} \times 100$$

11.3.6 Penetrometer Techniques

In Method 102E of B.S. 2782, a test is described for determining the '*Deformation under Heat of Flexible Polyvinyl Chloride Extrusion Compound*' with two specimens each 13 mm ($\frac{1}{2}$ in) in diameter and $1{\cdot}27 \pm 0{\cdot}07$ mm ($0{\cdot}050 \pm 0{\cdot}003$ in) thick. A specimen is measured accurately* for thickness and then placed on a flat horizontal base. A force is applied of 9·8 N (1000 gf) terminating in a vertical plunger 3 mm ($\frac{1}{8}$ in) in diameter and flat, and the whole assembly placed in an oven for 24 h at 70 ± 1°C. After this the assembly is removed from the oven and allowed to cool for one hour at room temperature. Finally the thickness of specimen in the deformed area is remeasured between three and five minutes after removal of the load.

The deformation under heat of the test specimen is taken from the difference between the initial and final thicknesses calculated as a percentage of the initial thickness, and the result is expressed as the mean of two readings.

Of far more importance and wide acceptance in penetrometer methods is the *Vicat softening point* test which in conventional form is Method 102D of

*Under a 3 mm ($\frac{1}{8}$ in) diameter foot exerting a pressure of $0{\cdot}24 \pm 0{\cdot}03$ N (24 ± 3 gf).

B.S. 2782, A.S.T.M. D.1525–65T, DIN.53460 and ISO R.306; in the A.S.T.M. Method an alternative (faster) rate of temperature increase is permitted whilst in B.S. 2782 and ISO R.306 other variants are offered.

In Methods 102D, 102F and 102J of B.S. 2782, the specimen is at least 10 mm ($\frac{3}{8}$ in) square and for sheet material the thickness thereof is used unless (a) it is more than 6·4 mm ($\frac{1}{4}$ in) in which case it is reduced to approximately 3 mm ($\frac{1}{8}$ in) by machining one face or (b) it is less than 2·5 mm (0·1 in) when two or three pieces are stacked together to give a thickness of

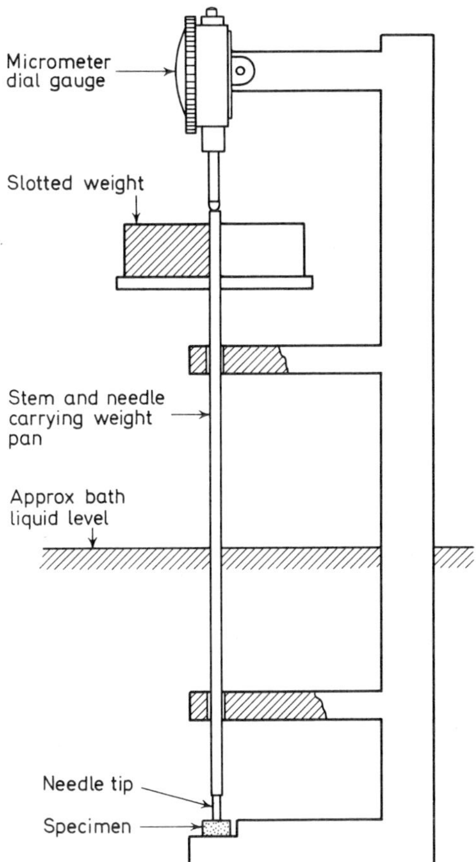

Figure 11.14. Vicat softening point apparatus (B.S. 2782). Note. It is recommended that the frame and indenting rod are constructed from low expansion alloy

at least 2·5 mm (0·1 in). If the specimen is moulded, the thickness is 3·8 mm ($\frac{1}{8}$ in). Two specimens are used.

A schematic arrangement for the apparatus is given in Figure 11.14. However, it is usual to arrange such units in banks of up to six.

The penetration of the rod into the specimen is measured by the dial gauge to 0·01 mm. The flat-ended tip of the rod is 3 mm ($\frac{1}{8}$ in) long, preferably hardened, of circular or square cross-section and area 1·000 ±

0·015 mm^2. The slotted weight is such that the total mass bearing on the specimen is 1000 to 1050 g for Methods 102D and 102F and 5000 to 5050 g for Method 102J, the combined weight of the rod and weight carrier being 100 g at maximum.

The materials of construction of the apparatus are such that, when the specimen is replaced by a piece of metal, the differential thermal expansion over the intended range does not exceed 0·02 mm.

The heating bath, containing a liquid with the same reservations as in Section 11.3.2, is fitted with heating controls such that its temperature can be raised at a constant rate of 50 ± 5 degC/h.

The Vicat Softening Point, Method 102D, or the Vicat Softening Point with 49 N (5 kgf) load, Method 102J, is taken as the temperature when the penetration reaches 1·0 mm, but if the $\frac{1}{10}$ Vicat softening point, Method 102F, is required then the reading is taken at a penetration of 0·1 mm. The respective uses of the Methods 102D and 102F are discussed by Willbourn and Stephenson[38].

11.3.7 Temperature Programming and Other Refinements

Automatic indication of the end points of Vicat softening point and deflection temperature under load tests is achieved by Graves and Loveless[44] using a lever actuating a mercury switch to cause a panel light to glow;

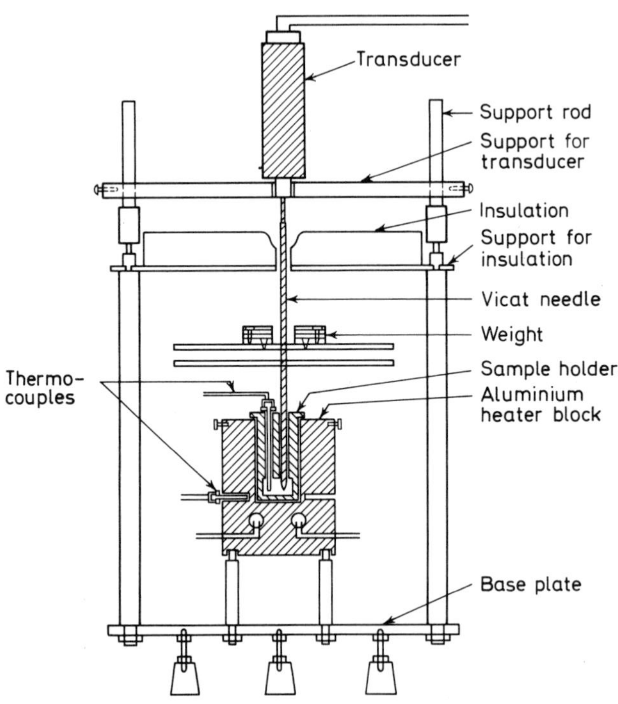

Figure 11.15. Automatic Vicat softening point apparatus (Ehlers and Powers)

they also describe V.S.P. tests at rates of temperature increase of both 50 degC/h and 120 degC/h (i.e. 2 degC/min), quoting results for seven thermoplastics to justify the alternative of the faster rate allowed in A.S.T.M. D.1525–65T (see Section 11.3.6 above).

Ehlers and Powers[45] use a transducer attached to the top of the penetrometer shaft of the Vicat apparatus to record depth of penetration against temperature fed to the recorder from a thermocouple (Figure 11.15).

For the automatic control of the rate of rise of the temperature bath, at either of the standard conditions (50 degC/h or 2 degC/min). Bestelink et alia[46] describe a set up in which the signal from the programming device is used to drive a motor which turns the contact of a variable transformer supplying the heater in the bath; thus the heat input into the bath is varied in relation to that required to maintain the correct heating rate.

11.4 LOW TEMPERATURE BRITTLENESS AND FLEXIBILITY TESTS

Most applications of rigid plastics make use of their stiffness, even if this is low by comparison with metals and stone; it is therefore logical that in the majority of quality control tests checking the thermal properties of such rigid plastics, it is loss of stiffness that is used as the parameter of measurement. Likewise, as many of the applications of flexible plastics utilise their characteristic 'give', the corresponding quality control for these follows their loss of flexibility with decrease in temperature. Amongst the standardised tests, however, will be found those which are slow in action, for example plotting torsional stiffness against temperature, and those which are fast acting, such as those which assess the temperature at which impact shock causes shattering. Early background to both types of test is described by Clash and Berg[47].

11.4.1 Torsional Stiffening Methods

Welding[48] reviewed torsion tests critically before some of them at least had been standardised and concluded they could be simple and accurate.

Method 104B of B.S. 2782: 1970 is a variant of the Clash and Berg apparatus[47]. Method 104D of the same standard simply uses the same test method after the test specimens have been heated at 100°C for 24 h over activated carbon, to assess resistance to ageing. The apparatus described in B.S. 2782 is shown in Figure 11.16 (plan and side elevations).

The method, for determining the 'Cold Flex Temperature' of flexible PVC extrusion compound, uses a test specimen 64 mm ($2\frac{1}{2}$ in) long, 6·4 ± 0·2 mm (0·250 ± 0·010 in) wide and 1·27 ± 0·08 mm (0·050 ± 0·003 in) thick. The test is carried out in duplicate.

The pulley of the apparatus is such that a torque of 0·057 Nm ($5{\cdot}7 \times 10^5$ cm dyn) can be applied to the specimen (e.g. by two 50 g weights acting on pulley 11·6 cm in diameter, if friction is negligible). The distance between the clamps, each of width at least sufficient to cover the full width of the specimen, is such as to leave a free length of specimen of 38 ± 0·5 mm (1·50 ± 0·02 in). The part of the apparatus carrying the specimen must be capable of being immersed into a liquid cooling bath, e.g. industrial methylated spirit and solid carbon dioxide, and a heating unit, such as an immersion heater, and stirrer must be provided.

In operation, the specimen is kept at room temperature for at least 12 h before test. Then, with the torque pulley clamped at the zero reading point, the specimen is clamped into the fixed (lower) and moveable (upper) clamps with neither tension nor slackness in the specimen. The assembly is then immersed in the bath held at a temperature, read by a thermometer

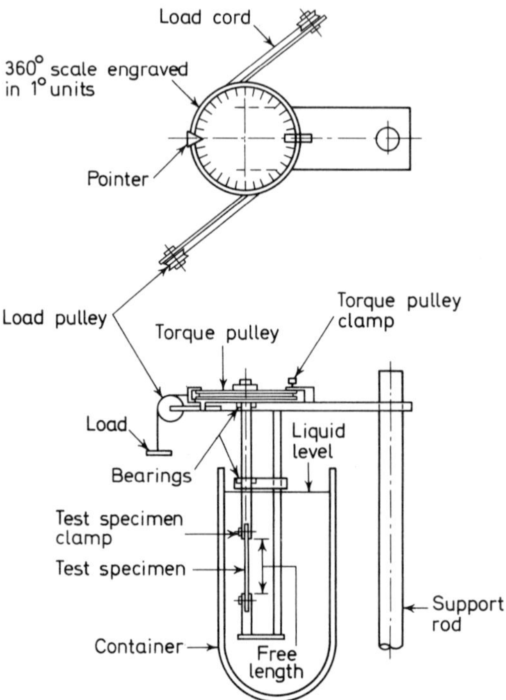

Figure 11.16. Cold flex temperature apparatus (B.S. 2782)

whose bulb is beside the specimen, just below that estimated to give a deflection of 30° with the compound under test. The stirrer is started at moderate speed and 5 min allowed to elapse at the steady temperature before the torque is applied. After this period the torque pulley is released for precisely 5 s, after which it is held stationary until a reading of the deflection to the nearest 1° is noted. The pulley is then returned to the zero position as soon as possible and clamped.

The heater, adjusted to give a temperature rise of approximately 2 degC/min is then switched on and the operation described above repeated at intervals of at most 5 degC until a deflection greater than 400° is obtained.

A graph is plotted of deflection against temperature and the temperature for a deflection of 200° is read off. 0·5 degC is deducted from this value for each 0·025 mm (0·001 in) of thickness of the specimen above 1·3 mm (0·050 in) and 0·5 degC added for each 0·025 mm (0·001 in) of thickness below 1·3 mm (0·050 in).

The mean of the results from the two specimens is recorded as the cold flex temperature.

A.S.T.M. D.1043–69, 'Stiffness Properties of Plastics as a Function of Temperature by Means of a Torsion Test', employs the same principle and defines a value T_f where the calculated apparent modulus of rigidity is 3160 kg/cm^2 (45 000 lb/in^2); the apparatus has been made automatic and recording by Scherr and Palm[49]. DIN.53445 also employs a torsion technique, but of a dynamic character by measuring the amplitudes of vibration, whilst DIN.53447 is the Clash and Berg method.

In view of their similarity to the above, it is worth mentioning the corresponding tests for elastomers. Thus in Part A.13 of B.S. 903: 1960 is described a somewhat similar test to B.S. 2782, with certain variations in specimen and conditions, but it does not define an end-point temperature. It is recommended that results are expressed as a plot of rigidity modulus, or log rigidity modulus, against temperature. Many product specifications call for the results to be expressed as the ratio of moduli at +20°C and −20°C. A.S.T.M. D.1053–65, the Gehman method, is a similar torsional test using calibrated torsion wires to measure the modulus of rigidity of elastomeric samples; this has been automated by Caspary[50].

Support for the Clash and Berg type of test (again before it was adopted by the B.S.I.) will be found in investigations described by Hayes and Lannon[51] who give results for PVC plasticised with various materials in a range of concentrations. Williamson[52], in an early paper, found the Clash and Berg type of apparatus too insensitive at low values of modulus, and described a simple torsion apparatus based on calibrated torsion wires, i.e. after the fashion of the Gehman technique; an automated version of this type of equipment has recently been reported by Wheeler[53].

11.4.2 Extensibility Method

Resistance to low temperature of flexible PVC sheet is measured by the extensibility in tension at specified temperature according to Method 104C of B.S. 2782: 1970.

A constant-rate-of-loading instrument is required, capable of increasing stress on a rectangular specimen (180 mm (7 in) long by 6·3±0·1 mm (0·250±0·005 in)×sheet thickness) uniformly from zero at a rate of 20·7±0·4 MN/m^2 (210±4 kgf/cm^2; 3000±50 lbf/in^2) per minute. The free (unclamped) length of specimen is 127 mm (5 in) and an extensometer device is needed to measure extension to the nearest one per cent. A cooling bath at −5±0·5°C, comprising one volume of industrial methylated spirit and three volumes of water, to which is added solid carbon dioxide as necessary, is provided for immersing that part of the apparatus which contains the clamped specimen.

The load required to give a stress of 10·3 MN/m^2 (105 kgf/cm^2; 1500 lbf/in^2) on the specimen is calculated from the width and thickness obtained gravimetrically (Chapter 5, Section 5.2.2). With the grips set 127±1·3 mm (5·000±0·050 in) apart, the specimen is mounted symmetrically in them, first in the upper one and then with a 20 g weight attached to its lower end to keep it taut, in the lower one. The mounted specimen is then completely immersed in the bath, for 30 or 60 s according to thickness, and after this the load is applied uniformly so that 10·3 MN/m^2 is reached in 30 s. When this stress is reached the extension of the specimen is read (all the extended specimen in the bath!) to the nearest 1·3 mm (0·05 in).

Three such determinations are carried out and their results averaged.

11.4.3 Flex Cracking Methods

The 'Cold bend temperature' of flexible PVC extrusion compounds, described as Method 104A of B.S. 2782: 1970, measures the lowest temperature, measured in multiples of 5 degC, at which none of a set of three test specimens fractures or cracks when wound onto a standard mandrel.

At least six specimens, each 100 mm (4 in) long and 4·8 mm ($\frac{3}{16}$ in) wide are cut from specially prepared sheet of thickness $1{\cdot}27 \pm 0{\cdot}08$ mm ($0{\cdot}050 \pm 0{\cdot}003$ in). They are tested in an apparatus of the type of Figure 11.17 in which all the essential features are given.

Three specimens are placed in the guides and one end of each is secured by clamp to the mandrel. Specimens and winder are then immersed in a cooling bath of industrial methylated spirit maintained at the required temperatures by solid carbon dioxide. After 10 min at the temperature, and whilst still immersed, the specimens are wound tightly around the mandrel for three complete helical turns at the rate of one revolution per second. They are then withdrawn from the bath and examined for signs of failure—fracture or surface cracking. The test is repeated, on fresh sets of specimens, at various temperatures until two temperatures, differing by 5 degC, are found such that at the higher no specimen fails and at the lower one or more fail.

The *cold bend temperature* is the higher of these two temperatures.

In the course of an investigation into low temperature failures of handbag materials, Wormald[54] developed a flexing tester operated at -5°C in which the test result was the number of flexings required to cause failure of a standard specimen when repeatedly flexed under controlled conditions, particularly incorporating in each cycle a rapid fold followed by an appreciable waiting period to simulate service conditions.

11.4.4 Impact Methods

A.S.T.M. D.746–64T, 'Brittleness Temperature of Plastics and Elastomers by Impact', determines the temperature at which 50% of the test specimens would fail under the conditions of the test (see also ISO/R.974).

Test specimens are die punched either as $6{\cdot}35 \pm 0{\cdot}51$ mm ($0{\cdot}25 \pm 0{\cdot}02$ in) wide pieces or as shown in Figure 11.18.

In either case they are $1{\cdot}91 \pm 0{\cdot}25$ mm ($0{\cdot}075 \pm 0{\cdot}01$ in) thick, of suitable length to be clamped in the apparatus as shown in Figure 11.19.

The alternative specimen is clamped so that the entire tab is inside the jaws for a minimum distance of 3·18 mm (0·125 in).

Sharp dies must be used. Hoff and Turner[55] have conducted an investigation into the effect of mode of specimen preparation on test results and

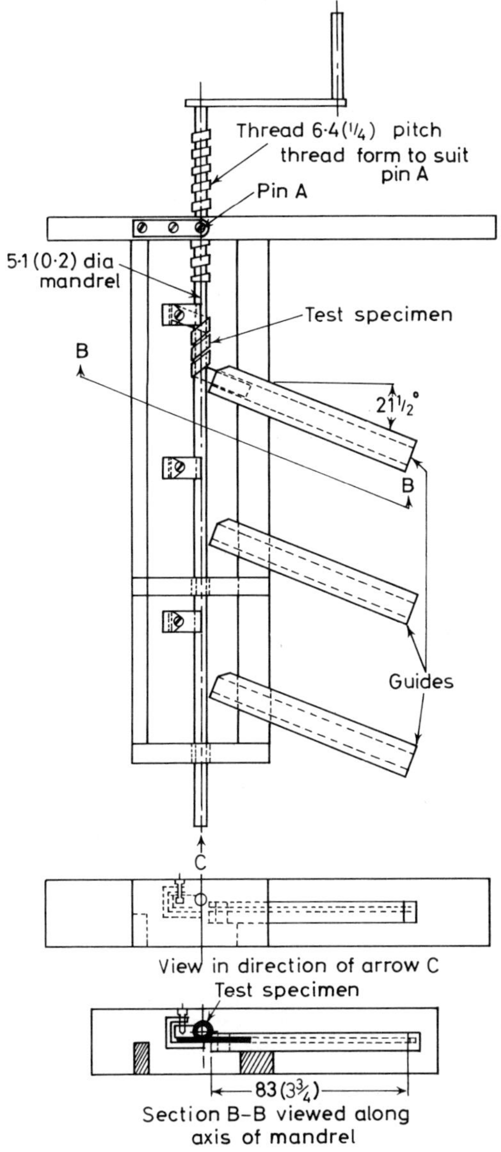

Figure 11.17. Cold bend test (B.S. 2782). Dimensions in millimetres with inch equivalents in parentheses

found up to 50 degC difference between razor cut and die cut specimens on low density polyethylene of melt flow index 20!

The striking edge moves relative to the specimen at a linear speed of 6·0 to 7·0 ft/s (1·8 to 2·1 m/s) at impact and for at least 6·4 mm (0·25 in) thereafter. A cooling bath is required, of liquid heat transfer medium without appreciable effect on the test specimens, and held within ±0·5°C of the desired value.

The specimen is mounted in the apparatus and immersed in the bath at the required temperature for 3±0·5 min. After this the specimen is impacted as described above and examined for failure—fracture or visible

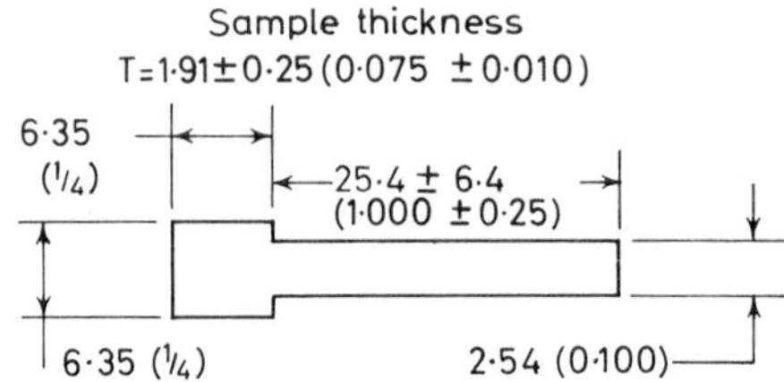

Figure 11.18. Alternative specimen for brittleness temperature test (A.S.T.M. D.746). Dimensions in millimetres with inch equivalents in parentheses

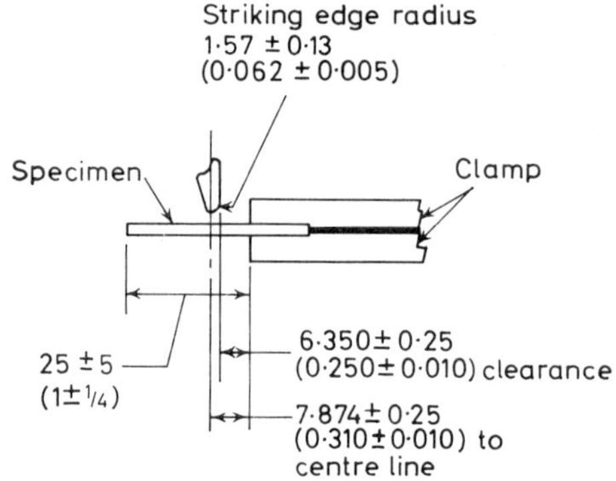

Figure 11.19. Clamp and specimen for brittleness temperature test (A.S.T.M. D.746). Dimensions in millimetres with inch equivalents in parentheses

crack (as shown by bending through 90°). A new specimen is used for each test.

The complete procedure is started at a temperature at which a 50% chance of failure is expected, and at least ten specimens are tested at this temperature. If all the specimens fail or do not fail, the bath temperature is increased or decreased (respectively) by 10 degC and the test repeated. When temperatures have been found where all failures and non-failures are recorded, the temperature is changed by steps of 2 or 5 degC, testing ten specimens each time and including the 'all fail' and 'no fail' temperatures.

Brittleness Temperature T_B

$$= T_h + \Delta_T\left(\frac{S}{100} - \frac{1}{2}\right)$$

where T_h = highest temperature at which failure of all the specimens occurs

Δ_T = temperature increment in °C

S = sum of the percentage breaks at each temperature (from a temperature corresponding to 'no fail' down to and including T_h)

Alternatively a graphic method may be used for determining T_B or, for quality control purposes, the test may be carried out at one standard temperature and the material deemed to pass if not more than five specimens fail out of the ten.

[See also B.S. 903: Part A25: 1968, 'Methods of Testing Vulcanised Rubber. Part A25. Determination of Impact Brittleness Temperature', where the test piece and equipment are essentially the same as above, but the brittleness temperature is defined differently.]

The same formula is used in A.S.T.M. D.1790–62, 'Brittleness Temperature of Plastic Film by Impact,' for obtaining T_B of film of thickness 0·25 mm (0·010 in) or less. Strip specimens 5·08±0·04 cm by 14·61±0·04 cm ($2\pm\frac{1}{64}$ in by $5\frac{3}{4}\pm\frac{1}{64}$ in) are die cut and tested by allowing an impact arm to fall onto a fold in the specimen.

The development and study of the A.S.T.M. D.746 method is described by Webber[56]. A number of authors[57–59] recommend notching the specimens to obtain more reproducible results, but this of course departs from the A.S.T.M. standardised procedure. The latter may be achieved conveniently by cooling banks of ten specimens simultaneously and testing all together[60]. Bata and Ito[61] have pointed out the importance of clamping pressure. Finally, another form of impact test, somewhat similar to that of A.S.T.M. D.1790, is described by Williams[62] for PVC sheet and coated fabrics. The apparatus is shown in Figure 11.20.

A is a brass vessel, insulated on sides and base, which forms an alcohol bath. B is a test piece holder pivot which with stop C and anvil D is carried on the base of the bath. Pillar E is also carried on the base (which projects to vessel A), through a strip of insulation to reduce heat transfer. Arm F on E is brazed to a housing G containing a ball bushing. Hardened steel plunger H, (with handle J) slides in the bushing and terminates in length of nylon rod K of the same diameter (also to reduce heat transfer). The nylon rod terminates in circular brass hammer L the lower face of which is parallel to the upper surface of the anvil. Between recessed collars M and N is fixed spring P of nominal strength 15 lbf/in. The position of M is such that the spring is free until the hammer is raised $\frac{1}{8}$ in above the anvil. When the hammer is raised, spring-loaded slide Q operated by knob R fixes it $1\frac{3}{8}$ in above the anvil, and pressing R releases it. Polyethylene film U, secured by hose clips, protects the steel parts from splashing and consequently rusting. Clip S holds the test piece and is screwed to an arm projected at right angles from the end of bent copper tube T. Over the upper part of the latter is slipped a piece of thick rubber tube to serve as a handle.

The test piece is $1\frac{1}{4}$ in long by $\frac{1}{4}$ in wide, is folded end to end and inserted $\frac{1}{8}$ in into the clip S. With the bath at the required temperature, the holder carrying the test piece is fitted over the pivot and rotated until the arm holding the clip contacts stop C, when the folded end of the test piece is located over the centre of the anvil and has its lower side just touching the anvil surface. After 15 s the hammer is released and then the piece withdrawn and examined. The result is recorded as a break if the coating of the test piece, or the test piece itself, is broken across the greater part of its width.

The end result may be in the form of the temperature at which 50% of specimens fail, as described above. This test, in essential details, now appears appears as Method 10, 'Determination of Cold Crack Temperature of

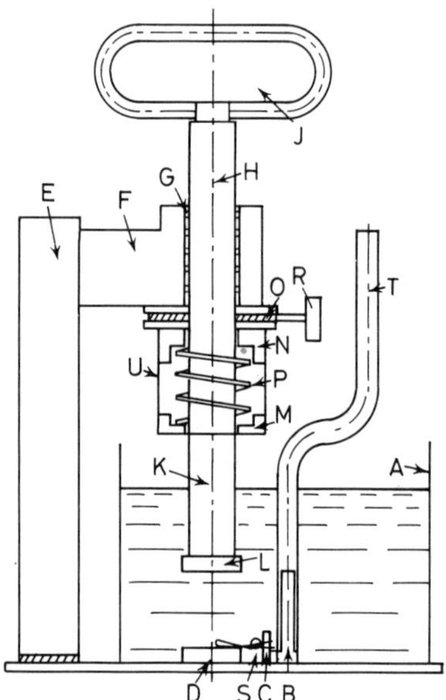

Figure 11.20. PVC brittle point (Williams)

PVC-Coated Fabrics', of B.S. 3424: 1961, 'Methods of Test for Coated Fabrics'.

11.5 NON-AMBIENT TEMPERATURE TESTS

If more meaningful data on temperature effects are required, more useful for design purposes than the arbitrary values obtained from the ad hoc thermal yield, brittleness and stiffening tests described in Sections 11.3 and 11.4, it is essential to perform relevant 'absolute' tests at the temperature or temperatures of interest. By and large, their execution follows the description of the tests at standard temperatures, the variation being in creating an atmosphere of the appropriate temperature around the specimens—and perhaps the whole apparatus—with the minimum of variation from point to point and at any one point. There may, of course, be practical difficulties in enclosing an apparatus, or even the essential part of it, in a temperature controlled atmosphere, or if this is successfully achieved, of actually operating it in such an enclosure. Carrying out impact tests of the swing hammer type is a particular case in point, but depending on the case in question, simple solutions are often available. Thus, impact testing of this type is a fast operation, and it has been found[63] that appropriate precooling of the test specimens followed by testing immediately after removal from the cooling cabinet, is a satisfactory alternative. There may also be unexpected pitfalls with otherwise simply accommodated tests, for instance ice formation on the guides of compression chucks (see Figure 8.30) causing apparently high resistance, or on the surface of specimens, leading to low results in resistivity measurements.

The profound influence of temperature on the properties of plastics is covered in many technical papers; by way of illustration, the reader is referred to Carswell and Nason[64], S.T.P.161[65] and Mathes[66].

11.5.1 Official Standards

There are no general methods recommended in British Standards, for instance, but the product specification B.S. 3534: 1964, 'Epoxide Resin Systems for Glass Fibre Reinforced Plastics. Part 2. Pre-Impregnating Systems', requires cross-breaking strength of prepared laminates to be measured, by Method 304B of B.S. 2782 (Chapter 8), at 10±2 degC below the 'lower classification temperature' after conditioning at that temperature for two hours; these temperatures range from 70°C to 200°C. No details or recommendations are given on how to carry out the test at elevated temperatures. The related B.S. 4045: 1966, 'Epoxide Resin Pre-Impregnated Glass Fibre Fabrics', replaces the high temperature test by one at room temperature after exposure for long periods at elevated temperature—and terms it a 'thermal endurance' test. The abilities to resist deterioration due to increased temperatures, as measured by the two tests, are of course quite different.

Three A.S.T.M. specifications give instructions for carrying out mechanical tests at above or below ambient temperature:

D.758–48	'Impact Resistance of Plastics at Subnormal and Supernormal Temperatures'
D.759–66	'Determining the Physical Properties of Plastics at Subnormal and Supernormal Temperatures'
D.2733–68T	'Methods of Test for Interlaminar Shear Strength of Structural Reinforced Plastics at Elevated Temperatures'.

All three cross-refer to A.S.T.M. E.197–66, 'Enclosures and Servicing Units for Tests above and below Room Temperature', with A.S.T.M. D.758 tying down the tolerances to ±2 degC near the test specimen and a maximum difference between inlet and outlet duct thermometers of 5 degC. A.S.T.M. E.197 itself gives no constructional details, but lays down requirements for temperature control in particular. In a similar general manner, DIN.53446, 'Testing of Plastics, Determination of the Temperature Limits' is of interest.

11.5.2 Elevated Temperature Testing

As far as short term mechanical testing is concerned, many of the suppliers of testing machines referred to in Chapter 8 market ovens and chambers suitable for enclosing all of the machine or, more often, that part immediately surrounding the test specimen. Klute and Mckee[67] describe a chamber for operation up to 500°F—a modest limit by present day standards—with interior working space 32 in by 16 in by 14 in, suitable for operation with 60 000 lb universal testing machine. There is a separate heating chamber and hot air is blown in at 100 ft^3/min; a 4 in thickness of glass fibre is used for insulation. Lavery et alia[68] have equipped a Scott tester for operation up to 650°F. Reference 65 describes techniques for measurement of electrical properties at elevated temperatures, but the reader is referred to Chapter 10 for details. If the examination is also required to involve contact with liquid media, the equipment described by Brunnberg et alia[69] may be used, at least for tensile testing.

11.5.3 Low Temperature Testing

Commercial units for low temperature testing are not available to the same degree as those for high temperatures, neither is it necessarily very satisfactory to use a cooler unit with the insulated chamber of a high temperature unit; many of the low temperature techniques use relatively expensive materials such as liquid nitrogen or liquid helium as coolants, and the consumption is unacceptably high with a high heat capacity insulant suitable for high temperature work. Expanded polystyrene, polyurethane or similar material is much more suitable for low temperature work.

Tests at certain 'critical' points may be carried out by contacting or surrounding the specimen and apparatus with the appropriate medium—for example solid carbon dioxide (−76°C), or boiling liquid oxygen (−183°C), liquid nitrogen (−196°C), liquid hydrogen (−253°C) or liquid helium (−269°C). Devices based on these fixed points are described by Watson and Christian[70] and McClintock and Warren[71], whilst the relative merits and disadvantages (and dangers!) of the various gases are summarised by Mathes[66].

For temperatures nearer ambient, straight refrigeration plant may be used, for example in the manner suggested by Klute and Mckee[67]. Use may also be made of synthetic mixtures of the type of methanol and solid carbon dioxide for instance. When contact with liquids is required, again reference 69 may be useful.

Following the ideas of Wessel and Olleman[72, 73], Ives and Mead describe a cryostat for low temperature mechanical testing, using the vapour from liquid nitrogen, which gives good temperature control up to near room temperature[63, 74].

The liquid nitrogen moves from its storage flask to the cold chamber of the apparatus through a dip tube which passes from the bottom of the vessel up to a well insulated pipe (as short as possible) and thence to the chamber; no valves or other restrictions are included in this line. The liquid nitrogen container is otherwise sealed, except for one vent pipe connected to the vapour above the liquefied gas; the increase in pressure of this vapour, caused by continuous slow evaporation, forces liquid or cold vapour into the working chamber.

Temperature is controlled by releasing or reducing the excess pressure in the storage flask and to this end the vent pipe is connected directly to an adjustable pressure relief valve (No. 1) and by way of a solenoid valve to relief valve No. 2. In operation, the settings of the valves are adjusted so that valve No. 1 permits the driving force in the flask to be high enough to ensure an adequate supply of coolant to the chamber, but the pressure is such that the supply is not excessive and overshooting is avoided. The second pressure relief valve is brought into action by the opening of the solenoid valve, which in turn is actuated by a temperature controller with a sensing element in the cold chamber. This second valve is adjusted to control the pressure at a

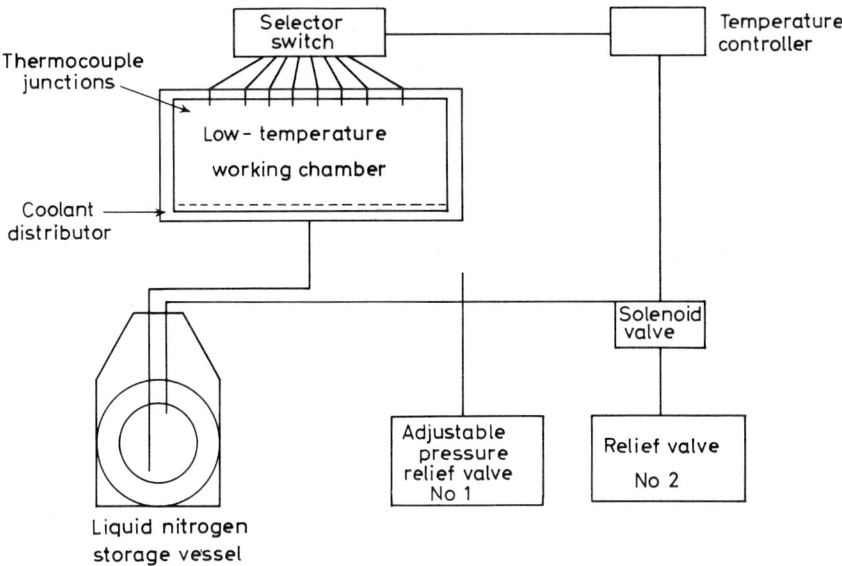

Figure 11.21. Schematic diagram of low temperature apparatus (Ives and Mead)

value slightly below that necessary to ensure a supply of coolant to the working chamber, and thus an enhanced demand for coolant, initiated by a rise in chamber temperature, can be met almost instantaneously.

The subject of testing at low temperatures has been reviewed by Lieb and Mowers[75].

11.6 COEFFICIENTS OF EXPANSION

Essentially, the determinations of the coefficients of linear and cubical expansion of plastics materials follow the classical methods of physics, but with the simplifying factor that *most* plastics materials have expansion coefficients several times larger than metals or glass for instance. Unfilled plastics materials have linear coefficients generally in the range 5 to 15×10^{-5}/degC compared to soda glass (1×10^{-5}), fused silica ($0{\cdot}045 \times 10^{-5}$), alumina ($2 \times 10^{-5}$) and iron ($1 \times 10^{-5}$). However, a value as low as $2{\cdot}1 \times 10^{-5}$/degC has been reported by Manfield[76] for an epoxide system filled with crushed fused quartz; there is also the complication that many plastics materials are not isotropic—laminates, filled mouldings, etc.—and many fabricated forms have moulded-in strains which the heat of an expansion test will tend to relieve, i.e. anneal.

For volume coefficients, the more accurate of the density measuring techniques described in Chapter 5, Section 5.1, may be used, carrying out the measurements at two or more temperatures of interest embracing a range wide enough that the coefficient is determined with sufficient accuracy, but not so wide as to span a transition and yield an average coefficient of meaningless value. Techniques based on density gradient tubes have been described by Gordon and Macnab[12] and on apparent loss in weight on immersion in liquid, by Ghanem[77].

If the material is isotropic, the linear coefficient may be obtained with sufficient accuracy for most purposes by dividing the cubical coefficient by three.

11.6.1 Standard Methods Based on Dilatometers

A.S.T.M. D.696–44 gives a method for determining the coefficient of linear thermal expansion using a quartz tube dilatometer (Figure 11.22), of outer tube length about 20 in.

The method is described as approximate and unsuitable for use with very soft plastics; the dial gauge, calibrated to 0·0001 in, should measure changes in length of the specimen (2–4 in length) with an accuracy of within $\pm 2\%$ over the temperature range being investigated (-30 to $+30$°C is recommended). The dial gauge, or equivalent, must not exert a pressure of more than 10 lbf/in^2 on the specimen.

Thin steel plates are cemented to each end of the cylindrical specimen to avoid indentation by the inner quartz tube (convex ended), contacting the dial gauge foot at its other end, and by the outer quartz tube (concave ended) contacting the lower end of the specimen with its convex inner surface. The length of the specimen (L) is measured with a scale or caliper capable of an accuracy of $\pm 0{\cdot}5\%$ of the specimen.

The specimen is mounted in the dilatometer and then placed in a temperature bath at $-30 \pm 0{\cdot}2$°C until the specimen reaches that temperature; this time is determined by preliminary 'blank' experiments. The dial gauge reading is recorded and the bath temperature is changed to $+30 \pm 0{\cdot}2$°C and the procedure repeated. Finally, the temperature is dropped to $-30 \pm 0{\cdot}2$°C again; the specification states 'If the change in length of the specimen due to heating, does not agree with the change in length due to cooling within 10% of their average (ΔL), the cause of the discrepancy shall be investigated and if possible eliminated'.

Average coefficient of linear thermal expansion

$$\alpha = \frac{\Delta L}{L \times T}$$

T = temperature difference in degree Centigrade over which the measurements were taken (i.e. 60 in the above).

An essentially similar method appears as Appendix A in the 'Recommendations for the Presentation of Plastics Design Data', of B.S. 4618,

Section 3.1, following the work of Committee PLC/36 (see Chapter 2).

The measurement of coefficient of cubical thermal expansion is described in A.S.T.M. D.864–52, using, in one variant, a glass dilatometer (Figure 11.23).

It consists of two pieces of Pyrex glass tubing sealed together as shown in Figure 11.23a. The lower piece, or bulb, is approximately 20 cm in length, 10 mm in outside diameter and 1 mm in wall thickness. The upper or capillary part is approximately 1 m in length, 6 mm in outside diameter and

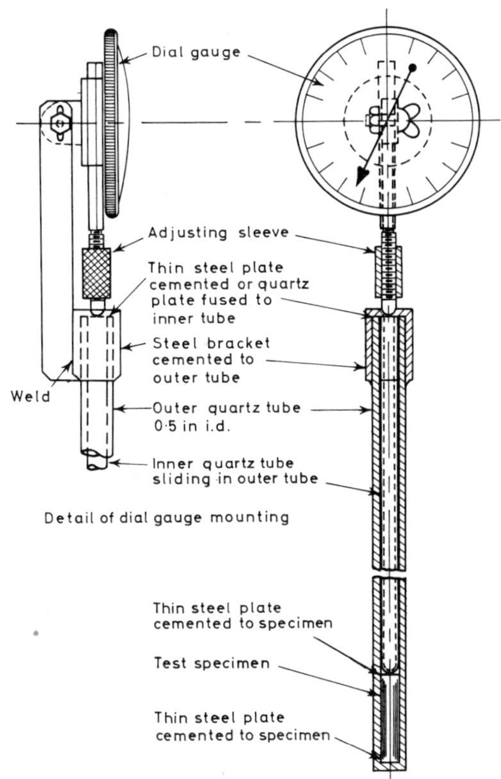

Figure 11.22. Quartz tube dilatometer (A.S.T.M. D.696)

about 0·7 mm in inside diameter. This latter is calibrated for uniformity and corrections applied if necessary. A scale, readable to the nearest 0·05 cm, is attached (see Figure 11.23b).

The test specimen is a rod approximately 15 cm in length by 6 mm by 6 mm, with rounded edges and free from voids, etc. The weight in grams, W, and volume in cm^3, V, are determined by direct measurements or via the specific gravity. The specimen is then sealed into the bulb of the dilatometer as shown in Figure 11.23, with the glass bubble below it to avoid excessive heating of the specimen during the sealing operation. The weight of the tube plus sealed in specimen is determined, W_1.

The tube is filled with mercury by vacuum and the meniscus level thereof is adjusted by piano wire so that level of the mercury is within the attached scale over the whole of the temperature-range of interest. The weight of the tube plus specimen plus mercury is determined, W_2.

The tube is immersed into a temperature bath ($\pm 0{\cdot}2^\circ$C control) and the temperature thereof adjusted to the lowest point on the range to be investigated. When the mercury meniscus level has been steady for 15 min,

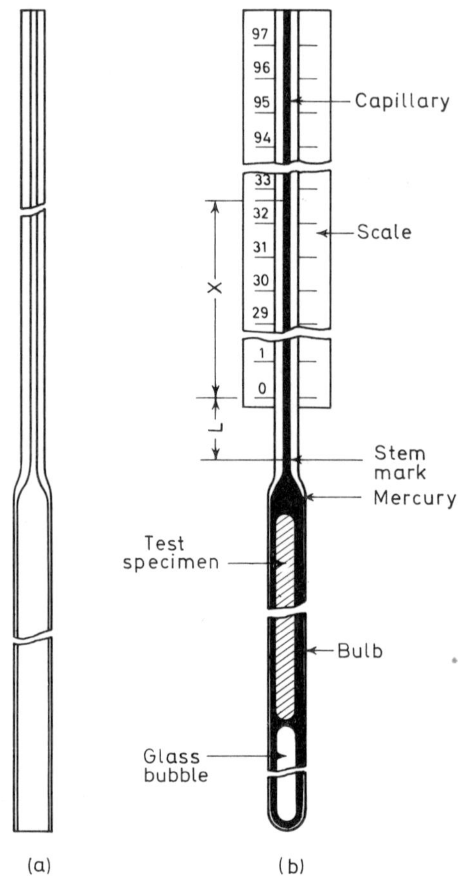

Figure 11.23. Glass dilatometer (A.S.T.M. D.864)

the position in the scale, X, is noted. At suitable increments, 5 or 10 degC, the procedure is repeated.

The whole procedure is carried out at least twice; if the results of the two increasing temperature operations do not agree, the procedure is repeated as often as necessary to obtain repeated sets of figures. The position of the mercury meniscus is plotted against temperature; generally two straight lines are obtained, one either side of the transition point. The slopes of these two lines are determined.

The internal volume, V_0, of the bulb below the stem mark is calculated from the formula:

$$V_0 = \frac{W_2 - W_1}{S_m} + V - (X_0 + L)A$$

where S_m = density of mercury (13·53 g/cm^3)
$X_0 + L$ = distance of mercury meniscus from stem mark, in cm, at time of weighing
and A = cross-section of capillary bore, in cm^2

The coefficient of cubical expansion

$$\alpha = \frac{A\Delta X}{V\Delta T} - \frac{V_0}{V}(\alpha_m - \alpha_g) + \alpha_m$$

where α_m = coefficient of cubical thermal expansion of mercury ($18{\cdot}2 \times 10^{-5}$/degC)
α_g = coefficient of cubical thermal expansion of Pyrex glass ($1{\cdot}0 \times 10^{-5}$/degC)
and $\frac{\Delta X}{\Delta T}$ = slope (s) of straight line portion of plot of mercury meniscus position against temperature.

11.6.2 Other Methods, Based on Direct Measurement

Dannis[78, 79] has described two methods, both essentially direct measurements of the changes of linear dimensions of specimens. In one a strip specimen is used (Figure 11.24) and its length changes are observed with a micrometer microscope which follows the relative movement of a mark,

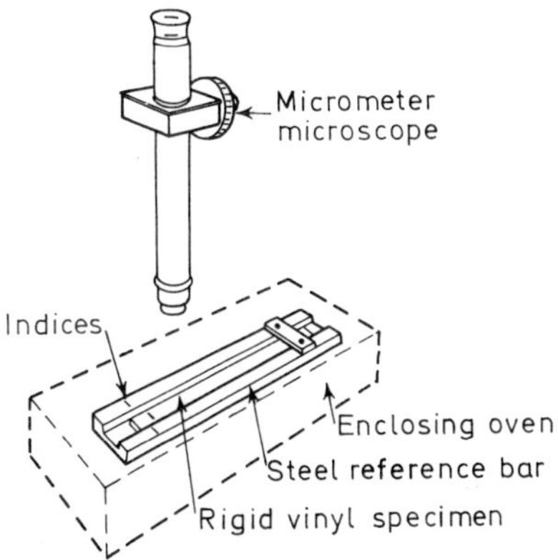

Figure 11.24. Linear expansion apparatus (Dannis)

ruled on the specimen, against an index mark on a steel reference bar. The specimen is held in a groove of the reference bar by being clamped at one end; the groove is polished and lightly lubricated to prevent sticking. The distance between clamp and mark is about 5 in (specimen length about 6 in) and the distance between the mark on the specimen and the reference mark is read to at least 0·0001 in accuracy. The temperature is recorded by a thermocouple inserted into a hole drilled into the reference bar.

The essential features of Dannis' second technique are shown in Figure 11.25.

One leg of a mirror tripod rests on the specimen under test, which latter is protected from indentation by a washer. The other two legs of the tripod rest on the rim of a stainless steel cell in which is contained the specimen. Movement of the specimen thus tilts the mirror and moves a galvanometer lamp spot. There is an auxiliary mirror on the rim of the cell, to check zero and allow accurate measurements of relative movement to be made.

A tilting mirror, moved by the expansion of a test piece relative to a comparator sample such as fused quartz, is used in a somewhat similar

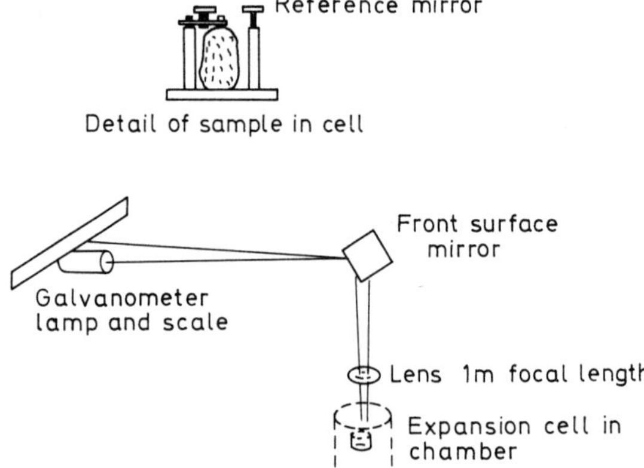

Figure 11.25. Linear expansion apparatus (Dannis)

technique described by Hamilton et alia[80]. In Reference 13, an Instron or similar tensile testing machine (see Chapter 8) is employed for linear expansion measurements.

11.7 SHRINKAGE

Closely related to the coefficient of linear thermal expansion is the property known as 'shrinkage'; this general-sounding term is normally taken to apply to the difference between the linear dimensions of a moulding and the corresponding dimensions of the mould cavity from which the moulding was produced.

There are five methods in B.S. 2782: 1970:

Method 106A Shrinkage of phenolic moulding material.

Method 106B Shrinkage of aminoplastic moulding material.

Method 106C After-shrinkage at 20°C of aminoplastic moulding material.

Method 106D After-shrinkage at 38°C of aminoplastic moulding material.

Method 106E After-shrinkage at 100°C of aminoplastic moulding material.

In Methods 106A and 106B the shrinkage is simply determined by subtracting the diameter of the cold moulding (prepared under the conditions specified in the relevant product B.S.) from that of the cold mould and dividing by the latter, using a marked diameter in case the specimen is not perfectly circular. In Method 106A the disc is 100–130 mm (4–5 in) in diameter and 3·18 mm ($\frac{1}{8}$ in) thick; in Method 106B it is 50·8 ± 1·0 mm (2·00 ± 0·4 in) in diameter and 3·18 mm ($\frac{1}{8}$ in) thick.

In Methods 106C, D and E the disc is as used in Method 106B, but the after-shrinkage is obtained by subtracting the diameter of the moulding, after subjecting it to one of the conditions (with stated tolerance) for an agreed period, from the corresponding diameter before exposure and dividing by the latter diameter. In the case of the 20°C conditions, a desiccator with calcium chloride therein is used.

A.S.T.M. D.955–51 covers the subject somewhat more comprehensively in making specific reference to compression and transfer moulding and injection moulding; for the latter two a longitudinal bar is used. The mode of expression of results is as for Methods 106A and B of B.S. 2782.

A.S.T.M. D.1299–55 is somewhat akin to Methods 106C, D and E of B.S. 2782, i.e. covering the 'Shrinkage of Moulded and Laminated Thermosetting Plastics at Elevated Temperatures'. Temperatures of 70°C, 90°C, 105°C, 130°C, 180°C or 230°C are used, as appropriate, with specimens 100 ± 1·0 mm (4 in) in diameter or 100 ± 1·0 mm square, both 3·2 ± 0·13 mm (0·0125 ± 0·005 in) in thickness.

In A.S.T.M. D.702–68, 'Cast Methacrylate Plastic Sheets, Rods, Tubes and Shapes', Section 6.1.8 gives a method for measuring shrinkage due to exposure for 30 min to 125°C or 140°C according to the grade of material. 30·5 cm (12 in) square sheets are used, and the separation of gauge marks placed thereon is measured to the nearest 0·25 mm (0·01 in). Also of relevance is A.S.T.M. D.2566–68, 'Linear Shrinkage of Thermosetting Casting Systems during Cure'.

Finally, DIN.53464, which makes specific reference to phenolic, urea formaldehyde and melamine formaldehyde moulding materials, lays down precise moulding conditions for these materials and uses test bars 120 ± 2 mm long by 15 ± 0·5 mm wide by 10 ± 0·5 mm thick.

11.8 THERMAL STABILITY AND DEGRADATION

So far in this chapter we have been dealing mainly with *reversible* processes, measurements designed to study or check the influence of change of temperature on plastics materials wherein any change of property so induced may be cancelled out by return to the original datum temperature, i.e. no permanent 'damage' has been done. We turn now to the tests for measuring the *irreversible* effects of change of temperature and, as a consequence, we are only concerned with the effect of heat, since lowering the temperature reduces the energy in a system and of itself causes no permanent change—though depending on circumstances the process of returning to the original state may be a slow one. Elevation of temperature will cause loss of volatile constituents such as an evaporation of plasticiser if present, as a first irreversible step; however, considering the polymer alone, since most are based on a wholly or partially carbon backbone they are suceptible to bond scission at relatively modest temperatures, perhaps as low as 100°C depending on polymer type, environment and duration.

The mode of polymer degradation varies widely with change in polymer type; for instance polyethylene in oxygen initially cross-links, PVC 'unzips' by losing hydrogen chloride and polymethylmethacrylate depolymerises into monomer almost quantitatively. It is not therefore surprising that standardised tests for thermal stability tend to differ from polymer to polymer and clearly, with an unknown material it is not sufficient, say, just to measure weight loss due to heating—there may be in fact a gain. The subject is a vast one and the reader is referred to References 81–85 for theoretical considerations and practical results.

A very general guide to the subject, in the form of a standard, is DIN.53446, 'Testing of Plastics. Determination of Temperature—Time limits'; Goldfein[86] has described some techniques for predicting degradation from 'master curves' obtained for certain plastics materials.

11.8.1 Standard Tests

METHODS BASED ON MECHANICAL OR OTHER PROPERTY DEGRADATION

The elevated temperature tests referred to in Section 11.5.1 come into this category in certain instances, in that some of them may be carried out at temperatures which cause irreversible changes. A.S.T.M. D.794–68, 'Determining Permanent Effect of Heat on Plastics', lays down general requirements for conducting any suitable standard tests at high temperatures; for example, A.S.T.M. D.700–68, 'Phenolic Molding Materials', under the requirements for 'Heat Resistance', requires determination of flexural strength after seven days ageing at the specified temperature followed by reconditioning to normal temperature. If the flexural strength so determined is less than 75% of the initial flexural strength the material is deemed to have failed.

Similarly, Methods 101A and 101B of B.S. 2782: 1970 determine the crushing strength after heating ('heat resistance') of, respectively, moulding material and laminated sheet. The test is basically a compression strength measurement (Chapter 8) after the test specimens have been cooled following heating in an air oven for 17 ± 1 h at 130–140°C (266–284°F), followed by 6 h ± 15 min at 165–175°C (329–347°F) and total immersion in fusible metal at 390–410°C (734–770°F) for 30 ± 2 min. The specimens are either cylinders (Method 101A) of length 9·5 ± 0·2 mm and similar diameter or cubes (Method 101B) of side of the same dimension; the latter may be a composite specimen made by building up thin sheet. The compression load on the (heat-treated) specimens is increased steadily so that failure occurs within 30 ± 15 s.

Method 108A of B.S. 2782: 1970, 'Ageing of Polythene by Hot Milling', requires 250 g of material to be milled for 3 h on an open two roll mill [recommended size: 150 × 300 mm (6 in × 12 in)—with speeds in the range 20–30 rev/min and a small differential speed between them]. The gap setting is 1·5 mm (0·060 in) and temperature 160 ± 5°C; cross-mixing is carried out at 5 min intervals. Appropriate standard physical tests are carried out after the ageing.

Method 108B of the same B.S., 'Ageing of Polythene by Hot Air Oven', uses a specimen approximately 76 mm × 76 mm × 1·3 mm (3 in × 3 in × 0·05 in) which is heated for 48 ± 2 h at 140 ± 1°C in an oven with fan circulation of the air. The specimen rests on a sheet of polytetrafluorethylene or a sheet of metal coated with that polymer. This method of ageing is particularly used for examination of polythene used for cable insulation or sheath and is usually followed by determination of power factor at 1–20 MHz (Chapter 10) or, for black compounds, by determination of melt flow index (Chapter 6).

In Method 110A of B.S. 2782: 1970, 'Extensibility after Heat Ageing of Flexible PVC sheet', the test is carried out as described in Method 104C (see Section 11.4.2 above), but at a temperature of 23 ± 1°C, after heating for 24 h at 100°C in the presence of activated carbon as described in Method 107F (see below).

METHODS BASED ON WEIGHT LOSS

There is a considerable number of 'loss in weight on heating' tests in the Method 107 series of B.S. 2782: 1970. Most, however, are not ageing tests, but rather control tests of *solids content* of resin solutions and monomer or resin content of laminating resins; as such they are not considered to fall within the scope of this monograph (being more analytical in character). Nevertheless two merit mention, Methods 107F and 107G.

Method 107F 'Loss in Weight on Heating of Plasticised Polyvinyl Chloride Compound' uses three specimens, each 101·6 ± 0·5 mm (4·000 ± 0·020 in) long, 6·35 ± 0·13 mm (0·250 ± 0·005 in) wide and 1·27 ± 0·08 mm (0·050 ± 0·003 in) thick. Each is weighed to the nearest 0·0001 g and placed in a cage constructed of 30 mesh bronze gauze. Each cage consists of a container 67 mm ($2\frac{5}{8}$ in) in diameter and 6·4 mm ($\frac{1}{4}$ in) deep, formed by soldering a 6·5 mm ($\frac{1}{4}$ in) wide strip of the gauze at right angles to the periphery of a 67 mm ($2\frac{5}{8}$ in) diameter disc of the gauze. The cage is provided with a lid 70 mm ($2\frac{3}{4}$ in) in diameter made in a similar manner. The assembled cage must not be greater than 9·5 mm ($\frac{3}{8}$ in) in depth and must be able to pass freely through the aperture of a metal can of external diameter approximately 90 mm ($3\frac{5}{8}$ in) and height approximately 115 mm ($4\frac{1}{2}$ in). The aperture of the can must be 76 mm (3 in) in diameter and in the lever type lid there is a 3 mm ($\frac{1}{8}$ in) diameter vent hole.

A layer of 120 ± 5 ml of activated carbon (5–14 mesh B.S. 410) is placed in the metal can and a loaded cage placed thereon, followed by another layer of 120 ml of activated carbon. The second cage is placed on the second layer of carbon and so on until the third cage is covered with a fourth layer of carbon. The lid is then placed firmly on the can and the whole assembly placed in a circulated oven or bath at 100 ± 1°C, in the latter case the vent hole being fitted with a vent pipe to permit total immersion of the can. After 24 h the can is removed, the specimens removed from the cages, cooled in a desiccator to room temperature and reweighed. The loss in weight is expressed as a percentage of the original weight. In A.S.T.M. D.1203–67, 'Loss of Plasticiser from Plastics (Activated Carbon Methods)', the principle is the same but, amongst other differences, the heating temperature is 70 ± 1°C. Two methods are described, one where the sample is in direct contact with the activated carbon, the other where the specimens are in a small wire-mesh cage. [See also DIN.53407 and ISO/R.176.]

Method 107G, 'Loss in Weight on Heating of Cellulose Acetate Moulding Material', employs two test specimens, each $50{\cdot}8 \pm 0{\cdot}06$ mm ($2 \pm 0{\cdot}025$ in) long, $25{\cdot}4 \pm 0{\cdot}06$ mm ($1 \pm 0{\cdot}025$ in) wide and $1{\cdot}52 \pm 0{\cdot}05$ mm ($0{\cdot}060 \pm 0{\cdot}002$ in) thick. They are placed on edge in a well-ventilated oven, of capacity not less than 3200 cm^3 (200 in^3), set to 70 ± 1°C. After 3 h the specimens are removed, cooled over calcium chloride in a desiccator and weighed (W_1). They are then replaced in the oven for a further 24 h, removed, cooled as before and reweighed (W_2). (All weighings are carried out in a weighing bottle.)

$$\text{Loss in weight on heating (\%)} = \frac{(W_1 - W_2) \times 100}{W_1}$$

Many of the various material specifications in Part 26 of the A.S.T.M. series contain weight loss tests as shown in Table 11.2.

Table 11.2 A.S.T.M. WEIGHT LOSS TESTS

A.S.T.M. No.	*Material*	*Specimens* (in)	*Conditions*
D.706–63	Cellulose acetate moulding compounds	3 each: $3 \times 1 \times \frac{1}{8}$ or 2 dia. $\times \frac{1}{8}$	72 h at 82 ± 1°C (180 ± 2°F)
D.707–63	Cellulose acetate butyrate moulding compounds	ditto	ditto
D.729–57	Vinylidene chloride moulding compounds	$3 \times 1 \times \frac{1}{8}$	ditto
D.786–49	Cellulose acetate plastic sheets	$3 \times 1 \times$ thickness of sheet	ditto
D.787–63	Ethyl cellulose moulding compounds	2 dia. $\times \frac{1}{8}$	ditto
D.1562–60	Cellulose propionate moulding compounds	ditto	ditto

In A.S.T.M. D.1457–69, 'TFE-Fluorocarbon Resin Moulding and Extrusion Materials', there is a 'Thermal Instability Test' the index of which is obtained by multiplying by 1000 the difference in specific gravity between (*a*) discs moulded by sintering compressed powder at 380°C (716°F) for 30 min followed by cooling to 300°C (572°F) at a rate of $1{\cdot}2 \pm 0{\cdot}5$ degC ($2{\cdot}2 \pm 0{\cdot}9$ degF) per minute in the oven and then to room temperature in air, and (*b*) discs similarly prepared but sintered for 2 h.

There is also a generalised technique in A.S.T.M. D.756–56, 'Resistance of Plastics to Accelerated Service Conditions'. Procedure II thereof involves heating for 72 h at 60 ± 1°C followed by measuring weight and dimensional changes and assessing any visual effects.

In DIN.53405 (nearly equivalent to ISO/R.177), weight loss of plasticised material at 70°C is determined when a sample is placed between specified contacting films etc. to simulate end uses.

METHODS SPECIFICALLY FOR VINYL CHLORIDE POLYMERS

Method 109A of B.S. 2782: 1970 and DIN.53381–Sheet 1 both assess the time taken for the vinyl chloride polymer or copolymer to evolve sufficient hydrogen chloride to change the colour of a piece of Congo Red paper placed with its lower edge 25 mm (1 in) above the top of the polymeric sample, the upper end of the paper being held in position by a plug of glass wool inserted in the mouth of the tube. The assembly is placed in a constantly-stirred oil bath at 180 ± 1°C to the level of the upper surface of the material in the tube. The heat stability (mean of two determinations) is taken as the time in minutes between insertion of the tube in the hot oil bath and the first signs of a change of the indicator paper from red to blue.

In B.S. 6746: 1969, 'PVC Insulation and Sheath of Electric Cables', (Appendix F), there is an ageing test for six types thereof which involves suspending samples in a tube about 100 mm in diameter and about 300 mm long, which tube is immersed vertically in an oven so heated and controlled that the samples are maintained at 82 ± 1°C; the samples must not be less than 20 mm from any other sample and from the wall of the tube. Clean air, preheated to approximately the same temperature, is supplied to the tube at the rate of 0·2 m^3/h, entering at the bottom and flowing upwards and out of the tube top. Air turbulence is maintained in the tube by a paddle blade rotating at about 60 rev/min, or by any suitable equivalent means. The ageing is carried out for 120 h and the loss in weight determined, which is then expressed as weight loss per unit area, in mg/cm^2, of the sample by measuring the exposed area of the sample, neglecting the cut ends. The number of samples to be used is specified and varies according to the dimensions of the cable.

For two types of insulation and sheath, the ageing is however carried out in an air oven for 168 h at 135 ± 2°C with an air flow sufficient to give between three and ten changes per hour. In this case elongation at break is determined and compared with the corresponding value before ageing.

A.S.T.M. D.793–49, 'Short-time Stability at Elevated Temperatures of Plastics Containing Chlorine', actually involves titration of the hydrogen chloride liberated when 10 g of material, cut so that no dimension is greater than $\frac{1}{16}$ in, is heated in the bottom of an Erlenmeyer flask. This latter is placed in a constant temperature bath at 180 ± 2°C to within 3 cm of the top of the flask and is fitted with a rubber stopper holding two glass tubes. One of the latter is a glass inlet tube and extends nearly to the bottom of the flask (Figure 11.26). The tube is connected to the gas preheater, in the bath, which in turn is connected to a nitrogen cylinder. The outlet tube is connected to an absorption tube containing 40 ml of 0·1 N sodium hydroxide solution.

The nitrogen gas flow is adjusted to a rate of 2–4 bubbles/second in the absorbing tube and allowed to flow for 30 min. The absorbent solution is washed into a beaker, acidified with nitric acid and the chloride present precipitated with 50 ml of 0·02 N silver nitrate solution. The excess silver nitrate is back titrated with 0·02 N ammonium or potassium thiocyanate solution, using ferric nitrate as indicator. A 'blank' is performed of 40 ml of the 0·1 N sodium hydroxide solution diluted with distilled water to approximately the same volume as used for the test.

$$M = [(A-B)-C] \times \text{normality of silver nitrate solution} \times 3{\cdot}65$$

where M = Short time stability expressed as milligrams of hydrogen chloride evolved per gram of test specimen (under the test conditions)

A = millilitres of silver nitrate solution

B = millilitres of thiocyanate solution

C = millilitres of silver nitrate solution consumed in the 'blank'.

PVC compositions discolour markedly when degradation sets in, due it is

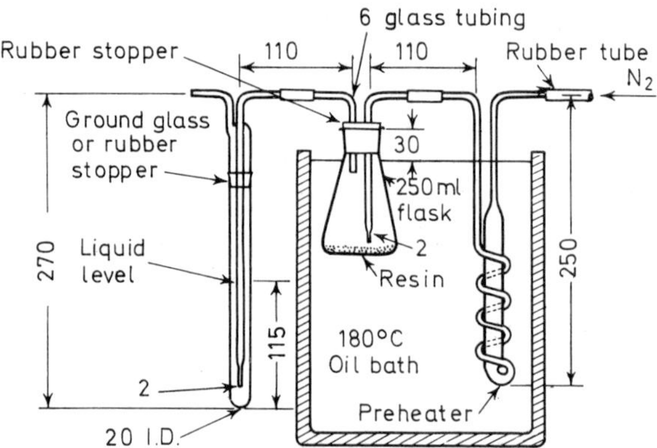

Figure 11.26. PVC stability at elevated temperature test (A.S.T.M. D.793). Dimensions in millimetres

believed to the creation of conjugated double-bond systems by loss of hydrogen chloride. Suitably specified oven ageing tests which depend on visual assessment of such degradation are therefore popular, and A.S.T.M. D.2115–67 and DIN.53381–Sheet 2 are two such standard tests. There is nothing particularly profound in these methods and the reader is referred to the original standards for details.

DIN.53381–Sheet 3 describes a test for the thermal stability of vinyl chloride polymers and copolymers in which a sample is heated in a glass tube, itself in a heated oil bath, and the hydrogen chloride evolved is carried in a gas stream into potassium chloride solution where the pH value is measured. The result is expressed in minutes as the time taken for the pH of the solution to fall to 3·8.

OTHER STANDARD METHODS FOR PLASTICS

A spectacular method is included in A.S.T.M. D.701–49, 'Cellulose Nitrate (Pyroxylin) Plastic Sheets, Rods and Tubes', and is called 'the fuming-off temperature'. Shavings not thicker than 0·025 mm (0·001 in) are prepared and heated for 2 h in an oven at 50°C to drive off moisture and solvents. 0·1 g approximately of the dried shavings are placed in a test tube, 15 cm by 1·3 cm, lightly stoppered with a perforated or notched cork and the tube immersed 5 cm into a liquid bath at 100°C. The bath is heated at a rate between 3 and 5 degC per minute and the temperature noted at which the sample decomposes explosively. At least five such tests are run.

In the dimensional stability test for 'Extruded Acrylic Plastic Sheet', in A.S.T.M. D.1547–61, the presence of bubbles is studied after specimens 6 in square are heated at either 143±6°C or 154±6°C for 15, 30 or 45 min according to, respectively, grade and thickness of material. The specimens, after heating at 50±6°C, are rested on a preheated glass plate sprinkled with talcum powder.

Finally, for testing the 'Thermal Oxidative Stability of Propylene Plastics' (A.S.T.M. D.2445–65T), a quantity of pellets or granules is placed in a U-tube which is partially immersed in an oil bath at 150±0·5°C. Oxygen is metered in at one end of the tube at a rate of 10 cm^3/min, and the pellets or granules are inspected at regular intervals for signs of failure as manifested by the appearance of surface crazing. Embrittlement is confirmed by removing a few of the granules or etc. and crushing them; if they are readily reduced to powder embrittlement is taken as complete.

RUBBER TESTING METHODS AND OTHERS

There is considerably more system in the standardised methods for ageing of rubber, in B.S. 903 for instance, and the subject as a whole has been comprehensively studied by Scott[87]. Part A.19 'Accelerated Ageing Tests' of B.S. 903: 1956 'Methods of Testing Vulcanised Rubber' describes three ageing apparatus:

Method A Cell Type Oven Method
Method B Oven Method
Method C Oxygen Pressure Method

In the last mentioned an oxygen pressure of 300±15 lbf/in^3 gauge (21±1 kgf/cm^2 gauge) is used. The effects of ageing are normally followed by any changes in tensile strength, modulus (see Section 8.4.6) elongation at break and hardness.

Also of interest, because of the use of plastics in electrical insulation applications, is the series B.S. 2011 'Basic Methods for the Climatic and Durability Testing of Components for Telecommunication and Allied Electronic Equipment' and, in the present context, particularly Part 2B: 1966 'Dry Heat' (which reconciles with the corresponding I.E.C. publication). Nine standard temperatures, with tolerances, are given and methods are described for ageing with and without thermal shock, that is with and without sudden temperature change respectively.

11.8.2 Non-Standard Methods

The technical literature is full of papers describing ad hoc tests for measuring the thermal degradation of polymers; it is neither practical nor usefully informative to attempt even a brief review. However, two instrumented techniques which, although by no means new in concept, have gained popularity in the study of polymers over the past few years, must be mentioned since although they are not standardised (yet) they are capable of standardisation and unambiguous description. Five other recent developments are also touched on.

DIFFERENTIAL THERMAL ANALYSIS (D.T.A.)

D.T.A. is an instrumented technique for following the effects of heat on a small sample; as the temperature of a 'furnace' surrounding the sample is increased in controlled manner, a thermocouple in the sample records the

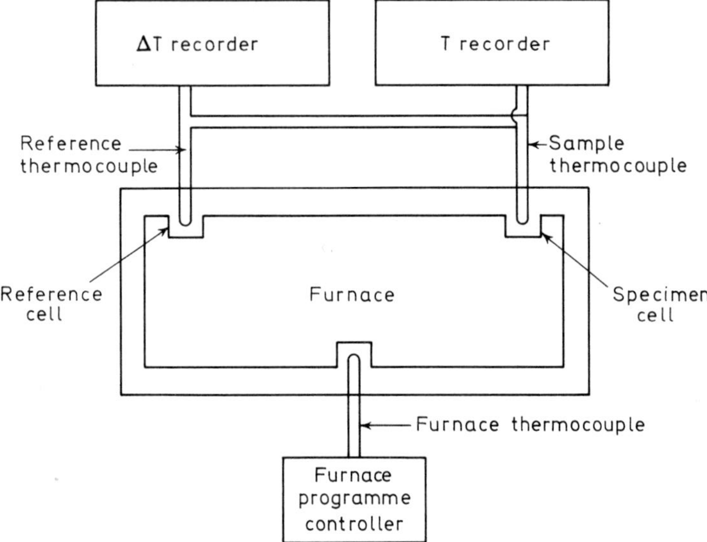

Figure 11.27. D.T.A. apparatus (diagrammatic)

response therein, whether the differential is zero (i.e. no effect other than that of heating), negative (i.e. endothermic reaction taking place) or positive (i.e. exothermic reaction taking place). D.T.A. apparatus is available commercially and it is only necessary here to describe the broad principles (see Figure 11.27).

Typically, the furnace must be without significant temperature differentials, have low heat capacity in the heating space so as to respond rapidly to (deliberate) variations in heat input and be provided with a programme controller to give uniform heating rate. It must be capable of rapidly transferring heat to the sample holder. This latter comprises two identical cells, one for the material under examination and the other for an inert material (e.g. alumina) to serve as reference. Thermocouples dip into the materials in the cells and the temperature (T) of the test sample, and the temperature differential (ΔT) between this and the inert material, are fed in e.m.f. form to a recorder to plot ΔT against T.

Figure 11.28[88] well illustrates the heat resistance conferred on cellulose by impregnation with melamine-formaldehyde resin, whilst Figure 11.29[88] shows the glass transition point of polystyrene.

Useful introductions to the subject of Differential Thermal Analysis are given by Kissinger and Newman (Chapter IV of Reference 83), Manley[86, 89],

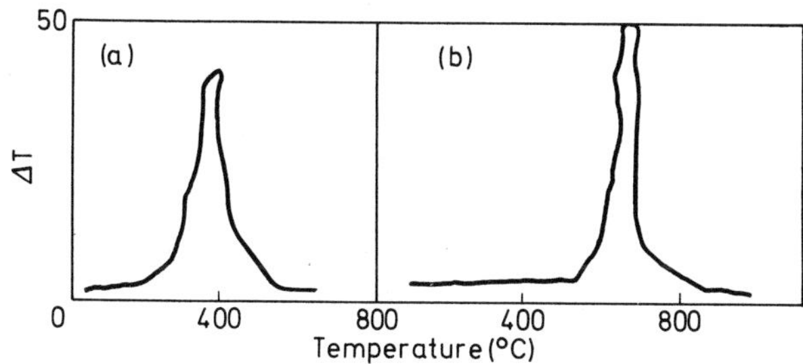

Figure 11.28. D.T.A. thermograms of (a) cellulose alone and (b) cellulose impregnated with melamine-formaldehyde resin (Manley) (Reproduced from S.C.I. Monograph no. 17, p. 175, 1963)

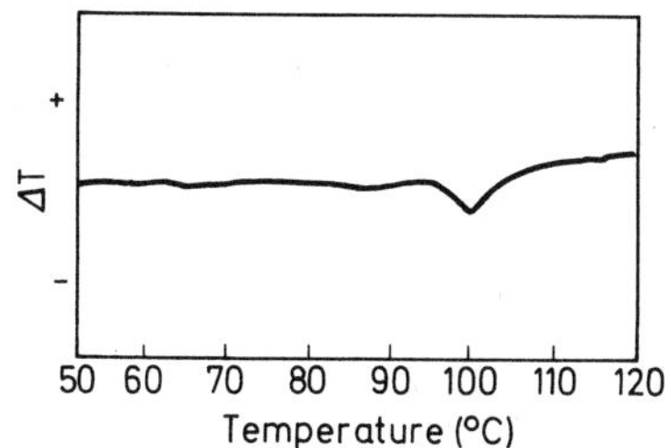

Figure 11.29. D.T.A. thermogram showing glass transition point of polystyrene (Manley) (ibid)

Double[90] and Smith[91, 92] and References 33 and 34 should also be consulted.

One particularly versatile and compact commercial unit is made by du Pont[93]. D.T.A. and T.G.A. (see below) are both extensively described in Reference 94.

THERMOGRAVIMETRIC ANALYSIS (T.G.A.)

T.G.A. is the technique of coupling a small furnace to an accurate balance so that weight change can be recorded continuously against temperature; T.G.A. equipment may conveniently form ancillary equipment to a D.T.A. apparatus[93] and, for instance, may identify a D.T.A. change as being physical or chemical according to whether or not there has been simultaneous weight loss. References 34 and 90, 91, 95 and 96 may be consulted for further information; the last-mentioned has a particularly useful bibliography. Chiu[97] has also described the coupling of T.G.A. to gas chromatographic equipment so that the volatile products produced by heating can be analysed.

THERMAL VOLATILISATION ANALYSIS (T.V.A.)

Somewhat similar is a method described by McNeill[98, 99] in which the volatile products from a sample heated in a continuously evacuated system are passed to the cold surface of a trap some distance away. A small pressure develops which varies with the rate of volatilisation of the sample and if this pressure is recorded as the temperature is increased linearly, a T.V.A. (thermal volatilisation analysis) thermogram is obtained. This type of thermogram is stated to have certain advantages over T.G.A. in characterising polymers, studying their degradation behaviour and detection of volatile impurities.

THERMOPARTICULATE ANALYSIS (T.P.A.)

Yet another thermal method is described, for instance, by Murphy et alia[100]; it is known as thermoparticulate analysis (T.P.A.) for which it is claimed that, amongst other advantages, the technique gives more reproducible results than D.T.A. As used by Murphy and his co-workers a small strip of

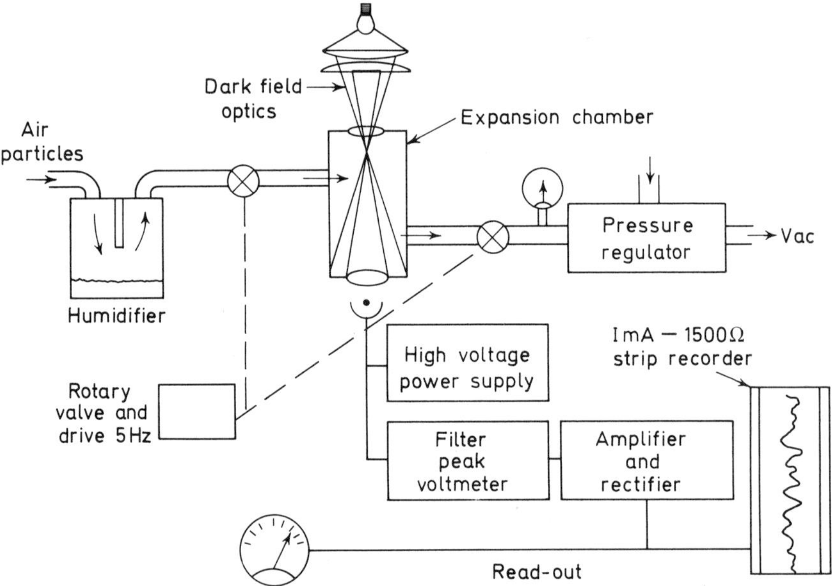

Figure 11.30. Schematic diagram of T.P.A. apparatus (Murphy et alia) (Reprinted from 'Plastics Design and Processing', July 1964 issue. Copyright 1964, Lake Publishing Corporation, Libertyville, Illinois, U.S.A.)

material is mounted in a copper tube with a thermocouple and the whole placed in a simple furnace programmed to rise in temperature by 50 degC per hour. Hydrogen is swept over the sample at a rate of 3 cm^3/s and after

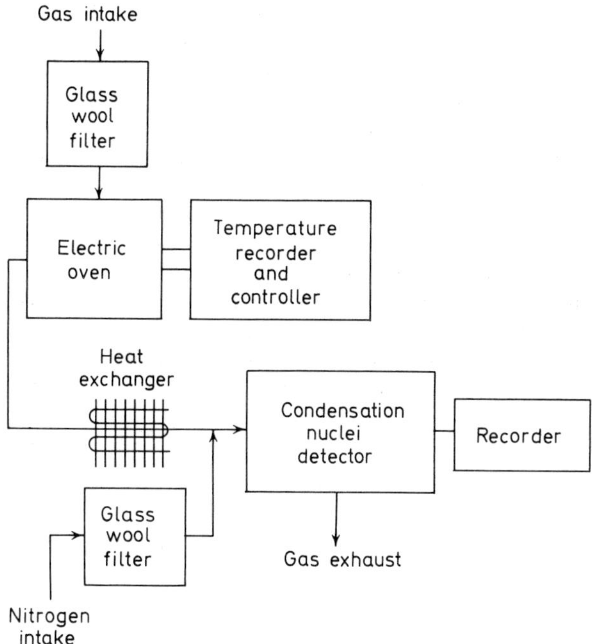

Figure 11.31. Schematic diagram of condensation nuclei counter (Murphy et alia) Reprinted from 'Plastics Design and Processing', July 1964 issue. Copyright 1964, Lake Publishing Corporation, Libertyville, Illinois, U.S.A.)

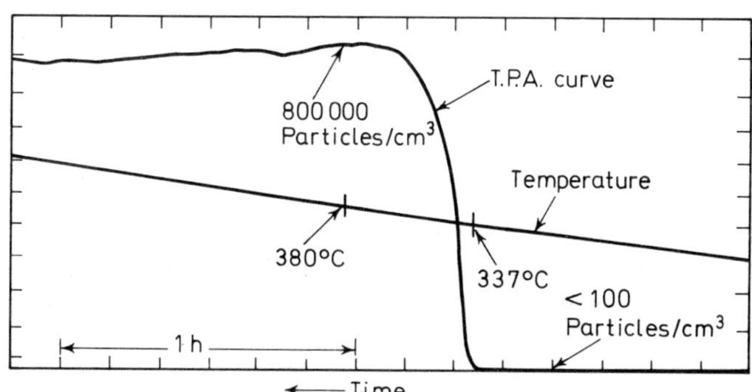

Figure 11.32. T.P.A. thermogram of polystyrene in hydrogen (Murphy et alia) Reprinted from 'Plastics Design and Processing', July 1964 issue. Copyright 1964, Lake Publishing Corporation, Libertyville, Illinois, U.S.A.)

the gas has passed through the furnace it is fed through a simple heat exchanger to a 'condensation nuclei detector'; however as the latter only functions at a gas flow of 100 cm^3/s, nitrogen is bled in at a rate of 97 cm^3/s after the heat exchanger. The 'condensation nuclei detector' is an effectively continuous method for measuring the number of particles (of radii 10^{-7}–10^{-5} cm) in the gas stream and this is achieved by adiabatic expansion of the gas sample after saturation with water vapour under such conditions that supersaturation is achieved and water condenses out on the walls of the vessel and on the particle nuclei but without autonucleation of water droplets. The droplets thus formed scatter light and their number is measured photoelectrically.

Condensation nuclei, i.e. particles, are generated from polymers on heating and their number increases rapidly, for instance, above a characteristic decomposition temperature (see Figures 11.30 and 11.31 for outline detail of the apparatus).

The thermogram of Figure 11.32 was produced with unfilled polystyrene and the characteristic temperature is clearly seen—even if one views the author's precise value of 337°C with a little scepticism. Polystyrene filled with silica gave the same result. The authors' comment that the D.T.A. of polystyrene in nitrogen has been reported to indicate an onset of decomposition at 375°C, but that this disparity could easily be explained by the much higher rate of temperature increase which would almost certainly have been used in the D.T.A. work.

For further information, see also Reference 101.

DIFFERENTIAL SCANNING CALORIMETRY (D.S.C.)

In Differential Thermal Analysis, as has already been explained above, the temperature differential between sample and 'control' is measured whilst the environment temperature round both is programmed. The result is meaningless in absolute terms though interesting in identifying transitions for example.

In Differential Scanning Calorimetry (D.S.C.), the temperature of the environment is still programmed, but the temperatures of the sample under test and the inert reference sample are continuously maintained at the same level by automatic balancing of the power fed to each holder. The differential power is thus measured and plotted against temperature. Amongst other claims made for this technique are measurements of specific heat, changes of specific heat with temperature, temperature and heat of fusion, degree of cure and quantitative evaluation of thermal stability. One apparatus commercially available is a standard accessory to the equipment described in Reference 93 (see also References 102–104).

Miller[105] has contributed an article on D.T.A., T.G.A. and D.S.C. and Lambert[106] gives details of the use of D.S.C. in measurements of glass transition temperature.

CONSTANT HEAT FLOW THERMAL ANALYSIS (C.F.T.A.)

In yet another variation on the theme, a constant heat flow is fed into the sample and the temperature of the latter simply measured as a function of time. The apparatus, and some results including transition point determinations, have been described by Steffens[107].

ELECTROTHERMAL ANALYSIS (E.T.A.)

The change in electrical conductivity of polymers as a parameter by which to measure thermal effects has been investigated by Pope[108] and Chiu[109] has described equipment for simultaneous D.T.A., T.G.A. and E.T.A. analyses.

11.9 FLAMMABILITY

The danger of fire is, or should be, an important consideration in the selection of any material for many applications, particularly those of a constructional nature. Plastics, being based mainly on carbon, hydrogen and oxygen, are not very fire-resistant as a class, though the risks inherent in their use are perhaps sometimes overplayed as a result of ignorance, gross misapplication or mental conditioning based on the fact that the earliest commercial 'plastic' was based on cellulose nitrate. Nevertheless, not all plastics are capable of being burnt and others, which in their normal state can be, may be effectively protected by suitable additives. It therefore follows that reliable methods of assessing fire resistance, or at least comparing materials for this characteristic, are required, always bearing in mind that results will depend to a considerable degree on the physical conditions of test selected. The results will also depend on the physical form of the material, particularly the surface (available to oxygen) to volume ratio; a classic example is that whilst unplasticised PVC is 'self-extinguishing', that is it ceases to burn when the source of heat is removed, 'foam' or 'expanded' unplasticised PVC burns fairly readily.

The subject is a large one and the reader is referred to Scott[110] and Allen and Chellis[111] for useful reviews of test methods and to Ashton[112] for a discussion on the use of plastics in buildings in the light of requirements for fire safety.

11.9.1 Terminology

The title of this section is '*flammability*' and this is now the widely accepted term meaning 'tendency to burn'; for many years the term '*inflammability*' was used in the U.K. synonymously with 'flammability', causing great confusion because usually the prefix '*in*' creates the antithesis. However, a few years back the B.S.I. standardised on the latter term, thereby bringing the U.K. terminology into line with that of the U.S.

Tests may be divided into two classes, '*ignition tests*' and '*burning tests*'. The former ascertain how readily and/or under what conditions a material will ignite whereas the latter measure how fast the material will burn when once ignited and perhaps also the associated effects (for example the amount and temperature of the gases evolved). Associated terms which may be

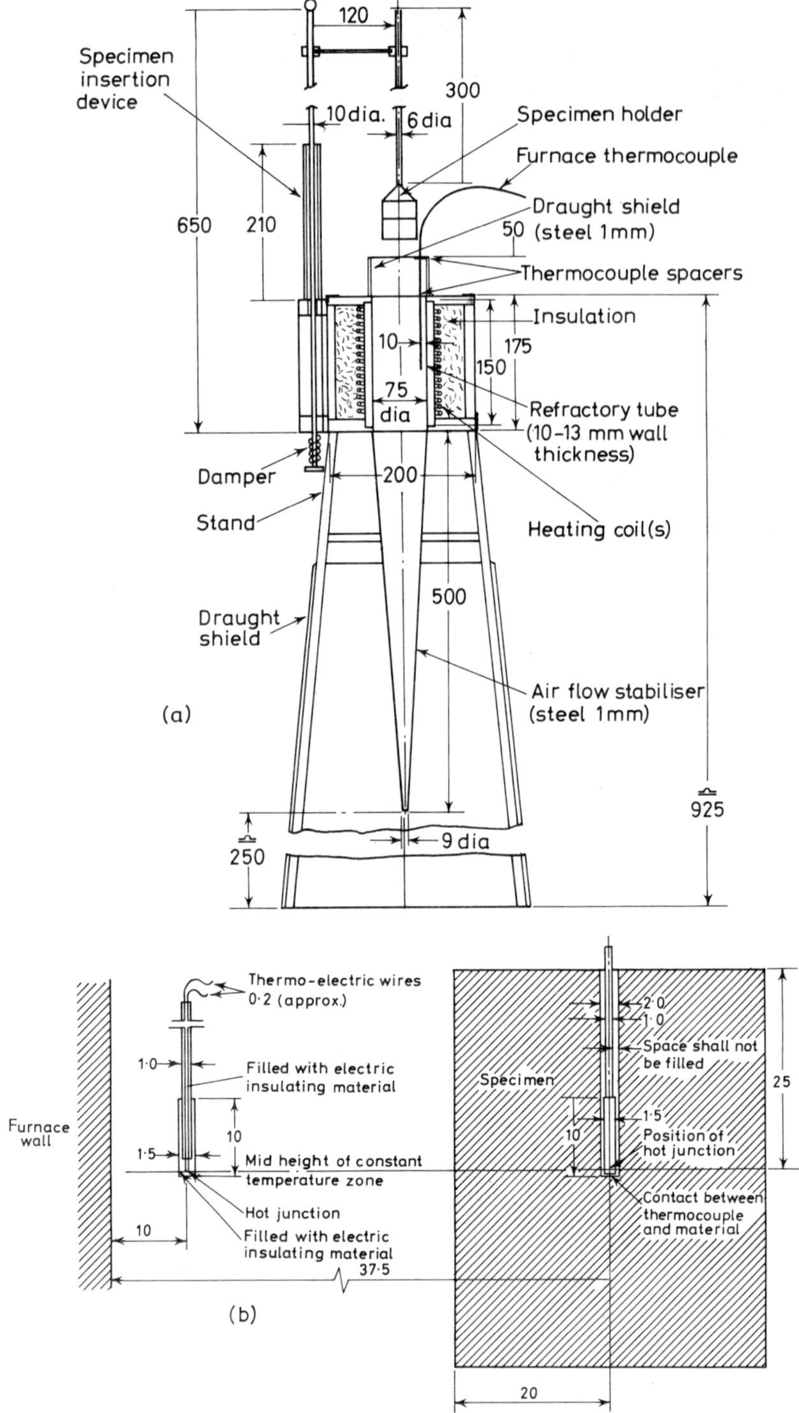

Figure 11.33. (a) Non-combustibility apparatus—general arrangement. (b) Non-combustibility apparatus—furnace and specimen thermocouple locations (B.S. 476: Part 4: 1970). All dimensions in millimetres

enountered in the description of tests include *'ignitability'*, *'combustibility'*, *'ignition temperature'*, *'flame resistance'* and variants thereon.

11.9.2 Ignition Tests

BRITISH STANDARD METHODS

A 'non-combustibility' test forms B.S. 476: Part 4: 1970, 'Fire Tests on Building Materials and Structures'. The specimens are 40 mm wide × 40 mm high × 50 mm high (with specified tolerances), layered up to this 'height' (thickness) if necessary, and three are used. They are dried by heating at 60 ± 5C for 24 h and then cooled to ambient temperature over a calcium chloride desiccator.

A special furnace (see Figure 11.33), is described in the B.S; it is heated to 750°C for at least 10 min and a specimen suspended centrally in the furnace

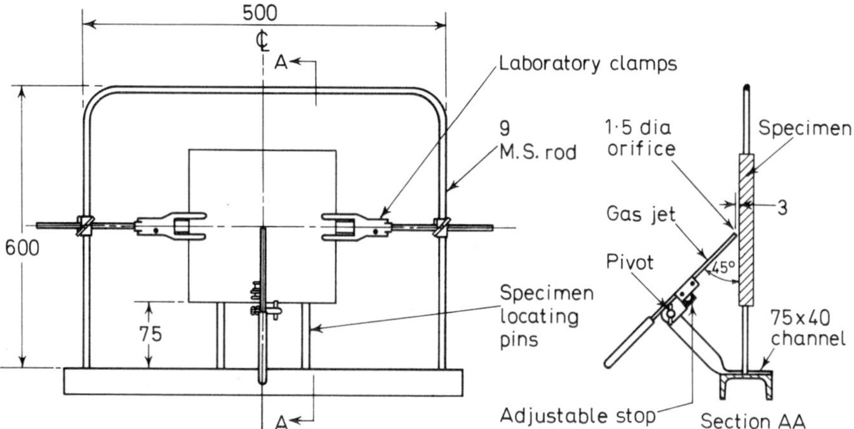

Figure 11.34. Ignitability test (B.S. 476: Part 5: 1970). All dimensions in millimetres

for 20 min. During this period a record is kept of the temperatures of the thermocouples in the specimen and in the furnace and the occurrence and duration of any flaming are noted.

A material is deemed 'non-combustible' if, during the 20 min period, none of the three specimens either:

1. causes the temperature reading from either of the two thermocouples to rise by 50 degC or more above the initial furnace temperature, or
2. is observed to flame continuously for 10 s or more inside the furnace.

Otherwise the material is deemed 'combustible'.

This test is not particularly suitable for plastics and another is available, as B.S. 476: Part 5: 1968, 'Ignitability Test for Materials'. Specimens 228 mm × 228 mm by the normal thickness are conditioned to equilibrium in air at 10–21°C and 55–65% relative humidity and then tested in an apparatus the framework of which is a U-frame of 9 mm mild steel fixed into a rigid steel base (see Figure 11.34).

Two laboratory clamps are attached centrally one to each vertical arm of the U-frame to hold the specimen and two supports of 9 mm mild steel are provided, as shown, so that the lower edge of the specimen rests on them.

A gas jet is provided from a copper tube having one end reduced to 1·5 mm diameter and pivoting on a fixed strap; an adjustable stop is provided so that the jet can be set 3 mm (+1·5 mm−0) from the face of the specimen when under test and at an angle of 45° to the vertical. The flame of the gas jet is adjusted to liberate 1060 cal/min ± 2% and is applied to the surface of the specimen for 10 s.

If any of three specimens 'flames' for more than 10 s after removal of the jet or if burning of the specimen extends to the edge within this period, the material is classified as 'easily ignitable' and its performance indicated by the letter '*X*'. If no specimen flames for more than 10 s and burning does not extend to the edge within this period the classification is 'not easily ignitable' and indicated '*P*'. If the surfaces of the material are different, they are tested as separate entities.

The test is conducted in a 'reasonably draught free atmosphere'.

A.S.T.M. METHODS

In A.S.T.M. D.1929–68, 'Ignition Properties of Plastics', three criteria can be measured:

flash ignition temperature—the lowest initial temperature of air passing around the specimen at which a sufficient amount of combustible gas is evolved to be ignited by a small external pilot flame.

self-ignition temperature—the lowest initial temperature of air passing around the specimen at which, in the absence of an ignition source, the self-heating properties of the specimen lead to ignition or ignition occurs of itself, as indicated by an explosion, flame or sustained glow.

self-ignition by temporary glow—a special case of self-ignition temperature where, in some cases, slow decomposition and carbonisation of the specimen results only in glow of short duration at various points therein without general ignition.

The test is carried out in a carefully specified furnace, the general design of which is shown in Figure 11.35.

Specimens (3 ± 0·5 g) are moulding pellets, or 2 cm × 2 cm pieces of sheet or moulding, bound together by fine wire. Initially, a first approximation of the flash ignition temperature is made. With *low air flow* (2·54 cm/s) and the specimen charged, the transformer controlling the current to the nichrome wire is adjusted to give a rise in thermocouple T_2 of approximately 600°C/h. The pilot light is lit and placed across the hole in the top of the furnace. The value of the air temperature (T_2) is noted at which the combustible gases are ignited; this is identified as the point at which there is a rapid rise in specimen temperature (T_1).

For *medium air flow* the procedure is the same but the air setting is 5·08 cm/s and for *high air flow* it is 10·16 cm/s.

To obtain first approximations of the self-ignition temperature, the same three procedures are repeated, but without the pilot flame. The temperature T_2 is noted at which the specimen flames, explodes or glows.

Second approximations of the flash ignition and self-ignition temperature are obtained using the air flow settings from the first tests which gave the lowest flash and self-ignition temperatures, with a temperature rise setting of 300 degC/h.

Finally, minimum flash ignition and self-ignition temperatures are determined, using air flow rates as for the second approximation tests. The transformer settings are adjusted until the air temperatures (T_1) are constant over a period of 15 min, the air temperatures being not more than 10 degC below the corresponding second approximation figures. The specimen is then placed in the furnace and ignition of gases or specimen flaming, exploding or glowing noted as appropriate. If these effects occur, the determination is repeated 10 degC lower still and so on until no ignition, etc.,

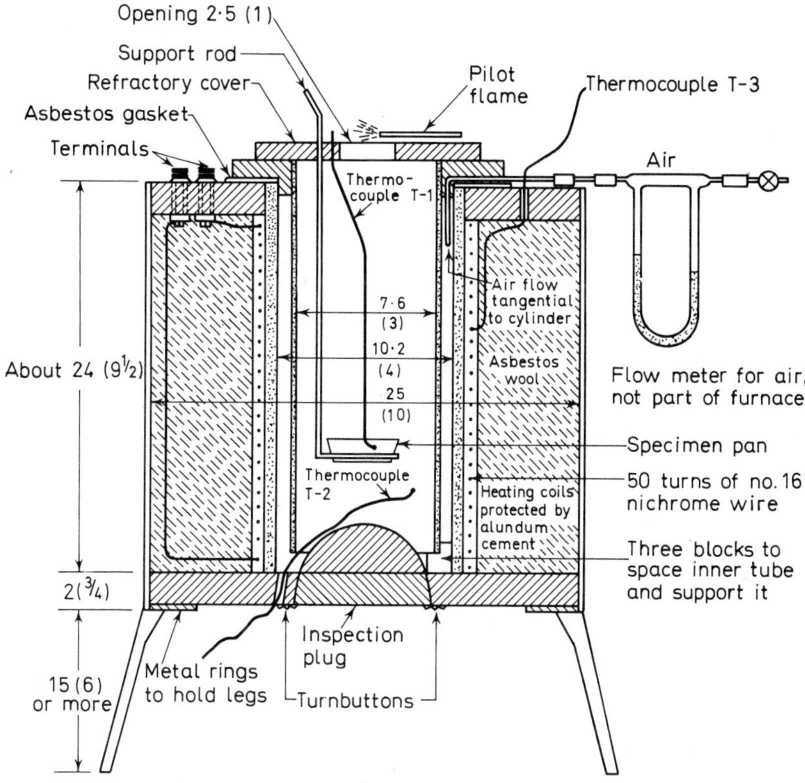

Figure 11.35. Hot air ignition furnace assembly (A.S.T.M. D.1929). Dimensions in centimetres with inch equivalents in parentheses

occurs in 30 min. The minimum flash ignition or self-ignition temperatures are the lowest air temperatures at which ignition, etc., occurs.

Similar in principle is ISO/R.871, 'Plastics. Determination of the Temperature of Evolution of Flammable Gases from Plastics', where however the *decomposition temperature* is defined as the lowest temperature at which the flash point type of igniting flame causes ignition of evolved gases to last at least 5 s.

A.S.T.M. D.229–69, 'Testing Rigid Sheet and Plate Materials Used for

Electrical Insulation', contains a test for flame resistance to be used for those materials found non-burning or self-extinguishing by a slightly modified version of the method of A.S.T.M. D.635 (see below). Under the test conditions, 'ignition time (I) is the period in seconds required to produce ignition' and 'burning time (B) is the period in seconds that the specimen burns after removal of the ignition source'.

The 'flame cabinet' and ancillary equipment are specified in detail; the essential principle is to place a weighed specimen, $13 \pm 0{\cdot}8$ mm ($\frac{1}{2} \pm 0{\cdot}036$ mm) thick by $13 \pm 0{\cdot}25$ mm ($\frac{1}{2} \pm 0{\cdot}01$ in) width by $254 \pm 1{\cdot}6$ mm ($10 \pm \frac{1}{16}$ in) long in an electric heating coil/specimen holder between arc electrodes which are simultaneously energised and whose function is to ignite gases emitted from the specimen. The ignition time (I) is determined and the coil and arc are de-energised 30 s after this. The total time T is noted when the specimen ceases to burn. Four specimens are examined and weight loss is also recorded.

$$\text{Burning Time } B = T - I - 30$$

11.9.3 Burning Tests

BRITISH STANDARD METHODS

A variety of ad hoc tests exist in B.S. 2782: 1970, 'Methods of Testing Plastics'. Method 508A, 'Rate of Burning', employs (three) specimens 150 mm (6 in) long, 13 mm ($\frac{1}{2}$ in) wide and $1{\cdot}5 \pm 0{\cdot}1$ mm ($0{\cdot}060 \pm 0{\cdot}005$ in) thick; pencil lines are drawn, one at 25 mm (1 in) and the other at 127 mm (5 in) from one end.

The specimen is rigidly clamped, in a draught-free atmosphere, with its long axis horizontal and its transverse axes at 45° to the horizontal, 6 mm ($\frac{1}{4}$ in) above a 130 mm (5 in) square piece of clean wire gauze (18 meshes to the linear inch)—see Figure 11.36.

Six mm ($\frac{1}{4}$ in) of the unsupported end of the specimen projects beyond the gauze. A non-luminous flame, 13–19 mm ($\frac{1}{2}$–$\frac{3}{4}$ in) high, from an alcohol lamp or Bunsen burner, is so placed that the top of the flame just touches the specimen. After 10 s the flame is removed and the specimen allowed to burn. The time taken for the edge of the flame to travel the 100 mm (4 in) between the two lines is measured and the rate of burning, in mm or in/min, is calculated.

If three consecutive specimens do not burn as far as the second mark the material is reported as 'resistant to flame propagation'. If three consecutive specimens do not burn to the first mark and show no flame or after-glow 5 s after removal of the burner, the material is reported as 'self-extinguishing'.

Method 508B of B.S. 2782, 'Degree of Flammability of PVC Extrusion Compound', employs (two) specimens 230 mm (9 in) long and 25 mm (1 in) wide, cut from moulded sheet $1{\cdot}27 \pm 0{\cdot}07$ mm ($0{\cdot}050 \pm 0{\cdot}003$ in) thick. 13 mm ($\frac{1}{2}$ in) squares are marked on the specimens.

The specimens are mounted one at a time in a special shield 305 mm (12 in) square and 760 mm (30 in) in height. There is a ventilating opening 25 mm (1 in) high around the bottom and a window of heat resisting glass forms one of the four sides, the other three being of any suitable sheet material. A door is provided on one side for insertion of the specimen.

A piece of celluloid, 25 mm (1 in) square and 0·25 ± 0·03 mm (0·010 ± 0·001 in) thick, compounded from cellulose nitrate of nitrogen content 11·00 ± 0·10% and containing 25% camphor, is fixed to one end of the specimen with the aid of acetone. The contact area is 160 mm^2 ($\frac{1}{4}$ in^2) and across the full 25 mm (1 in) width. After 2 h the specimen is hung in the centre of the shield so that it faces the window; the clip holding the specimen leaves 216 mm ($8\frac{1}{2}$ in) exposed below it, and the celluloid is at the lower end. The celluloid is ignited and the door quickly closed. The time required for the flame either to extinguish itself or completely burn the specimen is recorded as the 'time of burning'. The area of the test specimen that is burned or charred is measured to the nearest 160 mm^2 ($\frac{1}{4}$ in^2) and also recorded; any area which melts and drops is included.

The 'Degree of Flammability of Thin Polyvinyl Chloride Sheeting',

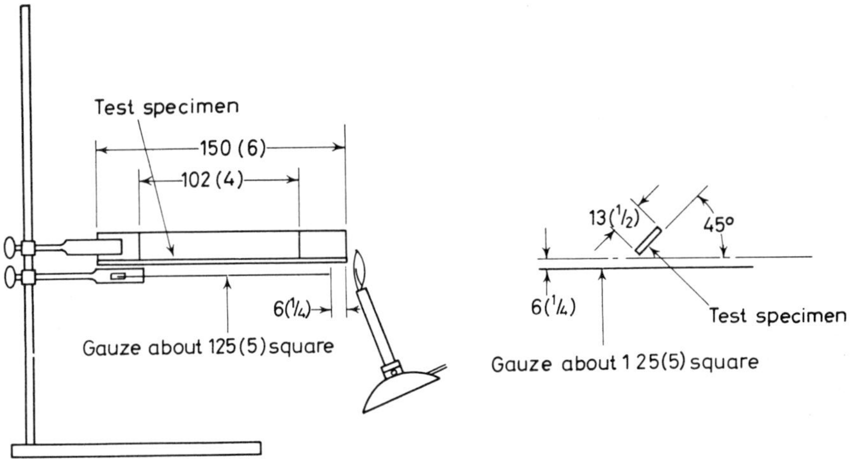

Figure 11.36. Rate of burning test (B.S. 2782). Dimensions in millimetres with inch equivalents in parentheses

Method 508C, involves stretching the test material over and inverted U-frame (Figure 11.37).

There are two of the 178 mm (7 in) radius supports, each L shaped in cross-section; they are made of brass about 1·3 mm (0·05 in) thick and kept apart by wire spacers so that the inner edges of the supports are 25·4 ± 0·8 mm ($1 \pm \frac{1}{32}$ in) apart. The specimen, 550 mm ($21\frac{1}{2}$ in) long and 35 mm ($1\frac{3}{8}$ in) wide, is held in position over the supports by two strips of spring steel. One of the supports is calibrated in 5 mm ($\frac{1}{4}$ in) intervals around its perimeter. The whole apparatus is operated in a draught free atmosphere.

Below one end of the specimen, by the 'start' mark on the calibration, a copper trough is placed on a platform on the base of the apparatus. 0·1 ml of ethanol is placed in the trough and ignited. After burning has ceased the distance is noted from the 'start' mark over which the specimen is burnt. Six specimens are used and, if the surfaces of the sheeting are different, three specimens are tested with one surface outwards and three with the same surface inwards. The degree of flammability is reported as the average of the six results, to the nearest 3 mm ($\frac{1}{8}$ in).

Method 508D, 'Flammability (Alcohol Cup Test)', is suitable for materials that cannot be tested by Methods 508A, B or C, but it is not suitable for materials that melt rather than char. The specimen is a 150 mm (6 in) square, by the natural thickness unless this be greater than 50 mm (2 in); three specimens, or possibly six, are needed. The apparatus is illustrated in Figure 11.38a and b.

The metal frame shown in Figure 11.38b is constructed of brass strip 12·5–13 mm wide and 3–3·5 mm thick. The wire grid is made of nickel-chrome resistance wire of about 0·15 mm diameter and stretches across the underside of the frame; the two projections shown in Figure 11.38a are to support the specimen centrally in the frame.

The frame rests on four supports at an angle of 45° to the horizontal, and the lengths of the four legs of the support are such that the centre of the underside of the frame is 152 mm above the surface of the base of the apparatus, which is covered with asbestos millboard of not less than 3 mm thickness. A flat bottomed metal cup, 17·5 mm in external diameter, 7·3 mm high and about 1 mm wall thickness, rests in a shallow depression on a piece of asbestos board, so that the base of the cup is 25 mm below the centre of

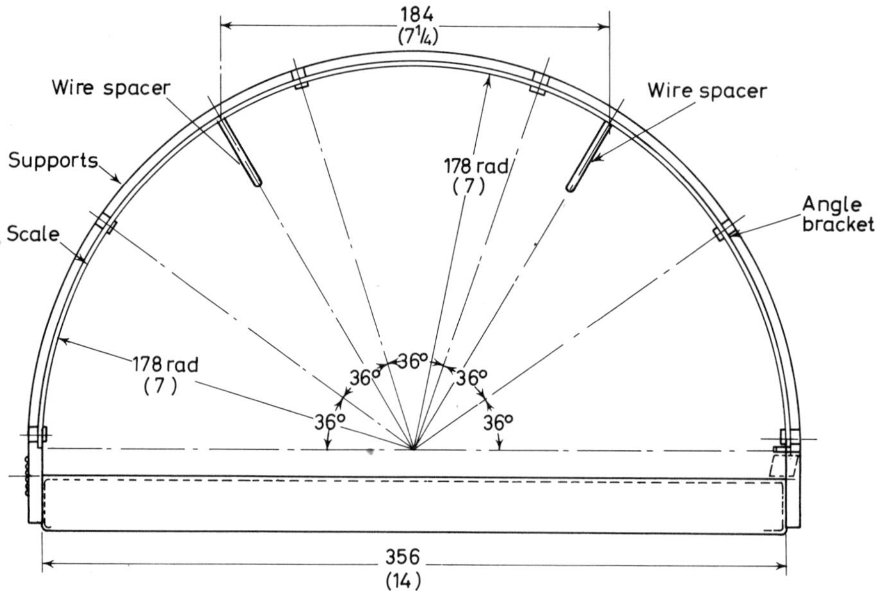

Figure 11.37. Flammability test—general arrangement (B.S. 2782). Dimensions in millimetres with inch equivalents in parentheses

the lower surface of the sample. The apparatus is placed in a draught free atmosphere.

Before testing, the cup is warmed by igniting 0·3 ml of ethanol in it. When the ethanol has burnt out, another 0·3 ml is placed in the cup and the test specimen is mounted in position. The second quantity of ethanol is ignited $2\frac{1}{2} \pm \frac{1}{4}$ min after the ignition of the first and when this second quantity has burnt out a note is made of the time that the specimen glows or flames. When the specimen has ceased to glow or flame a note is made (1) of any material that dropped from the specimen and continued to burn after reaching the base of the apparatus, (2) the percentage of the area of the underside of the specimen that is charred or scorched and (3) the length of that part of the edge (of the underside) that is scorched. (Areas that are covered with soot but not burnt are not deemed 'charred areas'.)

If the length of the scorched part of the edge of any of the three specimens exceeds 51 mm the material is designated 'flammable'. If this length is less

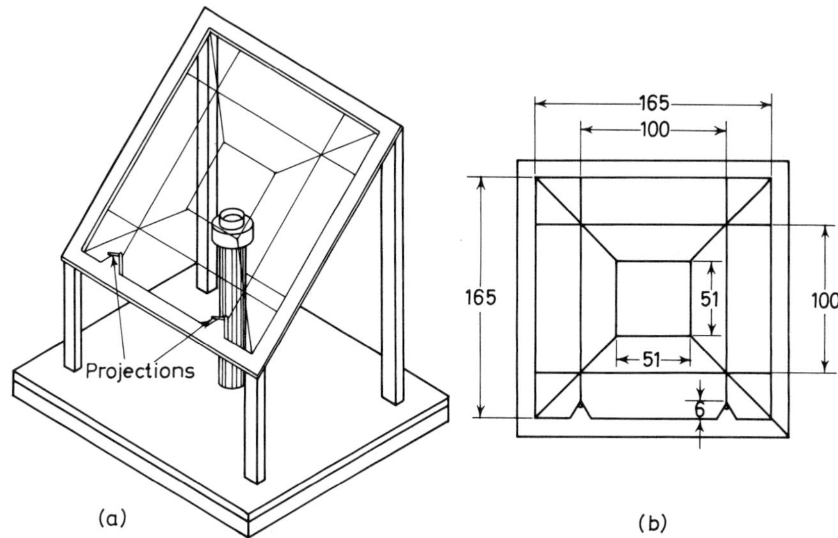

Figure 11.38. Flammability (alcohol cup test) apparatus (B.S. 2782). Dimensions in millimetres

than 51 mm for each of the specimens, the material is of 'low flammability'. If, in addition, each of the three specimens, or five out of six, meet the requirements (1) that flaming or glowing does not continue for more than 5 s after the alcohol has burnt out, (2) any material that drops from the specimen does not continue to burn after reaching the base of the apparatus and (3) charring or scorching does not extend over an area exceeding 20% of the underside of the specimen, the material is described as having 'very low flammability'.

Method 508E, 'Incandescence Resistance of Rigid Thermosetting Plastics', is identical with ISO Recommendation R.181; the apparatus is shown in Figure 11.39. [It also forms the method of DIN.53459 and is known as the Schramm and Zebrowski Method.]

The igniting bar D is a silicon carbide rod 8 mm in diameter, with a usable length of about 100 mm and metallised ends. The bar is held between metal grips mounted on ceramic or asbestos plate C hinged to base plate A by bearings B. Provision is made for electrically heating bar D to 950°C and controlling the temperature at this value. A counter weight E causes the igniting bar to exert a force of about 300 mN (30 gf) on the specimen H (80–130 mm long, $10 \pm 0{\cdot}2$ mm wide and $4 \pm 0{\cdot}2$ mm thick). The position of the igniting bar in relation to the test specimen and the length to which the bar can move along the specimen are controlled by adjusting screw F which can be brought into contact with removable plate G. Clamp K, mounted on column J, holds the test specimen. J is provided with a sliding base so as to accommodate specimens of different length. Metal rod M, 8 mm in diameter and 150 mm in length, can be moved into the exact position occupied by the igniting bar when the latter is vertically above its hinge and is used to locate the front end of the specimen before test.

The test is conducted in a draught free enclosure. The weighed specimen is clamped into position after swinging plate C with bar D attached, downwards away from clamp K and moving rod M into position vertically above the hinge. Clamp K and column J are adjusted to bring the face of the free end of the specimen into contact with M. The latter is then moved away and bar D swung back so as to be close to the end of the specimen. Screw F,

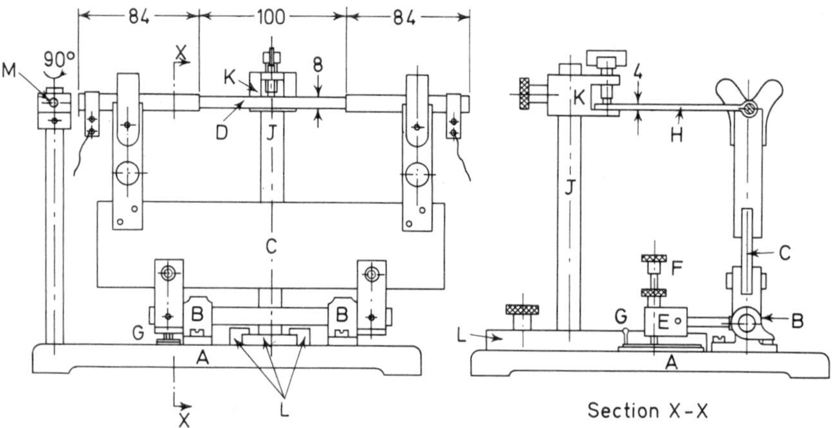

Figure 11.39. Incandescence resistance apparatus (B.S. 2782). Dimensions are in millimetres

while contacting plate G, is adjusted so that bar D just touches the end of the specimen. The bar is then swung away and plate G moved away from under screw F.

The bar D is heated to 950°C whilst it is tilted away from the specimen and when it is constant at this temperature it is brought back into contact with the end of the specimen and released. After 180±5 s, the bar is again swung away and any flame extinguished (without the use of liquid). The specimen is removed, cooled and reweighed. The outside of the specimen is then wiped with a dry cloth to remove any superficial charred material and 'polished' by any appropriate means so that the extent of the area affected by heat can be easily identified. The length of the part of the specimen that is not charred is measured to the nearest 0·1 mm along the middle of each of two 10 mm wide surfaces of the specimen. The difference, in millimetres, between the original length of the specimen and the length, after test, of the part not affected by heat is designated the flame spread S.

The loss in weight W, to the nearest 0·1 g, is expressed in milligrammes. The average value of W for three specimens and the average value of S, from the six results obtained from three specimens, are substituted in the equation:

$$\text{Incandescence Resistance} = \text{Log}_{10}\left(\frac{100\,000}{WS}\right)$$

Another test suitable for plastics forms B.S. 476: Part 6: 1968; it is known as the 'Fire Propagation Test for Materials' or, colloquially as the 'hot box test'. It is intended to follow the test for 'ignitability' already described from B.S. 476: Part 5: 1968, with the classification '*X*' or '*P*' as found.

A test sample comprises three specimens of each face. The specimens are 228 mm (+0–1·5 mm) square and of the normal thickness, but not exceeding 50 mm; if material thicker than this has to be tested, it is cut away from one surface to 50 mm (+0–3 mm) thickness.

The apparatus is very carefully specified in dimensions and is shown in outline in Figure 11.40. It is essentially a combustion chamber with all but one wall made from asbestos board of specified density, thermal conductivity and specific heat; the sixth wall is formed by the test specimen fixed in position by a holder made from the same asbestos board.

There is a row of 14 gas jets issuing from 1·5 mm diameter orifices at 12·5 mm centres in a horizontal stainless steel tube of 9 mm bore, arranged so that the jets impinge horizontally on the specimen 25 mm above the bottom of the exposed face. The holes in the tube are 3 mm (+1·5–0) from the face of the specimen. The gas supplied to the jets must liberate 7560 cal/min ±2%.

The two electric elements are each of 1000 W rating, are connected in parallel and the input is controllable at 1800 W (430 cal/s) ±2% and at 1500 W (358 cal/s) ±2%.

In operation a specimen (after conditioning as in B.S. 476: Part 5) is placed in the holder, with a recess suitable to the thickness of the specimen, and the holder is then mounted to form one vertical face of the combustion chamber, the seal being gas tight. The gas supply is turned on and the jets ignited and test timed from this point; after 2 min 45 s the electric current is turned on to give the higher input, then turned down to the lower input at 5 min. The duration of the test is 20 min. During the test the temperature difference between ambient conditions and that of the thermocouples at the exit of the cowl is recorded or measured at intervals of $\frac{1}{2}$ min for the first 3 min, at 1 min intervals from 4 to 10 min and 2 min intervals from 12 to 20 min.

The apparatus is calibrated similarly using a specified piece of asbestos board.

The mean temperature rise above ambient for the three specimens is plotted against time on the same graph as the curve obtained by the calibration operation.

The index of performance, *I*, for the material under test is determined as follows:

$$I = \sum_{\frac{1}{2}}^{3} \frac{\theta_{\mathrm{m}}-\theta_{\mathrm{c}}}{10t} + \sum_{4}^{10} \frac{\theta_{\mathrm{m}}-\theta_{\mathrm{c}}}{10t} + \sum_{12}^{20} \frac{\theta_{\mathrm{m}}-\theta_{\mathrm{c}}}{10t}$$

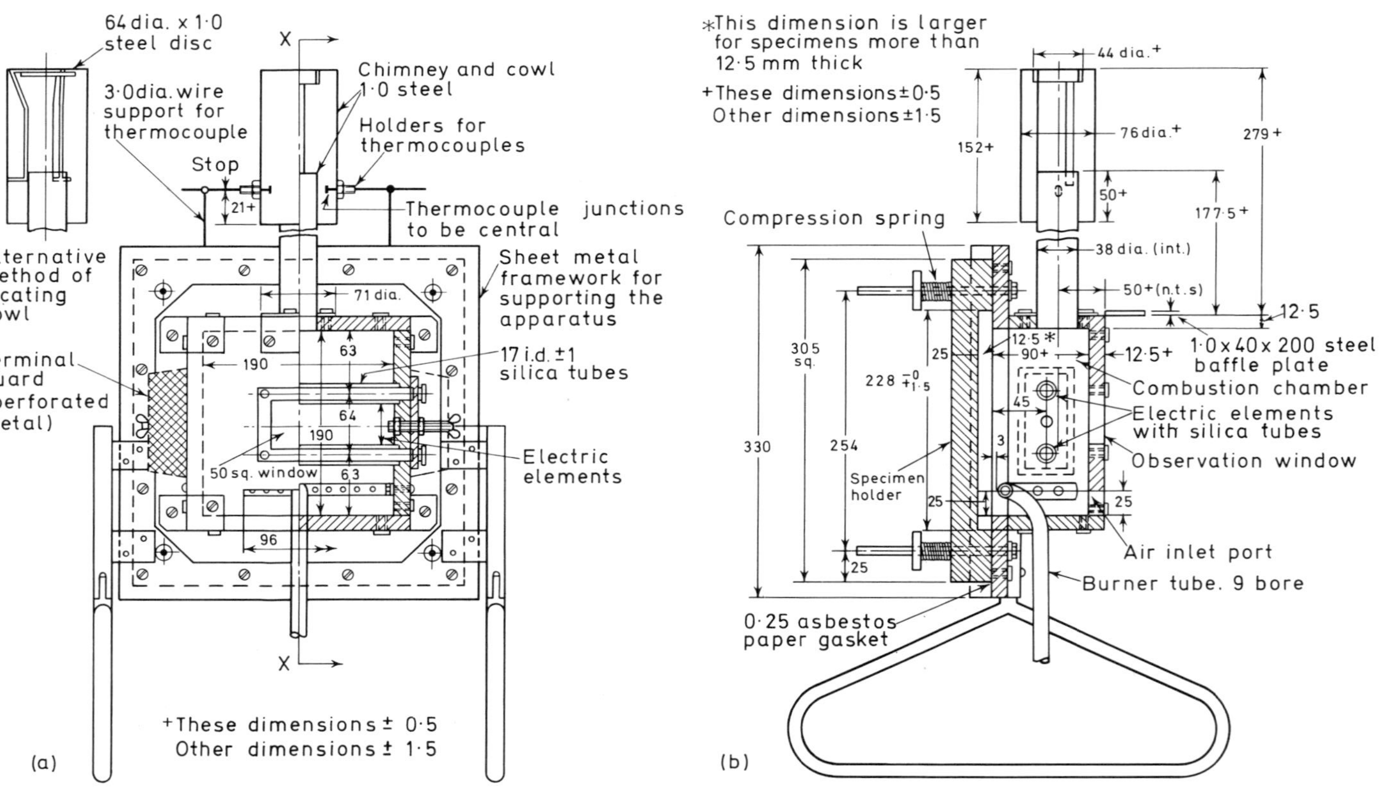

Figure 11.40. Fire propagation test B.S. 476: Part 6: 1970. (a) Part front elevation and section of apparatus, (b) section XX of apparatus. All dimensions in millimetres

where t = Time in minutes from the origin at which readings were taken.
θ_m = Temperature in degC of the mean curve for the material under test at time t.
θ_c = Temperature in degC of the calibration curve for the apparatus at time t.

'In computations only the positive value of $(\theta_m - \theta_c)/10t$ is used' (this is understood to mean that all negative values are ignored). As already stated, classification 'X' or 'P' is quoted with the result of this test.

It has been found to be very sensitive to small variations in certain variables and amendments to the standard have been proposed. It remains to be seen whether reproducibility between laboratories can be achieved.

In B.S. 476: Part 1: 1953, Section Two is entitled 'Surface Spread of Flame for Materials' and at the time of writing is certainly the most widely accepted (and specified) U.K. flammability test for constructional materials. (N.B. Tests for materials are not to be confused with tests for structures—See Section Three of the same standard and B.S. 476: Part 3: 1958 'External Fire Exposure Roof Tests'.) The standard test uses six specimens 9 in × 36 in (229·6 mm × 914·4 mm), each painted round the edges and $1\frac{1}{2}$ in (38·1 mm) therefrom with a specified sodium silicate composition. The specimens are exposed in turn to a source of radiant heat such that at one end a gold disc thermocouple in place of the specimen would record 500°C and the other 130°C, the intervening temperatures being also specified. Immediately the specimen is exposed to this radiant heat a specified gas flame is applied to the hotter end for one minute. At the same time measurements are commenced to record the time taken for flame to spread along the specimen surface, 3 in (76·2 mm) from the lower long edge; these are continued until the flames have died out or for 10 min, whichever is the longer time.

For each of the six specimens a curve is drawn of the distance of flame spread against time, from which is read the distance of spread during the first $1\frac{1}{2}$ min, the distance during the 10 min and the final distance of spread of flame.

Effective Spread of Flame = $\bar{x} + 1{\cdot}04[(x_1 - \bar{x})^2 \ldots + (x_6 - \bar{x})^2]^{\frac{1}{2}}$ where $x_1 \ldots x_6$ are the individual spreads of flame of each of the six specimens, and $\bar{x}$ the mean thereof.

Classifications according to this test are as follows:

Class I. Surfaces of very low flame spread. Those faces on which not more than 7·5 in (19·0 cm) effective spread of flame occurs.

Class II. Surfaces of low flame spread. Those faces on which the effective spread of flame neither exceeds 12 in (30·5 cm) during the first $1\frac{1}{2}$ min, nor exceeds a final value of 24 in (60·7 cm).

Class III. Surfaces of medium flame spread. Those faces on which the effective spread of flame neither exceeds 12 in (30·5 cm) during the first $1\frac{1}{2}$ min, nor exceeds 33 in (83·9 cm) after 10 min.

Class IV. Surfaces of rapid flame spread. Those faces on which the effective spread of flame either exceeds 12 in (30·5 cm) during the first $1\frac{1}{2}$ min, or exceeds 33 in (83·9 cm) after 10 min.

As a guide to performance, B.S. 476: Part 1: 1953 includes in an appendix an 'Apparatus for Preliminary Surface Spread of Flame Test'; this is simpler than that described above, but is *not* an alternative thereto. It uses six specimens each $3\frac{3}{4}$ in × 12 in (9·53 cm × 30·5 cm) and the apparatus is one third of the size of the full scale test; the temperature range is the same, but over the 12 in length instead of the 36 in. Classification is as above, with flame spread distances and time factors altered appropriately (see Reference 113).

At the time of writing, these two surface spread of flame tests are being revised and will appear ultimately as B.S. 476: Part 7 (Part 1 is to disappear altogether). The apparatus is essentially the same, but the methods of calibration and computing the test results are being improved.

(The Paint Research Station has described an even smaller simulation, in which a test specimen 9 in × 4 in is placed at an angle of 16–17° to a 10 in 1 kW electric fire element; one end of the specimen contacts one end of the element, through a mica guard.)

Before leaving British Standard tests, mention must be made of the methods to be found in the various product specifications, for cable, sheath and insulation in particular; the reader is referred to the individual specifications (see Chapter 2, Appendix 2).

A.S.T.M. METHODS

A.S.T.M. D.635–68, 'Flammability of Rigid Plastics over 0·127 cm (0·050 in) in Thickness', is very similar to Method 508A of B.S. 2782: 1970 described above, with slight differences in detail and an extensive discourse on judging the results. Somewhat similar is A.S.T.M. D.1692–68, 'Flammability of Plastics Sheeting and Cellular Plastics', which employs ten specimens 5·1 × 15·2 cm (2 × 6 in), with a normal thickness of 1·3 cm ($\frac{1}{2}$ in) or under, or cut to this value if over.

The specimen is rested on a 6·4 mm ($\frac{1}{4}$ in) mesh of 0·8 mm ($\frac{1}{32}$ in) steel wire. The mesh is 76 mm × 216 mm (3 × 8$\frac{1}{2}$ in) and 13 mm ($\frac{1}{2}$ in) of its length is bent upwards in a right angle. The mesh is supported over a 'batswing' bunsen burner (9·5 mm barrel with 48 mm maximum width wing) placed below one end (see Figure 11.41).

The specimens are each marked with a gauge mark across its width, 25·4 mm (1 in) from one end. A 38 mm (1$\frac{1}{2}$ in) blue flame is applied under the vertical part of the mesh, centrally spaced with respect to the width of the specimen, and in a draught-free atmosphere, and is removed after 60 s or when the flame front reaches the first line. The time, in seconds, to when the flame reaches the gauge mark is noted (t_g), if this happens, or to when the flame is extinguished before reaching the gauge mark if this occurs (t_e).

If the flame front reaches the gauge mark, the burning rate is $(762)/(t_g)$ cm/min $[(300)/(t_g)$ in/min]. If the gauge mark is not reached, the 'burning extent' is equal to 152 mm (6 in), minus the distance from the unburst end to the nearest evidence of the flame front along the upper surface of the specimen. The burning rate then equals 60 times the burning extent divided by t_e. If this happens with all ten specimens, the average burning rate, burning extent and extinguishment time of all specimens are calculated. If, on the other hand, one or more specimens have burnt past the gauge mark, the average burning rate is calculated for these only.

There is a vertical strip test in A.S.T.M. D.568–68, 'Flammability of Flexible Plastics', but unlike Method 508D of B.S. 2782: 1965, no cellulose nitrate fuse is used and the specimens are 45·7 cm (18 in) long. Inevitably, the bunsen burner is applied for a longer period to ignite the lower end of the specimen; otherwise the two methods are very similar.

A rather more elaborate test is described in A.S.T.M. D.1433–58, 'Flammability of Flexible Thin Plastic Sheeting'. Specimens are 3 in in width and 9 in in length and five are used in each material direction. The specimen is mounted in a special holder in a carefully specified cabinet with two nylon threads running across the specimen as gauge lines. These threads are connected to microswitches so that, when the flame front of the ignited specimen reaches the first thread and it burns through, the microswitch is

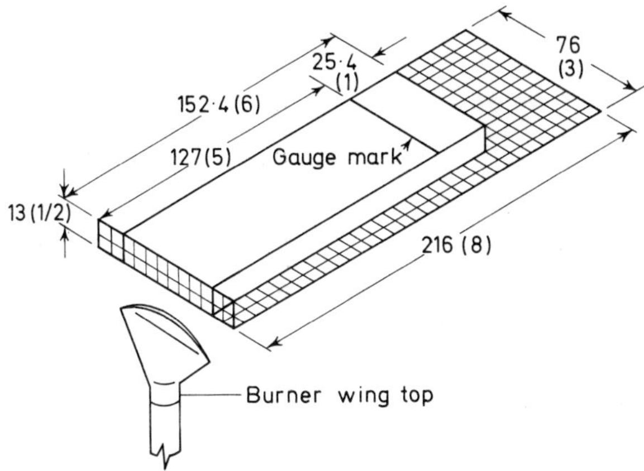

Figure 11.41. Flammability test (A.S.T.M. D.1692). Dimensions in millimetres with inch equivalents in parentheses

actuated to start a timing mechanism and when the second nylon thread is reached its microswitch stops the timer. By this means the average rate of burning between the gauge marks (6 in apart) is measured.

Materials found to be self-extinguishing by the method of A.S.T.M. D.635 are examined by A.S.T.M. D.757–65, 'Flammability of Plastics, Self-Extinguishing Type'; this is essentially identical to Method 508E of B.S. 2782: 1970 except that weight loss is not recorded.

On a larger scale A.S.T.M. (Book of Standards 14) provides:

E 84–68 'Surface Burning Characteristics of Building Materials'

E 162–67 'Surface Flammability of Materials Using a Radiant Heat Energy Source'

E 286–69 'Surface Flammability of Building Materials Using an 8 ft (2·44 m) Tunnel Furnace'.

Useful comments on these tests, and on some of the B.S. 476 tests, will be found in References 114 and 115, describing investigations using these methods.

GERMAN DIN METHOD

DIN.53459, 'Testing of Plastics. Determination of the Resistance to Incandescence by the Schramm and Zebrowski Method', is the ISO/R.181 and B.S. 2782: 1970, Method 508E, test. [See above.]

11.9.4 Other Tests

Quite apart from the considerable number of B.S. and A.S.T.M. tests referred to above, and those of other standardising bodies such as N.E.M.A. and D.T.D., there are many issued by official bodies such as the Air Registration Board, British Railways and the National Coal Board, to name but three in the U.K. The reader is referred to Scott[110] for further details and references. The destruction of the material under test, and its contribution to the provision of 'fuel' in a structure fire, are not the end of the story by any means. The toxic, corrosion and smoke hazards resulting from combustion of the material must not be overlooked; the contributions of Silversides[116], Gross et alia[117] and Rasbash[118] are particularly interesting in this respect.

The technical literature contains a plethora of articles on specially developed methods and the results obtained therefrom; it is not practicable to review these methods here but, by way of conclusion, reference is made to an article by Fenimore and Martin[119] containing a description of a new concept of test for flammability; it appears that a test of this type is to become an A.S.T.M. standard (D–2863). The essential principle is to determine the minimum volume fraction of oxygen, in a gaseous atmosphere of oxygen and nitrogen, which will sustain the candle-like burning of a stick of polymer (0·6 cm × 0·3 cm × 8 cm); the result is termed the 'limiting oxygen index'. Results for a wide range of plastics, and plastics compositions containing fire retardants, are presented (see also Reference 120), whilst further test data from a 'round robin' programme have been described by Goldblum[121] and by Isaacs[122, 123].)

REFERENCES

1. RITCHIE, P. D. (ed.) *Physics of Plastics,* Chapter 4, Iliffe Books Ltd., London (1965)
2. GORDON, M., *High Polymers—Structure and Physical Properties,* Chapter 4, 2nd edn, Iliffe Books Ltd., London (1963)
3. FRITH, E. M. and TUCKETT, R. F., *Linear Polymers,* Chapter 8, Longmans Green & Co., London (1951)
4. NIELSEN, L. E., *Mechanical Properties of Polymers,* Chapter 2, Reinhold Publishing Corporation, New York (1962)
5. BUECHE, F., *Physical Properties of Polymers,* Chapter 4, Interscience Publishers, New York (1962)
6. MEARES, P., *Polymers: Structure and Bulk Properties,* Chapter 10, D. Van Nostrand and Company Ltd., London (1965)
7. BRYDSON, J. A., *Plastics Materials,* 2nd edn, Iliffe Books Ltd., London, 37 (1969)
8. *Handbook on Polymer Assessment,* Appendix 10, Joint Services Non-Metallic Materials Advisory Board, Her Majesty's Stationery Office, London (1965)
9. BOYER, R. F. and SPENCER, R. S., Thermal Expansion and Second-Order Transition Effects in High Polymers, Part 1, Experimental Results, *J. appl. Phys.,* **15,** No. 4, 398 (April 1944)
10. MILLANE, J. J. and MCLAREN, S. M., 'The Volume-Time-Temperature Relationship of Polystyrene', *J. appl. Chem.,* **2,** No. 9, 554 (September 1952)
11. MEARES, P., 'The Second-Order Transition of Polyvinyl Acetate', *Trans. Faraday Soc.,* **53,** 31 (1957)
12. GORDON, M. and MACNAB, I. A., 'A New Diffusion Gradient Method for Thermal Expansion Studies with Applications to Polystyrene', *Trans. Faraday Soc.,* **49,** 31 (1953)
13. ROTHSTEIN, E. C. and SPECHLER, D., 'A Method of Rapid Determination of Thermal Expansion and Apparent Second-Order Transition Temperature of Polymer Films', *Polymer Engng Sci.,* **6,** No. 2, 112 (April 1966)

14. WOOD, L. A., BEKKEDAHL, N. and PETERS, C. G., 'Application of the Interferometer to the Measurement of Dimensional Changes in Rubber', *J. Res. nat. Bur. Stand.*, **23,** 571 (November 1939)
15. EISENBERG, A. and ROVIRA, E. 'A New Method for the Rapid Determination of Glass Transition Temperatures', *J. Polymer Sci.*, Part B, *Polymer Let.*, **2,** 269 (1964)
16. HELLER, J. and LYMAN, D. J., 'Measurement of Glass Transition Temperatures of Polymers by a Differential Pressure Transducer', *J. Polymer Sci.*, Part B, *Polymer Let.*, **1,** 317 (1963)
17. WILEY, R. H., 'Refractometric Determination of Second-Order Transition in Polyvinyl Acetate', *J. Polymer Sci.*, **2,** No. 1, 10 (February 1947)
WILEY, R. H. and BRAUER, G. M., 'Refractometric Determination of Second-Order Transition Temperatures in Polymers, II, Some Acrylics, Vinyl Halide, and Styrene Polymers', *J. Polymer Sci.*, **3,** No. 3, 455 (June 1948)
Also:
WILEY, R. H. and BRAUER, G. M., 'Refractometric Determination of Second-Order Transition Temperatures in Polymers, III, Acrylates and Methacrylates', *J. Polymer Sci.*, **3,** No. 5, 647 (October 1948)
WILEY, R. H. and BRAUER, G. M., 'Refractometric Determination of Second-Order Transition Temperatures in Polymers, IV, Butadiene-Acrylonitrile Copolymers', *J. Polymer Sci.*, **3,** No. 5, 704 (October 1948)
WILEY, R. H., BRAUER, G. M. and BENNETT, A. R., 'Refractometric Determination of Second-Order Transition Temperatures in Polymers, V, Determination of Refractive Indices of Elastomers at Temperatures from 25 to −120°C', *J. Polymer Sci.*, **5,** No. 5, 609 (October 1950)
18. SHELDON, R. P., 'A Simple Method for the Determination of Glass Temperatures of Amorphous Polymers', *J. appl. Polymer Sci.*, **6,** No. 24, S43 (Nov.–Dec. 1962)
19. GARRETT, R. R., 'Rapid Measurement of Glass Temperatures and Their Application to the Problem of Structure of EP Rubbers', *Rubb. Plast. Age*, **46,** No. 8, 915 (August 1965)
20. NIELSEN, L. E., BUCHDAHL, R. and LEVREAULT, R., 'Mechanical and Electrical Properties of Plasticised Vinyl Chloride Compositions', *J. appl. Phys.*, **21,** No. 6, 607 (June 1950)
21. SAUER, J. A. and WOODWARD, A. E., 'Transitions in Polymers by Nuclear Magnetic Resonance and Dynamic Mechanical Methods', *Rev. mod. Phys.*, **22,** No. 1, 88 (January 1960)
22. WILLBOURN, A. H., 'The Glass Transition in Polymers with the $(CH_2)n$ Group', *Trans. Faraday Soc.*, **54,** 717 (1958)
23. BARRETT, R. M. and GORDON, M., 'Applications of Rebound Resilience to the Cure of Polyester Resins', *Physical Properties of Polymers,* Society of Chemical Industry Monograph No. 5, S.C.I., London, 183 (1959)
24. CHEETHAM, I. C., 'Rolling Ball Loss Spectrometer', *I.R.I. Trans.*, **41,** No. 4, T.156 (August 1965)
25. FUOSS, R. M., 'Electrical Properties of Solids. VI. Dipole Rotation in High Polymers', *J. Amer. chem. Soc.*, **63,** No. 2, 369 (February 1941)
26. MEAD, D. J. and FUOSS, R. M., 'Electrical Properties of Solids, XIII, Polymethyl Acrylate, Polymethyl Methacrylate, Polymethyl-α-chloracrylate and Polychloroethyl Acrylate', *J. Amer. chem. Soc.*, **64,** No. 10, 2389 (October 1942)
27. REDDISH, W., 'Dielectric Study of the Transition Temperature Regions for Poly (vinyl Chloride) and Some Chlorinated Poly (vinyl Chlorides)', *J. Polymer Sci.*, Part C, *Polymer Symp.*, No. 14, Interscience Publishers, New York, 123 (1966)
28. WARFIELD, R. W., 'A New Method for Determining the Second Order Transition in Thermosetting Polymers', *S.P.E. J.*, **15,** No. 8, 625 (August 1959)
29. WUNDERLICH, B., 'Motion in Polyethylene, I, Temperature and Crystallinity Dependence of the Specific Heat', *J. chem. Phys.*, **37,** No. 6, 1203 (September 15 1962)
30. STEERE, R. C., 'Detection of Polymer Transitions by Measurement of Thermal Properties', *J. appl. Polymer Sci.*, **10,** No. 11, 1673 (November 1966)
31. GRISKEY, R. G., 'Use of Thermodynamic Properties in Polymer Characterisation', *Mod. Plast.*, **45,** No. 13, 215 (September 1968)
32. MANLEY, T. R., 'The Detection of Transition Points of Polymers by Comparative Thermal Diffusivity Measurements', *Polymer*, **10,** No. 2, 148 (February 1969)
33. KE, BACON, (ed.) *Newer Methods of Polymer Characterisation,* Chapter IX, Interscience Publishers, New York (1964)
34. SLADE, P. E. Jr. and JENKINS, L. T., *Techniques and Methods of Polymer Evaluation,* Volume 1, *Thermal Analysis,* Marcel Dekker Inc., New York (1966)
35. KEARNEY, J. J. and EBERLIN, E. C., 'The Determination of Glass Transition Temperatures

by Differential Thermal Analysis', *J. appl. Polymer Sci.*, **3,** No. 7, 47 (January–February 1960)

DANNIS, M. D., 'Transition Measurements Using Differential Thermal Analysis Techniques', *J. appl. Polymer Sci.*, **7,** No. 1, 231 (January–February 1963)

STRELLA, S., 'Differential Thermal Analysis of Polymers, I, The Glass Transition', *J. appl. Polymer Sci.*, **7,** No. 2, 569 (March–April 1963)

MARTIN, A. E. and RASE, H. F., 'Use of Differential Thermal Analysis in Studying Glass Transitions and Thermal Degradation of Polystyrene', *Industr. Engng. Chem. (Prod. Res. Develop.)*, **6,** No. 2, 104 (June 1967)

36. HOLROYD, L. V., CODRINGTON, R. S., MROWCA, B. A. and GUTH, E., 'Nuclear Magnetic Resonance Study of Transitions in Polymers', *J. appl. Phys.*, **22,** No. 6, 696 (June 1951)
37. TORDELLA, J. P., WEBBER, A. C. and COOPER, E. B., 'Measurement of the Effect of Temperature on Some Physical Properties of Plastics', *A.S.T.M. Special Technical Publication No. 132,* 14, American Society for Testing and Materials, Philadelphia (1953)
38. STEPHENSON, C. E. and WILLBOURN, A. H., 'The Vicat Softening Point Test Method: Correlations, Uses and Variants', *A.S.T.M. Special Technical Publication No. 247,* 169, American Society for Testing and Materials, Philadelphia (1959)
39. SHERR, A. E., 'Preliminary Evaluation of Polymer Properties', *S.P.E. J.*, **21,** No. 1, 67 (January 1965)
40. HORSLEY, R. A., 'The Limitations of Standard Tests', *Trans. Plast. Inst., Lond.*, **33,** No. 106, 119 (August 1965)
41. GOHL, W., 'Measurement of Dimensional Stability at Elevated Temperatures', *Kunststoffe*, **49,** No. 5, 228 (May 1959)
42. WALLHAUSER, H. and CHRISTMANN, E., 'Dimensional Stability under Heat, A Comparison of the Martens and ISO/R75 Methods', *Kunststoffe*, **56,** No. 8, 551 (August 1966)
43. SANDIFORD, D. J. H. and BUCKINGHAM, K. A., 'Softening Point Determination by The Tensile Deformation versus Temperature (T.D.T.) Test', *Brit. Plast.*, **34,** No. 11, 594 (November 1961)
44. GRAVES, F. L. and LOVELESS, H. S., 'Convenient Measurement of Deflection Temperature Under Load and Vicat Softening Point', *Mater. Res. Stand.*, **3,** No. 1, 33 (January 1963)
45. EHLERS, G. F. L. and POWERS, W. M., 'An Automatic Vicat-Type Heat-Distortion and Softening-Range Apparatus', *Mater. Res. Stand.*, **4,** No. 6, 298 (June 1964)
46. BESTELINK, P. N., MILLWOOD, R. and STEPHENSON, C. E., 'Temperature Control in Plastics Testing', *Plastics, Lond.*, **24,** No. 266, 510 (December 1959)
47. CLASH, R. F. Jr. and BERG, R. M., 'Stiffness and Brittleness Properties of Non-Rigid Vinyl Chloride-Acetate Resin Compounds', *Symp. Plast.*, 54, American Society for Testing and Materials, Philadelphia (1944)
48. WELDING, G. N., 'Low-temperature Flexibility of Rubber-like Materials', *Plastics, Lond.*, **20,** No. 214, 158 (May 1955)
49. SCHERR, H. J. and PALM, W. E., 'An Automatic and Recording Torsion Measuring Apparatus', *Mater. Res. Stand.*, **8,** No. 12, 13 (December 1968)
50. CASPARY, R., 'Automatic Apparatus for Measuring Torsion Modulus and Recovery as a Function of Temperature', *Kautsch. u. Gummi Kunststoffe*, **20,** No. 10, 587 (October 1967)
51. HAYES, R. and LANNON, D. A., 'The Low Temperature Flexibility of Plasticised Polyvinyl Chloride', *Brit. Plast.*, **26,** No. 291, 301 (August 1953)
52. WILLIAMSON, I., 'An Improved Instrument for the Evaluation of the Physical Properties of High Polymer Compositions', *Brit. Plast.*, **23,** No. 256, 87 (September 1950)
53. WHEELER, A., 'Evaluation of Flexible Materials by Means of an Automatic Torsion Pendulum', *Plast. & Polymers*, **37,** No. 131, 469 (October 1969)
54. WORMALD, D., 'Low Temperature Cracking Test for Plasticised P.V.C.', *Brit. Plast.*, **31,** No. 9, 392 (September 1958)
55. HOFF, E. A. W. and TURNER, S., 'A Study of the Low-Temperature Brittleness Testing of Polyethylene', *Bull. Amer. Soc. Test. Mat.* **No. 225,** 58 (October 1957)
56. WEBBER, A. C., 'Brittleness Temperature Testing of Elastomers and Plastics', *Bull. Amer. Soc. Test. Mat.* **No. 227,** 40 (January 1958)
57. BESTELINK, P. N. and TURNER, S., 'The Low-Temperature Brittleness Testing of Polyethylene', *Bull. Amer. Soc. Test. Mat.* **No. 231,** 68 (July 1958)
58. BIRKS, A. M. and RUDIN, A., 'Measurement of the Brittleness Temperature of Polyethylene', *Bull. Amer. Soc. Test. Mat.* **No. 246,** 49 (May 1960)
59. TURNER, S., 'The Low-Temperature Brittleness of Polymers', *Brit. Plast.*, **31,** No. 12, 526 (December 1958)

60. WHITEHEAD, ALAN D., The Ancient House, Ardleigh, Nr. Colchester, Essex.
61. BATA, G. L. and ITO, M., 'Additional Variables in the Brittleness Temperature Testing of Polyethylene', *Bull. Amer. Soc. Test. Mat.* **No. 248,** 55 (September 1960)
62. WILLIAMS, H. O., 'Measurement of PVC Brittle Point', *Brit. Plast.,* **31,** No. 3 107 (March 1958)
63. IVES, G. C. and MEAD, J. A., 'The Measurement of the Mechanical Properties of Plastics at Very Low Temperatures', S.C.I. Monograph No. 5, *The Physical Properties of Polymers,* Society of Chemical Industry, London, 80 (1959)
64. CARSWELL, T. S. and NASON, H. K., 'The Effect of Environmental Conditions on the Mechanical Properties of Organic Plastics', *A.S.T.M. Symposium on Plastics,* American Society for Testing and Materials, Philadelphia, 22 (1944)
65. 'Symposium on Temperature Stability of Electrical Insulating Materials', *A.S.T.M. Special Technical Publication No. 161,* American Society for Testing and Materials, Philadelphia (1954)
66. MATHES, K. N., 'Electrical and Mechanical Behavior of Polymers at Cryogenic Temperatures', *S.P.E. J.,* **20,** No. 7, 634 (July 1964)
67. KLUTE, C. H. and MCKEE, L. B., 'Plastics Testing at High and Low Temperatures', *Bull. Amer. Soc. Test. Mat.* **No. 202,** 50 (December 1954)
68. LAVERY, T. E., GROVER, F. S., SMITH, S. and KITCHEN, L. J., 'Equipment for High-Temperature Testing', *Rubb. Age, Lond.,* **80,** No. 5, 834 (February 1957)
69. BRUNNBERG, I., KUBAT, J. and SONDERLUND, G., 'Tensile Testing Device for Measurements on Polymeric Materials between −100 and +150°C, Mechanical Behavior of Cellulose in Liquid Media', *J. appl. Sci.,* **13,** No. 4, 571 (April 1969)
70. WATSON, J. F. and CHRISTIAN, J. L., 'Cryostat and Accessories for Tension Testing at −423°F', *Mater. Res. Stand.,* **1,** No. 2, 87 (February 1961)
71. MCCLINTOCK, R. M. and WARREN, K. A., 'Tensile Cryostat for the Temperature Range 4 to 300°K', *Mater. Res. Stand.,* **1,** No. 2, 95 (February 1961)
72. WESSEL, E. T. and OLLEMAN, R. D., 'Apparatus for Tension Testing at Subatmospheric Temperatures', *Bull. Amer. Soc. Test. Mat.* **No. 187,** 56 (January 1953)
73. WESSEL, E. T., 'Tension Testing Apparatus for the Temperature Range of −320°F–452°F', *Bull. Amer. Soc. Test. Mat.* **No. 211,** 40 (January 1956)
74. IVES, G. C. and MEAD, J. A., 'A Cryostat for Mechanical Testing of Materials', *Mater. Res. Stand.,* **1,** No. 3, 194 (March 1961)
75. LIEB, J. H. and MOWERS, R. E., 'Testing of Polymers at Cryogenic Temperatures', *Testing of Polymers* (Edited by J. V. Schmitz), Volume 2, Chapter 3, Interscience Publishers, New York (1966)
76. MANFIELD, H. G., 'A Universal Resin Casting System for Electronic Applications', *Brit. Plast.,* **34,** No. 10, 539 (October 1961)
77. GHANEM, N. A., 'A Simple Method for the Determination of the Coefficient of Cubical Expansion of Polyester Resins', *Paint Tech.,* **28,** No. 3, 48 (March 1964)
78. DANNIS, M. L., 'Dimensional Changes in Rigid Vinyls', *Mod. Plast.,* **31,** No. 7, 120 (March 1954)
79. DANNIS, M. L., 'Thermal Expansion Measurements and Transition Temperatures, First and Second Order', *J. appl. Polymer Sci.,* **1,** No. 1, 121 (Jan.–Feb. 1959)
80. HAMILTON, W. O., GREENE, D. B. and DAVIDSON, D. E., 'Thermal Expansion of Epoxies between 2 and 300°K', *Rev. Sci. Instrum.,* **39,** No. 5, 645 (May 1968)
81. GRASSIE, N., *Chemistry of High Polymer Degradation Processes,* Butterworths, London (1956)
82. 'High Temperature Resistance and Thermal Degradation of Polymers', *S.C.I. Monograph No. 13,* Society of Chemical Industry, London (1961)
83. WALL, L. A., 'Pyrolysis, Analytical Chemistry of High Polymers, Part II, Analysis of Molecular Structure and Chemical Groups', *High Polymers* (Edited by G. M. Kline), Volume XII, Chapter V, Interscience Publishers, New York (1962)
84. MADORSKY, S. L., *Thermal Degradation of Organic Polymers,* Interscience Publishers, New York (1964)
85. PINNER, S. H. (ed.) *Weathering and Degradation of Plastics,* Columbine Press, Manchester (1966)
86. GOLDFEIN, S., 'Prediction Techniques for Mechanical and Chemical Behavior', *Testing of Polymers* (Edited by W. E. Brown), Volume 4, Interscience Publishers, New York. 121 (1969)
87. SCOTT, J. R., *Physical Testing of Rubbers,* Chapter 11, Maclaren and Sons Ltd., London (1965)
88. MANLEY, T. R., 'Differential Thermal Analysis and Its Application to Polymer Science',

Techniques of Polymer Science, *S.C.I. Monograph No. 17,* Society of Chemical Industry, London, 175 (1963)

89. MANLEY, T. R., 'Characterisation of Filled Amino Resins by Thermal Analysis', *Trans. J. Plast. Inst. Lond.,* **35,** No. 117, 525 (June 1967)
90. DOUBLE, J. S., 'Differential Thermal Analysis of Polymers', *Trans. J. Plast. Inst. Lond.,* **34,** No. 110, 73 (April 1966)
91. SMITH, D. A., 'Thermal Analysis of Polymers, Part I', *Rubb. J.,* **150,** No. 3, 33 (March 1968)
92. SMITH, D. A., 'Thermal Analysis of Polymers, Part 2', *Rubb. J.,* **150,** No. 4, 21 (April 1968)
93. 'du Pont 900 Differential Thermal Analyzer' and 'du Pont 950 Thermogravimetric Analyzer'. E.I. du Pont de Nemours and Co. (Inc.), Instrument Products Division, Wilmington, Delaware 19898 or du Pont Company (United Kingdom) Ltd., Instrument Products Division, 64 Wilbury Way, Hitchin, Herts.
94. SCHWENKER, R. F. JR. and GARN, P. D., *Thermal Analysis, Volume 1: Instrumentation, Organic Materials and Polymers,* and *Volume 2: Inorganic Materials and Physical Chemistry,* Academic Press Inc., New York (1969)
95. CHIU, J., 'Applications of Thermogravimetry to the Study of High Polymers', *J. appl. Polymer Sci., Appl. Polymer Symp.,* No. 2, 25 (1966)
96. KEATTCH, C., 'An Introduction to Thermogravimetry', Heyden and Son Ltd. in co-operation with Sadtler Research Laboratories Inc., London (1969)
97. CHIU, J., 'Polymer Characterisation by Coupled Thermogravimetry–Gas Chromatography', *Analyt. Chem.,* **40,** No. 10, 1516 (August 1968)
98. MCNEILL, I. C., 'Thermal Volatilization Analysis: A New Method for the Characterization of Polymers and the Study of Polymer Degradation', *J. Polymer Sci.,* **4,** No. 10, Part A–1, 2479 (October 1966)
99. MCNEILL, I. C., 'Thermal Volatilization Analysis of High Polymers', *Europ. Polymer J.,* **3,** No. 3, 409 (August 1967)
100. MURPHY, C. B., VAN LIUK, F. W. JR. and PITSAS, A. C., 'T.P.A.: A New Method for Thermal Analysis of Polymeric Materials', *Plast. Des. Process.,* **4,** No. 7, 16 (July 1964)
101. MURPHY, C. B. and DOYLE, C. D., 'Thermoparticulate Analysis', *J. appl. Polymer Sci., Appl. Polymer Symp.,* No. 2, 77 (1966)
102. 'du Pont Differential Scanning Calorimeter Cell.', E.I. du Pont de Nemours and Co. (Inc.), Instrument Products Division, Wilmington, Delaware 19898, or du Pont Company (United Kingdom) Ltd., Instrument Products Division, 64 Wilbury Way, Hitchin, Herts.
103. 'Perkin-Elmer Model DSC-1B Differential Scanning Calorimeter', Perkin-Elmer, Instrument Division, Norwalk, Connecticut.
104. 'Differential Scanning Calorimeter', *Ind. Equip. News, N.Y.,* **18,** No. 1, 17 (January 1969). Also: Grant Instruments (Cambridge) Ltd., Barrington, Cambridge.
105. MILLER, G. W., 'The Thermal Characterisation of Polymers', *Appl. Polymer Symp. No. 10, Analysis and Characterisation of Coatings and Plastics,* **35,** Interscience Publishers, New York (1969)
106. LAMBERT, A., 'Glass Transition Measurements on Polystyrene by Differential Scanning Calorimetry', *Polymer,* **10,** No. 5, 319 (May 1969)
107. STEFFENS, E., 'Thermal Analysis by a Constant Heat Flow', *J. appl. Polymer Sci.,* **12,** No. 10, 2317 (October 1968)
108. POPE, M. I., 'A Method of Thermal Analysis of Polymers by Measurement of Electrical Conductivity', *Polymer,* **8,** No. 2, 49 (February 1967)
109. CHIU, J., 'Technique for Simultaneous Thermogravimetric, Derivative Thermogravimetric, Differential Thermal and Electrothermal Analysis', *Analyt. Chem.,* **39,** No. 8, 861 (July 1967)
110. SCOTT, K. A., 'The Flammability of Plastics: Methods of Testing', *Technical Review No. 27,* Rubber and Plastics Research Association of Great Britain, Shawbury (1965) See also: R.A.P.R.A. Bulletin, 168–173 (Nov.–Dec. 1965)
111. ALLEN, L. B. and CHELLIS, L. N., 'Flammability Tests', *Testing of Polymers* (Edited by J. V. Schmitz), Volume 2, Chapter 11, Interscience Publishers, New York (1966)
112. ASHTON, L. A., 'Fire Safety in Buildings and the Use of Plastics', *Trans. J. Plast. Inst. Lond.,* **32,** No. 98, J27 (April 1964)
113. 'Surface Spread of Flame Tests', *Insulation,* **11,** No. 2, 74 (March–April 1967)
114. 'Symposium on Fire Test Methods', *A.S.T.M. Special Technical Publication No. 301,* American Society for Testing and Materials, Philadelphia (1961)
115. 'Symposium on Fire Test Methods', *A.S.T.M. Special Technical Publication No. 344,* American Society for Testing and Materials, Philadelphia (1963)

116. SILVERSIDES, R. G., 'Measurement and Control of Smoke in Building Fires', *Fire Test Methods, Restraint and Smoke, S.T.P. 422,* 125, American Society for Testing and Materials, Philadelphia (1967)
117. GROSS, D., LOFTUS, J. J. and ROBERTSON, A. F., 'Method for Measuring Smoke from Burning Materials', *Fire Test Methods, Restraint and Smoke, S.T.P. 422,* 166, American Society for Testing and Materials, Philadelphia (1967)
118. RASBASH, D. J., 'Smoke and Toxic Produced at Fires', *Trans. J. Plast. Inst. Lond.*, Supplement No. 2, 55 (January 1967)
119. FENIMORE, C. P. and MARTIN, F. J., 'Candle-Type Test for Flammability of Polymers', *Mod. Plast.*, **44,** No. 3, 141 (December 1966)
120. 'Measuring Polymer Flammability', *Engineering,* **205,** No. 5320, 522 (April 5 1968)
121. GOLDBLUM, K. B., 'Oxygen Index: Key to Precise Flammability Ratings', *S.P.E. J.*, **25,** No. 2, 50 (February 1969)
122. ISAACS, J. L., 'The Oxygen Index Flammability Test', *J. Fire & Flam.*, **1,** No. 1, 36 (January 1970)
123. ISAACS, J. L., 'The Oxygen Index Flammability Test', *Mod. Plast.*, **47,** No. 3, 124 (March 1970)

12

Thermal Properties

12.1 THERMAL CONDUCTIVITY

12.1.1 Introduction

Thermal conductivity may be defined as the quantity of heat per unit time passing normally through unit area of a sheet of material of unit thickness, for unit temperature difference across the faces. In common with other properties involving heat, a proliferation of units exists, the most commonly used being

$$\frac{\text{Btu in}}{\text{ft}^2\text{h degF}}, \frac{\text{cal cm}}{\text{cm}^2\text{s degC}}, \frac{\text{J cm}}{\text{cm}^2\text{s degC}} \text{ or } \frac{\text{W}}{\text{cm degC}}$$

and, in SI units,

$$\frac{\text{J m}}{\text{m}^2\text{s degC}} \text{ or } \frac{\text{W}}{\text{m degC}}.$$

The relations between these units may be helpful to the reader searching the literature, and Table 12.1 gives the principle factors required for conversion (rounded off, but accurate to 1 in 400 at worst):

The importance of the thermal conductivity of plastics materials is best considered in relation to the two principal forms in which plastics are used in this context, namely the solid and cellular states. The thermal conductivity of substances covers a spectrum of 5 decades with cellular plastics near the low conductivity end of the scale, solid plastics an order higher and metals covering the top decade. Figure 12.1 illustrates some typical materials.

Since the thermal conductivity of plastics varies with temperature and density for both the solid and cellular forms (in the former case reflecting the crystallinity of the material and in the latter the amount of entrapped gas) these two parameters should always accompany a statement of conductivity.

SOLID PLASTICS

The thermal conductivity of plastics is obviously of great practical importance to the designer of processing equipment such as extruders, injection moulding machines, etc., in relation to the flow properties of the material and hence the throughput, but additionally it is a property which is intimately related to the structure of the polymer molecule itself and correlates with other important properties, such as transition temperatures and phase changes. In consequence a large proportion of published papers on the subject is concerned with the theoretical aspects in relation to structure. The German authors Eiermann, Hellwege and Knappe are particularly prolific in this field (References 1–7), amongst others (References 8–11).

The theory of the thermal conductivity of polymeric materials and its variation with temperature is not well understood and relatively little data are available over the wide temperature range necessary for the complete understanding of the behaviour of plastics in relation to their structure.

The mainly amorphous polymers such as natural rubber, polystyrene, PVC, PMMA and polycarbonate show relatively small changes in thermal

Table 12.1 THERMAL CONDUCTIVITY UNITS

		$\frac{\text{cal}}{\text{cm s degC}}$	$\frac{\text{W}}{\text{cm degC}}$	$\frac{\text{W}}{\text{m degC}}$	$\frac{\text{kcal}}{\text{m h degC}}$	$\frac{\text{Btu in}}{\text{ft}^2\text{ h degF}}$
$1\frac{\text{cal cm}}{\text{cm}^2\text{ s degC}}$ or $\frac{\text{cal}}{\text{cm s degC}}$	≡	1	4·19	419	360	2900
$1\frac{\text{J cm}}{\text{cm}^2\text{ s degC}}$ or $\frac{\text{W}}{\text{cm degC}}$	≡	0·239	1	100	86	693
$1\frac{\text{J m}}{\text{m}^2\text{ s degC}}$ or $\frac{\text{W}}{\text{m degC}}$	≡	0·00239	0·01	1	0·86	6·93
$1\frac{\text{kcal m}}{\text{m}^2\text{ h degC}}$ or $\frac{\text{kcal}}{\text{m h degC}}$	≡	0·00278	0·0116	1·16	1	8·06
$1\frac{\text{Btu in}}{\text{ft}^2\text{ h degF}}$	≡	0·000345	0·00144	0·144	0·124	1

conductivity between −180°C and +100°C; in general a constant or slightly increasing conductivity is observed up to the softening point and a slightly falling conductivity above[1, 5, 7, 8, 12, 13]. This behaviour is supposed to be dependent mainly on the nature of the Van der Waals bonds and is illustrated by the increase in conductivity in the strain direction and the accompanying decrease in a perpendicular direction when specimens are subjected to a tensile force[1, 3]. The thermal conductivity against temperature curves for amorphous polymers are also characterised by a break or an abrupt change in slope at the glass transition temperature, although precise measurement techniques are necessary to locate such changes[1, 7, 11].

On the other hand, the thermal conductivity of partly crystalline polymers may either decrease with temperature (polythene, polyoxymethylene) or increase (polyethylene terephthalate, PCTFE, PTFE, isotactic polypropylene)[1, 4, 5]. In the first class, measurements on polythenes of differing density and therefore (known) crystallinity suggest that the contribution of the crystalline regions to the overall conductivity is high and in fact conforms to the T^{-1} law for low molecular weight crystals, the thermal resistivity (reciprocal of thermal conductivity) against temperature curve being linear and passing approximately through the origin. The high contribution of the crystalline region combined with the relatively small one of the amorphous regions causes the curves for differing density polythenes to differ widely in shape, although all show significant falls of the order 20% from room temperature to the softening point. Measurements on a material in the

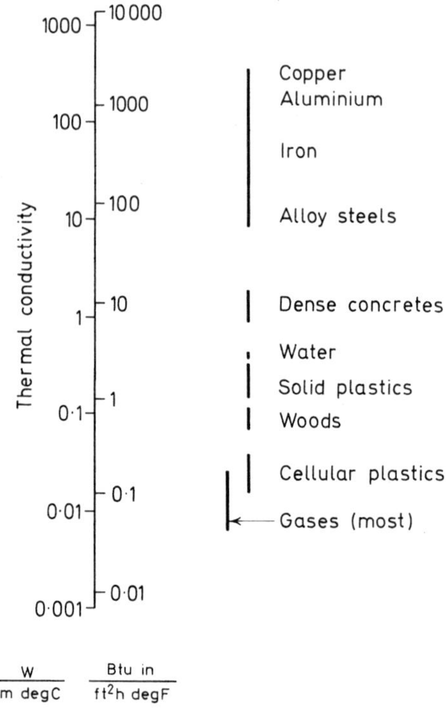

Figure 12.1. Thermal conductivity of various media at room temperature

second class (polyethylene terephthalate) show a slightly increasing conductivity over the whole of the temperature range from $-180°C$ to $+100°C$, but in this case the effect of varying crystallinity is much less marked than for polythene.

Some typical data for various solid plastics taken from the references already quoted are shown in Figure 12.2.

Data on thermosetting materials are rather sparse and rather less amenable to theoretical treatment, since thermosets are less commonly used in the virgin or unfilled state and they exist in a large number of varieties, cf. epoxides, phenolics; or are used in conjunction with inorganic reinforcements, e.g. paper, fabric and glass laminates. Ratcliffe[14], Schoenborn et al[15], report data on a number of laminates, and Knappe[5], also gives data on filled urea, phenolic and melamine resins. In view of the number of variables possible for a glass fibre reinforced laminate, even with a given resin system, the paper by Knappe and Martinez–Freire[16] is of some interest since it successfully attempts to correlate calculated and measured conductivity values for laminates with specified lay-ups.

The effect of fillers on thermal conductivity is well illustrated in Ratcliffe's excellent paper[17] on silicone rubbers and other elastomers, the fillers considered including silica, titania, calcium carbonate and zinc oxide at various levels. The paper includes a critical review of the relationship between conductivity of discrete and continuous phases in various two phase systems.

CELLULAR PLASTICS

The importance of cellular plastics in the field of thermal insulation needs no stressing and one may be sure that practically all the established plastics in current use have at some time or another been produced in the expanded form, albeit in many cases only experimentally. The function of the polymeric material itself in this context is a curious one in that it serves merely as an enclosure for entrapped air or gases and does not contribute to the insulating properties itself, in fact rather the reverse.

It is often possible to expand a given material in a wide range of densities and, for example, in the authors' laboratory expanded polystyrene has been produced in a controlled (more or less!) range of densities between 1 and

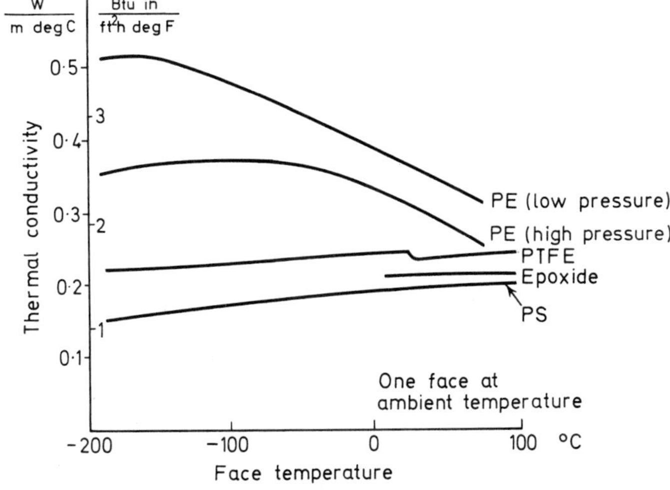

Figure 12.2. Thermal conductivity of typical solid plastics

46 lb/ft³ as compared with a value of 66 lb/ft³ for the solid material. Nevertheless, for most practical applications the density is usually at the low end of the range, for obvious reasons, and of the order of 1–5 lb/ft³. The conductivity falls steadily with decreasing density to a minimum value in the range quoted, and then begins to rise again due to increased convection effects caused by a higher proportion of open cells. It should be noted here that it is common practice to speak of conductivity of materials when convection and radiation are also involved. Strictly, one should say 'effective' conductivity, but the adjective is rarely used and is implied.

Closed cell materials give the lowest conductivities and this is the clue to the mechanism of heat transfer. For the low density range referred to, true heat conduction through the solid matrix is not a large proportion of the total, and at ambient temperatures radiation transfer is rather small. Most of the transfer is caused by inter-cellular convection currents in the gaseous phase and the reduction in overall conductivity resulting from replacing air or carbon dioxide by the much larger molecules of the chlorofluorocarbon gases ('Freon', 'Arcton') in cellular polyurethane for example is striking, being of the order two-fold. This improvement relates to freshly made material and a slow increase in conductivity with time subsequently occurs due, not to loss of 'Freon', but rather to permeation of air, and a stable conductivity approximately 50% greater than that of the fresh material finally results.

The relation between the structure of the polymer matrix, the cell size and orientation, proportion of closed cells, polymer conductivity, etc., and the properties of the gaseous phase, has been well worked out by a number of authors including Skochdopole[18, 19], Doherty et al[20], Guenther[21], Knox[22], Harding[23], Mittasch[24], and Küster[25]. In a theoretical paper Norman[26] gives heat transfer equations for a material in the process of expansion.

Figure 12.3 shows some typical data for various cellular and fibrous materials down to liquid nitrogen temperatures. The apparently anomalous

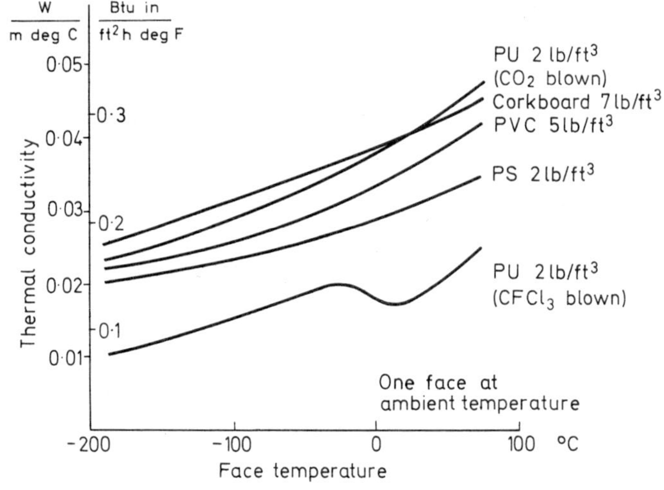

Figure 12.3. Thermal conductivity of cellular plastics

behaviour of the Freon blown polyurethane material in the vicinity of 0°C and below is believed to be caused by latent heat effects associated with the low boiling point of the Freon.

THERMAL DIFFUSIVITY

Thermal conductivity is a steady state phenomenon, by which is meant that it is mainly important in situations where temperatures are steady or only changing slowly. In applications where rapid temperature changes are involved a more useful property is the thermal diffusivity, a, which is a function of thermal conductivity, specific heat and density as defined by the equation:

$$a = \frac{k}{\rho C_p}$$

where k = thermal conductivity, ρ = density and C_p = specific heat at constant pressure.

A common error in applying this equation is to use a conductivity unit containing mixed length dimensions, e.g. Btu in/ft^2 h degF, instead of the consistent Btu ft/ft^2 h degF.

In this example, the density and specific heat must be expressed in appropriate units, such as lb/ft^3 and Btu/lb degF, the dimensions of thermal diffusivity in this instance being ft^2/h (the dimensions of thermal diffusivity in SI units are m^2/s).

Although diffusivity may be determined directly by suitable methods which sometimes yield the conductivity also, there are some circumstances where it can only be calculated from the relationship given above.

Experimental methods of measurement involve the determination of a rate of temperature change as distinct from the rate of heat flow in a steady state measurement of conductivity. Details of non-steady state methods for determining diffusivity and conductivity are given in Section 12.1.4.

12.1.2 Methods for Solid Plastics

INTRODUCTION

Thermal conductivity apparatus for solid plastics is usually based on specimens of simple geometry such as a disc or square, whose dimensions (diameter or length) are large relative to their thickness, so that the heat flow through their plane is large compared with possible lateral heat losses from the edges. One face is usually heated to a constant temperature and the other cooled, and the problem resolves into a measurement of the heat input and temperature difference and a knowledge of the extraneous losses from heater and specimen. These data together with specimen area and thickness are sufficient to give the conductivity from the defining equation of Fourier:

$$\frac{dq}{dt} = kA\frac{d\theta}{dx}$$

where k = thermal conductivity $\qquad \frac{d\theta}{dx}$ = temperature gradient

A = area $\qquad \frac{dq}{dt}$ = rate of heat flow

Under steady state conditions when temperatures are constant, the temperature gradient becomes θ/x, where θ is the temperature difference and x the specimen thickness, and the heat flow becomes q/t.

With the simple method described, heat is lost from both the sides and back of the heater, and an obvious improvement is to sandwich the heater between two specimens and cold plates. The side losses from heater and specimens are reduced by suitable insulation, but they may still form a significant proportion of the total heat input. They may be evaluated by using a pair of specimens of low, known conductivity material, for example a cellular plastic calibrated by an absolute method. If the losses are known to say 10% and only represent 10% of the heat input, the error due to this uncertainty is only 1%. The side loss calibration must be carried out over the range of specimen thickness employed and over the appropriate temperature range, and it is to be noted that it is a function of both hot and cold face temperatures and of the ambient temperature.

By using large area thin specimens and very thin heater plates the side losses may be reduced to negligible proportions, but of course the preparation of such specimens becomes increasingly difficult the larger the area and the smaller the thickness.

The most convenient specimen size lies usually in the range 2–8 in across by $\frac{1}{8}$–$\frac{1}{4}$ in thickness. The area of such specimens may easily be measured with the required precision, and the thickness similarly. The most important requirement of the specimens, however, is that the faces should be as flat as possible, since errors due to imperfect contact between specimens and heater and cold plates can be very large. As an example, an air film of thickness 0·001 in on one side of a 0·1 in thick specimen of a typical solid plastic, would cause an error approaching 10% if the resulting film temperature drops were neglected. It is customary to reduce such errors by coating the specimens with a film of relatively high conductivity liquid (compared with air) such as glycerol, or a suitable grease. Additionally a correction for the resulting small temperature drop in such films may be adduced from a measurement with hot and cold plates in contact and specimens removed, but the contacting medium present. Plates are invariably of high conductivity metal such as copper, brass or aluminium to ensure uniform surface temperatures.

Temperatures (invariably of the plates rather than the specimens) are normally measured with fine wire thermocouples in conjunction with a good quality vernier potentiometer reading to 1 μV, and heat flows by measurement of heater input wattage, both measurements being capable of high precision. A stable power source is a necessity and batteries are still sometimes used, although the many admirable transistorised power supplies now available are more convenient, and do not suffer from the long term drift associated with batteries. This is particularly important where measurements on low conductivity materials are involved, since steady state conditions may not be reached for many hours after energising the heater.

UNITED KINGDOM TESTS

The only standard test method available in the U.K. suitable for solid plastics is that of B.S. 874: 1965, 'Methods of Determining Thermal Properties', Clause 10c. This method is applicable to specimens in the form of thin sheet $\frac{1}{16}$–$\frac{1}{2}$ in thick approximately and 3 in diameter, at hot face temperatures up to about 100°C. The range of conductivity is 1–10 Btu in/ft^2 h degF.

Two specimens are required, one on each side of a 3 in brass or copper heater plate, the whole being assembled between two hollow cold plates of similar metal through which coolant is circulated. The heater is made from nichrome tape on a mica or laminated former enclosed between two recessed plates, the total thickness being $\frac{1}{4}$ in approximately. 36 s.w.g. nichrome/constantan thermocouples are mounted two or three to a face in holes drilled close to each surface of the heater and cold plates, these holes being drilled radially and to different depths. This ensures that any variations from an isothermal condition across a plate may be detected. The whole

Figure 12.4. 3 in disc apparatus of B.S. 874

apparatus when assembled is surrounded by an insulating powder or fibrous insulation although with the latter some care is needed in applying it uniformly each time so that the edge losses remain constant. These losses may be evaluated by calibration with known low conductivity specimens previously measured by a suitable method (see Section 12.1.3).

Figure 12.4 shows the general arrangement.

Although the basic apparatus is simple, considerable care is required as with all thermal conductivity measurements, to achieve a reasonable accuracy (say 5% in this case) and the descriptions of this and the other apparatus covered by B.S. 874 do not give much indication of the many pitfalls which can occur. Particular care must be observed that temperature measurements are accurate and relate to specimen surfaces, and that all sources of heat loss from heaters and specimens, including conduction loss along connecting wires and thermocouples, are properly taken into account or rendered negligibly small. In addition, cold face coolant temperatures must be maintained very constant and the attainment of temperature stability in the apparatus ensured before readings are taken.

The temperature limitations of the B.S. 874 apparatus are not mandatory and suitable modifications are described in References 17 and 27 which enable measurements to be made down to about −150°C and up to 250°C. In the latter reference the use of a cylindrical guard ring round the heater and specimen assembly made it possible to reduce side losses to zero at the expense of slight additional complications in operation.

Due to the difficulties in accurate measurement of small temperature differences with thermocouples, the minimum differential across the specimens is specified as approximately 15 degC.

It is interesting to note in passing that as far as is known no British Standard Specification invokes the use of the B.S. 874 3 in method (at least for plastics). This is, however, less surprising when it is considered that the essentially more important conductivity measurements on cellular plastics by the guarded hot plate method of B.S. 874 have only been required since the publication of B.S. 3837: 1965, 'Expanded Polystyrene Board for Thermal Insulation Purposes'.

UNITED STATES TESTS

The only U.S. test suitable for solid plastics is that of A.S.T.M. C.518–67, 'Thermal Conductivity of Materials by means of the Heat Flow Meter', which is essentially a comparative method stated to be suitable for temperatures between −45°C and +540°C approximately.

It is not uncommon with A.S.T.M. methods of test for their scope to be so broad as to admit practically any reasonable possibility. In this case, the size of the apparatus is not specified nor the conductivity range, only the conductance of the specimen. Thus it appears possible to test a 2 in thick specimen of conductivity up to 4 Btu in/ft^2 h degF, but the specimen would need to be 24 in square. At the other extreme a specimen $\frac{1}{4}$ in thick of conductivity up to 0·5 Btu in/ft^2 h degF is permitted, but the size is not stated, although by inference it is not less than 8 in square.

In essence the apparatus consists of a single cold plate and a plain hot plate with the material under test and a prepared specimen of known conductivity sandwiched in between. The prepared specimen has a multiway differential thermocouple attached to its surfaces so that the temperature difference across it may be easily measured. The heat flux through the composite is thus known, and additional couples measuring the temperature difference across the test specimen give the remaining data required (plus thicknesses).

The prepared specimen, or heat flow meter as it may be regarded, is itself calibrated by a preliminary comparison with a pair of standard slabs measured absolutely by a guarded hot plate apparatus. A minor advantage of the method is that no quantities of heat are required to be measured, only temperature differences. However, at best the method is a comparative one and the errors in measuring the standards and in calibrating the heat flow meter are cumulative.

An improvement given in the appendix to A.S.T.M. C.518 is to use two heat flow meters and two cold plates. By supplying the latter from two separate constant temperature baths, measurements at two different mean temperatures are obtained simultaneously.

OTHER METHODS

As far as national tests are concerned, few countries besides the U.K. and the U.S. seem to have adopted methods suitable for solid plastics as opposed to methods for cellular materials. This is not to say that no interest is evident, but that the need for data on solids is relatively small and presumably met by the fairly small number of investigators working in this field. Some of the methods they use will be briefly described.

Modifications to the unguarded two specimen disc or plate apparatus have been made by some authors. Hennig and Knappe[3, 28] used a very thin heater plate made of constantan foil and, in conjunction with thin, large area specimens the edge losses were made negligible. This apparatus was also used for measuring elastomers and polymers in the stretched state. By employing a specimen of known conductivity (polymethylmethacrylate) on one side of the heater, measurements could be made with only one specimen.

Another modification sometimes employed to avoid using two specimens is to have a sheet of uncalibrated low conductivity material on one side of the heater, backed by a second heater. This guard heater is adjusted until the temperature difference across the uncalibrated material is zero at equilibrium, whence all the energy in the main heater is directed through the test specimen (apart from the side loss which can be dealt with as previously discussed).

A longitudinal heat flow method is described by Powell, Rogers and Coffin[29] which is sufficiently versatile to be used for such diverse materials as PTFE and copper, but primarily it is intended for measurements at cryogenic temperatures down to a few degrees absolute. A long test bar is used, heated at one end and cooled at the other, the assembly being appropriately guarded to eliminate heat losses. Thermocouples mounted along the specimen measure the gradient, and the conductivities at different temperatures are obtained simultaneously. This method may be used at other temperatures and extended to a comparison technique by having a standard and unknown specimen joined end to end.

The use of a specimen with permanently mounted thermocouples as a heat flow meter has already been mentioned. Another type of heat flow meter, which relies on the semiconducting properties of certain elements to give an electrical output when heat passes through them, has been available for a number of years. Both types are discussed by de Jong and Marquenie[30].

Kline[8] used a concentric cylinder radial heat flow apparatus with a central heater and water jacketed specimen. End effects were reduced (and neglected) by using a relatively long specimen (up to 12 in). An unusual method reported by Schröder[31] uses two liquids of differing boiling points separated in suitable containers by a test piece whose conductivity is required. The lower liquid (of higher boiling point) is maintained at its boiling point and the heat passing through the specimen causes the upper liquid to boil. The heat flow is assessed by measuring the rate of evaporation and involves merely a stopwatch timing of a few minutes. The two faces of the specimen are of course at the boiling points of the liquids, which are chosen appropriately. If the boiling points and heat of vaporisation are known the method is absolute, otherwise the apparatus may be calibrated using specimens of known conductivity.

Another unusual method relates to a specific problem, that of measuring the conductivity of a wire coating. Hodgetts' solution[32] is to employ a radial heat flow method by surrounding a length of coated wire by a water filled tube and plotting wire temperature against wattage input as different currents are passed along it. The wire temperature is known from the resistance–temperature characteristics. This method is an inversion of the radial hot wire method commonly used for measuring liquids.

A useful, although early, review of methods of measuring solids is given by Cooper[33].

Guarded hot plate methods, widely used for low conductivity measurements (see Section 12.1.3) are not used extensively for solid plastics, partly because the guarding is less effective with higher conductivity materials and partly because the simpler plain heater plate methods are fairly satisfactory anyway. However, the paper by Hansen and Ho[9] describes briefly an 8 in guarded plate used with $\frac{1}{4}$ in thick specimens, and the method is otherwise noteworthy for the extremely low temperature differential employed, of the order of 2 degC. Eiermann et al also use a guarded method for the range -180°C to $+90$°C[2, 4].

Finally, as an example of an extremely simple and elegant method derived from everyday experience, the 'thermal comparator' devised by Powell[34] is cited. This is based on the tactile impressions of relative coldness experienced when materials of differing conductivity are handled. The greater the conductivity the cooler the 'feel'. The sensation depends on a complex function of conductivity, specific heat, density and surface texture. The method depends on measuring the differential rate of cooling of two metal spheres, previously heated to a constant temperature, when placed one in direct contact and one in near contact with a test piece. The cooling rate is simply measured by a differential thermocouple connected to a galvanometer or potentiometer. The apparatus is calibrated by means of specimens of known conductivity and the relationship between square root of thermal conductivity and the differential couple output after a fixed time interval (usually 10 s) is found to be linear. A modified single contact version with a thermocouple at the tip gives virtually an instantaneous reading related to the contact temperature and this version may be made direct reading in conductivity units.

12.1.3 Methods for Cellular Plastics

THE GUARDED HOT PLATE

Cellular plastics have the lowest thermal conductivity of any solid materials available at this time and the determination of this property involves the measurement of relatively small heat flows, which dictates the use of large area, thin specimens. A lower limit on thickness is imposed by the difficulty of measuring this property on lightweight compressible materials and the desire to have test thicknesses approaching practical user values; therefore it is usual for specimens to be of the order of 1–2 in thick. To ensure that the basic requirement of normal heat flow through the measurement area of a specimen is observed, the length or diameter to thickness ratio is usually in the region of 6 or 8 to 1 and this leads to specimens 8–12 in square or diameter.

With a similar assembly to that described in Section 12.1.2 for solids, comprising a heater plate and two specimens sandwiched between two cold plates, the heat losses from the sides of cellular specimens and heater could well approach the heat flow normal to the specimens and thus cause very considerable errors. By the use of the guard ring technique, widely applied to electrical measurements (see Chapter 10), these errors may be effectively eliminated and the guarded hot plate apparatus which results is probably the most widely used apparatus for the measurement of low conductivity materials. Its greatest virtue is that it is an absolute method capable of a good precision provided that a great deal of care is taken in both construction and operation.

The principle is simple. The heater plate is split into two portions, a central area and a surrounding annulus. These two parts have separately wound heaters to that the energy to the centre can be measured independently. The gap between them is bridged only by such part of the heater former as is necessary to provide sufficient physical support for the whole. The faces of the plate are usually copper or aluminium with thermocouples let into their surfaces on both central and guard areas. Two specimens and two liquid cooled cold plates complete the apparatus and the whole assembly is usually lagged.

In operation the cold plates and the central heater are maintained at the required temperatures and the guard heater is adjusted so that the guard and centre thermocouples indicate equality of temperature at equilibrium. Under these conditions the heat flux from the centre heater is assumed to be normal to the plane of the heater and specimens, and therefore there is no gain from, or heat loss to, the guard ring. The mean conductivity of the specimens is calculated in the usual manner, the relevant area being that of the centre heater based on the mid gap dimensions, and the wattage being that dissipated in the centre heater (note that only half the input energy flows through each specimen).

The attainment of complete equilibrium is even more important with low conductivity materials than with solid plastics, since the long time constant of cellular materials means that equilibrium times of the order of hours, or even a few tens of hours, are involved and the rate of change of temperature is slow and therefore difficult to detect. For this reason power supplies which are stable over long time periods are a necessity. Since temperature differences of the order of 0·05 degC between centre and guard heater plates have a significant effect on the measurement, the selection and construction of thermocouples is extremely important and it is necessary to determine the thermal e.m.f.s to within 1 μV.

As temperature differences across the specimens of as little as 10 degC or less may be used, cold plate temperatures must be maintained very constant and a thermostat bath temperature control of at least ±0·1 degC is required.

NATIONAL TESTS

The U.K. guarded hot plate method is described in outline in B.S. 874: 1965, Clause 10a. It is stated to be suitable for conductivities of 0–1 Btu in/ft^2 h degF and hot face temperatures of up to 80°C. However, this is by no means the limit of its capabilities and it has been successfully used at liquid nitrogen cold face temperatures (−196°C). The upper temperature limit is governed mainly by the choice of heater constructional materials and the necessity for the hot plate centre and guard surfaces to remain co-planar at all temperatures.

The cold and hot plates are 12 in square and two specimens of the same area and preferably $1\frac{1}{2}$–2 in thick are required. These are placed on each side of the hot plate, which has a central area 8 in square separated from the surrounding guard area by a $\frac{1}{16}$ in wide gap, and the combination clamped between two cold plates cooled by liquid circulation. The whole apparatus is insulated with 4–6 in of cork granules or layers of glass fibre insulation. This is the limit of the constructional details given in the standard.

Further details obtained from the National Physical Laboratory (to which the design is due) are as follows. 36 s.w.g. 80: 20 nichrome/constantan thermocouples are mounted in the surfaces of the plates, normally three or four on each side of the centre heater, and the same number on each side of the guarded area at corresponding positions, so that the temperature differences across the guard gap may be accurately measured. Three similar couples are mounted in each cold plate surface. When properly constructed it is possible for all the hot face thermocouples to agree to within 1 μV when the apparatus is assembled but not energised.

The thermocouple combination referred to is not widely used in this country and even less outside it. It is a much better choice than the well-known chromel (90 nickel: 10 chromium)/constantan couple because of its higher stability and consistency and it is preferred to copper/constantan because of the much lower thermal conductivity of nichrome, which reduces conduction errors. Copper wires are necessary for all connections from the reference junctions onwards to the measuring potentiometer. A stable reference temperature is essential and in addition thermocouple cold junctions need to be 6–8 in long to ensure constancy and consistency.

The hot plate is made of $\frac{1}{8}$ in thick aluminium sheets glued to the heater former, which is a thin laminate with nickel-chrome heating tape wound upon it. Holes are drilled round the former between centre and guard windings to coincide with the guard gap, so that heat conduction is reduced to a minimum.

Figures 12.5 and 12.6 show details of a typical 12 in guarded hot plate apparatus.

In operation it is tedious to balance exactly guard and centre temperatures and it is usually more convenient to know the effect of a given amount of off-balance on the answer so that an appropriate correction can be applied. To do this, small adjustments to the guard energy are made around the

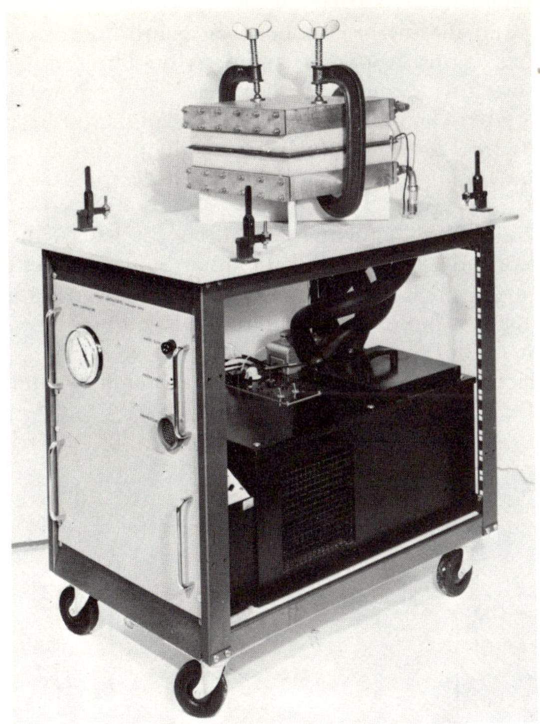

Figure 12.5. 12 in guarded hot-plate apparatus (B.S. 874)

Figure 12.6. 12 in guarded heater plate (B.S. 874)

balance point and the differences between guard and centre thermocouple readings plotted against apparent conductivity. The centre energy must be maintained constant throughout and after each guard adjustment the apparatus must be left for equilibrium to be re-established. A typical plot is shown in Figure 12.7 and the correct conductivity is given, of course, by the intercept at zero difference. Over small temperature differences the plot is linear.

The virtue of this refinement in technique is that, with experience, the slope of the plot indicates whether the apparatus is responding properly or not, and also it is a measure of the efficiency of the plate, a high slope being undesirable since it indicates a high lateral conductivity between centre and guard portions of the hot plate.

The extension of the apparatus and technique to measurements on insulators at liquid nitrogen temperatures is described by Hickman and Ratcliffe in Reference 35.

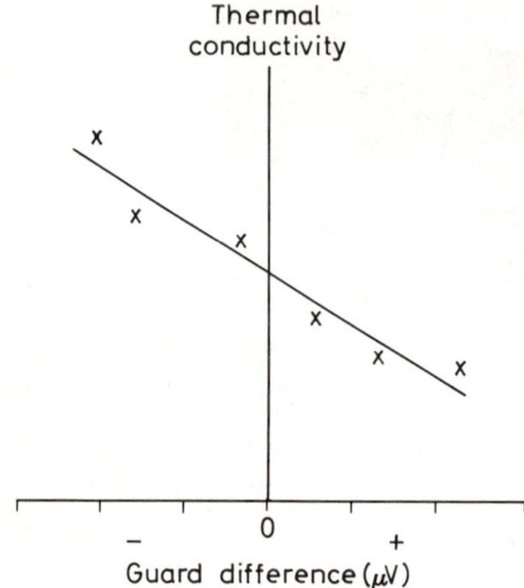

Figure 12.7. Thermal conductivity against guard difference

Typical temperatures of measurement given in B.S. 874 are −18/5°C or 0/20°C for refrigeration insulation, 10/27°C for building insulation and 20/82°C for hot water system insulation.

B.S. 2972: 1961, 'Methods of Test for Thermal Insulating Materials', is the main standard for thermal test methods (excluding plastics) invoking B.S. 874 for conductivity. B.S. Specifications for cellular plastics are currently B.S. 3837: 1965 (Expanded Polystyrene—see above), B.S. 3869: 1965, 'Rigid Expanded Polyvinyl Chloride for Thermal Insulation Purposes and Building Applications', B.S. 3927: 1965, 'Phenolic Foam Materials for Thermal Insulations and Building Applications' and B.S. 4370: 1968, 'Methods of Test for Rigid Cellular Materials'.

The U.S. version of the guarded hot plate apparatus is described in A.S.T.M. C.177–63, 'Thermal Conductivity of Materials by Means of the Guarded Hot Plate', in very general terms. Two types are described, one metal surfaced for the range −70°C–+260°C, the other with a refractory surface suitable for 100°C–700°C approximately.

Two well known variants of the metal version are known as the National

Bureau of Standards plate and the National Research Council plate.

The range of the apparatus is given in terms of specimen conductance, which should not exceed 10 Btu/ft^{2}h degF. The specimen thickness may be up to 4 in, but this latter requires a 24 in plate with 12 in square centre. Specimens up to 2 in thick require a 12 in plate with 6 in square centre.

Details regarding thermocouple type and position are vague, and it is permissible to fix them either in the plates or directly on the specimens. Guard balance is required to be sufficiently close to produce less than 0·5% error in the measured conductivity. When equilibrium is reached measurements are made at 30 min intervals until four successive sets of data give conductivity values differing by not more than 1%.

References 36–38 give details of refinements made by various experimenters and the results of intra-laboratory comparisons. A recent (1967) A.S.T.M. publication, STP 411, 'Thermal Conductivity Measurements of Insulating Materials at Cryogenic Temperatures', includes some papers on modified versions of the guarded hot plate apparatus suitable for very low temperature measurements.

A.S.T.M. C.420–62T, 'Thermal Conductivity of insulating Materials at Low Temperatures by means of the Wilkes Calorimeter', covers a method specifically designed for measuring low conductivity materials down to liquid nitrogen temperatures. This method has now reverted to a proposed method and the specification withdrawn (see A.S.T.M. Part 14, 1969). Briefly, a test specimen is placed on a heater plate which is maintained at the required hot face temperature, and a hollow calorimeter filled with liquid nitrogen or other appropriate liquefied gas contacts its upper surface. The body of the calorimeter is surrounded by a guard calorimeter and a second guard calorimeter also surrounds the outlet tube of the measuring calorimeter, which is connected to a gas meter. These guard calorimeters are likewise filled with liquefied gas. The rate of evolution of gas is suitably timed and from the thermal properties of the liquefied gas the heat flux through the guarded area of the specimen is calculated. Specimen surface temperatures are measured with thermocouples.

The heat flow meter apparatus of A.S.T.M. C.518–67, which is also suitable for cellular plastics, has already been described in Section 12.1.2.

A.S.T.M. Specifications covering particular materials are C.534–64T, 'Preformed Flexible Elastomeric Cellular Thermal Insulation in Sheet and Tubular Form', C.578–65T, 'Preformed Block-Type Cellular Polystyrene Thermal Insulation', and D.2341–65T, 'Rigid Urethane Foam'.

The German guarded hot plate apparatus described in DIN.52612 is unusually large, specimens 20 in square being required. The hot plate itself is 30 in square and the guard ring section falls outside the specimens. The cold plates are also 30 in square and the space between guard and cold plates is filled with granular or powder insulation, which also completely surrounds the apparatus.

An alternative arrangement utilises the same basic apparatus in conjunction with a single specimen. The top specimen and cold plate are removed and replaced by a backing heater, which eliminates losses from the upper surface of the hot plate.

In the U.K., guarded hot plate apparatus to B.S. 874 and A.S.T.M. C.177 is commercially available from Yarsley Laboratories[39], and to A.S.T.M.

C.177 from D. A. Pitman Ltd.[40]. In the U.S. the A.S.T.M. apparatus is also available from Custom Scientific Instruments Incorporated[41].

OTHER METHODS

Relatively few methods for cellular plastics exist other than the guarded hot plate and the non-steady state methods described in Section 12.1.4, which is a reflection of the difficulties involved, and only one method will be given here (and strictly this is a quasi-steady state method).

This is a simple heat flow meter technique, due to Lang[42]. The meter itself is 9 in × 9 in × 0·09 in and it is placed with a surrounding spacer in the centre of two specimens 18 in square, which are sandwiched between cold and hot plates 14 in square. The apparatus is calibrated with known, low conductivity specimens and it is found that the conductivity is proportional to the product of specimen thickness and heat flow meter output. A very high precision is claimed for this measurement and also it is stated that equilibrium times are as short as half to one hour with cellular materials. An analysis of the above method by Norris and Fitzroy[43] seems to substantiate the claims made by Lang, but it is based on the proviso that the two specimens should have a uniform conductivity and also that the heat flow meter is at the exact mid-plane of the specimens.

This method has been investigated and developed over the last decade in a number of countries; currently it is already covered by an A.S.T.M. standard and is being considered as a secondary method in the U.K. by the B.S.I. Committee revising B.S. 874. In addition an ISO working party on cellular plastics is proposing a similar method.

Typically, the apparatus is designed round specimens 12 in square (300 mm) and an inch or two in thickness. The single specimen arrangement consists of an assembly, usually horizontal, of hot plate (liquid circulated or electrically heated) specimen, heat flow meter and cold plate (liquid circulated). Sometimes the heat flow meter is sandwiched between two identical specimens and sometimes the single specimen version has an added cold plate and heat flow meter on the reverse side of the heater so that two simultaneous measurements can be made on different specimens. This modification is described in A.S.T.M. C.518–67, 'Thermal Conductivity of Materials by Means of the Heat Flow Meter'. The limitation imposed by Lang[42] that the meter should be exactly at the mid-plane of the specimens no longer seems to be valid.

Usually, the apparatus is employed at fairly low temperatures (heat flow temperature below say 100°C), but there is no upper limit theoretically, nor practically if the heat flow meter itself is mounted near the cold-plate. A.S.T.M. C.518 suggests use up to 500°C approximately. The heat flow meters themselves are available commercially as fine wire multi-junction thermocouple units based on a thin sheet plastics material such as rigid PVC, although the A.S.T.M. document gives constructional details for a meter based on thin work sheet.

The apparatus has to be calibrated by means of known conductivity specimens previously measured by the guarded hot plate method, which at best can give an accuracy of about 2%. Since the precision of the apparatus is approximately 3% only, the uncertainty of a determination may be of the order 5%, which is acceptable for most practical purposes.

Commercial versions of the apparatus are available from Yarsley Laboratories[39] and D.A. Pitman Ltd.[40].

The accuracy of guarded hot plate sets has been assessed at international level by the Institut International du Froid[44] and the results published suggest that the best national laboratories can achieve agreement within 1% and that a precision of 0·5% is achievable, as witness the NPL results on the reference material, a resin bonded glass fibre.

12.1.4 Diffusivity and Non-Steady State Methods

Diffusivity and non-steady state methods generally have attracted much interest in recent years. Their simplicity and speed offer obvious advantages over the time consuming equilibrium methods, but with inorganic building materials in particular the chief attraction is the possibility of testing samples in the moist condition. This is not usually possible with the conventional methods, since the moisture distribution is altered by the heat flow and water is driven towards the cold face of the specimen.

Before discussing diffusivity methods proper, which are absolute and which depend upon a particular solution of the heat conduction equation, an example of a quasi-dynamic method for determining conductivity may be of interest, since it is essentially a comparative method and no theoretical treatment is required. The method uses a single specimen and a heat sink technique and is described by Goldfein and Calderon[45].

The specimen is sandwiched between a heated (or cooled) plate and a heat sink, the temperature of which is plotted against time from when the plate is applied. The apparatus was calibrated with NBS measured samples and the conductivity of these plotted against the product of heating (or cooling) rate and specimen thickness. For cellular materials between 0·12 and 0·38 Btu in/ft^2 h degF, a precision of 1% is claimed. Specimen thickness between $\frac{1}{2}$ and 1 in is recommended.

The two general ways in which non-steady state methods can be applied have been described by Clarke and Kingston[46, 47]. The first way is that in which a constant heat input is suddenly applied to one face of a slab of material and in this case either the rate of temperature rise of the heated face is measured with the opposite face maintained at a constant temperature, or alternatively the temperature rise of the non-heated face is measured when it is perfectly insulated. In this way the conductivity and specific heat are given simultaneously with diffusivity.

In the second, one or both faces of the slab are suddenly brought to a constant temperature(s) (dissimilar), and the rate of temperature rise of either the opposite perfectly insulated face in the former case, or some interior point in the specimen in the latter case, determined.

The specimen is normally of a very simple geometry, such as a plane sheet or a cylinder, otherwise the mathematical solution of the heat flow equation becomes too unwieldy. With such a one-dimensional heat flow system the problem consists in solving the equation

$$\frac{\partial \theta}{\partial t} = \frac{k}{\rho C_p} \cdot \frac{\partial^2 \theta}{\partial x^2}$$

for given boundary conditions, where θ, t and x are respectively temperature, time and distance, and k, C_p and ρ the thermal conductivity, specific heat and density respectively.

Clarke and Kingston's method is based on a pile of test specimens interleaved with a strip heater of very thin steel. This arrangement ensures that, effectively, the central specimen is subjected to a constant heat flux on both sides and a centrally mounted thermocouple of very fine gauge wires monitors the temperature via a d.c. amplifier and output meter.

The test assembly is placed in a constant temperature cabinet before test, specimens being of the order of 3–5 in square and an inch thick. The temperature rise employed is of the order of 2 degC and it is claimed that a test takes about two minutes to complete. However, initially, before energising the heater, the specimen temperatures must be very constant and their drift less than 1 millidegreeC per minute.

The accuracy of both conductivity and specific heat measurement is claimed to be high, but even with thermocouple wires as fine as 41 s.w.g. errors become significant with materials of conductivity less than 0·6 Btu in/ft^2 h degF, so that measurements on cellular plastics are ruled out.

The Building Research Station's transient method, based on a simplification of the Clarke and Kingston method, is claimed to be suitable even for cellular plastics, providing thin specimens are used ($\frac{1}{2}$ in for expanded polystyrene, $\frac{3}{4}$ in for higher conductivity materials) in order to keep the side loss relatively low. Theoretically it is claimed that there is no upper limit to thermal conductivity measurement although practically the heater power requirements may set a limit. However, there is a limitation on diffusivity set by low thermal capacity samples, i.e. cellular plastics, since the finite thermal capacity of the heater becomes relatively significant, although the resultant errors may be corrected. The method is described by Ball[48].

In its simplest form the apparatus consists of two cold sinks (metal plates backed by concrete) and two specimens sandwiching a heater made from conducting paper. A set of fine wire thermocouples in contact with the specimens' faces are so connected as to measure the mean temperature differential which is usually of the order one or two degrees Celsius during a run. The temperature differential is plotted against time from energising the heater at a constant rate. From the curve produced, sets of data at time t and $2t$ are used to calculate both conductivity and diffusivity. The steady state is normally achieved within $\frac{1}{2}$–1 h, depending on specimens, when the usual Fourier equation can be used.

One of the most important features of the apparatus is that because of the small temperature differential employed measurements are possible on moist materials with a minimum disturbance to the moisture distribution. It is anticipated that the method will be incorporated into B.S. 874 during its current revision.

The apparatus is available commercially from Yarsley Laboratories[39] in a simplified form.

Another example of the second class is the method used by Braden[49] for measurements on dental polymers. Here the specimen is a rectangular prism with a centrally embedded thermocouple and its temperature is plotted against time after immersing it suddenly in a constant temperature bath. The method appears only to give diffusivity. Data for polythene, polypropylene, polymethylmethacrylate and silicone and polysulphide rubbers are given. Chung and Jackson[50] similarly used a cylindrical specimen initially maintained at 180°C, then suddenly cooled by pumping water through an external jacket. Data are given for polymethylmethacrylate and polystyrene. The precision is stated to be 1% and accuracy 2 to 3%. Hattori[51] also used a cylindrical specimen and reports data on PVC, polystyrene, polythene (low pressure and high pressure), PTFE and PCTFE. A review of methods and data for diffusivity is given by Berlot[52] with particular emphasis on polymers.

The most popular methods at present for cellular plastics are those based on line heat sources. If a thin wire is embedded in a homogeneous medium and heated, the rate of temperature rise at some nearby point is a function of the diffusivity and may be derived from the heat flow equation

$$\frac{\partial\theta}{\partial t} = \frac{k}{\rho C_{p}}\left(\frac{\partial^{2}\theta}{\partial r^{2}}+\frac{1}{r}\cdot\frac{\partial\theta}{\partial r}\right)$$

where the symbols have their usual meaning and r is the radial distance from the wire.

In practice, the heater wire is usually mounted inside a long narrow metallic tube together with thermocouples which monitor the rate of temperature rise, and the whole instrument is known as a heated probe. The principles are described by Hooper and Lepper[53] and Hooper and Chang[54], and experimental details worked out by Eustachio and Schreiner[55] and Mann and Forsyth[56]. The solution of the heat flow equation used by the above authors is

$$\Delta\theta = \frac{Q}{4\pi k}\log_{e}\frac{t_2}{t_1}$$

where Q = constant heat input per unit time, per unit length of heater
$\Delta\theta$ = temperature rise in the time interval $t_1 \rightarrow t_2$
t_1, t_2 = time intervals after heater energised

$\Delta\theta$ plotted against $\log_e t_2/t_1$ gives a straight line of slope $Q/4\pi k$ whence k is derived.

Alternatively, by standardising on the two time intervals t_1, t_2 and on the heater input Q, k becomes proportional to $(1/\Delta\theta)$ and hence a single calibration with a material of known conductivity is sufficient to determine the constant of proportionality. This method now appears in A.S.T.M. D.2326–64T, 'Thermal Conductivity of Cellular Plastics by means of a Probe', and Figure 12.8 is taken from this document.

Although extremely useful as a device for production control testing and comparative tests on similar materials, the heated probe seems liable to give anomalous results from time to time on cellular materials (see for example the data for cork and expanded polystyrene in Reference 56). It appears that this is connected with the small size of the probe so that its behaviour is more a function of local variations and discontinuities in cell structure than is that of the conventional plate methods which give average values over large areas and thicknesses of specimen. In the case of soft, easily penetrable materials, the insertion of the probe must alter the local density and hence conductivity, and with rigid materials, in which it is necessary to make a prepared hole, the tightness of fit must obviously be an important factor in the determination.

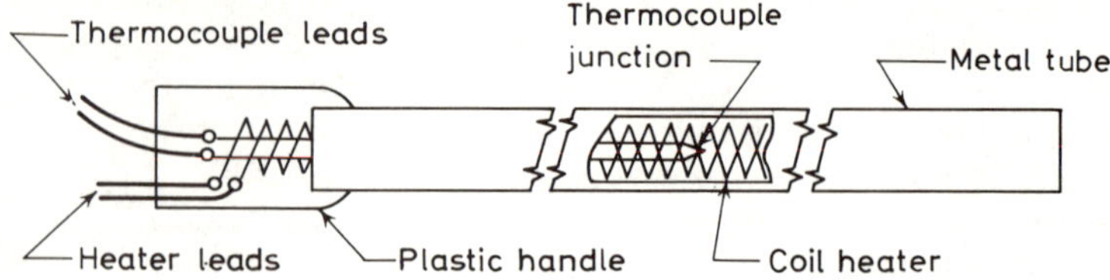

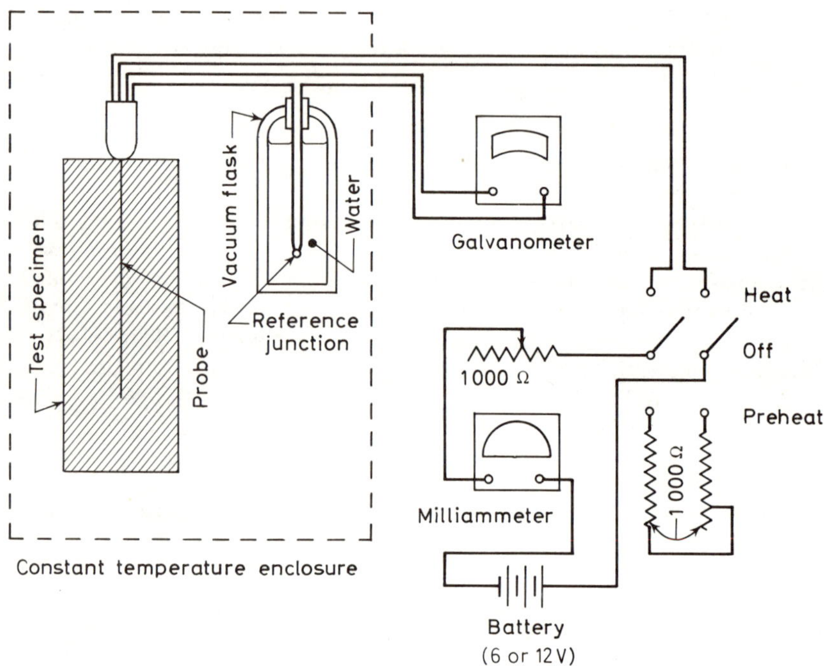

Figure 12.8. Probe and electrical circuit (A.S.T.M. D.2326)

Commercial models of probe type instruments are available from Custom Scientific Instruments Incorporated[41] and H. Tinsley & Co. Ltd[57].

12.2 SPECIFIC HEAT

12.2.1 Introduction

When a crystalline solid is heated its temperature rises, becomes constant at the melting point, and rises again when the material is all liquid. This behaviour is conveniently represented by a curve of heat content per unit mass (enthalpy) plotted against temperature. Figure 12.9 shows an idealised plot of a crystalline substance. The change in enthalpy at the melting point is the latent heat.

The slopes of the initial and final parts of the curve give the heat content per unit mass per unit temperature rise or *specific heat,* which may be more formally defined as the amount of heat required to raise the temperature of unit mass by one degree.

Polymeric materials do not behave quite in this way of course, and their melting points (or rather softening points) are not sharply defined since crystallinity is either absent completely (e.g. polystyrene, PVC) or only partial (polyethylene, PTFE). Figure 12.10 shows two curves illustrating an amorphous and a crystalline polymer, from which it can be noted that the specific heat (slope of plot) of the crystalline polymer tends to increase much more rapidly below the transition temperature than that of the amorphous polymer, and above the transition point the specific heat remains fairly constant with temperature for both polymers, albeit at a higher level for the crystalline material.

A knowledge of specific heat is important in the processing of plastics and is required also in calculating thermal diffusivity from thermal conductivity measurements. From the theoretical viewpoint a curve of specific heat against temperature is extremely valuable in the calculation of fundamental thermodynamic properties (enthalpy, entropy, free energy, for example) and changes in such properties throw valuable light on such physical processes as quenching and cooling, crystallisation, melting, and the various transitions found in polymers.

Warfield, Petree and Donovan[58] claim that specific heat derminations are the most accurate way of measuring second order transitions and Melia[59] in his excellent, comprehensive review essay covers theoretical aspects in some detail (including 96 references). A recent review by Hands[60] outlines present theory on polymers and gives 60 references.

Figure 12.11 shows typical specific heat data for some common thermoplastics. Since the property is a function of unit mass of material, the specific heat of cellular plastics is the same as for the solid material.

The present unit of heat is the joule, and the unit of specific heat is J/g degC although cal/g degC is still widely used (the British unit Btu/lb degF still thrives in engineering practice, particularly in the U.S., but the size of the unit is identical to the cal/g degC unit).

The calorie as unit quantity of heat was defined originally by assuming the specific heat of water to be unity at some convenient temperature, but now that energy and work can be measured so precisely in terms of the fundamental electrical units (and time) it has been found expedient to re-define the quantity of heat in these units.

The range of specific heat is small and many solids, including plastics, have values between 0·4 and 4 J/g degC. Liquids generally have values well below 4 but water is anomalous in this respect.

There are two general ways of measuring the specific heat of solids; one is to apply a measured quantity of heat (usually relatively small) to a sample under adiabatic conditions and measure the resulting temperature rise, and the other is to employ some variation on the method of mixtures and measure the temperature rise in a suitable liquid when the sample, pre-heated to an

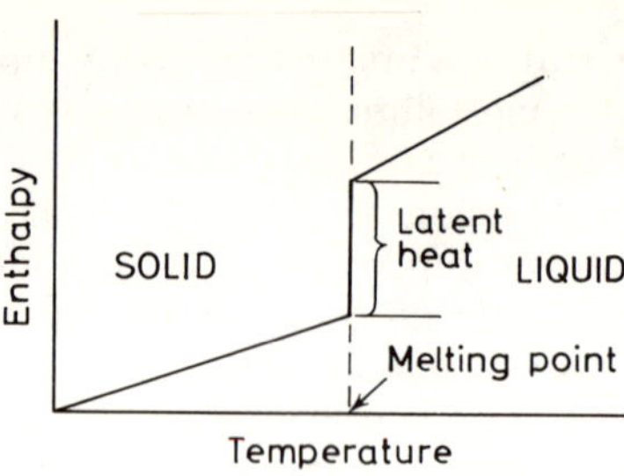

Figure 12.9. Enthalpy of ideal crystalline material

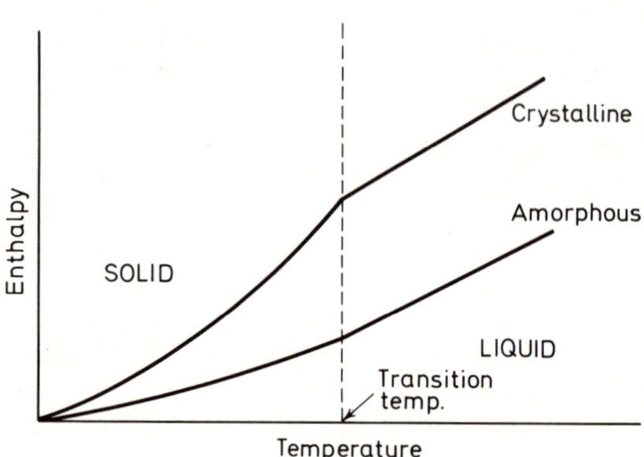

Figure 12.10. Enthalpy of amorphous and crystalline polymers

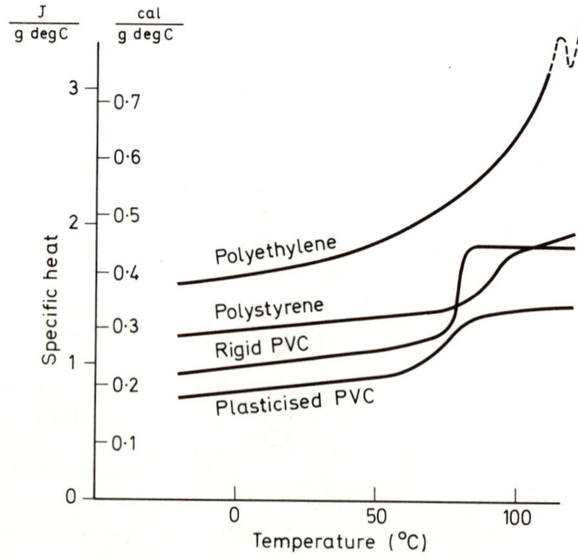

Figure 12.11. Specific heat of some thermoplastics

elevated temperature, is immersed in it. These latter techniques are generally referred to as 'drop' methods and usually they give specific heat over a fairly large range of temperature only, so some care is needed in their application.

12.2.2 Adiabatic Calorimeters

One of the most precise ways of measuring specific heat is by means of an adiabatic calorimeter; it is also the most relevant and direct method since a small measured amount of heat is applied to the specimen and its resultant temperature rise recorded. It is adiabatic since the temperature of the system is allowed to rise, but no heat exchange with the surroundings is permitted and this is normally contrived by surrounding the calorimeter proper by a

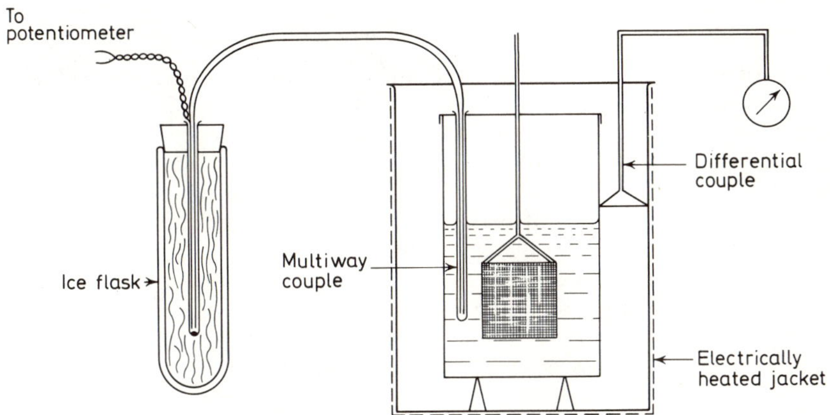

Figure 12.12. Simple adiabatic calorimeter (NPL)

jacket maintained at the same temperature to within very close limits (of the order of 0·01 degC or less). All the heat applied to the specimen, less that used in raising the temperature of the calorimeter itself, is therefore employed in raising the specimen temperature; the specific heat is simply calculated. The thermal capacity of the calorimeter is determined in a separate experiment.

Figure 12.12 shows the elements of a simple adiabatic calorimeter. The inner calorimeter containing the specimen sits inside a larger vessel, with a minimum of solid insulation between the two. The facing surfaces of the inner and outer containers are highly polished to reduce radiation heat transfer. The outer vessel has heater windings on all surfaces and a differential thermocouple is affixed to both inner and outer calorimeter walls and connected to a galvanometer.

The centre calorimeter has a heater supplied by a stable battery or power supply and suitable means of monitoring the current and voltage. For liquids a stirrer is essential also. In the case of solids sometimes the heater is embedded in the material or wound directly on it, but in either case liquid immersion is essential. A sensitive thermometer is also included in the calorimeter capable of measuring small temperature rises of a degree or two.

Initially the system must be in equilibrium and the jacket and calorimeter temperatures identical as indicated by zero output from the differential couple. Beginning at ambient temperature, a constant current is passed through the calorimeter heater for a measured finite time and as the temperature rises the jacket heater current is adjusted so that the differential couple output remains substantially at zero. When equilibrium is re-established at the new temperature the specific heat S is given by

$$EIt = mS\theta + H\theta$$

where E, I = voltage and current applied to calorimeter heater (V, A)
t = duration of current application (s)
m = mass of specimen (g)
S = specific heat (J/g degC)
θ = temperature rise (degC)
H = water equivalent of calorimeter, stirrer, etc.

H is determined in a separate experiment with the test specimen removed.

The following description of the NPL Adiabatic Calorimeter for determinations in the range 25–75°C is taken from an NPL leaflet[61].

'The apparatus consists of a thin-walled copper calorimeter, chromium-plated on the outside, measuring about 8 cm in diameter and 13 cm high. This is supported inside an electrically-heated copper jacket, about 1·5 mm thick, which is chromium-plated on the inside. The calorimeter is about half-filled with oil—we use transformer oil—and contains an electrical heater, a copper gauze basket, which acts as a stirrer, and also a five-way differential thermocouple with one set of junctions in melting ice.

'When the apparatus is at a constant temperature electrical energy is dissipated at a constant measured rate for a known time, say, 300 s, and the temperature rise of the calorimeter and contents is measured by the five-way couple. To prevent heat losses from the calorimeter during the test run the outer jacket is heated to maintain the calorimeter and jacket always at the same temperature. A differential thermocouple, connected to a galvanometer, and attached to the two units, enables this to be done accurately.

'The experiment is then repeated with the copper-gauze basket stirrer filled with the material to be tested. Sheet materials are placed vertically on edge about $\frac{1}{16}$ in apart to allow the oil to pass between. The thermal capacity of the test sample is then calculated from the difference in electrical energy required to give the same temperature rise of the calorimeter with and without the test sample.'

The techniques of adiabatic calorimetry particularly with regard to measurements on plastics are described by Aukward et al[62], Tautz et al[63] (down to −150°C), Hellwege, Knappe and Wetzel[64] and Wunderlich and Dole[65]. The difficulties of making precision measurements and the complications of the adiabatic calorimeter are described by Dole and co-workers[66] in their paper on automatic control of adiabatic jackets.

Details of the construction and operation of an adiabatic calorimeter together with an analysis of errors, are given in a useful paper by Bowring, Garton and Norris[67].

12.2.3 Drop Calorimeters

Drop calorimeters are widely used on account of their simplicity. A specimen, often contained in a metal capsule, is heated to some appropriate constant temperature in an oven or furnace and allowed to drop into liquid in a stirred calorimeter. The temperature of the calorimeter is plotted against time and from the curve the temperature rise, allowing for heat losses or gains to or from the environment, is adduced. The specific heat is calculated from this rise after applying corrections for the water equivalent of the calorimeter, etc., determined separately.

A.S.T.M. C.351–61, 'Mean Specific Heat of Thermal Insulation', is typical, although not intended specifically for plastics. The calorimeter fluid is normally water, but other fluids are allowed and the specimen is heated to approximately 100°C in a tubular heater, although again other

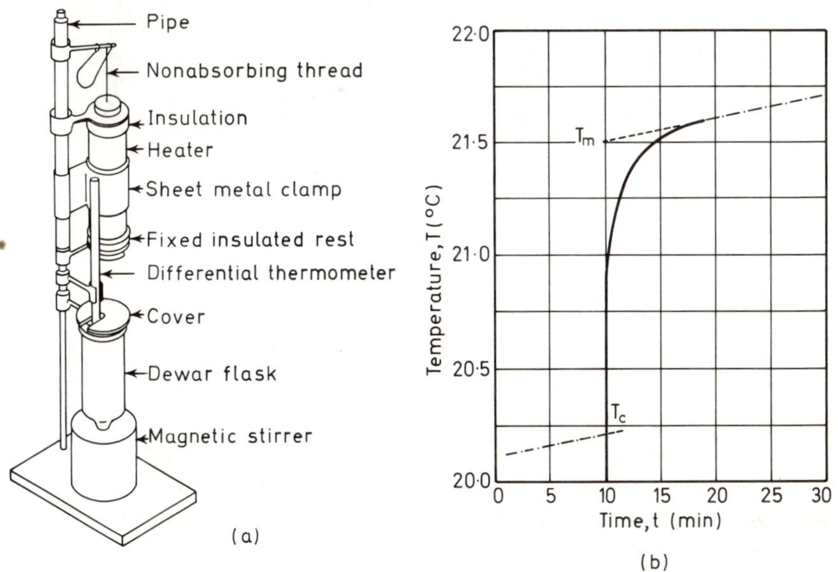

Figure 12.13. Drop calorimeter and temperature—time curve (A.S.T.M. C.351)

temperatures may be used. Figure 12.13 shows the overall set-up and a typical time–temperature curve.

The beauty of drop calorimeters is that the same basic calorimeter is used whatever temperature is of interest (within reason) and the only problem is heating (or cooling) the specimen. For example Wilkes[68] describes a method which he used for temperatures above ambient down to −180°C, the specimen being appropriately heated or cooled in a test tube as shown in Figure 12.14 and then 'poured' into the calorimeter, the latter comprising simply a brass capsule with a tripping lid actuated by the falling specimen, contained in a Dewar flask. The temperature rise of the capsule was measured with thermocouples.

Sometimes the drop calorimeter and adiabatic methods are combined, the heated capsule being dropped into an adiabatic container, and Griskey

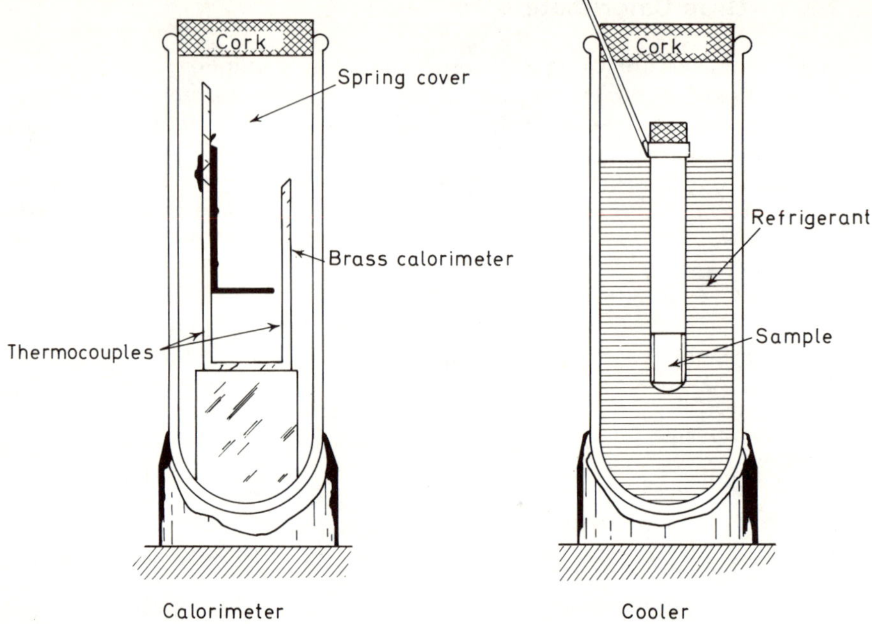

Figure 12.14. Low temperature drop calorimeter (Wilkes)

and Hubbell[69] describe such a method as applied to measurements on methacrylic polymers, measurements being made in the range 120–130°C.

12.2.4 Other Methods

One other method is worthy of brief mention, called the 'thin heater calorimeter' by its originator Hager[70]; results obtained with it on PVC are reported by Dunlap[71].

It is intended for thin flexible sheet material or film and a layer of such material is wound with a metal foil into a tight roll. The foil is heated by the passage of a current and a thermocouple buried in the assembly measures the resulting small temperature rise (a few degF). It is claimed that, by virtue of the construction of the roll and the low thermal mass of the heater, the heating effect is in fact adiabatic, since the measurement of temperature rise is made before heat can escape.

REFERENCES

1. EIERMANN, K., 'Thermal Conductivity of High Polymers', *J. Polymer Sci., Polymer Symp.* Pt. C, No. 6, 157 (1963)
2. EIERMANN, K., HELLWEGE, K-H and KNAPPE, W., 'Quasi-steady State Measurement of Thermal Conductivity of Plastics in the Temperature Range −180°C to +90°C', *Kolloidzschr.*, **174,** No. 2, 134 (February 1961)
3. HENNIG, J. and KNAPPE, W., 'Anisotropy of Thermal Conductivity in Stretched Amorphous Linear Polymers and in Strained Elastomers', *J. Polymer Sci., Polymer Symp.*, Pt. C, No. 6, 167 (1963)

4. EIERMANN, K. and HELLWEGE, K-H, 'Thermal Conductivity of High Polymers from −180°C to +90°C', *J. Polymer Sci.*, **57,** No. 165, 99 (March 1962)
5. KNAPPE, W., 'The Steady-state Absolute Measurement of Thermal Conductivity of Commercial High Polymers from 20–100°C', *Kunststoffe,* **51,** No. 11, 707 (November 1961)
6. EIERMANN, K., 'Measurement of Thermal Conductivity of Amorphous and Crystalline Polythene', *Kolloidzschr.*, **180,** No. 2, 163 (February 1962)
7. EIERMANN, K., 'Thermal Conductivity of Plastics and its Dependence on Structure, Temperature and Pre-Treatment', *Kunststoffe,* **51,** No. 9, 512 (September 1961)
8. KLINE, D. E., 'Thermal Conductivity Studies of Polymers', *J. Polymer Sci.*, **50,** No. 154, 441 (1961)
9. HANSEN, D. and HO, C. C., 'Thermal Conductivity of High Polymers', *J. Polymer Sci.*, Pt. A, **3,** 659 (1965)
10. RITCHIE, P. D. (ed.) *Physics of Plastics,* Chapter 4, Iliffe Books Ltd., London (1965)
11. ANDERSON, D. R., 'Thermal Conductivity of Polymers', *Chem. Rev.*, **66,** No. 6, 677 (December 1966)
12. SHELDON, R. P. and LANE, SISTER K., 'Thermal Conductivities of Polymers, I, Polyvinyl Chloride', *Polymer,* **6,** No. 2, 77 (February 1965)
13. SHELDON, R. P. and LANE, SISTER K., 'Thermal Conductivities of Polymers II, Polyethylene', *Polymer,* **6,** No. 4, 205 (April 1965)
14. RATCLIFFE, E. H., 'The Thermal Conductivities of Resin/Glass Laminates', *Plastics, Lond.*, **22,** No. 233, 55 (February 1957)
15. SCHOENBORN, E. M., ARMSTRONG, A. A. and BEATTY, K. O., 'Thermal Properties of Certain Laminated Plastics', *Bull. Amer. Soc. Test. Mat.* **No. 174,** 54 (May 1951)
16. KNAPPE, W. and MARTINEZ-FREIRE, P., 'Measuring and Calculating the Thermal Conductivity of Glass Fibre Reinforced Plastics', *Kunststoffe,* **55,** No. 10, 776 (October 1965)
17. RATCLIFFE, E. H., 'Thermal Conductivity of Silicone Rubber and some other Elastomers', *I.R.I. Trans.*, **38,** No. 5, 181 (1962)
18. SKOCHDOPOLE, R. E., 'The Thermal Conductivity of Foamed Plastics', *Chem. Eng. Prog.*, **57,** No. 10, 55 (October 1961)
19. PATTEN, G. A. and SKOCHDOPOLE, R. E., 'Environmental Factors in Thermal Conductivity of Plastic Foams', *Mod. Plast.*, **39,** No. 11, 149 (July 1962)
20. DOHERTY, D. J., HURD, R. and LESTER, G. R., 'The Physical Properties of Rigid Polyurethane Foams', *Chemy Ind.*, **No. 30,** 1340 (28 July 1962)
21. GUENTHER, F. O., 'Cellular Materials—Composition, Cell-size, Thermal Conductivity', *S.P.E. Trans.*, **18,** No. 7, 243 (July 1962)
22. KNOX, R. E., 'Insulation Properties of Fluorocarbon Expanded Rigid Urethane Foam', *A.S.H.R.A.E. J.*, **7,** No. 4, 43 (October 1962)
23. HARDING, R. H., 'Heat Transfer through Low-density Cellular Materials', *I. and E.C. Proc. Design and Development,* **3,** No. 2, 117 (April 1964)
24. MITTASCH, H., 'Dependence of the Thermal Conductivity of Plastic Foams on the Degree of Moisture', *Plaste u. Kautschuk,* **16,** No. 4, 268 (April 1969)
25. KUSTER, W., 'The Thermal Conductivity of Foamed Plastics', *Kunststoffe,* **60,** No. 4, 249 (April 1970)
26. NORMAN, R. H., 'Heat Conduction in an Expanding Material', *I.R.I. Trans. & Proc.*, **37,** No. 1, 30 (February 1961)
27. RATCLIFFE, E. H., 'Thermal Conductivities of Fused and Crystalline Quartz', *Brit. J. appl. Phys.*, **10,** No. 1, 22 (January 1959)
28. KNAPPE, W., 'Measurement of Thermal Properties of Poor Conductors with a 2-plate Apparatus without Guard-ring', *Z. Angew Phys.*, **12,** No. 11, 508 (November 1960)
29. POWELL, R. L., ROGERS, W. M. and COFFIN, D. O., 'An Apparatus for Measurement of Thermal Conductivity of Solids at Low Temperatures', *J. Res. nat. Bur. Stand.*, **59,** No. 5, 349 (November 1957)
30. DE JONG, J. and MARQUENIE, L., 'Heat Flowmeters and their Applications', *Instrum. Pract.*, **16,** No. 1, 45 (January 1962)
31. SCHRODER, J., 'Apparatus for Determining the Thermal Conductivity of Solids in the Temperature Range 20 to 200°C', *Rev. Sci. Instrum.*, **34,** No. 6, 615 (June 1963)
32. HODGETTS, G. B., 'A Method of Measuring the Thermal Conductivity of Wire Coatings', *Brit. J. appl. Phys.*, **13,** No. 7, 310 (July 1962)
33. COOPER, G. L., 'Methods of Measuring the Thermal Conductivity of Solids', *Atomic Energy Research Establishment Unclassified Report,* A.E.R.E. Inf/Bib., 104 (1956)

34. POWELL, R. W., 'Thermal Conductivity Measurements by the Thermal–comparator Method', *Proc. Black Hills Summer Conf. on Transport Phenomena, S. Dakota School of Mines and Technology, Rapid City,* Report Office of Naval Research Contract No. Nonr (G)—000 64—62, 95 (October 15 1962)
35. HICKMAN, M. J. and RATCLIFFE, E. H., 'The Measurement of the Thermal Conductivity of Heat Insulators at Low Temperatures', *Proc. 9th Int. Congress Refrig.,* **1,** 2149 (September 1955)
36. ROBINSON, H. E. and WATSON, T. W., 'Interlaboratory Comparison of Thermal Conductivity Determinations with Guarded Hot Plates', *S.T.P. 119,* American Society for Testing and Materials (Philadelphia), 36 (1952)
37. GILBO, C. F., 'Experiments with a Guarded Hot-plate Thermal Conductivity Set', *S.T.P. 119,* American Society for Testing and Materials (Philadelphia), 45 (1952)
38. ZABAWSKY, A., 'An Improved Guarded Hot-plate Thermal Conductivity Apparatus with Automatic Controls', *S.T.P. 217,* American Society for Testing and Materials (Philadelphia), 3 (1957)
39. Yarsley Laboratories, Chessington, Surrey.
40. D. A. Pitman Ltd., Jessamy Road, Weybridge, Surrey.
41. Custom Scientific Instruments, Inc., 541 Devon Street, Kearny, New Jersey, U.S.A.
42. LANG, D. L., 'A Quick Thermal Conductivity Test on Insulating Materials', *Bull. Amer. Soc. Test. Mat.* **No. 216,** 58 (September 1956)
43. NORRIS, R. H. and FITZROY, N. D., 'A Quick Thermal Conductivity Test on Insulating Materials', *Mater. Res. Stand.,* **1,** No. 9, 727 (September 1961)
44. 'Thermal Insulating Measurements', Report by Commission II of Institut International du Froid (Paris–17) (1968)
45. GOLDFEIN, S. and CALDERON, J., 'Apparatus for Determining Thermal Conductivity of Insulation Materials', *J. appl. Polymer Sci.,* **9,** No. 9, 2985 (September 1965)
46. CLARKE, L. N. and KINGSTON, R. S. T., 'Equipment for the Simultaneous Determination of Thermal Conductivity and Diffusivity of Insulating Materials using a Variable-state Method', *Aust. J. appl. Sci.,* **1,** No. 2, 172 (June 1950)
47. CLARKE, L. N. and KINGSTON, R. S. T., 'Further Investigation of Some Errors in a Dynamic Method for the Determination of Thermal Conductivity and Diffusivity of Insulating Materials', *Aust. J. appl. Sci.,* **2,** No. 2, 235 (June 1951)
48. BALL, E. F., 'A Simple Transient-Flow Method of Measuring Thermal Conductivity and Diffusivity', *Building Research Station (Garston), Current Papers, Research Series 65* (1967)
49. BRADEN, M., 'Measurement of the Thermal Diffusivities of Polymers', *Trans. J. Plast. Inst. Lond.,* **33,** No. 103, 17 (February 1965)
50. CHUNG, P. K. and JACKSON, M. L., 'Thermal Diffusivity of Low Conductivity Materials', *Industr. Engng. Chem.,* **46,** No. 12, 2563 (December 1954)
51. HATTORI, M., 'Thermal Diffusivity of Some Linear Polymers', *Kolloidzschr.,* **202,** No. 1, 11 (March 1965)
52. BERLOT, R., 'Mesure de la Diffusivité et de la Conductivité Thermique des polymers: Variation en fonction de leur etat structural et de la temperature', *Publications Scientifiques et Techniques du Ministere de L'Air, Notes Techniques,* No. N.T.154, Paris (1966)
53. HOOPER, F. C. and LEPPER, F. R., 'Transient Heat Flow Apparatus for the Determination of Thermal Conductivities', *Heat. Pip. Air Condit.,* **20,** 129 (August 1950)
54. HOOPER, F. C. and CHANG, S. C., 'Development of the Thermal Conductivity Probe', *Heat. Pip. Air Condit.,* **22,** 125 (October 1952)
55. EUSTACHIO, D. D. and SCHREINER, R. E., 'A Study of a Transient Heat Method for Measuring Thermal Conductivity', *Heat. Pip. Air Condit.,* **22,** 113 (June 1952)
56. MANN, G. and FORSYTH, F. G. E., 'Measurement of the Thermal Conductivity of Samples of Thermal Insulating Materials and of Insulation in situ by the Heated Probe Method', *Mod. Refrig.,* **59,** No. 699, 188 (June 1956)
57. H. Tinsley & Co. Ltd., Werndee Hall, South Norwood, London S.E.26.
58. WARFIELD, R. W., PETREE, M. C. and DONOVAN, P., 'The Specific Heat of High Polymers', *S.P.E. J.,* **15,** No. 12, 1055 (December 1959)
59. MELIA, T. P., 'The Specific Heats of Linear High Polymers', *J. appl. Chem.,* **14,** No. 11, 461 (November 1964)
60. HANDS, D., 'The Thermophysical Properties of Polymers, Part I, Specific Heat', Rubber and Plastics Research Association of Great Britain, *Technical Review* (Shawbury), No. 51 (February 1970)

61. Private communication

62. AUKWARD, J. A., WARFIELD, R. W., PETREE, M. C. and DONOVAN, P., 'Technique for Measuring Specific Heat of Thermosetting Polymers', *Rev. Sci. Instrum.*, **30,** No. 7, 597 (July 1959)

63. TAUTZ, H., GLUCK, M., HARTMANN, G. and LEUTERITZ, R., 'The Specific Heat of High Polymers in the Temperature Range −150°C to +180°C', *Plaste u. Kautschuk*, **10,** No. 11, 648 (November 1963)

64. HELLWEGE, K-H, KNAPPE, W. and WETZEL, W., 'Specific Heat of Polyolefins and High Polymers between 30°C and 180°C', *Kolloidzschr.*, **180,** No. 2, 126 (February 1962)

65. WUNDERLICH, B. and DOLE, M., 'Specific Heat of Synthetic High Polymers, VIII, Low Pressure Polyethylene', *J. Polymer Sci.*, **24,** No. 106, 201 (April 1957)

66. DOLE, M., HETTINGER, W. P., LARSON, N., WETHINGTON, J.A. and WORTHINGTON, A. E., 'Calorimetry of High Polymers, I, Automatic Temperature Recording and Control of Adiabatic Jackets', *Rev. Sci. Instrum.*, **22,** No. 11, 812 (November 1951)

67. BOWRING, R. W., GARTON, D. A. and NORRIS, H. F., 'Measurement of the Specific Heats of Santowax "R", Para-, Meta- and Ortho-Terphenyl, Diphenyl and Dowtherm "A" ', *U.K.A.E.A. Report A.E.E.W.-R.38,* H.M.S.O. (London) (1960)

68. WILKES, G. B., 'Thermal Conductivity, Expansion and Specific Heat of Insulators at Extremely Low Temperatures', *Refrig. Engng,* **52,** No. 1, 37 (July 1946)

69. GRISKEY, R. G. and HUBBELL, D. O., 'Calorimetric Behavior of Methacrylic Polymers', *J. appl. Polymer Sci.*, **12,** No. 4, 853 (April 1968)

70. HAGER, N. E., 'Thin Heater Calorimeter', *Rev. Sci. Instrum.*, **35,** No. 5, 618 (May 1964)

71. DUNLAP, L. H., 'Specific Heats of Poly (vinyl chloride) Compositions', *J. Polymer Sci.*, Pt. A–2, **4,** No. 5, 673 (1966)

13

Permanence Tests

13.1 INTRODUCTION

The miscellaneous collection of subjects considered in this chapter have one factor in common—they are generally test methods designed to evaluate the resistance of plastics materials to change as a result of exposure to some environment other than normal ambient. Many are accelerated tests and invariably these have at least one shortcoming in common, a very result of their aim to provide a forecast of durability within a finite period of time. Unless they are 'accelerated' in nature they offer nothing over a field trial, but to achieve this acceleration the stringency of test conditions may be so enhanced as to scale an energy barrier and create an effect which will never occur in normal usage. At least, if this does not happen the equation of the results of accelerated ageing tests into likely service life is difficult if not impossible and/or the accelerated conditions are so idealised that they may not truly simulate the sequence of events in the field; the latter, and the 'energy barrier' consideration, may lead to orders of merit from the accelerated ageing tests which are not fulfilled in practice.

This chapter does not deal with the effects of elevated temperature alone; this subject has been considered in Chapter 11 and, as far as degradative effects specifically are concerned, in Section 11.8 thereof. Nor can it be claimed that all the degradative influences that plastics materials are likely to encounter, are covered, but it is hoped that at least the more important ones have been touched on. For a brief general discussion of the behaviour in adverse environments, see Eshenaur[1].

13.2 WATER ABSORPTION AND WATER SOLUBLE MATTER

13.2.1 General Considerations

The resistance of plastics materials to water varies significantly according to their chemical nature and, if filled or reinforced, the nature of additives or reinforcements present. The sensitivity to moisture can well be imagined to be quite different for the relatively hydrophilic secondary cellulose acetate and the decidedly hydrophobic polystyrene and their take-up of water (*w/w*) confirms this:

Water Absorption (24 h)	Cellulose Acetate: 1·9–5·7% according to grade G.P. Polystyrene: 0·03–0·05%

(Data according to A.S.T.M. D.570—see below[2])

In practical terms of influence on mechanical properties, reference has already been made (Chapter 3, Section 3.1) to the 20 fold increase in impact strength of nylon 6 between dryness and moisture saturation.

A particularly interesting effect is described by Uijlenburg[3] who studied the water absorption of unplasticised PVC at various temperatures up to 140°F and then measured certain mechanical properties, particularly related to the material in pipe form. It was found that at 140°F effects were produced by water absorption which were not encountered at more normal temperatures and therefore a proposal for a speedy measurement of the long term bursting pressure of unplasticised PVC pipe, by testing at 140°F, was likely to yield very misleading results as to performance at more normal ambient temperatures.

Braden[4] has analysed the processes involved in the absorption of water by plastics materials, which he concluded depend on only two parameters: the diffusion coefficient and the equilibrium uptake. The former governs the kinetics of water absorption and is highly temperature dependent. Equilibrium uptake is virtually unaffected by temperature. Blank[5] examined five equations proposed for relating water absorption to time, concluded that none were universally applicable and deduced an equation for obtaining equilibrium absorption (Q) from values of absorption (q_1, q_2, and q_3) measured at equally spaced short intervals:

$$Q = \left[\frac{q_2^4 - q_1^2\ q_3^2}{2q_2^2 - (q_1^2 + q_3^2)}\right]^{\frac{1}{2}}$$

Further experimental study of this equation is described by Blank[6].

The effect of fillers is well illustrated by the differing requirements for maximum water absorption specified in B.S. 771: 1959, 'Phenolic Moulding Materials', for a high shock resistant PF, probably rag filled, and a heat resistant PF, probably asbestos filled:

HS : 95 mg ⎫ $\bar{x}$ values (i.e. maxima for averages of several months
HR: 32 mg ⎭ production)

(Data according to B.S. 2782: 1970, Method 502A—see below.)

The standard water absorption test is designed as no more than a quality control test, for use purely and simply as a guide to water sensitivity, by measuring water take-up under a specified but arbitrary set of experimental conditions with no pretence to absolute significance or even to an assessment of the effect of moisture attack on, say, electrical or mechanical properties for instance (for this see, for example, Hauck[7]). Even so, great care is needed to obtain reproducible and comparative test data from such water absorption tests. Control of temperature is obviously of paramount importance in a diffusion or rate process. The presence of grease, even finger marks, on the surface, may influence results. The physical state of the actual surface must be specified, for instance whether 'as moulded' (with a resin skin) or as left after machining. For hydrophobic materials in particular the precise mode of removal from the water immersion bath and of drying and the time taken to weigh may all significantly influence the test results. The head of water could have an effect and, since *all* plastics materials are relatively resistant to water, water absorption tests generally must take heed of surface to volume ratio of test specimens. Practically all standard tests, being essentially short term, do not allow equilibrium to be attained throughout the specimen thickness (this could take months, or even years!) and therefore the effect measured is basically a surface one. Thus, the total surface area of the specimen is of paramount importance in a test measuring water absorption by simple weight increase, as is the surface/volume ratio in tests measuring water absorption on a wt/wt basis. Finally, a constituent may be present which is significantly soluble in water, i.e. it will be leached out in the test, or the material may even hydrolyse. A quantity of the test specimen lost by solution in the water environment will obviously affect the results of a test which is designed to assess attack of water by weighing water uptake; the loss of soluble matter may cause a much lower nett gain in weight, may cancel out the gain or even exceed it. If water solubility is anticipated it should be measured in the test.

The work of Braden[4] demonstrates that two materials of the same equilibrium uptake of water could, if their diffusion coefficients were different, give quite dissimilar water absorption test results after some arbitrarily selected short test period of, say, 24 h.

13.2.2 Standard Techniques

BRITISH STANDARD METHODS

B.S. 2782: 1970, 'Methods of Testing Plastics', contains no less than ten methods for determining water absorption:

Method 502A Water absorption of phenolic moulding material
Method 502B Cold water absorption of aminoplastic moulding material
Method 502C Water absorption and water soluble matter of polyvinyl chloride extrusion compound
Method 502D Water absorption of laminated sheet
Method 502E Water absorption of laminated tube and rod
Method 502F Water absorption. Procedure A of ISO Method
Method 502G Water absorption. Procedure B of ISO Method
Method 503A Boiling water absorption of aminoplastic moulding material
Method 503B Boiling water absorption. Procedure A of ISO Method
Method 503C Boiling water absorption. Procedure B of ISO Method

Methods 502F with 502G and 503B with 503C are identical in experimental details with ISO/R.62, Procedures A and B, and R.117, Procedures A and B.

The essential features of the various methods are summarised in Table 13.1:

All methods require that the specimens shall not make contact over any substantial area with one another or with the container during immersion.

A.S.T.M. METHODS

Seven procedures are included in A.S.T.M. D.570–63, 'Water Absorption of Plastics', which differ in their severity of duration and temperature but

Table 14.1 CREEP DATA (References 2 and 24)

Method	*Specimen details*	*No. of specimens*	*Test*
502A	Moulded disc 52·1±2·0 mm (2·05±0·08 in) in diameter and 11·94±0·76 mm (0·470±0·030 in) thick. Machined and glass paper finished on all surfaces to 48·36±0·20 mm (1·900±0·008 in) in dia., and 10·16±0·20 mm (0·400±0·008 in) in thickness.	2	Dry for 1 h at 50±2°C, cool in desiccator and weigh immediately after removal (W_1). Immerse for 168±2 h in distilled water at 25±0·5°C. Remove, dry with clean cloth or filter paper and reweigh 5–15 min after removal from the water (W_2). Water absorption = (W_2-W_1) mg.
502B	Moulded disc 50·8±1·00 mm (2·00±0·04 in) in diameter and 3·2±0·18 mm (0·125±0·007 in) thick	2	Dry for 24±1 h at 50±3°C, cool in desiccator and weigh immediately after removal (W_1). Immerse for 24±1 h in distilled water at 23±0·5°C. Remove, dry with clean cloth or filter paper and reweigh not more than 1 min after removal from the water (W_2). Heat for 24±1 h at 50±3°C, cool and reweigh (W_3). If $W_3<W_1$ Water absorption = (W_2-W_3) mg If $W_3>W_1$ Water absorption = (W_2-W_1) mg
502C	Knife-punched disc from sheet 1·27±0·07 mm (0·050±0·003 in) thick. Diameter of punch at cutting edge: 50·8±0·2 mm (2·00±0·01 in)	2	(All weighings carried out in a weighing bottle). Dry over $CaCl_2$ or other desiccant at room temperature and pressure for 24±1 h. Weigh immediately (W_1). Immerse in distilled water at 50±1°C for 48±1 h. Remove, cool for 30±15 min at 20°±5°C, dry with clean cloth or filter paper and reweigh within 5 min of removal from water (W_2). Dry over $CaCl_2$ or etc. as before and weigh at daily intervals until consecutive weights do not differ by more than 2 mg (W_3). Water absorption = (W_2-W_3) mg. Water soluble matter = (W_1-W_3) mg.
502D	38·10−0·000+0·51 mm (1·500−0·000+0·020 in) square by thickness of sheet unless this exceeds 25·4 mm (1 in) when it is reduced to 25·40±0·25 mm (1·000±0·010 in) by machining one face only.	2	Weigh specimen (W_1), immerse in distilled water at (usually) 23±0·5°C for 24±1 h. Remove, dry with clean cloth or filter paper and reweigh within 2 min of removal from water (W_2) Water absorption = (W_1-W_2) mg. (with thickness quoted)

Table 13.1—*continued*

Method	*Specimen details*		*No. of specimens*	*Test*
502E	Tube or rod 38·10±0·51 mm (1·500±0·020 in) long		2	As 502D, with dimensions of cross-section quoted
502F and 502G	Moulding material, Extrusion compound	50±1 mm diameter by 3±0·2 mm thick	3	Dry for 24±1 h at 50±2°C, cool in dessiccator and, for 502F only, weigh to nearest mg (W_1). Immerse in distilled water at 23±0·5°C for 24±1 h. Remove, dry with clean dry cloth or filter paper and reweigh within 1 min of removal from water (W_2). For 502G only, redry for 24±1 h at 50±2°C, cool in desiccator and reweigh (W_3). 502F: Water absorption = (W_2-W_1) mg 502G: Water absorption = (W_2-W_3) mg
	Sheet, Casting resin, Laminating resin	50±1 mm square by 3±0·2 mm thick		
	Tube, Rod	50±1 mm long		
503A	As 502B		2	As 502B except that boiling distilled water is used and the specimen is cooled for 15±1 min in water at 20±5°C before drying and weighing.
503B and 503C	As 502F and 502G except that casting and laminating resins are not mentioned		3	As 502F and 502G respectively (except that for some reason the tolerance on 50°C is ±3°!) but boiling distilled water is used and the specimen is cooled for 15±1 min in water at 20±5°C before drying and reweighing.

are not, mercifully, altered by the chemical nature of the material under examination. Inevitably, however, different specimens are used according to the physical form thereof:

Moulded plastics: disc 2 in (50·8 mm) diameter and $\frac{1}{8}$ in (0·32 mm) thick. The thickness tolerance is ±0·007 in for hot moulded materials and ±0·012 in for cold moulded and cast materials.

Sheet: bar 3 in (7·62 cm) long by 1 in (2·54 cm) width by the thickness of the material. For most materials the tolerance on thickness is ±0·008 in.

Rod: 1 in long for rods 1 in diameter or under and $\frac{1}{2}$ in long for rods of greater diameter.

Tube: 1 in long for tubes of internal diameter less than 3 in. For tubes of internal diameter 3 in or more, the specimen is 3 in length cut in the circumferential direction and 1 in width cut lengthwise.

Three specimens are used. For materials the water absorption characteristics of which may be affected by temperatures in the region of 110°C, conditioning is carried out 50±3°C for 24 h, after which the specimen is cooled in a desiccator and then immediately weighed. For materials unaffected by a temperature of 110°C the conditioning is at 105–110°C for 1 h.

The seven procedures are:

1. $24-0+\frac{1}{2}$ h in distilled water at 23 ± 1°C on edge.
2. As 1, but for 120 ± 4 min.
3. As 2 followed by 1, total immersion period being 24 h.
4. After 1, reimmerse, weigh at end of first week and thereafter at two weekly intervals until equilibrium is substantially obtained (as defined).
5. 120 ± 4 min in boiling distilled water, on edge; cool in distilled water at room temperature for 15 ± 1 min.
6. As 5, but for 30 ± 1 min.
7. As 5, but in distilled water at 50 ± 1°C for 48 ± 2 h.

For comparing materials, it is specified that both procedures 1 and 4 shall be used.

After all immersion treatments the specimens are wiped off with a dry cloth and weighed immediately. With specimens $\frac{1}{16}$ in or less in thickness, a weighing bottle must be used.

DIN METHODS

DIN.53471 and DIN.53475 follow ISO/R.117 and R.62 respectively. DIN.53472 offers a variety of standard methods somewhat after the style of A.S.T.M. D.570, but using specimens of the ISO/R.62 and R.117 type. DIN.53473 employs similar specimens again, but the absorption is from an atmosphere of 92–93% relative humidity at $20° \pm 2$°C for 1, 2, 4, 7 or a multiple of seven days.

13.3 CHEMICAL RESISTANCE

13.3.1 General Considerations

Water absorption and water solubility are in reality but one aspect of the general subject of *chemical resistance* and probably the only excuse for treating the effect of water as a separate entity—and before all others—is the profusion of the medium, even if mainly saline. By their very nature the basic polymers as such would not be expected to be subject to chemical attack except by specific compounds or classes of compounds. The greatest source of aggressive media will be the organic solvents since the polymers themselves are mainly organic and, subject to the influence of molecular weight and crystallinity, non-polar polymers are generally susceptible to the attack of non-polar solvents and polar polymers to polar solvents. However, this is a specialised subject (see, for instance, Chapter 8 of Gordon[8]) and in this chapter we will be concerned mainly with the extractive effects of organic solvents.

Organic plastics are generally resistant to the attack of aqueous media, for the reasons briefly outlined in Section 13.2 above, but if the polymer chain is prone to oxidative attack, concentrated nitric acid for instance may have a pronounced degradative effect. Much may be gained from the ad hoc approach of immersion of small strip specimens in the medium, preferably at various temperatures, and observing the effects—discoloration, crazing, cracking, complete degradation, etc. Partial immersion has much to recommend it, as the attack at an air–liquid interface, with oxygen accessibility, may be much more severe. 'Monitoring' the attack, by measurement of the change of some important physical property such as hardness and transparency, is a valuable refinement; Jessup[9] used flexural strength, hardness and electrical properties when examining the effects of dilute sulphuric acid and certain organic solvents on a range of thermosetting plastics.

A review, based on practical experience, of the effect on certain plastics and rubbers of attack by chemicals used in synthetic fibres manufacture, particularly sulphuric acid, is provided by Evans[10].

13.3.2 Standard Methods

There is only one method for chemical resistance evaluation, decidedly specific, in B.S. 2782: 1970, 'Methods of Testing Plastics'; it is Method 505A, for determining 'Resistance to Concentrated Sulphuric Acid of Rigid Polyvinyl Chloride Compounds'. At least the one method has the merit of being quantitative because it is based on comparison of shear strength of discs 0·43 mm (0·017 in) thick, before and after immersion in concentrated sulphuric at 95–100°C for 24 h, the shear strength test being essentially that of Method 305A of the same standard (see Chapter 8). The results are expressed as the percentage retention of shear strength.

On the other hand, B.S.I. Committee PLC/36 (see Chapter 2, Section 2.2.1) is known to be working on a recommendation for the presentation of data on chemical resistance (of plastics) to liquids; about 50 liquids will probably be suggested. It will form part of B.S. 4618.

A.S.T.M. D.543–67, 'Resistance of Plastics to Chemical Reagents', which also includes a useful bibliography of chemical resistance of plastics, lays down general procedures for the examination of 50 specified reagents and solvents, by following weight and dimensional changes and alteration in mechanical properties. With regard to the latter it is noted that tensile tests are generally preferred, though for rigid materials flexural tests may be particularly sensitive in detecting the development of surface cracks.

A.S.T.M. D.2299–68 describes how to carry out tests for 'Determining Relative Stain Resistance of Plastics', not so much by chemicals as the hazards of everyday life such as coffee, tea, lipstick, shoe polish and so on. The specific hazard of sulphide staining is covered by A.S.T.M. D.1712–65.

The 'Resistance of Plastics Films to Extraction by Chemicals' is specified in A.S.T.M. D.1239–55, using distilled water, soap solution, cottonseed oil, mineral oil, kerosine, 50% ethyl alcohol or any of the other chemicals described in A.S.T.M. D.543. Specimens, squares of side $50 \pm 0{\cdot}25$ mm, are first conditioned, then weighed and afterwards immersed by vertical suspension in 400 ml of the test liquid which has been already maintained for at least 4 h at the test temperature. After the period of immersion (24 h at 23°C in the standard), the specimens are rinsed in certain cases, gently wiped with soft cloth or absorbent tissue and reweighed. (For non-volatile media with good adhesion to the film, a technique is provided for estimating the 'blank' absorption.)

DIN.53476 follows ISO/R.175 in laying down procedures for measuring the resistance of plastics to some 50 liquids.

13.4 ENVIRONMENTAL STRESS CRACKING

Hulse (see Section 3.7 of Reference 11) suggests that the term 'stress cracking' (often termed 'environmental stress cracking') be limited 'to the delayed brittle cracking that can occur when polyethylene test pieces are stressed usually multiaxially, in an "active environment". The effect of the "active environment" is either to induce cracking where none would occur in its absence, or to cause cracking to appear at lower stresses or in shorter times than would be experienced in an "inactive environment".' (Nowadays it is acknowledged that polypropylene at least will also show this behaviour.) Hulse considers that all ductile or non-brittle cracking, all cracking or rupture resulting instantaneously from applied stress and all fatigue or normal long term failures should be excluded from the 'stress rupture' heading.

The references selected below have been generally chosen to reconcile with this distinction.

The subject has received a considerable amount of attention in the literature and Howard, in particular, has produced useful reviews[12–14]. In the first of these articles he mentions the variables which have been found to affect the results obtained from polyethylene and describes the test methods in use at that time (1959); the last mentioned reference in particular applies fracture theory to the phenomenon. Another useful review is provided by Hopkins and Baker[15] whilst the mechanism and theory have been further considered by Gaube[16], O'Connor and Turner[17, 18], Isaksen, Newman and Clark[19] and Tung[20]. Suffice to state here that quite apart from temperature, time, environment, stress level (inter alia) all having their separate effects, results are dependent on density (i.e. crystallinity), melt flow index, mode of fabrication, etc. (see also References 21 and 22).

There are two official test methods of which one is the so-called 'bent strip' or Bell Telephone Laboratories method, as appears in A.S.T.M. D.1693–66, 'Environmental Stress-Cracking of Type 1 Ethylene Plastics'. ('Type 1' ethylene plastics, according to A.S.T.M. D.1248, have densities in the range 0·910–0·925 g/cm^3.) The history, development and improvement of this method have been described by Kaufmann[23, 24] and Rudin and Birks[25] for example; Fulmer[26] has correlated it with stress relaxation tests in an environment of 'Igepal' solution (see below).

The A.S.T.M. method uses a conditioned specimen blank cut from sheet 0·125±0·005 in thick (Figure 13.1) into which is cut the central imperfection shown by the jig of Figure 13.2, the use of which is mandatory; it may readily be imagined that test results are heavily dependent on the mode of preparation of the specimens and particularly the care with which the central cut is made.

The conditioning is carried out prior to 'nicking' and consists of 1 h in water or steam at 100°C followed by between 5 and 24 h at 23±1°C before starting the test. Ten such specimens are bent through 135° and placed in the specimen holder (Figure 13.3) by means of (a) a specified bending clamp assembly and (b) a transfer tool assembly (to transfer the specimen from the bending clamp to the holder).

Not more than 10 min after the first specimen, and not less than 5 min after the last specimen has been bent into the holder, the whole assembly is inserted into a hard glass tube 200 mm long and of o.d. 32 mm. Immediately afterwards, the tube is filled, to 0·5 in above the top specimen, with fresh

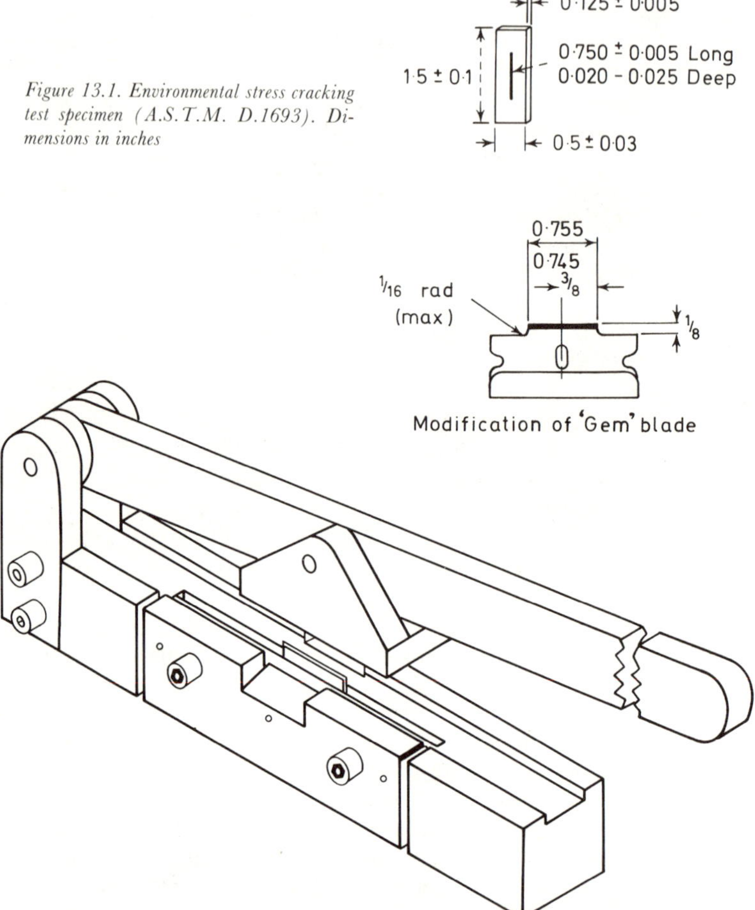

Figure 13.1. Environmental stress cracking test specimen (A.S.T.M. D.1693). Dimensions in inches

Figure 13.2. Environmental stress cracking test jig for nicking specimen (A.S.T.M. D.1693). Dimensions in inches

reagent solution at 23 ± 1°C, stoppered with a foil wrapped cork and placed in a constant temperature bath at 50 ± 0·5°C. (The reagent is described as 'Igepal CO–630', an alkyl-aryl polyethylene glycol.) The nick in the specimen must not touch the tube wall. The specimens are inspected regularly for 'failure', e.g. at 48 h intervals, and it is suggested that quality specifications be drawn up such that, for a given period of test, acceptance can be made on the basis of not more than five specimens failing.

It will be seen that the test as standardised is a 'go-not-go' method; Rudin and Birks[25] describe refinements which they claim enable the test to be used reliably as a quantitative test, by assessing the time to produce 50% of failures in the specimens.

Various other methods have been suggested, and early techniques have been described as already stated by Howard[12] and by Elbers and Fischer[27]; in the latter a tensile method is described for determining the critical short term load in a non-ionic surface active agent environment. Lander[28] advocated a similar test using a tensile impact specimen (Chapter 8, Section 8.11) cut or machined from sheet 0·040 in thick; under tensile loading, 20 specimens at a time are examined in the presence of a surface-active agent such as 'Igepal'.

This method now forms A.S.T.M. D.2552–66T, 'Environmental Stress Rupture of Type III Polyethylenes under Constant Tensile Load'. Results

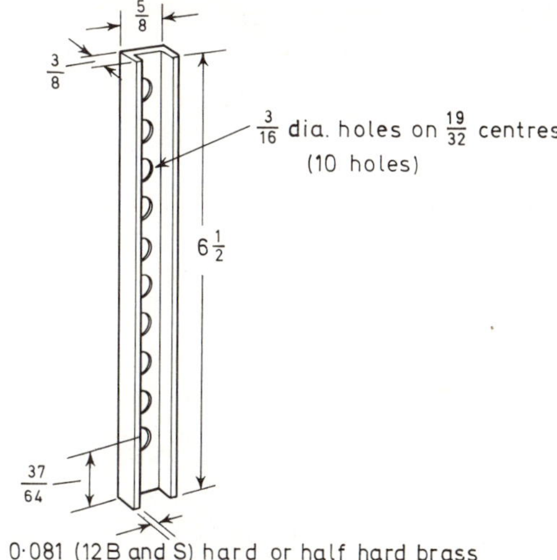

Figure 13.3. Environmental stress cracking test specimen holder (A.S.T.M. D.1693). Dimensions in inches

are expressed as 'F_{50}' values, the probable time for 50% of the specimens to fail in a brittle manner under a constant tensile load of between 8 and 10 $\times 10^7$ dyne/cm^2 (selected so that not more than three failures out of twenty are of a ductile nature) in 'Igepal CO–630' environment at 50°C. Figure 13.4 shows a suitable general arrangement for the test.

It is claimed that this tensile method is more reproducible than the bent strip technique, amongst the objections to which are that the stress induced in the specimen is dependent on the inherent stiffness of the polymer (especially if the test is applied over a range of polyethylenes—Larsen[29] refers to its application to Type III polyethylenes, i.e. of densities 0·941–0·965 g/cm^3), the stress and strain changes during the test (and changes different from one grade of polymer to another) and that no control is exercised over the rate of stressing the test specimens in the holder (Figure 13.3).

Bauer[30] has also advocated the tensile test in a paper which, in addition, described tests on moulded cups and bottles. Nisizawa[31] gives very brief details of an apparatus (also involving tensile stress) where the environment can be liquid or gaseous.

Bartoo[32] describes correlation of stress cracking with product life, the tests being applied to whole articles, such as bottle crates in high density polyethylene, by total immersion in liquid environment. A.S.T.M. D.2561–67T,

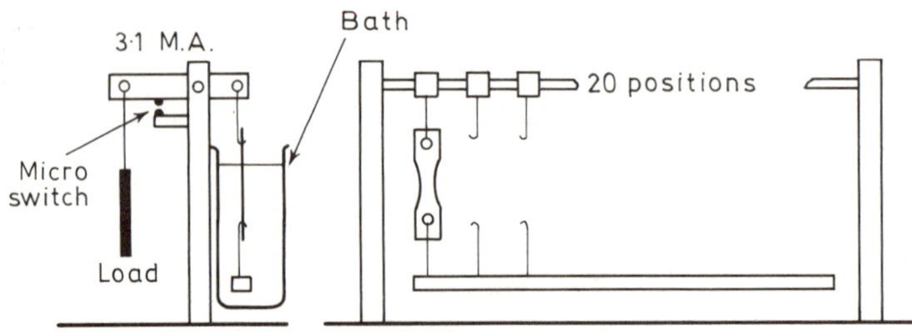

Figure 13.4. Environmental stress rupture test apparatus (A.S.T.M. D.2552 and Larsen)

'Environmental Stress–Crack Resistance of Blow-Moulded Polyethylene Containers', endeavours to standardise procedures for certain types of finished articles.

13.5 CRAZING

The appearance of crazing in this chapter is somewhat dubious but it is felt that there is more justification in terming it a 'permanence property' (of appearance) than a 'mechanical property' (of surface fracture). Crazing is an assembly of very fine cracks generally on the surface of a moulding or etc.; it is most commonly associated with polystyrene (see Section 3.7.6 of Chapter 3 of Reference 11). Crazing may result from the application of external strain, probably as a long term effect (Chapter 14), when cracks are formed without complete fracture, or from the effects of internal stresses resulting from the moulding or forming process, or both. Appropriate liquid environments influence the progress of crazing; thus kerosine has been used[33] to assess the propensity of polystyrene mouldings to develop craze patterns, the kerosine being held to accelerate the normal process of development of crazes with time in a strained article. It has been suggested[34] that the effect of solvents such as kerosine is to 'plasticise' the polymer and reduce the critical strain at which crazing starts such that if that strain is already present in the moulding, crazing occurs. Therefore the value of the kerosine test as a diagnostic method can be seen, but its serious limitations as a means of forecasting liability to craze may be readily imagined and no doubt account for it not being adopted as a standardised method.

For the record, the essential details of the kerosine test[33] were as follows:

The moulding, or a representative section thereof, is immersed in kerosine for 1 min at 20 ± 5°C and removed without wiping. If at the end of the 30 min the moulding is free from crazes it is deemed satisfactory. (In actuality, the test as described was used as a check on the effectiveness of annealing).

Ligroin has also been used in this type of test and another variation is to bend strips in cantilever fashion and 'paint' the surface in tension with ligroin, watching for the appearance of craze patterns. In Ministry of Technology Aerospace Material Specification DTD 5592, polymethyl methacrylate cast sheets are tested similarly using acetone, but actual fracture of the bent strips is deemed as failure.

In B.S. 1493: 1967, 'Polystyrene Moulding Materials', impact strength test specimens are injection moulded and then notched (B.S. 2782: 1970, 'Methods of Testing Plastics', Method 306A using Specimen C—see Chapter 8 Section 8.10.2). In view of the hazards of this operation, if the test results are considered uncharacteristically low, additional specimens are completely immersed in n-heptane at 23 ± 2°C for 10 min and then air dried for 20 min. Satisfactorily moulded and notched specimens show little or no evidence of crazing.

Sauer and Hsiao[35] have reviewed the subject from its root causes to methods of prevention. Sherman and Axilrod[36] examined polymethyl methacrylate at various temperatures by tensile tests conducted so that the onset of crazing could be readily observed in the specimens; incidentally they concluded that the conception of a critical-strain as propounded by Maxwell and Rahm (e.g. see Reference 34) was not applicable to polymethyl methacrylate. A detailed study of the 'solvent' immersion test, examining the effects of temperature, time, solvents and other liquid environments (particularly domestic products) has been undertaken by Ziegler[37] who gives a useful early bibliography. More recent studies include those of Spurr and Niegisch[38], who examined polycarbonate as well as polystyrene and polymethyl methacrylate, Kambour[39] who added styrene-acrylonitrile

Table 13.2 STRESS CRACKING AND CRAZING

Variable	*Environmental stress cracking*	*Stress crazing*
Materials affected*	Polyethylene, polypropylene	Polystyrene, polycarbonate, polymethyl methacrylate
Environment	Environment required	Environment not required but will accelerate crazing if present
Stress state	Occurs only in polyaxial stress field	Occurs in uniaxial or polyaxial stress field
Experimental observation	Brittle fracture	Loss of material transparency by formation of craze lines
Controlling factor	Time duration of stress and stress magnitude	Stress or strain magnitude
Material parameters	Molecular weight, crystallinity, orientation, melt index, thermal history	Orientation, molecular weight, thermal history, surface condition, impurities, cross linking.

*Other materials are also affected but these are the materials commonly investigated.

copolymer to those three and measured the void contents of crazes—and Leghissa and Salvatore[40] who described a method for actually measuring the extent of crazing by a photometric method. Knight[41] has examined the phenomenon from a mathematical point of view, particularly from the aspect of stability of the crazes, i.e. their lacking in tendency to develop into wholesale cracks. Dempsey[42] measured the critical strain associated with crazing. The relationship between orientation and crazing of injection moulded polystyrene has been studied by Murphy[43]; a ball indentation test for assessment of 'environment stress cracking', which study included polystyrene, polymethyl methacrylate and polycarbonate inter alia, has been described by Pohrt[44] and correlated with the method of DIN.53449 (draft). Stalki and Haslett, in a recent paper[45], have described further work on the bent strip method, using carefully defined fibre stresses (see Chapter 8 Section 8.7), with a view to producing an A.S.T.M. Method.

It may be imagined that there is considerable confusion, especially in the literature, as to what constitutes 'crazing' and 'environmental stress cracking'; the comparison produced by Broutman[46] is helpful (see Table 13.2).

One may criticise some of the details of this table; for instance, environmental stress cracking may not lead to brittle fracture—ductile failure has been observed. Again crazing is probably never wholly uniaxial.

Broutman also considers the fracture mechanics of stress cracking and crazing.

Andrew's recent book on fracture of polymers[47] contains a number of references to the subject.

13.6 DAMP HEAT AND CLIMATIC CYCLING TESTS

There are certain ageing tests available which, in the words of the Scope of A.S.T.M. D.756–56 'Resistance of Plastics to Accelerated Service Conditions':

> '. . . cover procedures for determining the weight and shape changes occurring in plastics under various conditions of use, not where exposure to direct sunlight, weathering, corrosive atmospheres or heat alone is involved, but where changes in atmospheric temperature and humidity are encountered. This embraces the interior of buildings, and the interior of transport facilities such as motor vehicles, airplane cargo spaces or wing interiors, holds of ships, and railroad cars'.

Notwithstanding this explanation, Procedure II of A.S.T.M. D.756 involves dry heat ageing only and has therefore been mentioned in Chapter 11 (Section 11.8.1). Others specified are:

Procedure I. 24 h at 60 ± 1°C at a relative humidity of 85–89% followed, within 2 h, by 24 h at 60 ± 1°C in an oven.

Procedure III. 24 h at 70 ± 1°C at a relative humidity of 70–75% followed, within 2 h, by 24 h at 70 ± 1°C in an oven.

Procedure IV. 24 h at 80 ± 1°C over distilled water in an oven followed, within 2 h, by 24 h at 80 ± 1°C in an oven.

Procedure V. 24 h at 80 ± 1°C at a relative humidity of 70–75% followed, within 30 min by 24 h at −40 ± 2°C or −57 ± 2°C, then within 2 h, by 24 h at 80 ± 1°C in an oven and, finally, within 30 min by 24 h again at −40 ± 2°C or −57 ± 2°C.

Procedure VI. 24 h at 38 ± 1°C at a relative humidity of 100% followed, within 2 h, by 24 h at 60°C in an oven.

Procedure VII. As Procedure VI except that the temperature is 49 ± 1°C in both humid and dry heat periods.

Desiccator cooling or warming up is used between each part of the various cycles.

Specimens are weighed, their dimensions measured and visual changes noted after each of the various parts of the various cycles.

The relevance of B.S. 2011 to plastics testing has already been mentioned in Chapter 11 (Section 11.8.1). Part 2C of this British Standard describes a test for 'Damp Heat; (Long Term Exposure)' which involves exposure to a temperature of 40 ± 2°C and a relative humidity of 90–95%; no mist or condensed water must reach the components under test. Part 2D, 'Damp Heat, Cyclic', specifies a 24 h cycle of the following:

1. 25 ± 10°C ('laboratory temperature') to 55 ± 2°C in $1\frac{1}{2}$–$2\frac{1}{2}$ h during which period the relative humidity must be between 80 and 100 per cent and condensation must occur on the components.
2. 55 ± 2°C for 16 h, during which a periodical excursion of 2–3°C variation in temperature, must occur at least four times per hour. The relative humidity during this part of the cycle is 95–100% and condensation will occur.
3. Cool to 'room temperature', the relative humidity remaining at 80–100%, to complete the 24 h cycle. During this period droplets of water must not appear on the components.

These tests have been derived from earlier specifications issued by the U.K. Ministry of Supply, particularly specifications K 114 and RCS 11.

Similarly interesting, from a related industry, is Part F2: 1966, 'Resistance to Humidity under Condensation Conditions' of B.S. 3900, 'Methods of Test for Paints' (Basically formed out of Ministry of Defence Specification DEF.1053). Here the cycle involves the temperature cycling continuously from 42°C to 48°C and back to 42°C in 60 ± 5 min, the heating and cooling periods being approximately equal. Heating is effected through an open water bath so that 100% relative humidity is achieved and copious condensation occurs on the test panels.

B.S. 4618: 1970, 'Recommendations for the Presentation of Plastics Design Data', Recommendation 4.2, 'Presentation of Data on the Change of Linear Dimensions with Moisture Absorption,' to be published soon.

13.7 U.V. RESISTANCE, OZONE RESISTANCE AND WEATHERING RESISTANCE

Weathering effects are a combination, inter alia, of those due to heat (cf. Chapter 11) including infrared radiation, to damp (see Sections 13.2 and 13.6 above), to ultraviolet radiation, to oxidation, to ozone formed by the action of u.v. on atmospheric oxygen and perhaps to radiation from active fallout or emitting sources. The last mentioned is conveniently separated for consideration (see Section 13.9) and those already discussed, of heat and damp, may well be encountered as singular or combined degradative influences. However, u.v., oxidation and ozononlysis are unlikely to be hazards except as constituents of general weathering attack and are therefore dealt with together, though tests for the effects of u.v. or ozone alone are available and are mentioned; the reason for this non-classical approach is that in general weathering the separate degrading influences affect the changes or rate of change brought about by the others; put simply, the action of heat and damp may well be far more severe than that of the sum of the separate actions of heat and damp and likewise, say, the separate and combined effects of u.v. and heat. To this synergistic phenomenon must be added the varying intensity of the degradative influences, which very variation may be more severe in action than a continuous high level of the influence. Again, to take a simple example, repeated drying out caused by occasional saturation with water in a high temperature environment may be far more damaging than continuous immersion at the same temperature.

In seaboard or polluted atmospheres, salt spray, sulphur dioxide, hydrogen sulphide and similar additional influences may have to be considered, but as somewhat special cases these are discussed in Section 13.8. Similarly, microbiological attack is best taken separately (Section 13.12).

Estevez[48] analyses the weathering of plastics into its separate components and lists those major objections to accelerated tests: (1) the degree of acceleration vis-a-vis 'natural' ageing is unknown, (2) the order of merit of materials produced by an accelerated test may be quite different from that obtained in use and (3) phenomena may occur in accelerated tests which never occur naturally—and vice versa. [(2) could result at least in part from (3).] With regard to point (1), the very variability of the weather at any one geographical location, quite apart from the vast differences in average climate between various parts of the world, obviously militates against the chances of any direct correlation between accelerated weathering test results and the results of natural weathering.

The problem of predicting weathering resistance of plastics, the methods therefor, and many test results, are to be found in a symposium edited by Kamal[49].

13.7.1 Tests for Ozone Resistance

The 'ozonolysis' of unsaturated carbon–carbon linkages is a well known phenomena and analytical tool; it is to be expected that polymers containing such unsaturated bonds will be prone to ozone attack and indeed most elastomers are subject to such degradation to a greater or lesser degree. Saturated polymers, however, are quite resistant to ozone and for this reason there are no standard tests for plastics as such, which specifically examine this property, that is separately from its contribution to the general degradative influence of weathering. There is, however, A.S.T.M. D.574–62, 'Ozone-Resisting Insulation for Wire and Cable'.

The subject has been generally reviewed by Weiss[50] and, for elastomers specifically, by Scott[51] who points out that ozone concentration varies from a few parts per 100 million in normal outdoor air to about 90 p.p.m. in Los Angeles 'smog'; Jones[52] reports even higher figures in a paper on the distribution of ozone at the earth's surface. It may be much higher still in the vicinity of high voltage electric discharges. It is necessary to carry out the test in the absence of light to avoid confusing the effects with those due to the production of an oxidised surface layer which may protect the elastomer from ozone attack. Strain is an essential component of the conditions required to produce ozone cracking; the nature, especially whether static or dynamic, and intensity of the strain, the concentration of ozone and the temperature all having pronounced effects on the rate of crack formation, contribute to making it incredibly difficult to standardise a test giving reproducible results.

For the reader interested in elastomer testing there is, for example, B.S. 903: Part A.23: 1963, 'Methods of Testing Vulcanised Rubber. Determination of Resistance to Ozone Cracking under Static Conditions', in which strip specimens are extended 10, 20, 30, 50, 75, 100, 125 or 150% (at least four values thereof should be used) at a carefully controlled specified temperature in an atmosphere containing ozone, produced from air of relative humidity between 50 and 80% at 20°C, of concentration:

25 ± 5 parts per 100 million at 30 ± 1°C,
50 ± 5 parts per 100 million at 50 ± 1°C, or
$15\,000 \pm 1000$ parts per 100 million at 30 ± 1°C.

At the end of the test period the specimens are examined at between $\times 5$ and $\times 10$ magnification for the presence of cracks—ozone cracking is characterised by running in a direction orthogonal to that of the stress. By using an annular specimen, Amsden[53] claims to be able to examine for ozone cracking resistance at all strains between 0 and 50% simultaneously. Eagles[54] provides a very useful description of ozone resistance apparatus, dealing in detail with the production of the ozone and its control, and makes reference to commercial apparatus described elsewhere[55].

13.7.2 Natural Ageing and Weathering

Caryl[56] has covered the subject generally, if briefly, in a chapter of a recently published text book.

STANDARD METHODS

If one can afford the time to wait, evaluation of resistance to weathering, and of associated effects such as fading, is most reliably undertaken using the natural elements. Even then, of course, the results obtained should be carefully related to the geographical location, time of the year and climatic conditions prevailing at the time. As a compromise, the conditions may be 'monitored' in some way by measuring the effect on some standard and the results used on a comparative basis. This 'standard' might be a similar material or product of proven performance (over a prolonged period) or it might be something which changes by known degrees according to the amount of 'degrading influence' which has fallen on it. Such is the principle behind Method 507A, 'Colour Fastness to Daylight', of B.S. 2782: 1970, 'Methods of Testing Plastics'.

Specimens, conveniently 130 mm (5 in) long by 50 mm (2 in) wide, are mounted in a carefully specified 'exposure case' which faces south, with no obstruction in an easterly, southerly or westerly direction subtending a vertical angle greater than 20° nor any northerly obstruction subtending an angle greater than 70°. The 'exposure case' consists essentially of an open bottomed shallow square box, mounted at 45° to the horizontal and supported on legs so that its lower front edge is 760 mm (30 in) from the ground. A framed wire screen fits snugly into the 'case' box and on the upper surface of the wire mesh rests a rack which also fits snugly in the box. The rack comprises a frame with spaced horizontal battens across it, there being a 51 mm (2 in) gap between each batten; on the battens 51 mm (2 in) wide wooden flaps are hinged so that the lower 51 mm (2 in) of the depth of each batten may be covered. Over this rack rests a lid, kept at a distance of 25 mm (1 in) from the upper surface of the rack by a distance piece mounted in the centre of the rack. The lid consists of a framed sheet of 3 mm (24 oz) glass and again fits snugly into the box.

Each specimen is mounted so that its upper 51 mm (2 in) is under the flap of one batten, the next 51 mm spans the gap between adjacent battens and the remaining 25 mm rests on the next, lower batten. At least 6·4 mm ($\frac{1}{4}$ in) must separate each specimen laterally from its neighbour or from the edge of the rack frame. Dyed wool standards (see below) are exposed on the same rack in an essentially similar fashion (under the glass frame), until Standard No. 5 of B.S. 1006 changes colour such that the exposed portion of that standard has a contrast with the unexposed portion equal to Grade 4 on Geometric Grey Scale of B.S. 2662, 'Grey Scale for Assessing Change in Colour'. The standards are examined sufficiently often to ensure that the specified degree of colour change is observed. After this exposure, the specimens are cleaned with soap and cold water, dried and examined in a good north light against a white background in comparison with the dyed wool standards.

The colour fastness to daylight of the material is the number of the dyed wool standard that has changed colour to the same degree as the test specimen.

B.S. 1006: 1961, 'The Determination of Fastness to Daylight of Coloured Textiles', describes the dyed wool standards (which are available from the British Standards Institution). They range from a Fastness Rating of 1 (very low light fastness) to 8 (very high light fastness), each standard being approximately twice as light fast as the one below it. No reliable correlation can be given between B.S. 1006 standards and exposure to daylight in this country and the method must be used on this essentially comparative basis.

The compass directions apply only in the Northern Hemisphere, of course.

A similar procedure is followed by DIN.53388, Bl 1, 'Testing of Plastics: Determination of Resistance to Colour Change upon Exposure to Daylight', and by ISO/R.105 (for Textiles).

Full exposure to the natural elements is allowed in A.S.T.M. D.1435–65, 'Outdoor Weathering of Plastics'. The general design of the weathering racks and their siting are covered and the method of storage of 'control' samples is laid down (23 ± 1°C and 50 ± 2% relative humidity) so that final 'control' values for the material under test may be determined as well as initial values. Neuman[57] has stated that the relevant A.S.T.M. Sub-committee has now concluded that only the 45° exposure variant, of those suggested, need be used.

There are many articles in the literature describing the results of weathering trials, but that by Rugger[58], although qualitative, is a particularly helpful summary, covering a wide range of thermosets and thermoplastics in up to 15 years of exposure.

MEASURING THE EFFECTS OF WEATHERING

In many instances, change of physical appearance may be a sufficient criterion upon which to assess the effect of weathering (or other ageing); obviously this is particularly true if the intended application is basically decorative. When it comes to standardising techniques, however, degrees of colour change, darkening or fading, and changes in opacity need less subjective techniques. Alternatively if it is a mechanical property (or several), or electrical performance which is of importance, changes in these as a result of weathering should be determined, following for example the guide lines laid down by A.S.T.M. D.1435 (see above).

A number of techniques described in the literature are mentioned in Section 13.7.3 (below) because they refer to data obtained at least in part from accelerated tests. The work of Gouza and Bartoe[59], however, specifically relates to weathering trials in the U.S. on plastics, carried out in Bristol, Pennsylvania. The exposure was actually carried out on specimens under a cantilever bending stress, deflection being measured midway between the fulcrum and the point of loading. Different outer fibre stresses (see Chapter 8) were used on acrylic plastics and two end points noted, at the onset of crazing and at rupture; creep data are also reported. Further use of the cantilever, to assess the effects of weathering (and heat ageing), has been described by Kelleher et alia[60].

Recently[61], the techniques of attenuated total reflectance infrared (ATR) spectroscopy has been suggested; the chemical degradation of the surface molecules can be followed and the changes induced by weathering thus 'monitored' at a very early stage, without waiting for the 'massive' effects of usual change, mechanical property degradation, absorption spectroscopy pattern alteration or etc. to become manifest. In the article referred to, the technique is termed 'internal reflectance spectroscopy'.

In studying the weathering of glass fibre reinforced plastics, Crowder and Majumdar[62] have used electron microscope scanning.

ACCELERATING NATURAL WEATHERING

One way of accelerating natural weathering is to employ exposure sites which 'get a lot of weather' as it were. This apparently naive concept has a basis in fact, however, and Ellinger[63] describes the use of the Florida climate for accelerating the weathering of paints in particular, without bringing into play any elements or factors of an artificial test which are not encountered naturally (see above). South Florida has a consistent climate of a combination of warm temperature, high daily relative humidity, heavy annual rainfall, heavy dews and high values of solar radiation. There are undoubtedly other locations* enjoying similar advantages, but apart from the difficulty of accessibility to many would-be investigators, the degree of acceleration is probably not that great and necessarily indeterminate vis-a-vis any particular user condition.

The idea described by Garner and Papillo[64] is very intriguing; a device, termed 'EMMA', is a combination of ten aluminium mirrors so arranged as to intensify natural sunlight ten-fold, whilst the specimens are prevented from becoming unnaturally hot by a stream of cooled air. Results obtained indicated acceleration of degradation ranging from three-fold in unsaturated polyesters to nine-fold in rigid PVC. The intensification of energy, and

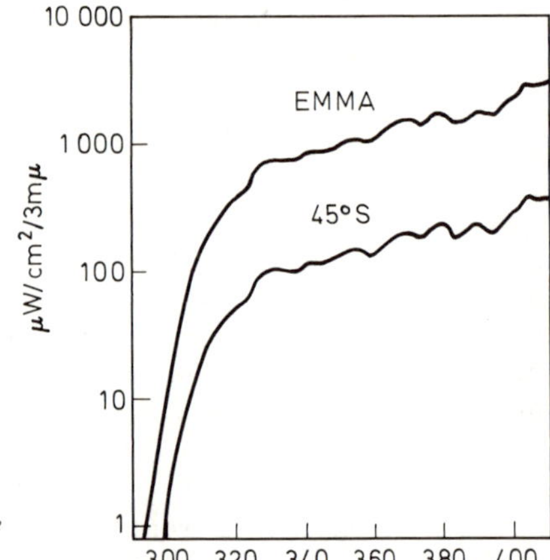

Figure 13.5. U.V. energy spectrum: EMMA and 45°S (From Garner and Papillo[64] by courtesy of the American Chemical Society)

correlation with 'typical sunlight energy spectrum' are shown in Figure 13.5.

Further description, including the addition of water spray (EMMAQUA) and results from such accelerated tests have been provided by Caryl[65].

13.7.3 Accelerated Weathering and Ageing

THE PROBLEM

With the possible exception of the two approaches described immediately above, natural weathering and ageing takes too long to achieve results in an industry as fast developing as plastics. The desire to capitalise on research and development as quickly as possible, and before one's competitors step in, eliminates in most cases any idea of waiting for natural weathering to take its course, for unless extrapolation of the effects of time is hazarded, the

*The now extinct U.K. Ministry of Supply at one time had a station in West Africa, and issued a number of reports describing the results of prolonged exposure on plastics.

results will only predict service life for just as long as the weathering trials were allowed to run. The temptation to speed things up is all too great for mortals to resist.

The difficulties encountered in doing this have already been briefly mentioned in the opening paragraphs of this Section and King[66] has considered the problem in some detail; it has been also discussed by Langshaw[67] who suggested that surface changes should be measured in the exposed specimens and correlated with recorded quantities of light intensity, temperature, etc., over short exposure periods so that a general equation could be built up, with known constants of correlation, and used anywhere the climatic data could be recorded. The idea sounds acceptable in theory but is probably very complicated in practice, bearing in mind that full account must be taken of possible interactions between climatic factors and the exercise would have to be undertaken for every different plastics composition; however, Goldfein[68] has attempted the prediction for four plastics materials.

To assess what has to be simulated—and accelerated—it is desirable to know how to measure the climatic conditions prevailing in different areas of the world. A useful introduction to the measurement of these factors is provided by Haynes[69].

Undoubtedly the greatest source of error in results from accelerated weathering and ageing derives from the radiation sources used. There is not only the factor of intensification which may yield sufficient energy to initiate phenomena which would never occur at the lower levels of intensity experienced naturally; equally important is how close the wavelength emission pattern simulates that of sunlight reaching the earth's surface—anywhere! Thus, although the sun radiates down to 2000 Å in the u.v., in passing through the earth's atmosphere only about two thirds of the sun's energy reaches the surface and all wavelengths below about 2900 Å are filtered out, through absorption, reflection and scattering. This minimum wavelength varies according to the season, being usually above 3000 Å in winter.

Basically four types of lamp source have been used, and in many countries all are still being used, for accelerated ageing and weathering tests on plastics. Low pressure mercury vapour lamps give an emission comprising a series of sharply defined bands including a very strong emission at 2537 Å; it is not difficult to imagine that such a radiation might easily cause a degradation or chemical change that would never occur in sunlight. High pressure mercury arc lamps give a better spectral distribution and are reasonably suitable if the lower wavelengths are filtered out. Carbon arc lamps are widely used and specified (see below), but their spectrum bears no resemblance to that of sunlight. Xenon arc lamps have become popular only recently and are now being considered for national standards; with an appropriate filter they do simulate sunlight reasonably closely, though the degree of acceleration compared with daylight may not be too great. Just how unsuitable carbon arc lamps are, and how much better filtered xenon lamps should be, is demonstrated by Hirt et alia[70], Bikales[71], Grum[72] and by Brighton[73]; Figure 13.6 is taken from this last reference.

As if to prove the point by exception, though, Howard and Gilroy[74] claim to have established a correlation between natural weathering of low density polyethylene and accelerated weathering incorporating exposure to the radiation of a carbon arc!

UNITED KINGDOM STANDARD TESTS

B.S. 2782: 1970, 'Methods of Testing Plastics', only contains one artificial (accelerated?) test of the type under discussion: Method 507B. It follows the same general lines as Method 507A (Section 13.7.2 above) and leaves the

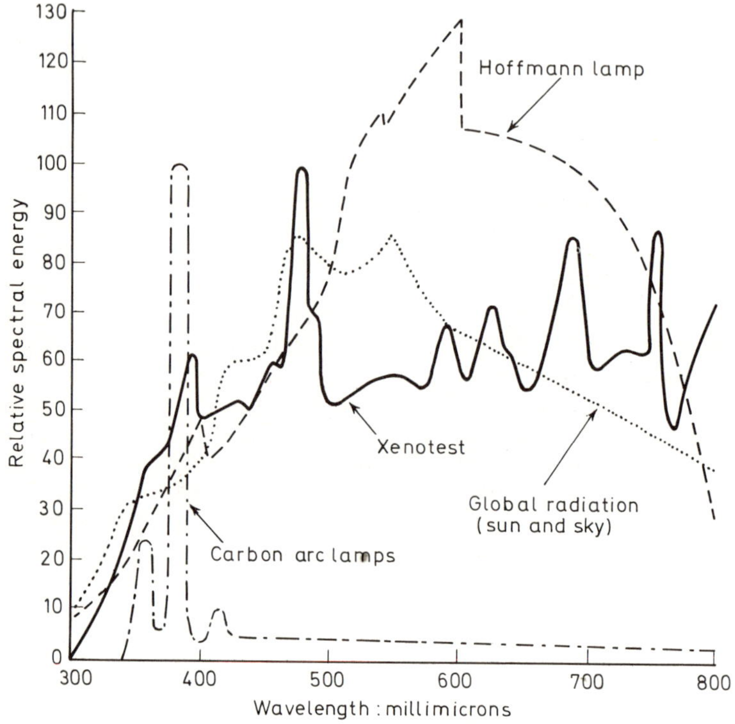

Figure 13.6. Spectral energy distribution in various types of illumination (Brighton) (Reproduced from 'Weathering and Degradation of Plastics', edited by Dr. S. H. Pinner and published by Columbine Press, Buxton)

artificial light source to be specified in the relevant material specification. It states that xenon arc lamps are to be preferred, admits that carbon arc lamps are used and implies that most mercury vapour lamps are not acceptable. It is interesting to note that B.S. 2487: 1969, 'Denture Base Polymer', in its test for colour stability, uses a 400 W S–1 bulb, 'a combination tungsten filament mercury-arc enclosed in a glass that has a low transmission below 280 nm'.

No accelerated weathering tests as such are standardised for plastics in the U.K. In Part F3: 1966, 'Resistance to Artificial Weathering (Enclosed Carbon Arc)', of B.S. 3900, 'Methods of Test for Paints', a much used method is described, derived essentially from Method No. 26, 'Resistance to

Accelerated Weathering', of Ministry of Defence Specification DEF.1053, but using 1600 W lamp instead of a 900 W one and including dry periods in the cycle.

The specimens (painted panels) are mounted as indicated in Figure 13.7, on the 10°–15° inclined frame, comprising two galleries, on the inner wall of a 4 ft diameter drum which rotates once every 20 min. The enclosed carbon arc lamp, of 1600 ± 50 W power consumption, is carefully specified. The atomisers produce a spray, by discharging 1500 ml/h each of water of specified purity at an air pressure of 7 lbf/in^2 (0·5 kgf/cm^2), such that the whole surface of each specimen is evenly wetted with fine spray approximately a quarter of a revolution before it reaches closest proximity to the lamp. Ambient temperature is 20 ± 2°C.

The 24 h cycle is as follows:

1. 4 h with atomisers on and fan off.
2. 2 h with atomisers off and fan on
3. 10 h with atomisers on and fan off.
4. 2 h with atomisers off and fan on.
5. 5 h with atomisers on and fan off.
6. 1 h with atomisers off and fan off and apparatus stopped.

An indication of correlation between results of this test on oil modified alkyd and oleoresinous paints with natural weathering in the United Kingdom is given and the comment is made that modification of the cycle may be necessary for correlation with specific environments, e.g. inclusion of moist sulphur dioxide for industrial pollution and salt spray for marine atmospheres. Indeed Chaplin[75] has recommended, for evaluation of exterior paint systems to be used in the London area, a cycle of two weeks comprising accelerated weathering (as above), oven at 68°C, sulphur dioxide cabinet (see below, Section 13.8), refrigeration at −7°C, salt spray (see below, Section 13.8) and humidification (see above, Section 13.6).

A.S.T.M. TESTS

A.S.T.M. D.795–65T, 'Exposure of Plastics to S–1 Mercury Arc Lamp', is a method involving the use of a mercury arc lamp under which specimens are rotated on a turntable at 30–40 rev/min, the temperature at the mid-point of the latter being 55–60°C. In an alternative procedure, periods of this exposure are interrupted with periods in a 'fog chamber' produced by atomising distilled water.

A.S.T.M. D.1501–65T, 'Exposure of Plastics to Fluorescent Sunlamp', covers three procedures all involving five 61 cm (2 ft) 20 W fluorescent tubes, again with specimens mounted on a turntable as above. Procedure 'A' involves exposure to the lamps only with the temperature at the centre of the turntable 21–25°C. 'B' is similar but with the temperature raised to 55–60°C by blown hot air and 'C' interrupts the heat and light exposure with fog. (Somewhat related is the commercial 'Climatest' marketed in the U.K. by Mectron (Frigistor) Ltd.[76] which incorporates 16 pairs of fluorescent lamps and a water spray; this, however, is not a standard technique—at least as yet.)

The monitoring of carbon arc lamp exposure tests, as used in A.S.T.M. E.188–69T, 'Operating Enclosed Carbon-Arc Type Apparatus for Artificial

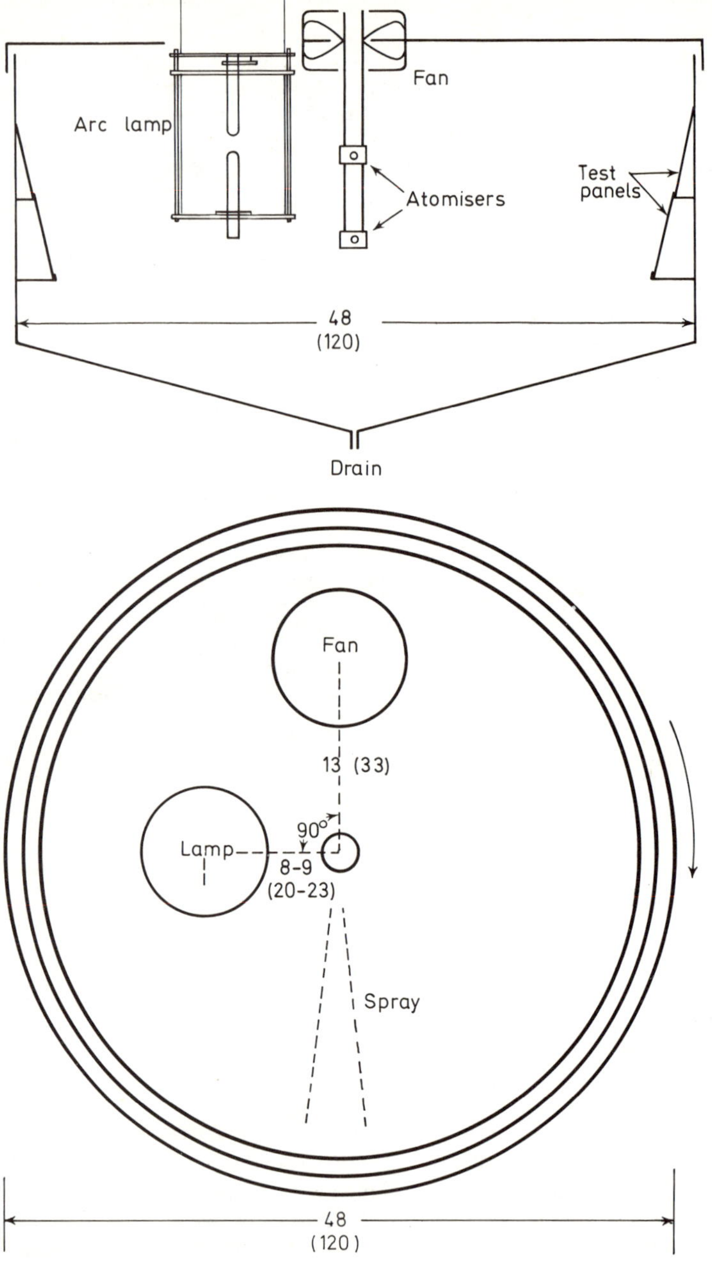

Figure 13.7. Artificial weathering apparatus for paints (B.S. 3900). Dimensions in inches with centimetre equivalents in parentheses

Light Exposure of Non-Metallic Materials', is described in A.S.T.M. D.1920–66T, 'Determining Light Dosage in Carbon-Arc Light Ageing Apparatus'. Standard specimens of yellow polymethyl methacrylate are injection moulded, containing no u.v. absorber but sufficient 'Solvent Yellow 33C.1.47 000' to produce a $17 \pm 1\%$ light transmission at 420 mμ when using a specimen $3{\cdot}15 \pm 0{\cdot}04$ ($0{\cdot}124 \pm 0{\cdot}0015$ in) thick. A double plaque of this thickness is produced and broken; one half is retained in a dark enclosure for reference purposes to check the original light transmission each time the exposed half specimen is measured. Light transmission at 420 mμ is plotted for the exposed specimens against exposure time on log–log paper and the slope of the plot characterises the performance of the carbon-arc apparatus.

For artificial weathering, there is the procedure laid down in A.S.T.M. D.1499–64, 'Operating Light- and Water-Exposure Apparatus (Carbon-Arc Type) for Exposure of Plastics', which in fact just specifies some variables of the equipment covered by A.S.T.M. E.42–65, 'Operating Light- and Water-Exposure Apparatus (Carbon-Arc Type) for Exposure of Non-Metallic Materials'. Herein are described no less than twelve different types of equipment, all utilising one or two carbon arc lamps and *generally* on the lines described above for Part F3 of B.S. 3900.

In the like manner, A.S.T.M. D.2565–66T, 'Operating Xenon Arc-Type (Water-Cooled) Light- and Water-Exposure Apparatus for Exposure of Plastics', covers procedures applicable when this type of equipment, as described in A.S.T.M. E.239–69T, is used to evaluate plastics materials. A.S.T.M. E.239–69T, entitled 'Operating Light- and Water-Exposure Apparatus (Water-Cooled Xenon Arc Type) for Exposure of Non-Metallic Materials', allows the use of four different types of apparatus, two each employing 2500 W or 6000 W xenon arc lamps, with water spray, etc., *generally* after the style of lay-out shown in Figure 13.7.

Mention should also be made of A.S.T.M. E.240–69T, 'Operating Water-Cooled Xenon-Arc Type Apparatus for Light Exposure of Non-Metallic Materials', where the effects of radiation alone, i.e. without simulated rainfall and other humidifying effects, are studied. (There is also a draft DIN.53389 in existence entitled 'Testing of Plastics; Determination of Resistance of Plastics to Colour Change upon Exposure to Light of a Xenon Lamp'.)

MISCELLANEOUS ACCELERATED WEATHERING APPARATUS AND ITS EVALUATION

Useful reviews of accelerated weathering apparatus have appeared from time to time, for example by Merrill and Myers in 1941[77], Sawyer in 1952[78] and Kinmonth in 1964[79]. The last mentioned includes the 'Xenotester' based on the xenon arc lamp (see above). Further description of this equipment will be found in the contribution by Brighton[73] and in Reference 80.

Results of tests on a wide range of thermoplastics and thermosets carried out in a 'Weather-Ometer'* (A.S.T.M. D.1499) are given by Chottiner and

*Manufactured by Atlas Electric Devices Co. of Chicago (U.S.A.). For the record, the same company's exposure lamp is known as a 'Fade-Ometer'. The term 'Fugitometer' will be found used in the U.K. for certain proprietary exposure lamps.

Bowden[81] who draw attention particularly to the different manner in which the various plastics deteriorate and report optimistically on the correlation between the data and use in the field. Brighton[73] has examined the relationship between accelerated and natural weathering in some detail and concludes that, whilst accelerated ageing tests can be used with confidence in predicting short term outdoor behaviour, the situation is not satisfactorily resolved for long-term service.

A summary of U.S. natural and accelerated test methods can be found in a report on polyolefines by Titus[82].

MEASURING THE EFFECTS OF ACCELERATED WEATHERING

Reference has already been made above to the desirability of obtaining some quantitative measure of degradation; an example was given of following the effect of natural weathering by cantilever bending tests. Hunt and Bauman[83] examined polyester resins, subjected to artificial weathering, by deterioration of clarity and increase in yellowing, the latter being taken as the difference between absorption at 4500 Å and at 6000 Å. A different type of optical method, again for polyesters, has been described by Gray and Wright[84] who measured the colour developed on treating polyesters with a solution of N,N-dimethyl-p-phenylenediamine in benzene-methanol solution. Weathered polyesters produced a pronounced colour whereas unexposed polyesters did not. Cipriani, Giesecke and Kinmonth[85], in connection with a description of a novel ageing device, used flexing tests for polypropylene, yellowing for PVC and nitrocellulose (lacquers) and brittle point and carbonyl content for polyethylene. Neuman[86] found a falling weight impact test (see Chapter 8 above) useful in following the degradation of rigid plastics by 'Weather-Ometer' and 'Fade-Ometer'.

The whole subject of weathering of plastics—the contributing factors, natural weathering, accelerated tests and techniques to assess the influence of weathering—has been well summarised by Scott[87].

13.8 SALT SPRAY RESISTANCE, STAINING ETC.

13.8.1 Salt Spray Resistance

In marine atmospheres, salt spray may be present in the atmosphere to a significant degree. It is not reckoned a serious hazard to plastics generally and there is no B.S. or A.S.T.M. test, for instance, designed to assess degradation of plastics by salt spray. When it must be checked, reference may be made to Test 2K, 'Salt Mist', of B.S. 2011: 1963, 'Basic Methods for the Climatic and Durability Testing of Components for Telecommunication and Allied Electronic Equipment', or Method No. 24, 'Resistance to Continuous Spray', or Method No. 36, 'Resistance to Intermittent Salt Spray', of Ministry of Defence Specification DEF/1053, 'Standard Methods of Testing Paint, Varnish, Lacquer and Related Products'; the former of the two DEF/1053 methods corresponds to B.S. 3900: Part F4: 1968. The salt solution of the B.S. 2011 method comprises:

Sodium chloride	27 g
Magnesium chloride (anhydrous)	6 g
Calcium chloride (anhydrous)	1 g
Potassium chloride	1 g
Distilled water to 1 litre	

and that of the two DEF/1053 methods is similar, but a little more complex. The B.S. 2011 salt mist chamber, constructed in corrosion resistant metal, is shown in Figure 13.8 (dimensions presumably in inches); the air pressure

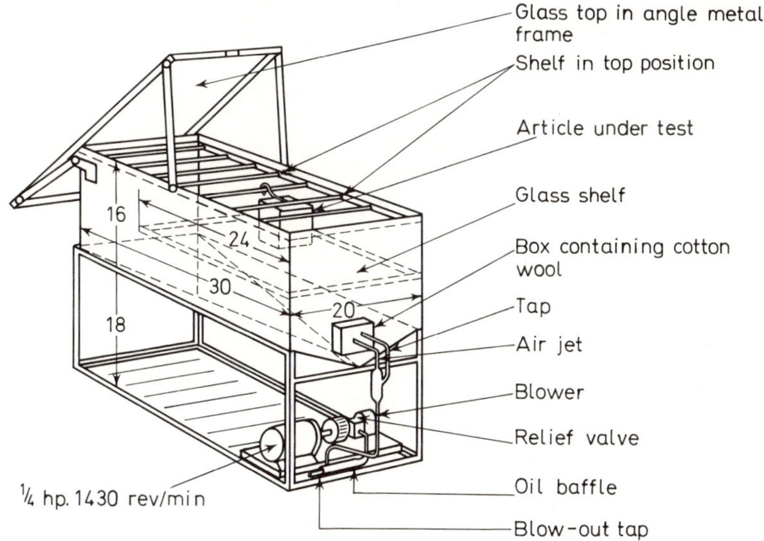

Figure 13.8. Salt mist chamber (B.S. 2011)

for atomising the salt solution is 4–5 lb/in^2 and the diameter of the nozzle $0{\cdot}048 \pm 0{\cdot}001$ in.

13.8.2 Sulphur Dioxide Resistance

In industrial atmospheres particularly there is likely to be a significant concentration of sulphur dioxide, derived from the burning of coal or fuel oil. Again, as far as is known, no standard test exists for plastics, but the C.R.L. test forming Part 3 of B.S. 1391: 1952, 'Performance Tests for Protective Schemes Used in the Protection of Light-Gauge Steel and Wrought Iron against Corrosion', may be useful. Specimens are suspended in the vapour above distilled water at 45°C, to which has been added $\frac{1}{100}$ th of its volume of a solution of 0·3 g sulphur dioxide in 100 ml distilled water. Further similar amounts of the sulphur dioxide solution are added twice daily.

13.8.3 Stain Resistance

Very general instructions are given in A.S.T.M. D.2299–68, 'Determining Relative Stain Resistance of Plastics'; basically the method is no more than physical contact of appropriate staining agents with the test surface—immersion of the specimen in low viscosity liquids, application of a thin film of viscous liquid or paste to the specimen or, for solid staining materials, applying a uniform coating to the test area.

More specifically, sulphides are a frequent hazard—from rubber, fumes, food, kraft paper—and many plastics are prone to staining thereby because of the frequent incorporation of compounds of lead, antimony, copper and cadmium for instance, as stabilisers, pigments and fillers. A.S.T.M. D.1712–65 requires that the specimen be half immersed in a saturated solution of hydrogen sulphide at approximately 23°C for 15 min. B.S. 1763: 1967, 'Thin PVC Sheeting (Flexible, Unsupported)', and B.S. 2739: 1967, 'Thick PVC Sheeting (Flexible, Unsupported)', both contain a sulphur stain resistance test based on immersion for 30 min in sodium sulphide solution acidified with hydrochloric acid.

13.9 IRRADIATION EFFECTS

The term 'irradiation' is a general one and any reference to its effect on plastics should include changes induced by naturally occurring radiation as well as by the man-made products of nuclear fission, for example. The naturally occurring hazards are summarised in Reference 88 and the effect of u.v. radiation has been discussed in Section 13.7.3.

The subject of testing polymers for radiation resistance has been comprehensively described by Metz[89] and, being somewhat out of the ordinary run of plastics evaluation, only a brief mention will be made here. With the advent of nuclear power for peaceful purposes, however, there is a need for constructional and insulation materials of known resistance to the products of nuclear fission (and combination).

The radiation concerned includes X-rays, fast electrons, gamma rays, alpha particles, fast protons and fast neutrons[90], all characterised by very high energy, certainly in relationship to the energies required to cause scission of atomic bonds. As far as polymer technology is concerned, high energy radiation may be used 'constructively' in polymerisation, curing and cross-linking generally[90–92]. Irradiation of polyethylene has been used commercially to modify its properties with advantage—increasing stiffness and heat resistance for example—though it is probably true to state that this development has suffered as a result of the appearance of high density polyethylenes, polypropylene and ethylenic copolymers. The effects of nuclear radiation on materials generally have been described by Reinsmith[93]. Polymers may be cross-linked (as polyethylene) or degraded (polymethyl methacrylate for example); Manley[94], Ballantine[95], Black[96] and Fischer et alia[97], amongst others, have discussed the irradiation of plastics materials and Charlesby[98] has devoted a complete book to the subject.

Broad principles for evaluation of plastics are laid down in A.S.T.M. D.1672–66, 'Exposure of Polymeric Materials to High Energy Radiation', but the reader is referred to Metz[89] for detail. Also of relevance is A.S.T.M. D.2568–68, 'Calculation of Absorbed Dose from X or Gamma Radiation', which gives a recommended practice applicable to materials containing hydrogen, carbon, nitrogen, oxygen, fluorine, silicon, sulphur, chlorine and phosphorus (in combinations of some or all).

13.10 ABLATION

'Ablation' is the term applied to the decomposition of reinforced plastics in particular when used as nose cones for rockets and missiles generally. The heat generated on entry of the missile into a planet's atmosphere is considerable and some way must be found to dissipate this heat in order to protect the functional parts of the equipment and ensure their satisfactory operation. One such way is by ablative cooling whereby certain reinforced plastics are employed in a sacrificial manner, their heat absorption and insulation during decomposition being sufficient to protect the units behind the nose cone. The subject is described in References 99–101.

As may be imagined, the evaluation of materials for ablation resistance is a highly specialised subject. As such, any description of the techniques is beyond the scope of this handbook; generally the accepted methods aim to simulate field conditions, for instance the 'Arc Wind Tunnel' in which an electric arc jet exhausts into a chamber at reduced pressure or, a little more simply, where the material under evaluation is placed in the exhaust stream of a rocket motor. The interested reader is referred to References 100, 102 and 103.

13.11 EROSION

The term 'erosion' is usually taken to mean the destructive effect of water droplets impinging on articles at high speeds; it is particularly applied to the hazards suffered by leading edges of airscrews and wings in aircraft. A test method described by Hoff and Langbein[104] consists essentially of mounting the specimens on the end of a rotating arm of radius 120 cm, which is capable of peripheral speeds of up to 470 m/s. The arm rotates under a set of eight sprays. Evaluation is by means of weight loss and depth of erosion.

'Erosion' is also applied to the more conventional wearing of a surface caused by flow of water. Newman, Toner and Achhammer[105] have described the effect of 50 washings in hot detergent solution on the visual appearance and mechanical properties of a wide range of reinforced plastics and moulded thermosetting compositions.

The subject has been examined in detail by Brunton[106].

13.12 MICROBIOLOGICAL ATTACK (ALSO 'BIOLOGICAL ATTACK')

13.12.1 What are Organisms?

The subject has been well summarised by Heap[107]:

'*Bacteria* are unicellular organisms ranging from 10 μ to 0·5 μ in length and lacking a definitely organised nucleus as is found in cells of higher plants and animals. Some forms possess fine, flexible hair-like projections called flagellae, others have a surrounding capsule and some forms have both. Normal reproduction is by simple fission, that is by a mature cell dividing into two identical cells, but under adverse conditions each cell produces a spore which is able to lie dormant for extensive periods of time until favourable conditions return. The spores may be airborne over great distances but the bacteria themselves occur mainly in the soil, in natural bodies of water and internally as agents of disease in plants and animals.

'The term *fungus* covers a large and heterogeneous group of organisms in the plant kingdom having in common a lack of chlorophyll so that they are unable to synthesise a food supply using sunlight. That group of fungi which is of interest as a cause of deterioration of commercial products is generally known as *mould* (or *mildew*). This term refers to small non-parasitic (saprophytic) fungi which consist of thread like filaments, known as hyphae; a collection of hyphae is known as a mycelium. Two types of spore are usually produced, the one for multiplication under favourable conditions and the other, which is very resistant to desiccation, for maintaining the species when its environment changes. These are distributed by the air in the same way as the bacterial spores until they reach a suitable environment such as soil, water or suitable bacterial material.

'Both groups of organism require for their growth a source of carbon and, for all except a few species of bacteria, this must be in the form of an organic material which can be broken down enzymatically to simple substances, for use as food. It is for this reason that bacteria and fungi act as destructive agents, since the breaking down of carbon-containing molecules, in for example cotton, paper and leather, is reflected in deterioration of physical properties and eventual disintegration of the material. Since rubbers and plastics are largely composed of organic materials they are obviously potential sources of nutrition for micro-organisms. Besides a source of carbon, other requirements for growth are the presence of other elements such as hydrogen, oxygen, nitrogen, sulphur, magnesium, phosphorus and traces of certain metals; atmospheric oxygen for fungus and aerobic bacteria (other bacteria, anaerobic, only grow in the absence of atmospheric oxygen); absence of ultra-violet light for fungi; high temperatures, 30°C being very favourable; and high humidities, relative humidities (r.h.) of 95 to 100% being very favourable for fungi whereas bacteria frequently need the presence of liquid water for growth.'

Generally, naturally occurring polymers are consumed by micro-organisms (bacteria and fungus) whereas synthetic polymers tend not to be. However, the picture is much confused by the presence of additives, reinforcements etc., which may themselves be attacked. The resistance, or otherwise, of plastics materials have been described by, inter alia, Heap[107], Leutritz[108], Witt, Chapman and Raskin[109], Stahl and Pessen[110], Hueck[111], Wessel[112], MacLachlan[113], Kulkarni[114], MacLachlan, Heap and Pacitti[115], Heap and Morrell[116], Kühlwein and Demmer[117] and Dolezel[118].

13.12.2 Methods of Test

The variations of test conditions, which can be used to assess the resistance of plastics materials to microbiological deterioration, may well be imagined to be vast—choice of single cultures of bacteria or fungi, or combinations thereof (as occurs usually in nature), the choice of temperature and humidity conditions, use of a natural environment (e.g. earth burial or tropical sites). Evaluation may be by visual examination and/or measurement of properties, the latter selected to highlight most significantly any likely deterioration, for example flexibility of a PVC composition where susceptibility of the plasticiser is suspected.

There are not very many relevant British Standard Tests. B.S. 1203: 1963, 'Synthetic Resin Adhesives (Phenolic and Aminoplastic) for Plywood', B.S. 1204: Part 1: 1964, 'Synthetic Resin Adhesives' (Phenolic and Aminoplastic) for Wood. Part 1. Gap-Filling Adhesives' and B.S. 1204: Part 2: 1965, 'Synthetic Resin Adhesives (Phenolic and Aminoplastic) for Wood. Part 2. Close-Contact Adhesives', all contain essentially the same test for measuring resistance to micro-organisms. This consists of measuring the strengths of bonds, prepared as specified, after the test pieces have been aged for four weeks at 25 ± 2°C in a culture medium and washing in cold water after removal. The culture medium is prepared by moistening the sawdust of the sapwood of any timber, or the sapwood *and* heartwood of perishable species of timber, such as ash, beech, birch, poplar and willow, with water containing $\frac{1}{2}$ oz (14 g) of domestic sugar per quart (1·1 litre) so that the sawdust is saturated but not so wet that water can be squeezed out by the hand (about three times the weight of dry sawdust). No preservatives, insecticides or added substance that inhibit or retard mould growth shall be present.

A 1 in (2·5 cm) layer of this moist sawdust is placed in a non-corrodible flat dish, covered with a sheet of glass and sealed with modelling clay or similar material to prevent loss of moisture. The dish is kept at 25 ± 2°C for a week whilst the prepared bond test pieces are immersed in cold water, which is changed daily, to remove free formaldehyde and mould inhibiting materials from the wood (which is bonded by the test adhesive). At the end of the week the wet test pieces are pressed into the surface of the sawdust, and the dish resealed and the ageing started.

B.S. 1982: 1968, 'Methods of Test for Fungal Resistance of Manufactured Building Materials Made of or Containing Materials of Organic Origin', examines with (1) wood-rotting basidiomycetes (fungi), (2) cellulose-attacking microfungi and (3) moulds or mildew. For the first the following are used:

Merulius lacrymans	Dry Rot fungus
Poria monticola	Pore fungus
Coniophora cerebella	Cellar fungus

Full details of how to grow the cultures are given. Evaluation of the effects, of 12 weeks incubation at not more than 22°C, and preferably not below 20°C, is by visual examination and weight change, against control experiments designed to ensure that the cultures are active and also to measure the effects of moisture alone.

The above test (1) is for, inter alia, plywood and chipboards (which will in all probability contain urea-formaldehyde or phenol-formaldehyde binder). Test 2 is described as being suitable for 'any material derived or partly derived from plant or animal sources' and need not concern us here. Test 3 is 'suitable for materials that are required to present a decorative finish or appearance and which may be disfigured by a superficial growth of moulds. Plasterboard, sheets made of plastics . . . may suitably be tested by this method'.

The moulds used are:
Chaetomimium globosum
Cladosporium cladosporioides
Paecilomyces varioti
Stachybotrys atra
Penicillium funiculosum
as a single suspension in water, for four weeks at a relative humidity of 95–100% and a temperature of 25±2°C. Evaluation of the effects is by visual examination under a ×10 magnification lens.

Part 2J, 'Mould Growth', of B.S. 2011: 1967, 'Methods for the Environmental Testing of Electronic Components and Electronic Equipment', employs a suspension of eight cultures of which three:
Aspergillus terreus
Paecilomyces varioti
Trichoderma viride
particularly are said to attack plastics. Experimental conditions are carefully specified and the period of ageing is 28 or 84 days at 28–30°C and relative humidity greater than 95%.

Similarly A.S.T.M. D.1924–63, 'Determining Resistance of Plastics to Fungi', specifies the test conditions using six test fungi, at a temperature of 28–30°C and relative humidity of not less than 95% for at least 21 days.

Unless otherwise agreed, A.S.T.M. practice is to use only one test organism for assessing 'Resistance of Plastics to Bacteria' (A.S.T.M. D.2676–67T); this is 'Pseudomonas aeruginosa'. Two procedures are described, one in which the specimens are laid on the surface of the nutrient agar containing bacteria and the other where the specimens are actually immersed in the agar.

Burgess and Darby[119] have described two methods for quantitative measurement of microbiological decay. In one, weight loss measurements are used to follow the attack on plasticised compositions; in the other the activity of the organisms is measured by oxygen absorption in a differential manometer against a sterile control flask. Otherwise the basic techniques are as before, using similar fungi.

As far as permeability to bacteria is concerned, Ronsivalli et alia[120] studied a range of plastics films and found them to be impermeable under the test conditions.

13.13 ATTACK BY INSECTS AND RODENTS

The examination of resistance to attack by insects and rodents is more a subject for the entymologist and the zoologist. The interested reader is referred to Hueck[111], Wessel[112], Ott[121], Becker[122, 123], Pacitti[124] and MacNulty[125].

Finally, for a general treatise on plastics degradation (and that of elastomers), the reader is referred to Lightbody et alia[126].

REFERENCES

1. ESHENAUR, R. E., 'Polymers in Adverse Environments', *Plast. & Polymers,* **37,** No. 131, 401 (October 1969)
2. Technical Data on Plastics, Manufacturing Chemists' Association Inc., Washington (U.S.A.) (1957)
3. UIJLENBURG, H., 'Water Absorption of Unplasticised P.V.C.', *Plastics, Lond.,* **25,** No. 275, 359 (September 1960)
4. BRADEN, M., 'The Absorption of Water by Plastics', *Trans. Plast. Inst. Lond.,* **31,** No. 94, 83 (August 1963)
5. BLANK, F., 'The Problem of Water Absorption of Plastics', *Plaste u. Kautschuk,* **9,** No. 8, 391 (1962)
6. BLANK, F., 'Water Absorption of Plastics', *Plaste u. Kautschuk,* **12,** No. 11, 657 (1965)
7. HAUCK, J. E., 'How Water Affects Plastics', *Mat. Des. Engng.,* **64,** No. 5, 93 (November 1966)
8. GORDON, M., *High Polymers, Structure and Physical Properties,* 2nd edn, Iliffe Books Ltd., London (1963)
9. JESSUP, J. N., 'Chemical and Thermal Resistance of Thermosetting Moulding Compounds', *Mod. Plast.,* **44,** No. 7, 174 (March 1967)
10. EVANS, L. S., 'The Chemical Resistance of Rubber and Plastics', *Rubb. & Plast. Age,* **44,** No. 11, 1349 (November 1963)
11. RITCHIE, P. D. (ed.) *Physics of Plastics,* Iliffe Books Ltd., London (1965)
12. HOWARD, J. B., 'A Review of Stress-Cracking in Polyethylene', *S.P.E. J.,* **15,** No. 5, 397 (May 1959)
13. HOWARD, J. B., 'Why do Plastics Stress-Crack?' *S.P.E. Trans.,* **4,** No. 3, 217 (July 1964)
14. HOWARD, J. B., 'Why do Plastics Stress-Crack?', *Polymer Engng. & Sci.,* **5,** No. 3, 125 (July 1965)
15. HOPKINS, I. L. and BAKER, W. O., 'Stress-Cracking of Polyethylene', *Kunststoffe,* **49,** No. 11, 621 (November 1959)
16. GAUBE, E., 'Creep-Rupture Strength and Stress Cracking of Low Pressure Polyethylene', *Kunststoffe,* **49,** No. 9, 446 (September 1959)
17. O'CONNOR, A. and TURNER, S., 'Environmental Stress Cracking of an Injection Moulding Grade of Polythene, Part I', *Brit. Plast.,* **35,** No. 9, 452 (September 1962)
18. O'CONNOR, A. and TURNER, S., 'Environmental Stress Cracking of an Injection Moulding Grade of Polythene, Part 2', *Brit. Plast.,* **35,** No. 10, 526 (October 1962)
19. ISAKSEN, R. A., NEWMAN, S. and CLARK, R. J., 'Mechanism of Environmental Stress Cracking in Linear Polyethylene', *J. appl. Polymer Sci.,* **7,** No. 2, 515 (March–April 1963)
20. TUNG, L. H., 'Stress-Cracking of Polyethylene Examined from the Viewpoint of Critical Strain', *J. Polymer Sci.,* Part A, **3,** 1045 (1965)
21. SUEZAWA, Y., HOJO, H., IKEDA, T. and OKAMURA, Y., 'Effects of Environment and Temperature on the Environmental Stress-Cracking of High-Density Polyethylene', *Mater. Res. Stand.,* **5,** No. 2, 55 (February 1965)
22. HOWARD, J. B. and GILROY, H. M., 'Some Observations on Stress-Cracking', *S.P.E. J.,* **24,** No. 1, 68 (January 1968)
23. KAUFMANN, K. A., 'Polyethylene Environmental Stress Cracking by the Proposed Bent Strip Test Method', *Bull. Amer. Soc. Test. Mat.* **No. 233,** 32 (October 1958)
24. KAUFMANN, K. A., 'Polyethylene Environmental Stress-Cracking by the Bent Strip Method', *Mod. Plast.,* **36,** No. 7, 146 (March 1959)
25. RUDIN, A. and BIRKS, A. M., 'Measurement of Environmental Stress-Cracking of Polyethylene', *Bull. Amer. Soc. Test. Mat.* **No. 245,** 60 (April 1960)
26. FULMER, G. E., 'Time, Temperature and Molecular Weight Effects in Environmental Stress-Cracking', *Trans. Soc. Rheol.,* **9,** Part 2, 121 (1965)
27. ELBERS, F. and FISCHER, F., 'Methods for the Practical Testing of Stress-Cracking Resistance of Various Types of Polyethylene', *Kunststoffe,* **50,** No. 9, 485 (September 1960)
28. LANDER, L. L., 'Environmental Stress Rupture of Polyethylene', *S.P.E. J.,* **16,** No. 12, 1329 (December 1960)
29. LARSEN, H. R., 'Environmental Stress Rupture Testing of Type III Polyethylene', *Mater. Res. Stand.,* **6,** No. 7, 335 (July 1966)
30. BAUER, P., 'Stress Cracking of Polyethylene', *Kunststoffe,* **57,** No. 11, 881 (November 1967)
31. NISIZAWA, M., 'A Modified Environmental Stress-Cracking Apparatus for the Liquid and Gas Phases', *J. appl. Polymer Sci.,* **12,** No. 7, 1785 (July 1968)

32. BARTOO, O. F., 'Stress Cracking and Product Life', *Plastics, Lond.,* **31,** No. 341, 294 (March 1966)
33. BAILEY, J., 'Annealing of Styrene and Related Resins,' *Mod. Plast.,* **24,** No. 2, 127 (October 1946)
34. MAXWELL, B. and RAHM, L. F., 'Factors Affecting the Crazing of Polystyrene', *S.P.E. J.,* **6,** No. 9, 7 (November 1950)
35. SAUER, J. A. and HSIAO, C. C., 'Stress-Crazing of Plastics', *Trans. A.S.M.E.,* **75,** No. 5, 895 (July 1953). Also: *S.P.E. J.,* **9,** No. 9, 13 (November 1953)
36. SHERMAN, M. A. and AXILROD, B. M., 'Stress and Strain at Onset of Crazing of Polymethyl Methacrylate at Various Temperatures', *Bull. Amer. Soc. Test Mat.* **No. 191,** 65 (July 1953)
37. ZIEGLER, E. E., 'The Crazing of Polystyrene', *S.P.E. J.,* **10,** No. 4, 12 (April 1954)
38. SPURR, O. K. JR. and NIEGISCH, W. D., 'Stress Crazing of Some Amorphous Thermoplastics', *J. appl. Polymer Sci.,* **6,** No. 23, 585 (Sept.–Oct. 1962)
39. KAMBOUR, R. P., 'Refractive Indices and Compositions of Crazes in Several Glassy Polymers', *J. Polymer Sci.,* Part A, 2, No. 9, 4159 (September 1964)
40. LEGHISSA, L. and SALVATORE, O., 'Photometric Method for Measuring Crazing in Transparent Plastics Materials', *Polymer Engng & Sci.,* **6,** No. 2, 127 (April 1966)
41. KNIGHT, A. C., 'Stress Crazing of Transparent Plastics, Computed Stresses at a Nonvoid Craze Mark', *J. Polymer Sci.,* Part A, 3, 1845 (1965)
42. DEMPSEY, L. T., 'Measurement of the Critical Strain Associated with the Environmental Stress-Crazing of Polymers', *Polymer Engng & Sci.,* **7,** No. 2, 86 (April 1967)
43. MURPHY, B. M., 'The Effect of Orientation on the Crazing of Injection Moulded Polystyrene Test Pieces', *Chemy Ind.,* No. 10, 289 (March 8 1969)
44. POHRT, J., 'Assessment of Tendency to Environmental Stress Cracking of Thermoplastics with the Aid of a Ball Indention Test', *Kunststoffe,* **59,** No. 5, 299 (May 1969)
45. STOLKI, T. J. and HASLETT, W. H., 'An Improved Variable Strain Bending Form for Determining the Environmental Craze Resistance of Polymers', *Mater. Res. Stand.,* **9,** No. 12, 32 (December 1969)
46. BROUTMAN, L. J., 'Fracture Mechanics of Stress Cracking and Crazing', *S.P.E. J.,* **21,** No. 3, 283 (March 1965)
47. ANDREWS, E. H., *Fracture of Polymers,* Oliver and Boyd Ltd., Edinburgh (1968)
48. ESTEVEZ, J. M. J., 'Some Thoughts on the Weathering of Plastics', *Trans. Plast. Inst. Lond.,* **33,** No. 105, 89 (June 1965)
49. KAMAL, M. R., *Weatherability of Plastic Materials, Applied Polymer Symposia No. 4,* Interscience Publishers, New York (1967)
50. WEISS, E., 'Testing for the Ozone Resistance of Polymers', *Testing of Polymers,* Chapter 9, (Edited by J. V. Schmitz), Volume 2, Interscience Publishers, New York (1966)
51. SCOTT, J. R., *Physical Testing of Rubbers,* 305, Maclaren & Sons Ltd., London (1965)
52. JONES, K. P., 'The Distribution of Ozone on the Earth's Surface', *J. I. R. I.,* **2,** No. 4, 194 (August 1968)
53. AMSDEN, C. S., 'A Novel Method for Testing the Static Ozone Resistance of Rubber', *Trans. I. R. I.,* **42,** No. 3, T.91 (June 1966)
54. EAGLES, A. E., 'The Development of Tests for the Determination of the Resistance to Ozone of Natural and Synthetic Rubbers', *Proc. I. R. I.,* **13,** No. 3, P.94 (June 1966)
55. 'Fully Automatic Ozone Tester', *Rubb. & Plast. Age,* **46,** No. 12, 1383 (December 1965)
56. CARYL, C. R., 'Methods of Outdoor Exposure Testing', *Testing of Polymers,* (Edited by W. E. Brown), Volume 4, Interscience Publishers, New York, 379 (1969)
57. NEUMAN, R. C., 'Does the Angle of Exposure to the Sun Make a Difference?' *Mater. Res. Stand.,* **9,** No. 6, 38 (June 1969)
58. RUGGER, G. R., 'Fifteen Years of Weathering Results', *S.P.E. Trans.,* **4,** No. 3, 236 (July 1964)
59. GOUZA, J. J. and BARTOE, W. F., 'Weathering of Plastics, A Method of Studying the Resistance of Plastics to Outdoor Exposure', *Mod. Plast.,* **33,** No. 9, 157 (March 1956)
60. KELLEHER, P. G., MINER, R. J. and BOYLE, D. J., 'Cantilever Beam Test Measurement of Plastics' Ageing', *S.P.E. J.,* **25,** No. 2, 53 (February 1969)
61. 'New Test Quickly Predicts How Well a Plastic Material Resists Weathering', *Mod. Plast.,* **45,** No. 16, 109 (December 1968) See also: CHAN, M. G. and HAWKINS, W. L., 'Internal Reflection Spectroscopy in the Prediction of Outdoor Weatherability', *Polymer Preprints,* American Chemical Society, Division of Polymer Chemistry, **9,** No. 2, 1638 (September 1968)
62. CROWDER, J. R. and MAJUMDAR, A. J., 'Weathering of G.R.P. Scanning Electron Microscope Studies', *Plastics, Lond.,* **33,** No. 371, 1012 (September 1968)

63. ELLINGER, M. L., 'Accelerated Natural Weathering Tests', *Paint Technol.*, **27,** No. 12, 40 (December 1963)
64. GARNER, B. L. and PAPILLO, P. J., 'Accelerated Outdoor Exposure Testing in the Evaluation of Ultraviolet Light Stabilisers for Plastics', *American Chemical Society Division of Organic Coatings and Plastics Chemistry,* Atlantic City Meeting, **22,** No. 2, 110–113 (September 9–14)
65. CARYL, C. R., 'Accelerated Outdoor Exposure Testing', *S.P.E. J.*, **23,** No. 1, 49 (January 1967)
66. KING, A., 'Ultra-Violet Light: Its Effect on Plastics', *Plast. & Polymers,* **36,** No. 123, 195 (June 1968)
67. LANGSHAW, H. J. M., 'The Weathering of High Polymers', *Plastics, Lond.*, **25,** No. 267, 40 (January 1960)
68. GOLDFEIN, S., 'Prediction of Mechanical Behavior of Plastics Undergoing Decomposition from the Combined Effects of Environmental Exposure and Stress', *J. appl. Polymer Sci.,* **10,** No. 11, 1737 (November 1966)
69. HAYNES, B. C., 'Fundamentals of Atmospheric Elements', Symposium on Conditioning and Weathering, *Special Technical Publication No. 133, Amer. Soc. Test. Mat.*, (Philadelphia), 3 (1953)
70. HIRT, R. C., SCHMITT, R. G., SEARLE, N. D. and SULLIVAN, A. P., 'Ultraviolet Spectral Energy Distributions of Natural Sunlight and Accelerated Test Light Sources, *J. Opt. Soc. Amer.*, **50,** No. 7, 706 (July 1960)
71. BIKALES, N. M., 'Take a Look at New Weathering Data', *Plast. Technol.*, **13,** No. 4, 11 (April 1967)
72. GRUM, F., 'Artificial Light Sources for Simulating Natural Daylight and Skylight', *Appl. Opt.*, **7,** No. 1, 183 (January 1968)
73. BRIGHTON, C. A., 'Correlation of Accelerated and Natural Weathering Tests', *Weathering and Degradation of Plastics,* Chapter 4, (Edited by S. H. Pinner), Columbine Press, Buxton (1966)
74. HOWARD, J. B. and GILROY, H. M., 'Natural and Artificial Weathering of Polyethylene Plastics', *Polymer Engng & Sci.*, **9,** No. 4, 286 (July 1969)
75. CHAPLIN, C. A., private communication
76. 'Climatest', Mectron (Frigistor) Ltd., Canal Estate, Station Road, Langley, Bucks. See also: *Paint Technol.*, **32,** No. 4, 34 (April 1968)
77. MERRILL, L. K. and MYERS, C. S., 'Accelerated Testing of Plastics for Weathering Resistance', *Bull. Amer. Soc. Test. Mat.* **No. 113,** 19 (December 1941)
78. SAWYER, R. H., 'Accelerated Weathering Devices', Symposium on Conditioning and Weathering, *Special Technical Publication No. 133, Amer. Soc. Test. Mat.* (Philadelphia), 91 (1953)
79. KINMONTH, R. A. JR., 'Weathering of Plastics', *S.P.E. Trans.*, **4,** No. 3, 229 (July 1964)
80. 'New Xenotest Accelerated Weathering Instrument', *Rubb. & Plast. Age,* **43,** No. 5, 469 (May 1962)
81. CHOTTINER, J. and BOWDEN, E. B., 'How Plastics Resist Weathering', *Mat. Des. Engng.*, **62,** No. 4, 97 (October 1965)
82. TITUS, J. B., 'The Weatherability of Polyolefins', Plastics Technical Evaluation Center, Picatinny Arsenal, Dover, New Jersey, *Plastec Report No. 32* (March 1968)
83. HUNT, W. C. and BAUMAN, R. P., 'Measuring Yellowing and Darkening in Weathered Plastics', *Mod. Plast.*, **32,** No. 2, 156 (October 1954)
84. GRAY, V. E. and WRIGHT, J. R., 'Colorimetric Method for Measuring Polyester Degradation Due to Weathering', *J. appl. Polymer Sci.*, **7,** No. 6, 2161 (November 1963)
85. CIPRIANI, L. P., GIESECKE, P. and KINMONTH, R., 'New Way to Peg Plastics Ageing', *Plast. Technol.*, **11,** No. 5, 3 (May 1965)
86. NEUMAN, A. C., 'Development of an Impact Test for Evaluation of Weatherability of Rigid Plastics', *Polymer Engng & Sci.*, **6,** No. 2, 124 (April 1966)
87. SCOTT, K. A., 'The Weathering of Plastics, I, Weathering Factors and Exposure Tests, A Review', *Technical Review 34,* Rubber and Plastics Research Association of Great Britain (Shawbury) (1966)
88. 'Space Radiation Effects on Materials', *A.S.T.M. Special Technical Publication No. 330,* Amer. Soc. Test. Mat. (Philadelphia) (1962)
89. METZ, D. J., 'Testing Polymers for Radiation Resistance', *Testing of Polymers,* Chapter 5, 151, (Edited by J. V. Schmitz), Volume 2, Interscience Publishers, New York (1966)
90. CHARLESBY, A., 'Atomic Radiation and Polymer Science', *Trans. I. R. I.*, **34,** No. 4, 175 (August 1958)

91. PINNER, S. H., 'Radiation Chemistry and Polymer Technology, Part I', *Brit. Plast.*, **34,** No. 1, 30 (January 1961)

92. PINNER, S. H., 'Radiation Chemistry and Polymer Technology, Part II', *Brit. Plast.*, **34,** No. 2, 76 (February 1961)

93. REINSMITH, G., 'Nuclear Radiation Effects on Materials', *Bull. Amer. Soc. Test. Mat.* **No. 232,** 37 (September 1958)

94. MANLEY, T. R., 'The Irradiation of Plastics', *Res. Appl. Chem.*, **12,** No. 2, 42 (February 1959)

95. BALLANTINE, D. S., 'Irradiation of Plastics', *S.P.E. J.*, **12,** No. 7, 27 (July 1956)

96. BLACK, R. M., 'Irradiation of Polymeric Materials', *Trans. Plast. Inst. Lond.*, **29,** No. 81, 98 (June 1961)

97. FISCHER, H., HELLWEGE, K-H and LANGBEIN, W., 'Change in Tensile Strength and Elongation of Some Plastics Caused by High-Energy Radiation', *Kunststoffe,* **58,** No. 9, 625 (September 1968)

98. CHARLESBY, A., *Atomic Radiation and Polymers,* Pergamon Press, London (1960)

99. SCHMIDT, D. L., 'Behaviour of Plastics in Re-Entry Environments', *Mod. Plast.*, **38,** No. 3, 131 (November 1960)

100. 'Reinforced Plastics for Rockets and Aircraft', *A.S.T.M. Special Technical Publication* **No. 279,** Amer. Soc. Test. Mat. (Philadelphia) (1961)

101. BASHFORD, V. G., 'Ablation', *Chemy Ind.*, No. 6, 224 (February 5 1966)

102. SCHMIDT, D. L. and SCHWARTZ, H. S., 'Evaluation Methods for Ablative Plastics', *S.P.E. Trans.*, **3,** No. 4, 238 (October 1963)

103. LISTON, E. M., 'Arc-Image Ablation Testing of Ablation Materials', *J. Macromol. Sci.-Chem.*, **A3,** No. 4, 705 (July 1969)

104. HOFF, G. and LANGBEIN, G., 'Rain Erosion of Plastics', *Kunststoffe,* **56,** No. 1, 2 (January 1966)

105. NEWMAN, S. B., TONER, S. D. and ACHHAMMER, B. G., 'Surface Erosion of Filled Plastics', *Mod. Plast.*, **36,** No. 4, 135 (December 1958)

106. BRUNTON, J. H., 'Cavitation Erosion Testing of Polymers', *Testing of Polymers,* Chapter 10, (Edited by J. V. Schmitz), Volume 1, Interscience Publishers, New York (1965)

107. HEAP, W. M., 'Microbiological Deterioration of Rubbers and Plastics', *Information Circular No. 476,* Rubber and Plastics Research Association of Great Britain (Shawbury) (September 1965)

108. LEUTRITZ, J. JR., 'The Effect of Fungi and Humidity on Plastics', *Bull. Amer. Soc. Test. Mat.* **No. 152,** 88 (May 1948)

109. WITT, R. K., CHAPMAN, J. J. and RASKIN, B. L., 'Effect of Moisture and Fungus on Plastic Insulating Materials', *Mod. Plast.*, **30,** No. 1, 119 (September 1952)

110. STAHL, W. H. and PESSEN, H., 'Funginertness of Internally Plasticised Polymers', *Mod. Plast.*, **31,** No. 11, 111 (July 1954)

111. HUECK, H. J., 'The Biological Deterioration of Plastics', *Plastics, Lond.*, **25,** No. 276, 419 (October 1960)

112. WESSEL, C. J., 'Biodeterioration of Plastics', *S.P.E. Trans.*, **4,** No. 3, 193 (July 1964)

113. MACLACHLAN, J., 'Microbiological Deterioration of Rubber and Plastics', *Rubb. J.*, **147,** No. 6, 58 (June 1965)

114. KULKARNI, R. K., 'Brief Review of Biochemical Degradation of Polymers', *Polymer Engng & Sci.*, **5,** No. 4, 227 (October 1965)

115. MACLACHLAN, J., HEAP, W. M. and PACITTI, J., 'Attack of Bacteria and Fungi on Rubber and Plastics in the Tropics', *S.C.I. Monograph No. 23,* Microbiological Deterioration in the Tropics, Society of Chemical Industry (London), 185 (1966)

116. HEAP, W. M. and MORRELL, S. H., 'Microbiological Deterioration of Rubbers and Plastics', *J. appl. Chem.*, **18,** No. 7, 189 (July 1968)

117. KUHLWEIN, H. and DEMMER, F., 'The Microbial Corrosion of Plastics', *Kunststoffe,* **57,** No. 3, 183 (March 1967)

118. DOLEZEL, B., 'The Resistance of Plastics to Microorganisms', *Brit. Plast.*, **40,** No. 10, 105 (October 1967)

119. BURGESS, R. and DARBY, A. E., 'Two Tests for the Assessment of Microbiological Activity of Plastics', *Brit. Plast.*, **37,** No. 1, 32 (January 1964)

120. RONSIVALLI, L. J., BERNSTEINS, J. B. and TINKER, B. L., 'Method for Determining the Bacterial Permeability of Plastics Films', *Food Technol.*, **20,** No. 8, 98 (August 1966)

121. OTT, D. J., 'Insect Resistance Test Methods', *Amer. Dyestuff Rep.*, **52,** No. 26, 26 (December 23 1963)

122. BECKER, G., 'Testing Results on Termite Resistance of Plastics', *Bull. Rilem,* **25,** 93 (December 1964)
123. BECKER, G., 'Resistance of Plastics to Termites', *Materialprufung,* **5,** No. 6, 218 (1963)
124. PACITTI, J., 'Attack by Insects and Rodents on Rubber and Plastics', *Information Circular 475,* Rubber and Plastics Research Association of Great Britain (Shawbury) (July 1965)
125. MACNULTY, B. J., 'The Testing of Materials Against Fungi and Termites in the Tropics', *S.C.I. Monograph No. 23,* Microbiological Deterioration in the Tropics, Society of Chemical Industry (London), 135 (1966)
126. LIGHTBODY, A., ROBERTS, M. E. and WESSEL, C. J., 'Plastics and Rubber', *Deterioration of Materials,* Chapter 9, (Edited by G. A. Greathouse and C. J. Wessel), Reinhold Publishing Corporation, New York (1954)

BIBLIOGRAPHY

ROSATO, D. V. and SCHWARTZ, R. T., *Environmental Effects on Polymeric Materials, Volume I, Environments,* Interscience Publishers, New York (1968)

ROSATO, D. V. and SCHWARTZ, R. T., *Environmental Effects on Polymeric Materials, Volume II, Materials,* Interscience Publishers, New York (1968)

14
Long-Term Mechanical Properties

14.1 INTRODUCTION

About two centuries ago the traditional materials of construction were stone, bricks and mortar and timber, with iron and steel as relative newcomers for consideration in stressed structural applications. The first iron bridge, in England at any rate (at Ironbridge near Coalbrookdale, in Shropshire) is dated 1779, and it is not difficult to imagine the difficulties the proposers of this innovation experienced in gaining acceptance of their idea. The bridge is still in use, to some extent an excellent testimonial to the original design, but it is doubtful if the builders could have predicted such a 'useful service life' from the data available to them at the time of construction. Much more recently it has been necessary to show that aluminium had acceptable long term properties before its use for durable, stressed components could be accepted. This initial prejudice against the use of aluminium was justified, because its strength is affected by its service environment and recent events, particularly in the aircraft industry, have shown that repeated applications of apparently moderate stresses can lead to catastrophic failures (fatigue).

Plastics materials have now reached a similar stage in their development, when many potential users require to be convinced of their suitability for long term use in stressed applications. It is generally known that 'plastics' toys have a short life in the hands of children, that a polythene washing-up bowl lasts for several years if it is treated with proper respect and that melamine/paper laminates used as kitchen working surfaces are virtually indestructible. More quantitative data can be obtained by applying the tests described in earlier chapters of this book (particularly Chapter 8), and various short term 'strength' properties can be measured using standard tests. However none of this information solves the problems of, for example, an aircraft designer who wants to promote a plastics material from its menial use as a luggage rack to a more significant role in the main structure of an aircraft. Our hypothetical designer has got to be convinced, by numerical data, that any material he uses will not deform to an unacceptable extent or fracture in use at the stress level and in the environment proposed (creep, stress rupture). As it is possible that the stress levels will vary frequently and systematically during the life of the proposed component the designer will require information to advise him of the likely results of such fluctuations.

From the results of tests devised to yield data on the long-term mechanical properties of materials it should be possible to assess the suitability of such materials for many types of use. A reasonable life for a stressed component in an aircraft might be 30 000 h (i.e. about $3\frac{1}{2}$ years in the air) although many other critical components would be acceptable with a much shorter working life, whilst at the other end of the scale water supply installations are generally designed on the assumption that they will be in service for 50 years. In the most general terms, a designer requires to be supplied with data from which he can predict the behaviour of a material throughout the designed life of the product.

In producing these data the plastics technologist is faced with a task of considerable magnitude. There are many more applications to be considered for plastics than was the case for aluminium, there are many more types and grades of plastics than there are of aluminium (or indeed of metals generally) and the range of the properties exhibited by plastics materials is greater. This is especially true in the field of deformation, where at room temperature some materials fracture virtually without deformation whilst others will elongate by hundreds of percent, and performance in this respect (and others of course) can be significantly affected by relatively trivial changes in temperature. Failure can be caused by a continuously applied load, or by fatigue or a combination of both. First we will consider continuous loading.

14.2 EFFECTS OF CONTINUOUSLY APPLIED STRESS

In many situations materials or components will be subjected to stresses which are:

1. steady and continuous, or
2. steady for relatively long periods (for example 8 h out of 24) with intervening periods at reduced or negligible stress, or
3. fluctuating, but by a relatively small amount.

In considering the effects of such stresses it is usually sufficient to sum the periods at the high stress to get a 'working life' for materials in situations such as 2 (but see Reference 1), and to assume that the highest stress is applied continuously in situations of the type 3. There are then three possible 'standard' test approaches open to the investigator.

a. Creep measurement in which, usually, a constant load is applied to a specimen, whilst its deformation is recorded, or

b. stress relaxation measurement—in which a specimen is deformed to a constant extent, whilst the load required to achieve this deformation is recorded, or

c. stress rupture measurement—in which a specimen is submitted to a constant load and the time to break is recorded.

The technique most suitable will, of course, depend on the material, its end use (and perhaps on the equipment available). Techniques do exist for estimating performance in one condition from the results of tests of a different type[2,3] although the mathematical processes may be formidable.

The most commonly used experimental method is creep measurement, and the most simple is stress rupture.

14.2.1 Creep

Creep is formally defined in A.S.T.M. D.674–56 (1961)[4] as 'The time dependent part of the strain which results from the application of a constant stress to a solid at a constant temperature. That is, the creep at a given elapsed time is equal to the total strain at the given time minus the instantaneous strain on loading. The creep extension is usually expressed a percentage of the initial unstretched length'. This 'Recommended Practice' also notes that it is more usual to apply a constant *load* to a creep specimen (e.g. to make no allowance for the decrease in cross-section that occurs as a tensile test specimen increases in length, a simplification almost universally applied in all testing procedure) and warns that (some) time dependent strain occurs rapidly, even during the application of the load. Other relevant definitions are also given.

A.S.T.M. D.674 is a recommendation intended to give general guidance for creep testing and it avoids specification of a precise method of test. Instead it offers advice on the general design of creep tests, and variables which can influence their results, such as temperature (the most important), humidity, vibration, anisotropy of the material; ageing (which can cover continued, slow polymerisation, crystallisation, plasticiser loss, change in water content) etc. Similar advice is given in Reference 5 which specifically relates to thermal history. DIN.53444[6] gives advice on creep testing very much as A.S.T.M. D.674. Recommendation 1.1. of B.S. 4618: 1970 'Recommendations for the Presentation of Plastics Design Data' considers the design of creep tests in a most comprehensive way (B.S.I. Committee PLC/36).

Creep tests may involve stressing specimens in almost any fashion (tension, compression, flexure, shear, etc.) depending on the proposed use of the material, and of course it may be necessary to devise tests to apply a combination of the above stress modes. However most measurements are made in tension[6, 7] probably because the stress is more uniformly distributed (as noted by Chasman[8]). In this text it is proposed only to consider tests in tension and to a lesser extent compression, as these are the most generally applicable. They are at the same time more difficult experimentally as deformations are smaller, applied loads are larger and alignment is more critical than with other test modes.

Thus for most practical purposes the measurement of creep can be regarded as a simple matter of applying a tensile load to a specimen, and measuring its change in length. A.S.T.M. D.674 recommends that the load should be known within 1% and that the grips and gripping technique should be designed to minimise eccentric loading of the specimen. Requirements for the measurement of change in length (usually and correctly change in gauge length) are much more critical, and the sensitivity of the extensometer should never be less than 0·001 in/in and may need to be as high as 0·000 001 in/in, with measurements being consistent and reproducible over periods of weeks, months or even years in extreme cases. It is always necessary to carry out the tests in a constant temperature room or at least enclose the specimens in some sort of controlled temperature chamber.

The simplest possible experimental technique is described by Findley[9] who studied creep in cellulose acetate specimens at 77 ± 1°F and 50 ± 2% r.h. at periods up to 7000 h (about ten months). The stresses involved ranged between 500 lbf/in^2 and 2700 lbf/in^2 (based on original cross sectional area) and these were achieved by the application of simple dead loads of 30 lb–150 lb to the specimens. Under these conditions extensions in the range 1–45% were produced, and measurements of adequate accuracy were obtained with a travelling microscope, which was used to determine the distances between pairs of marks clamped to the specimens. As would be expected the specimens decreased in thickness significantly during the

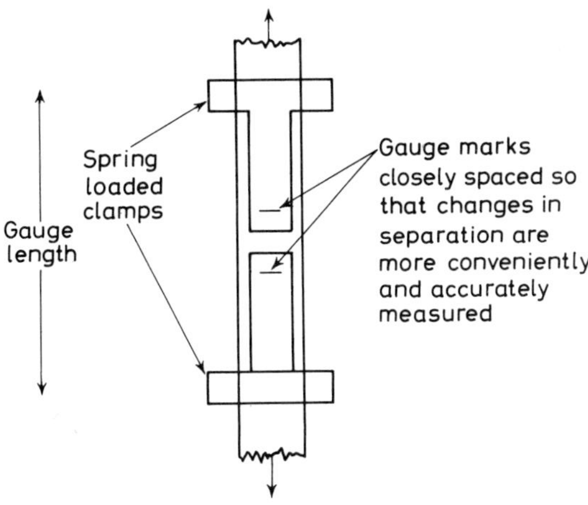

Figure 14.1. Arrangement of gauge marks (Findley)

progress of the tests, and it was necessary to spring load the clamps to accommodate this reduction (see Figure 14.1).

Results of this experimental work are shown in Figure 14.2.

As can be seen, this technique is capable of demonstrating the creep characteristics of this material over a range of stresses and deformations which might be relevant to consumer goods and non-critical applications, though the low creep region, of interest for serious engineering projects, is less adequately covered.

Thermosets generally creep less than thermoplastics and Telfair and his co-workers[10] used a similar loading technique for their work on various filled phenolics, but improved the sensitivity of strain measurement by the use of resistance strain gauges. Two such gauges were bonded to each test specimen, one on each of the two wider faces (presumably to counteract the effect of fortuitous bending as well as increase the sensitivity of the measurement) and two similarly connected gauges, bonded to an unstressed specimen, were included in the measuring circuit to minimise the effect of minor temperature changes. With this system, strains as small as 0·005% (0·000 05 in/in) were 'easily detected'. Various test temperatures and

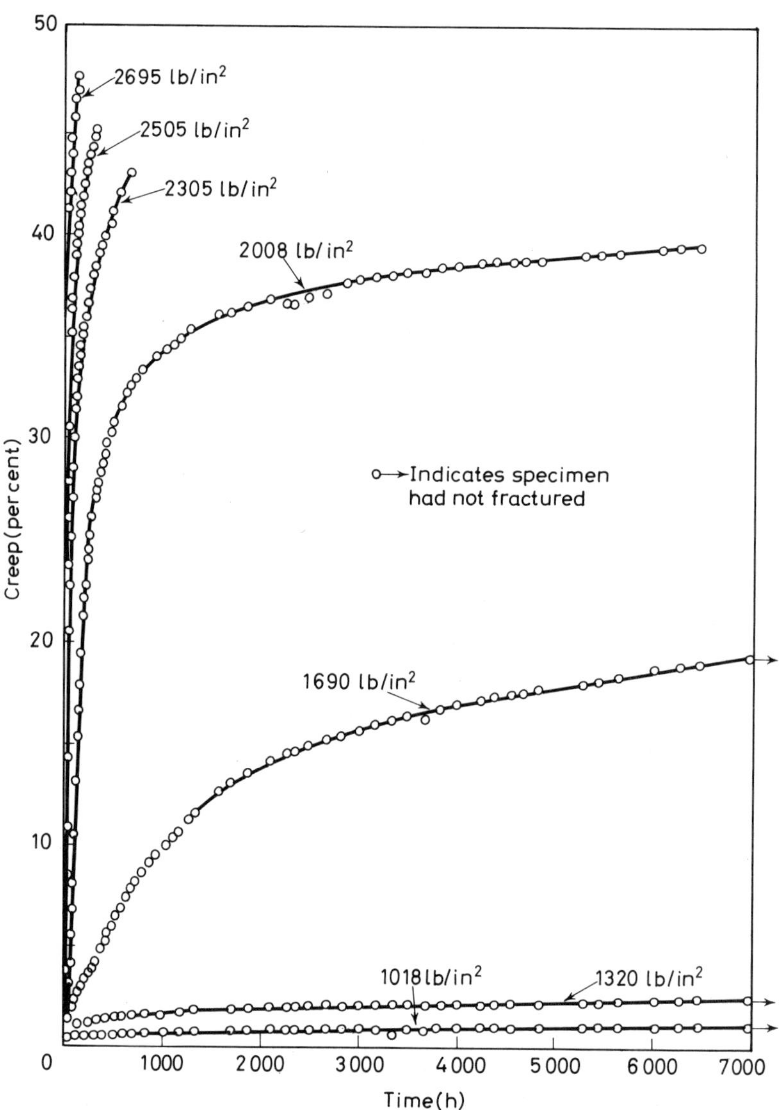

Figure 14.2. Creep-time diagram for cellulose acetate at 77°F (Findley[9]) (Reproduced by permission of McGraw-Hill, Inc.)

stresses were investigated, resulting in creep extensions of less than 1% at times up to 1 000 h. The work also paid attention to the recovery of the specimens after the removal of the stressing loads (just after completion of 1 000 h), and a typical creep and recovery curve is shown in Figure 14.3.

Subsequently Staff and others[2] reported their examination of filled phenolics, using techniques similar to those described above, but extending the time scale to 5 000 h at least, and 14 500 h in some cases. The description of their work is distinguished by their practical comments on the use of strain gauges and the precautions they took to avoid erroneous results from the gauges. They also comment on the prediction of performance of the materials they examined for service of up to five years.

More elegant techniques of creep determination involve the use of extensometers based on the optical lever principle and lever loading systems.

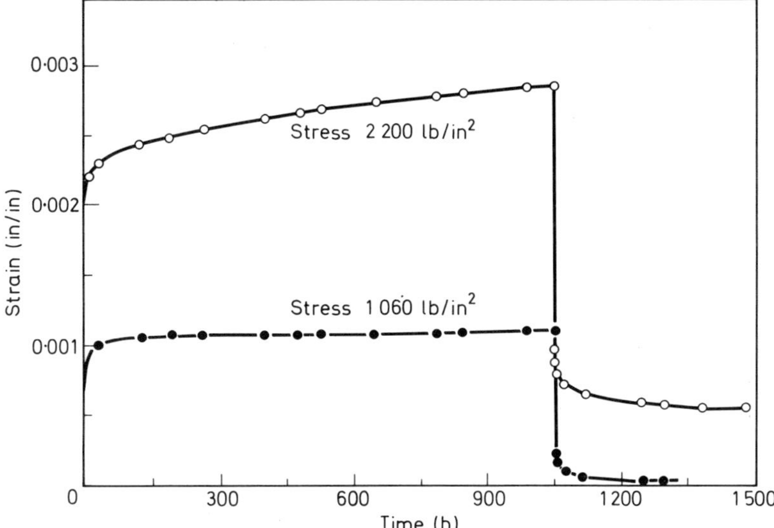

Figure 14.3. Creep and recovery of wood filled phenolic at 25°C (Telfair et alia[10]). (Reproduced by permission of McGraw-Hill, Inc.)

The two basic optical levers, both developed for conventional modulus measurements, are those of Lamb and Martens and they are adequately described by Fenner[11] who indicates that they will detect extensions of a few millionths of an inch. Dunn et alia[12] modified the conventional Lamb extensometer so that it was more suitable for use with thermoplastics (which suffer extensions rather greater than those of metals, but are less able to support the weight of an extensometer or the stresses which have to be applied to locate the knife edges). A working diagram of this modified extensometer is shown in Figure 14.4.

As can be seen, the instrument consists of two pairs of knife edges affixed one to each of four light rigid members. As the pairs of knife edges move apart one pair from the other, due to extension of the gauge length of the specimen, two rollers are caused to rotate by relative motion of the rigid members. The rollers carry small plane mirrors, their rotation resulting in deflection of a light beam reflected by both mirrors in turn on to a circular scale which is centred at the second mirror. The components of the extensometer are held together, and the knife edges are held in contact with the specimen, by sprung members approximately in line with the pairs of knife edges. It is said that when the scale is 50 cm from the final mirror and the gauge length is 8 cm a strain of 0·01 in/in results in a 12·8 cm displacement of the reflected light beam.

In the same paper there are descriptions of two more extensometers for use with materials which could not support the weight of the type described above—one for polythene which, whilst being optical in principle, does

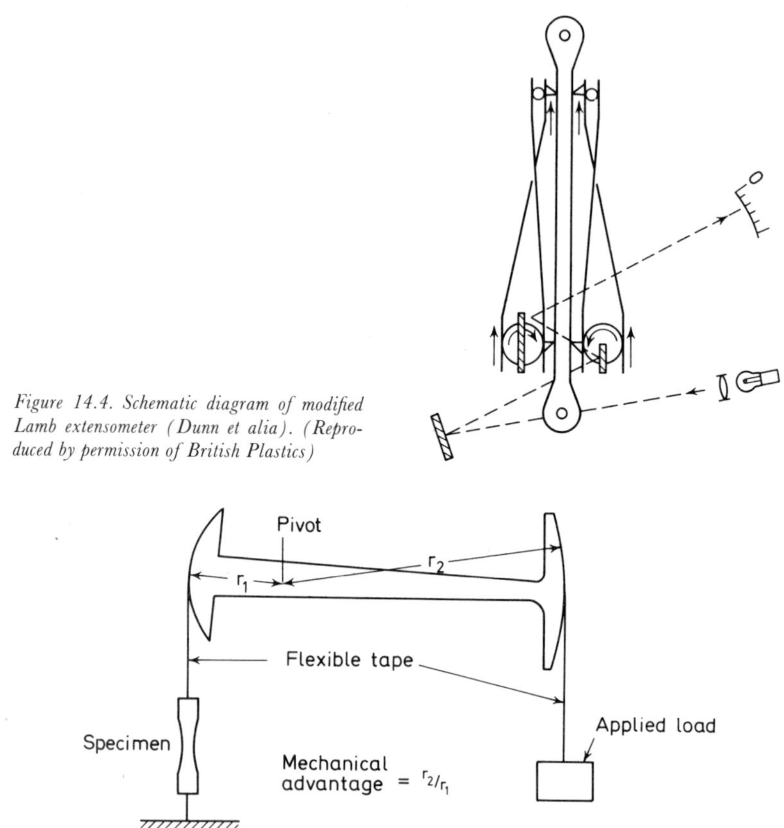

Figure 14.4. Schematic diagram of modified Lamb extensometer (Dunn et alia). (Reproduced by permission of British Plastics)

Figure 14.5. Typical single lever loading system

involve the insertion of locating pins (which actuate levers) into the test specimen—and the other for films or monofilaments, which is completely optical in principle, and involves the minimum of disturbance of the specimen. All three extensometers are used with single lever loading machines of the basic type shown in Figure 14.5.

Considerable ingenuity has been applied to the problems of measuring creep extension and three further examples are cited. Mullen and Dolch[13] describe apparatus devised for use with Portland cement, but which could well be modified for use with low creep plastics. It is suitable for use with specimens immersed in liquids, or which are inaccessible for other reasons, as the indication is remote from the specimen itself. In essence the extensometer consists of two concentric lengths of fine hypodermic needle tubing, arranged and coupled to the specimen as shown in Figure 14.6.

The fiducial marks, shown inset, are cut with a single edge razor blade whilst the tube is rotated in a watch maker's lathe and it is claimed that if

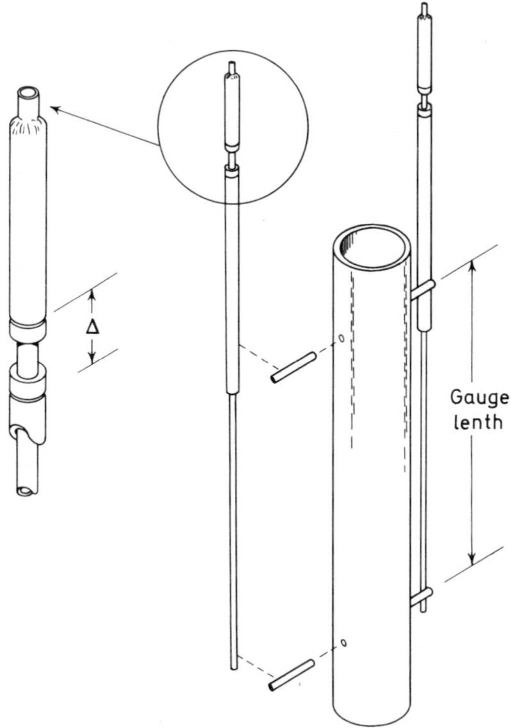

Figure 14.6. Periscope strain gauge (Mullen and Dolch)

they are illuminated in a standard fashion their separation could be determined to an accuracy of 0·000 040 in (0·000 010 in/in on a 4 in gauge length) with an appropriate linear traverse microscope.

Creep apparatus operated in the authors' laboratories (but designed and constructed by B.T.R. Industries Limited)[14] was devised to enable materials to be investigated whilst they were immersed in various liquids at both above and below ambient temperatures, using horizontal specimens. Light Invar structures are mounted at the gauge marks and the separation of these is measured by a portable extensometer, consisting of two micrometers reading in units of 0·000 01 in mounted end to end, as shown in Figure 14.7.

The micrometer drums are rotated until fine foil contacts on the ends of the spindles just touch vanes mounted on the Invar structures, this end point being indicated electrically by a suitable simple circuit. Because the extensometer is portable one of these relatively expensive units can serve several specimens, making a useful reduction in the capital costs of a creep test facility.

An automatic recording technique is described by Scherr and Palm[15] which relies on a shutter fixed to the moving grip of a tensile creep machine, and partially obstructing a beam of light directed at a photosensitive tube. The output of the tube is continuously recorded, and suitable calibration converts this record to one referring to the grip movement. The technique as described can only be expected to give approximate results, as grip

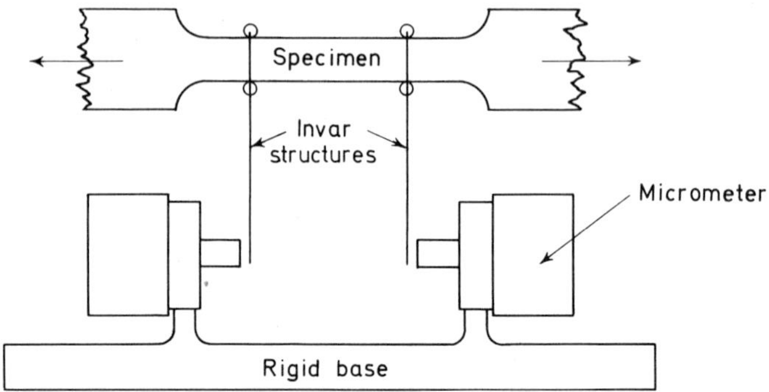

Figure 14.7. Micrometer extensometer

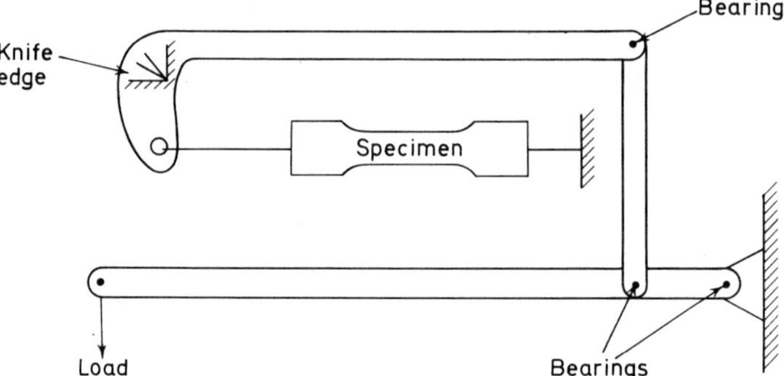

Figure 14.8. Two lever creep machine

separation only is determined, but a modification of the technique, to permit the monitoring of two gauge marks, would be possible.

The method of providing the stress has generally received less attention, as it presents less difficulty. As has already been seen, the problems associated with the large masses required to apply large stresses has been overcome by the use of lever systems. There is a limit to the mechanical advantage that can be achieved with a single lever, but multiple lever machines have been used, as shown in Figure 14.8.

The above diagram represents the system used in the authors' laboratory. The machine has a mechanical advantage of approximately 1:125, but because of friction introduced by the many parts, and the use of ball bearings instead of the theoretically more acceptable knife edge, it is necessary to calibrate each machine using a previously calibrated steel specimen fitted with strain gauges and a range of dead weights.

Special difficulties do arise however if it is necessary to determine creep at constant stress—that is if it is necessary to reduce the applied load in step with the reduction in area of the specimen as its extends. Andrade[16] stressed specimens by means of an appropriately shaped weight (hyperboloid of revolution) which was partially immersed in a liquid. As the specimen extended and decreased in cross-section the load was decreased by buoyancy of the liquid. Ward and Marriott[17] used a single lever machine, but the circular arc at the 'dead-weight' end of the lever in the conventional machine (as shown in Figure 14.5) was replaced by a specially profiled cam. In computing the shape of the cam it was assumed that Poisson's ratio* was 0·5, which is approximately true for many plastics. No allowance appears to have been made for the fact that elongation and section reduction of the parts of the specimen immediately adjacent to the grips will be less than occurs in the central part of the specimen.

The application of the creep load has been given considerable attention. The investigator is required to apply the load rapidly (to avoid the theoretical embarrassment of a stress which increased over a significant period of time) but without shock or vibration. Findley[9] supported his dead weights 'on planks, blocked up in such a way that they could be used as levers to lower the weights quickly but gently until they were supported by the specimen'. He then 'immediately' recorded extension and time. Dunn and his fellow workers[12] describe a system for automatic shock free loading which can apply loads up to 30 kg without vibration, in times considerably below $\frac{1}{2}$ s. Telfair[10] increased the difficulties of interpretation by adding weights in 'convenient increments' taking strain readings at each increment so that modulus data could be obtained. The advantages of knowing the modulus for the *actual creep test specimen* may be considerable, but this technique must increase the difficulties of mathematical interpretation of the results considerably.

It is appropriate to refer to a technical simplification that can be advantageous in certain circumstances. A number of specimens can be strung, end to end in series, and be stressed by one dead weight. This technique has obvious advantages if a number of specimens can be stressed by the same large weight, but little is gained if there is any significant risk of fracture, as the failure of one specimen would preclude the continuation of tests on other specimens in the 'string'.

The examination of specimens in compression is subject to the difficulties which apply to compression tests in general, described in Chapter 8 (Section 8.5). Of these the tendency of the specimen to buckle is the most serious, as this would obviously invalidate the measurements of small creep deformations. Dunn, Mills and Turner[12] and Thomas[18] suggest the use of a hollow cylindrical specimen to prevent or reduce buckling, whilst Jones, Koo and O'Toole[19] suggest the use of a square section specimen, which is almost totally enclosed in a close fitting tube. (This method is unsuitable for large

*When a material is stretched its cross-sectional area is reduced. Poisson's ratio is given by

$$\frac{\text{change in width per unit width}}{\text{change in length per unit length}}.$$

deformations.) In all three cases the intention is to use a specimen which is sufficiently slender to be stressed without the use of excessively large loads. In the work in References 12 and 19 decrease in specimen length is assumed to be directly related to the relative movement of the compression platens which is recorded by dial gauge but Dunn and his co-workers[12] do indicate that this assumption is of doubtful validity, and recommend the use of extensometers, whilst Thomas[18] shows that for some materials at least, the error associated with the use of platen mounted dial gauges is unacceptably large.

Creep tests may be made to yield simple factual data describing the performance of a material under stress conditions, or with a view to achieving a more fundamental understanding of the mechanism of the phenomenon. In the latter case it will be necessary at least to distinguish between the instantaneous deformation and the creep of the specimen, and it may be necessary to go into much greater detail—for example Findley[20] suggests that it might be legitimate to divide the creep curve into

1. primary creep
2. secondary (or steady-state) creep and
3. tertiary creep

In this text it is not proposed to discuss theoretical aspects further; additional information can be obtained from References 1, 3, 5, 7, 12, 21, 22, 23.

In presenting results it is usual to plot strain against time, producing separate curves for different stress levels and temperatures. Depending on the material, test temperature and duration, these may be most informative

Table 14.1 CREEP DATA (References 2 and 24)

Description of Material, etc.	*Initial applied stress dyne/cm²*	*Strain after* 100 s	*Strain after* 10^7 s (116 days approx.)
Polypropylene homopolymer 0·909 g/cm³ 20°C	2×10^8	0·031	—
	1×10^8	0·008	—
	5×10^7	0·004	0·013
Polythene, 20°C			
0·9127 g/cm³, MFI 70	2×10^7	0·02	
0·9245 g/cm³, MFI 20	2×10^7	—	0·03
0·9253 g/cm³, MFI 03	2×10^7	0·008	0·02
Acetal copolymer 20°C	2×10^8	0·008	0·023
	1×10^8	0·004	0·010
Unplasticised PVC 20°C	2×10^8	0·007	0·013
	1×10^8	0·004	0·006 (Estimated)
Nylon 66, 20°C at approx. 0% r.h.	2×10^8	0·007	0·015
As above, glass fibre filled	2×10^8	0·002	0·003
Woodflour/Phenolic 25°C	1×10^8	—	0·001
	3×10^8	—	0·005
Cloth/Phenolic 25°C	1×10^8	—	0·001
	3×10^8	—	0·004

on either linear, semi-logarithmic or fully logarithmic plots. From these factual displays of practically determined data it is common to prepare isochronous stress against strain curves (stress against strain at some constant time) or isometric stress against time curves (stress against time at some constant strain) and examples of these are shown in Reference 23.

Finally, it is perhaps worth mentioning that a number of more commonplace tests, such as hardness tests (load applied for perhaps 30 s) indentation tests on floor tiles, compression set tests on upholstery foams (load applied for perhaps 24 h) etc. are really elementary creep tests, insofar as they require the application of a constant load to a specimen and the measurement of the resulting deformation.

Typical creep data taken from published literature, and converted as far as possible into comparable units, etc. are shown in Table 14.1. (Conversions only approximate.)

14.2.2 Stress Relaxation

A.S.T.M. D.674[4] defines stress relaxation as 'The time-dependent change in the stress which results from the application of a constant total strain to a specimen at a constant temperature. The stress relaxation at a given elapsed time is equal to the instantaneous stress resulting when the strain is applied, minus the stress at the given time'. Throughout a given test the specimen is usually essentially constant in shape and size and thus the distinction between 'constant stress' and 'constant load' which has to be considered in creep tests, has no analogy in stress relaxation experiments. However the warning given in this A.S.T.M. (mentioned earlier in Section 14.2.1), concerning the rapidity of initial changes, applies equally to both creep and stress relaxation measurements, and the variables which affect creep testing are of similar importance in the second method of evaluating the effects of continuously applied stress. Continuing the analogy, stress relaxation tests can be devised to submit specimens to tensile, compressive or any other type of forces, but again simple tensile tests are the most commonly used for the evaluation of fundamental properties. The essential difference between the two types of test is the measured parameter—force instead of deformation (time is, of course, recorded in both cases) and at first sight it might appear that the measurement of force would be the simpler, as the difficulties of long time extensometry are avoided. The A.S.T.M. suggests that 'For relaxation tests a suitable device for applying the load is a screw, driven by a worm-wheel and worm, whose action is controlled automatically by the strain in the specimen. A spring dynamometer may be used to weigh the load during the relaxation test. This dynamometer with its indicator should permit measurement of load within 1% and the zero should not drift with time'.

Nielsen[25] describes a simple, basic test equipment for the measurement of stress relaxation, and indicates that simple instruments are suitable for use with low modulus (and by inference highly extensible) polymers. Rigid polymers present much greater difficulties, and for accurate work the apparatus must be very stiff because otherwise its deformation may be comparable to that of the polymer. Nielsen cites the case of a rigid specimen, 1 in long, stretched by 0·001 in in a relaxation experiment. The stress measuring device (and the apparatus) must not deform by more than 10 micro-inches if stress measurement accurate to 1% is to be achieved.

A more detailed description of a specific instrument is given by Curran, Andrews and McGarry[26]. This consists essentially of a rectangular aluminium frame, which is several orders of magnitude more rigid than the test specimens, with the specimen stretched along its longer axis. The requisite stress is applied to the specimen via one grip which is positioned by a keyed micrometer screw passing through one short frame member, and the (initial) strain in the specimen is indicated during the setting up process by a clip-on extensometer, which is subsequently removed. The stress is recorded by a group of four resistance strain gauges which are bonded to a

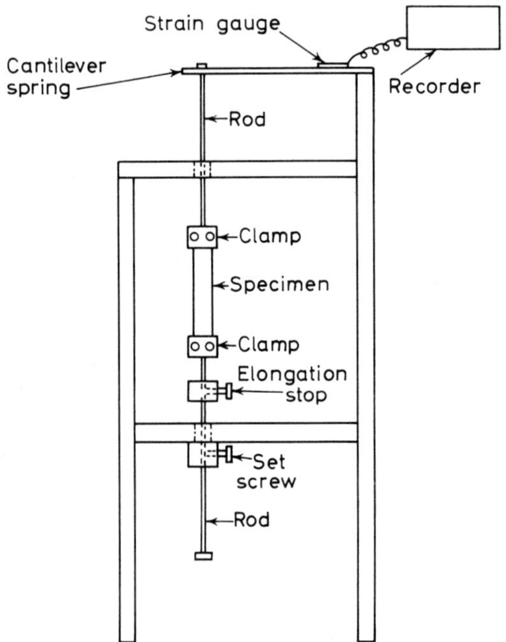

Figure 14.9. Simple apparatus for measuring stress relaxation (Nielsen)

thin strip steel tensile member, which restrains the lower grip. Typical results are shown in Figure 14.10.

This method is suitable for tests involving liquid environments, and a suitable environmental cell is described by the authors.

For the most accurate work this technique would not be acceptable because there are two mechanisms whereby the strain in the specimen could change during the course of the test. The strip steel load weighing member will decrease in length, if only marginally, as the stresses in it (and the specimen) decrease. More significantly, no attention is paid to the changes in strain distribution within the specimen as the test progresses. If one considers a typical dumb-bell specimen it is obvious that the stress will be greater in the narrow portion of the length. Under certain conditions of stress, this narrow portion will creep, permitting the larger sectioned parts of the specimen to retract, and thus invalidating the concept of constant strain. This type of mechanism is also possible in parallel sided specimens, as their stress distribution is far from uniform because of the disturbance caused by the gripping members, although in this case a simple common-sense analysis is not possible. Findley[3] refers briefly to apparatus 'in which

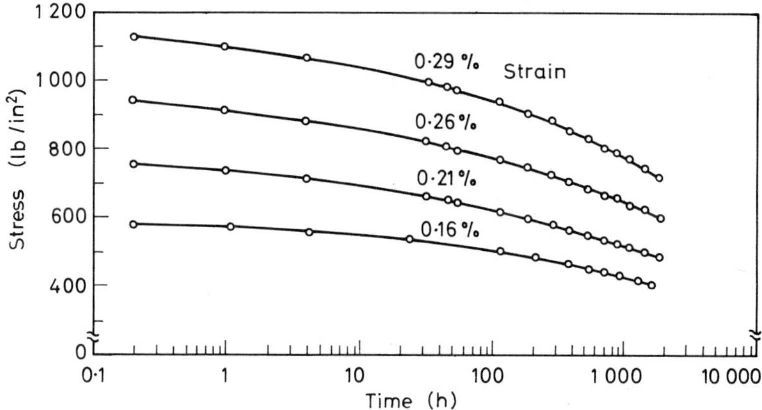

Figure 14.10. Typical stress relaxation data for 'multiphase' thermoplastics (Curran, Andrews and McGarry). (Reproduced by permission of Modern Plastics, McGraw-Hill, Inc.)

means are provided to make those adjustments, required as a result of the decreasing load, to maintain a constant strain in the specimen'.

14.2.3 Stress Rupture

Stress rupture tests are by far the most simple of the three types of long term continuous stress tests to apply, and they are in fact creep tests divested of the onerous task of determining the associated (usually) small elongations. Specimens are subjected to a stress, often defined as some fraction of their ultimate, short term failing stress, and the time which elapses between application of load and ultimate failure is recorded. For short time tests it is usually sufficient for the specimens to be observed by an operator with a stop watch or process timer; in the other extreme a calendar may be sufficient! Tests of intermediate duration, days, weeks or months, may involve the use of more elaborate devices to ensure that the time of failure is recorded with sufficient accuracy. Stress rupture test results are influenced by temperature, vibration, etc. as are creep and stress relaxation tests.

Examples of this type of test occur only infrequently in standard specifications. It is perhaps appropriate that two 'stress rupture' tests described in British Standards are relevant to plastics pipes, as it was necessary to justify the use of plastics in this field by extensive tests of this nature. B.S. 3505: 1968, 'Unplasticised PVC Pipe for Cold Water Services' and B.S. 3506: 1969, 'Unplasticised PVC Pipe for Industrial Uses,' both describe a 'Long term hydrostatic test'. A specimen of pipe is maintained at a constant (internal) pressure so selected that the specimen will burst after between one and ten hours. Throughout the test the internal pressure is kept constant to $\pm 2\%$ and the specimen is immersed in a water bath maintained at 20 ± 1°C. The time to burst is recorded. A second specimen is tested in a similar fashion, except that the applied pressure is selected to give a burst after between 100 and 1000 h. The minimum wall thickness and mean outside diameter of each specimen are determined before test, and the circumferential stress in the specimen is derived from:

$$P = \frac{2\delta t}{D-t}$$

where P = Pressure to be applied (bars)
δ = circumferential stress (bars)
t = minimum wall thickness (mm)
and D = mean outside diameter (mm)

As a quality control procedure on a pipe production unit, the pipe manufacturer is required to record data for all pipe produced to the above mentioned specifications grouped according to size and class (pressure rating) of pipe.

Subsequently circumferential stress is plotted against time on a log/log scale, and a regression line drawn through all the production data is extrapolated to yield a 50 year (438 000 h) value for circumferential stress. Even on a logarithmic scale this is a large extrapolation but in this case there are data supporting its validity[27] for unplasticised PVC. Both B.S. 3505 and B.S. 3506 require that the 50 year value of burst (circumferential) stress should exceed 230 bar (3340 lbf/in^2) for 7 in diameter and smaller pipes, and 260 bar (3770 lbf/in^2) for larger pipes.

A similar stress rupture test method (better insofar as tests are continued beyond 10 000 h) is, however, described in Appendix 'B' of A.S.T.M. D.2239–68, 'Polyethylene (PE) Plastic Pipe (SDR-PR)'; A.S.T.M. D.1598–67, 'Time to Failure of Plastic Pipe Under Long-Term Hydrostatic Pressure', also applies. Another A.S.T.M. method, D.2552–66T, 'Environmental Stress Rupture of Type III Polyethylenes Under Constant Tensile Load', is also worthy of mention. This describes a standardised method for submitting dumb-bell specimens to continuously applied forces, whilst the specimens are immersed in a surface active agent which is maintained at a constant temperature. The time taken for specimens to fracture is recorded, and specimens showing ductile failures, involving 'a noticeable degree of uniform cold drawing' are not considered in the assessment of results. The 'end point' of a given test (at one test/temperature combination) is defined as the 'probable time required for 50% of the specimens to fail in a brittle manner'. Finally, whilst specifications are being considered one should remember that the guidance given in A.S.T.M. D.674–56 and DIN.53444, directed specifically towards creep tests and referred to in Section 14.2.1, is almost all equally applicable to 'stress rupture' tests, and the 'Tentative Recommended Practice for Measuring Time-to-Failure of Plastics Under Tension in Various Environments' (A.S.T.M. D.2648–67T) is of general use in listing certain essential test requirements.

For most situations apparatus and techniques used for creep tests are equally applicable to stress rupture measurements. Two papers which specifically refer to the *measurement* of stress rupture are mentioned in this section. A number of others which discuss the results of such tests are invoked in Section 14.4.

Carey[28] describes a simple laboratory apparatus in which horizontally mounted tensile specimens (of polyethylene) are stressed by a simple multiplying lever. Elongations of up to about 20% are possible in the apparatus, and the specimen is immersed, if required, in a temperature controlled liquid. Numerous results are shown graphically.

Millane[29] has examined a series of glass reinforced plastics laminates, based individually on melamine, epoxide, phenolic silicone and polyester resins. As an essential preliminary, the flexural strengths of groups of three specimens, 6 in long, and $\frac{1}{8}$ in thick were determined at a 5 in span. Subsequently single specimens of similar size were stressed in the same configuration, under various applied loads, in air or water at various temperatures. Stress levels were selected to give flexural stress–rupture time curves extending to 1000 h; examples are shown in Figure 14.11.

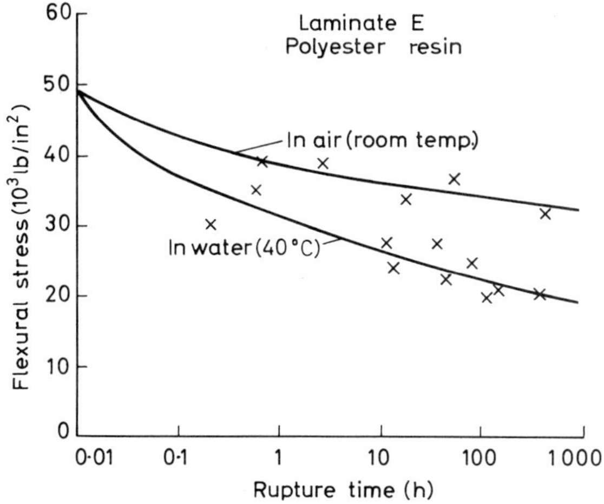

Figure 14.11. Typical stress rupture data for glass cloth reinforced laminates (Millane). (Reproduced by permission of British Plastics)

Millane's paper describes some practical aspects of the requisite apparatus, including an electro-mechanical timing device. However simple elapsed time counters are probably the most suitable equipment for this duty, and it is relatively simple to arrange for a counter to be switched off at time of fracture of the specimen.

14.3 FATIGUE

A.S.T.M. D.671–63T[30], which is specifically applicable to plastics, defines fatigue as 'The process of progressive localised permanent structural change occurring in a material subjected to conditions which produce fluctuating stresses and strains at some point or points and which may culminate in cracks or complete fracture after a sufficient number of fluctuations'. The tentative method continues to describe other terms relevant to the subject, and indicates that the fluctuating stresses and strains need not be of constant amplitude or frequency as they can result from random variation. In some cases the stress (or strain) may oscillate about a zero value, as in the case of a vibrating component, where any given elementary volume may be alternately subjected to tensile and compressive stresses of equal amplitude, but this is not an essential condition, and it is easy to visualise a component subjected to a constant tensile stress together with a smaller fluctuating stress; our elementary volume would then be subjected to a stress of oscillating value, which would not necessarily pass through zero. A.S.T.M. D.671 describes two experimental techniques both of which involve subjecting a cantilevered specimen to flexural stresses which vary cyclically.

Method A involves the use of an apparently specific 'fixed-cantilever, repeated constant deflection type of fatigue testing machine'. In essentials an 11·43 cm long, 1·90 cm wide and usually 0·76 cm thick specimen is held in a vice at one end, whilst its other end is oscillated sinusoidally by a connecting rod driven by a rotating eccentric wheel. The specimens may be rectangular with a hole 0·30 cm in diameter at the centre; or rectangular with a notch cut across the centre of one face, or of an elementary dumb-bell shape, the object of the departure from the simple rectangle being to cause a stress concentration, although the specimens are in all other respects machined and finished to avoid stress raising deformations. The machine described is fitted with a dynamometer to indicate stress in the specimen, and this is used in setting up for a test to show when the apparatus is adjusted to produce deformations of the required amplitude in the specimen. This presupposes an accurate knowledge of the load–deflection characteristics of the test piece, and appears to be unnecessarily complicated and subject to error. It is possible to adjust the specimen gripping vice so that the mean stress (or strain) is not zero.

The description of the method discusses the choice of specimen shape and defines three types of failure, namely (a) the formation of a spreading crack which, once formed, rapidly results in complete fracture of the specimen, (b) the occurrence of a crack or cracks which progress into the specimen rapidly at first, then more slowly, and rarely result in a complete fracture of the material and (c) the generation of excessive heat resulting from internal friction. The 'end point' of type (a) failures is easily detected, and guidance is given in the case of (b), where it is suggested that fatigue failure should be said to have occurred when the stiffness of the specimen has been reduced by a specified amount. Specimens which tend to 'overheat' may suffer progressive cracking (a) before a large temperature rise occurs or failure of type (b). Alternatively it may be necessary to reduce the testing temperature or the speed of testing. Finally, among other points of interest, method A warns that specimens liable to creep will indeed do so if the mean stress is other than zero.

Method B of A.S.T.M. D.671 describes the *use* of a specific 'fixed cantilever, repeated constant load type fatigue testing machine'. However the machine itself is described only in the most cursory manner, and is not referred to by name, and thus the description of the test method is of no practical value to a worker intending to carry out the test without other knowledge.

There is no British Standard devoted to the investigation of fatigue in plastics, but the five parts of B.S. 3518 (:1962 onwards), 'Methods of Fatigue Testing', have a certain general relevance although they apply to the testing of metals.

Very few plastics material specifications require fatigue tests to be made. A.S.T.M. D.1565[31] describes a simple compression fatigue unit for flexible vinyl chloride polymer or copolymer foam. Specimens 12 in square and perferably of 'original' thickness, but with the 12 in square faces parallel, are compressed between slightly larger platens. The upper platen is connected to a reciprocating mechanism, operating at 60 cycles per minute, which is adjusted so that in each cycle the specimen is compressed to 50% of its original thickness (75% in some cases) and allowed to relax completely. The effects of 250 000 such cycles are assessed visually, and by any change in thickness which results. A.S.T.M. D.1564[32] describes a total of six methods for the investigation of the fatigue of flexible urethane foam although two would perhaps be better classified as compression set tests. Of the remainder one is similar in principle to the test described in A.S.T.M. D.1565, two subjects a specimen to repeated traverses by an offset roller, which both depresses and shears the foam (one to a constant deflection and the other under a constant load) and the last method involves rotating an 11 in diameter disc of the foam, which is covered in a 'standard weight duck material', under a pair of weighted rollers whose axes of rotation are canted across two opposite radii at 15°. Again the specimen is subjected to both repeated compressive and shear forces. A.S.T.M. D.2406[33] is similar but less comprehensive.

All of the few standard methods require that a specimen is subjected to a stress or strain cycle in which one of these variables alternates between constant limits. To evaluate a material test schedules are usually designed so that groups of specimens are subjected to cycles in which either the mean or the maximum stress (or strain) changes progressively and results are plotted in the form of a curve showing the function of stress against cycles to failure, from which it may be possible to predict a fatigue limit, a value of the stress function for which the number of cycles approaches infinity. This is, in practical terms, a safe working cycle for the *specimen* in question, although it may still be difficult to interpret such data in terms of the performance of a component. This approach is obviously very slow, and a more rapid test technique was devised by Prot, who proposed that specimens should be cycled in equipment which increased the maximum stress at a rate which was kept constant for any one specimen. (In practice this can be achieved by allowing a stream of water or lead shot to flow into a loading container or by adding small weights, by hand or machine, at equal cycle increments.) The principle is adequately described in Reference 34 which also gives a useful summary of terms and mathematical techniques applicable to fatigue.

The most convenient technique for the application of a cyclic stress or strain is the 'rotational' one, in which an essentially cylindrical specimen is rotated about its axis of symmetry while its ends are constrained by two bearings which are deliberately and symmetrically misaligned. These bearings are kept in the position required to stress the specimen by small weights or springs as shown in Figure 14.12.

In use each elementary volume of the 'working' part of the specimen is subjected to stress which varies sinusoidally between nominally similar maxima of compression and tension. It is of course possible to vary the frequency of the test, the overall applied stress, etc., etc. Lazar[35] describes tests made on polymethyl methacrylate, polystyrene and nylon using this technique. Romualdi[36] and his co-workers describe a similar technique, but with the additional refinement that the specimens can be subjected to a superimposed tensile load, so that alternating stresses varying from complete reversal to pure tension can be developed by varying the magnitudes of the axial and bending stresses. Findley et al[37] developed another machine with the special merit that it was suitable for use at temperatures down to −320°F, and involved the use of very small specimens.

Another approach to fatigue testing, particularly suited to flexible materials, involves the repeated severe bending of thin sheet specimens. An appropriate technique is described by Carey[38], and the De Mattia Tester, which is described in A.S.T.M. D.430[39] is widely used for plastics coated fabrics, although its use is more common in the rubber industry. Repeated flexing, without rotation of the specimen, can also be used to study the fatigue

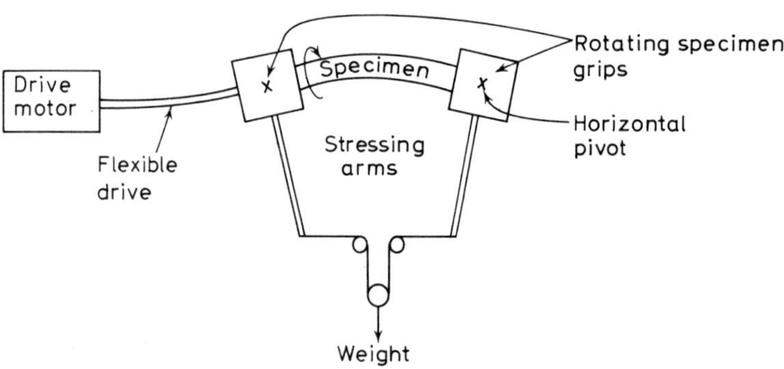

Figure 14.12. Basic design of rotational fatigue apparatus

of rigid specimens. Appropriately shaped specimens can be flexed as cantilevers, as described by Gotham[40] or Cessna and his co-workers[41]; specimens can be subjected to repeated three point bending, as described by Lavengood and Gulbransen[42] or they can be flexed in a four point manner, as by Hutchinson and Benham[43]. Finally, axially applied compressive and tensile stresses are considered by Lazar[35], briefly, and Boller[44] who describes in general terms an extensive programme of tests on reinforced plastics, which are reported in detail elsewhere.

A.S.T.M. D.671 discusses the special problems associated with the generation of heat within the specimens under test. The problem is perhaps special to plastics, as metals are usually fatigue tested at stress or strain levels within their elastic limits, which means that very little energy is lost in the loading/unloading cycle, and little heat is generated in the specimens. This more or less ideal situation does not exist when plastics specimens are tested, as cyclic deformation is usually accompanied by a significant loss of energy resulting in heating of the specimen. This heating is minimised by reducing the speed of testing, but there is an obvious practical limit—one cannot afford to occupy a test position in an expensive test machine for too long with one specimen. Cessna et alia[41], amongst others, alleviate this problem by monitoring the temperature of the surface of the specimen, arranging for a coolant to be supplied to the specimen when the temperature exceeds a predetermined limit. This cannot, however, be the complete solution, as the temperature is measured at, and the coolant is applied to, the surface of the specimen. Most plastics materials are good thermal insulators, whilst the heat is generated within the body of the specimen. Consequently when a coolant is applied the centre of the specimen will almost certainly be significantly hotter than the surface, resulting in enhancement of stress resulting from thermal expansion.

There is, of course, a further way of assessing the effects of cyclic loading on materials, by considering such a treatment as an accelerated ageing procedure (equivalent to heat or exposure to carbon arc radiation) and measuring the strength of a material 'before' and 'after'. Oberbach[45] briefly describes an investigation in which specimens were taken from the fatigue test before the expected failure occurred, and subjected to short term tests. He concludes that deterioration did not occur in the thermoplastic materials studied until immediately before fracture would have occurred in the fatigue test, whereas the glass reinforced plastics specimens decreased in ultimate strength progressively throughout the 'fatigue' process.

14.4 PREDICTION OF LONG TERM BEHAVIOUR

The prediction of long term (mechanical) behaviour is a fairly broad heading, and could be said to include consideration of the effects of accelerated weathering, normal ageing and many other factors on the strength of plastics materials. However it is not proposed to consider the subject in such an all embracing manner (although the reader is referred to a paper by Raphael[46] who considers the effects of oxidative degradation, catalytic degradation, volatilisation and cross linking on physical properties at various temperatures over periods of up to 4000 h), but to limit the field to predictions which can be based on 'continuous' tests of the types described in Sections 14.2 and 14.3. Even here, prediction is usually a hazardous business, and the wise prophet is well advised to qualify his statements with a careful exposition of the assumptions he has been forced to make. Any hypothesis relating test results to long term performance is at risk until sufficient time has elapsed for it to have been justified fully by subsequent assessment of the long term performance itself; by this time the more speedy tests may well be only of academic interest. In old established fields of investigation this still leaves considerable scope for the development and improvement of methods of prediction, because the occurrence of new materials is by now relatively infrequent, and most hypotheses can be supported by proven results for similar materials. This is not the situation in the Plastics Industry, which is in a state of rapid development, and prediction without *complete* justification is sometimes essential if new materials are to be properly exploited without intolerable delays.

For prediction to be useful it must indicate the condition of a material or component at some time in the future, and it is not uncommon to expect a test extending for 1000 h (about seven weeks) to yield information on which performance for 10 000 h, 100 000 h (11 years) or even longer, is predicted. To achieve this it is usually necessary to (a) extrapolate from the results of a relatively short term test by assuming that whatever relationship has been found to exist between the property or properties of interest and time holds over much longer periods or (b) accelerate the test by increasing stress, frequency of stress application, temperature or etc. in such a way that the working life of the component is compressed into the period of test. Extrapolation is more suitable for 'steady load' situations, whereas acceleration is more appropriate when fatigue is involved.

The most thoroughly investigated long term application of plastics is their use in cold water pipes, and the extrapolation of 1000 h data to 50 years advocated in British Standards 3505 and 3506 (cited in Section 14.2.3) is adequately justified by work undertaken by Richard et alia[27] on PVC pipes. More recent work by Gill[47] and Lloyd[48] is also relevant, and the latter, in particular, justifies the use of tests lasting between 0·001 h (3·6 s) and 10 h as a quality control procedure for use with batches of pipe made from material which has previously been shown to have satisfactory properties by longer term tests. (It is interesting to note that this work predicts service life by extrapolation of a failure stress–time curve which is not a straight line.) However other materials and applications have rarely been investigated in any depth.

A number of workers have considered the theoretical justification of the extrapolations described above, usually basing their arguments on the use of Arrhenius type equations. Typical mathematical treatments are described in References 49, 50 and 51.

At first sight it would appear that the 'accelerated' approach to the problem of prediction is much more easily justified, although second thoughts may lead to a different conclusion. If one considers the simple 'pounding' (repeated compression) test used for flexible polyurethane foam it is easy to see a logical justification for it. Presumably someone has calculated that cushions in an armchair will be compressed (by being sat on) about ten times per day, 365 days per year, and that a reasonable life is ten years, during which the cushion will be compressed about 36 500 times. A.S.T.M. D.1564 (mentioned in Section 14.3 above) specifies that a specimen should be compressed 250 000 times, and it is not unreasonable to suppose that the extra 'compressions' were added to allow for minor movements of the sitter, or as a safety factor, or to give a round number. Thus the test appears to have a fairly satisfactory practical basis. However, it makes no allowance for the hysteresis loss which occurs when the foam is compressed and released. The heat generated in one compression–relaxation cycle is trivial, but repetition of such cycles, at a rate of one per second, is a different matter. Similarly the specimen subjected to repeated flexing cannot at the same time be subjected to continuous compression, another practical and deleterious condition of normal use. In spite of all these limitations the test is useful and successfully sorts foams into good and bad types, possibly because the increase in temperature causes an appropriate degree of acceleration of the normal degradation processes other than flexing. A.S.T.M. D.671 warns that specimens may be softened by rapidly repeated deformations, and recommends a reduction in testing speed if this occurs. At a less dramatic level, however, heat can still cause spurious results in fatigue tests. The acceleration due to increase in repetition rate of flexing may be significantly different to the acceleration resulting from increase in temperature.

A typical investigation of fatigue properties is described by Molt[52] who relates experimental data to the life of a glass reinforced plastics boat hull.

Test stresses three times greater than those expected in practice were repeatedly applied to the boat hulls tested, and the effect of this increase was determined by reference to previously published data relating strain and fatigue life for similar materials. Regrettably the boats had not been in use long enough to justify the prediction of their life at the time the paper was published. Zarek[53] discusses methods of assessing the fatigue limit (the limiting stress which can be applied to a specimen for a number of times which approaches infinity, without causing failure) for specimen tested using the Prot 'progressive load' method (see Section 14.3). He concludes that plotting stress at failure against the square root of rate of increase of stress is most suitable.

REFERENCES

1. WILLIAMS, M. L. and HOWARD, W. H., 'Creep of High Polymers', *S.P.E. Trans.*, **2,** No. 1, 74 (January 1962)
2. STAFF, C. E., QUACKENBOS, H. M. and HILL, J. M., 'Long Time Tension and Creep Tests of Plastics', *Mod. Plast.*, **27,** No. 6, 93 (February 1950)
3. FINDLEY, W. M., 'Stress-Relaxation and Combined Stress Creep of Plastics', *S.P.E. J.*, **16,** No. 2, 192 (February 1960)
4. *A.S.T.M. D.674-56,* 'Testing Long Time Creep and Stress Relaxation of Plastics Under Tension or Compression Loads at Various Temperatures', American Society for Testing and Materials (Philadelphia)
5. RITCHIE, P. D. (ed.) *Physics of Plastics,* Iliffe Books Ltd., London (1965)
6. *D.I.N. 53444:* 'Testing of Plastics, Tensile Creep Test', Deutscher Normenausschuss (Berlin).
7. BROWN, W. E., *Testing of Polymers,* Volume 4, Chapter 1, Interscience Publishers, John Wiley and Sons, New York (1969)
8. CHASMAN, B., 'Creep and Time—Fracture Strength of Plastics Under Tensile Stresses', *Mod. Plast.,* **21,** No. 6, 145 (February 1944)
9. FINDLEY, W. N., 'Creep Tests on Cellulose Acetate', *Mod. Plast.,* **19,** No. 12, 71 (August 1942)
10. TELFAIR, D., CARSWELL, T. S. and NASON, H. K., 'Creep Properties of Molded Phenolic Plastics', *Mod Plast.,* **21,** No. 6, 137 (February 1944)
11. FENNER, A. J., *Mechanical Testing of Materials,* Philosophical Library Inc., New York (1965)
12. DUNN, C. M. R., MILLS, W. H. and TURNER, S., 'Creep in Thermoplastics—Review of Apparatus for Creep Measurements', *Brit. Plast.,* **37,** No. 7, 386 (July 1964)
13. MULLEN, W. G. and DOLCH, W. L., 'Periscope-Type Strain Gage Measures Creep in Immersed Specimens', *Mat. Res. & Stand.,* **6,** No. 4, 191 (April 1966)
14. CAMERON, J. B., 'The Temperature Limitations of Reinforced Plastics in Aggressive Environments', *Trans. Plast. Inst. Lond.,* **35,** No. 119, 681 (October 1967)
15. SCHERR, H. J. and PALM, W. E., 'A Recording Apparatus for the Measurement of Longitudinal Creep', *J. appl. Polymer Sci.,* **7,** No. 4, 1273 (July-August 1963)
16. ANDRADE, E. N. DA C., 'On the Viscous Flow in Metals and Allied Phenomena', *Proc. Roy. Soc.,* A.LXXXIV, 1 (June 1910)
17. WARD, A. G. and MARRIOTT, R. R., 'A Constant Stress Apparatus for the Study of the Creep Properties of Plastics', *J. Sci. Instrum.,* **25,** No. 5, 147 (May 1948)
18. THOMAS, D. A., 'Uniaxial Compressive Creep Studies', *Plast. & Polymers,* **37,** No. 131, 485 (October 1969)
19. JONES, E. D., KOO, G. P. and O'TOOLE, J. L., 'A Method for Measuring Comprehensive Creep of Thermoplastic Materials', *Mat. Res. & Stand.,* **6,** No. 5, 241 (May 1966)
20. FINDLEY, W. N., 'Mechanism and Mechanics of Creep of Plastics', *S.P.E. J.,* **16,** No. 1, 57 (January 1960)
21. GOLDFEIN, S., 'Creep of Glass Reinforced Plastics', *Bull. Amer. Soc. Test. Mat.* **No. 225,** 29 (October 1957)
22. Discussion of A.S.T.M. Bulletin **No. 225:** 'Creep of Glass Reinforced Plastics', *A.S.T.M. Bulletin No. 229,* 65 (April 1958)
23. POWELL, P. C., 'The Prediction of Long Term Deformation in Thermoplastic Components', *I.C.I. Plast. Today,* No. 31 (October 1967)

24. TURNER, S., 'Creep in Thermoplastics', *Brit. Plast.*, **37,** No. 8, 440 (August 1964); No. 9, 501 (September 1964); No. 11, 607 (November 1964); No. 12, 682 (December 1964); **38,** No. 1, 44 (January 1965)
25. NIELSEN, L. E., 'Stress Relaxation', *Mechanical Properties of Polymers,* Chapter 4, Reinhold Publishing Corp., New York (1962)
26. CURRAN, R. J., ANDREWS, R. D. and MCGARRY, F. J., 'Device for Measuring Stress-Relaxation of Plastics', *Mod. Plast.*, **38,** No. 3, 142 (November 1960)
27. RICHARD, K. and DIEDRICH, G., 'Standfestigkeitseigenschaften von einigen Hochpolymeren', *Kunststoffe,* **45,** No. 10, 429 (October 1955)
28. CAREY, R. H., 'Creep and Stress—Rupture Behaviour of Polyethylene Resins', *Industr. Engng Chem.*, **50,** No. 7, 1045 (July 1958)
29. MILLANE, J. J., 'Long Term Stress Resistance of Resin/Glass Laminates in Different Environments', *Brit. Plast.*, **33,** No. 5, 199 (May 1960)
30. *A.S.T.M. D.671-63T:* 'Tentative Methods of Test for Repeated Flexural Stress (Fatigue) of Plastics', American Society for Testing and Materials (Philadelphia)
31. *A.S.T.M. D.1565-66:* 'Standard Specifications for Flexible Foams made from Polymers or Copolymers of Vinyl Chloride', American Society for Testing and Materials (Philadelphia)
32. *A.S.T.M. D.1564-64T:* 'Tentative Method of Testing Slab Flexible Urethane Foam', American Society for Testing and Materials (Philadelphia)
33. *A.S.T.M. D.2406-68:* 'Standard Methods of Testing Molded Flexible Urethane Foam', American Society for Testing and Materials (Philadelphia)
34. *A.S.T.M. Special Technical Publication No. 91-A,* 2nd edn, 'A Guide for Fatigue Testing and the Statistical Analysis of Fatigue Data', American Society for Testing and Materials (Philadelphia) (1963)
35. LAZAR, L. S., 'Accelerated Fatigue of Plastics', *Bull. Amer. Soc. Test. Mat.* **No. 220,** 67 (February 1957)
36. ROMUALDI, J. P., CHANG, CHIAO-LIN and PECK, C. F. JR., 'A Fatigue Testing Machine for Range of Stress', *Bull. Amer. Soc. Test. Mat. No. 200,* 39 (September 1954)
37. FINDLEY, W. N., JONES, P. G., MITCHELL, W. I. and SUTHERLAND, R. L., 'Fatigue Machines for Low Temperature and for Miniature Specimens', *Bull. Amer. Soc. Test. Mat.* **No. 184,** 53 (September 1952)
38. CAREY, R. H., 'Fatigue Testing of Nonrigid Plastics', *Bull. Amer. Soc. Test. Mat.* **No. 206,** 52 (May 1955)
39. *A.S.T.M. D.430-59:* 'Standard Methods of Dynamic Testing for Ply Separation and Cracking of Rubber Products', American Society for Testing and Materials (Philadelphia)
40. GOTHAM, K. V., 'A Formalised Experimental Approach to the Fatigue of Thermoplastics', *Plast. & Polymers,* **37,** No. 130, 309 (August 1969)
41. CESSNA, L. C., LEVENS, J. A. and THOMSON, J. B., 'Flexural Fatigue of Glass-Reinforced Thermoplastics', *Polymer Engng & Sci.*, **9,** No. 5, 339 (September 1969)
42. LAVENGOOD, R. E. and GULBRANSEN, L. B., 'The Effect of Aspect Ratio on the Fatigue Life of Short Boron Fiber Reinforced Composites', *Polymer Engng & Sci.*, **9,** No. 5, 365 (September 1969)
43. HUTCHINSON, S. J. and BENHAM, P. P., 'Low-Frequency Plane-Bending Fatigue of P.V.C.', *Plast. & Polymers,* **38,** No. 134, 102 (April 1970)
44. BOLLER, K., 'Fatigue Characteristics of R.P. Laminates Subject to Axial Loading', *Mod. Plast.*, **41,** No. 10, 145 (June 1964)
45. OBERBACH, K., 'The Behaviour of Plastics in Fatigue Testing', *Kunststoffe,* **55,** No. 5, 356 (May 1965)
46. RAPHAEL, T., 'Predicting Service Life of Plastics', *Plast. Technol.*, **8,** No. 10, 27 (October 1962)
47. GILL, D. A., 'Long Term Burst Testing for Plastics Pipe', *Brit. Plast.*, **34,** No. 3, 126 (March 1961)
48. LLOYD, P. F. V., 'Stress Rupture Testing of Thermoplastic Pipes', *Plastics, Lond.*, **29,** No. 322, 39 (August 1964)
49. CAREY, R. H. and OSKIN, E. T., 'The Prediction of Long Time Stress Rupture Data from Short Time Tests', *S.P.E. J.*, **12,** No. 3, 21 (March 1956)
50. GOLDFEIN, S., 'Long Term Rupture and Impact Stresses in Reinforced Plastics', *Bull. Amer. Soc. Test. Mat.* **No. 224,** 38 (September 1957)
51. GOLDFEIN, S., 'General Formula for Creep and Rupture Stresses in Plastics', *Mod. Plast.*, **37,** No. 8, 127 (April 1960)

52. MOLT, R. P., 'Service Life Estimates for F.R.P. Structures', *S.P.E. J.*, **17,** No. 9, 977 (September 1961)
53. ZAREK, J. M., 'Accelerated Fatigue Testing of Polymethyl Methacrylate', *Brit. Plast.*, **30,** No. 9, 399 (September 1957)

15

Non-Destructive Tests

15.1 INTRODUCTION

The general heading 'non-destructive tests' covers a very broad field of activities, ranging from simple visual inspection, through weighing and confirming that dimensions are correct, up to ultrasonic and X-ray examination of finished products. The simplest of these tests are in every day use, and indeed some are so obvious and instinctive that they are never thought of as 'tests' as such. At the other extreme the more complex techniques are still a matter for the specialist and the research laboratory.

15.2 VISUAL EXAMINATION, WEIGHING AND MEASURING

If one is buying a pound of tomatoes or a car (or, for that matter, taking a wife) it is almost certain that the transaction will be preceded by a visual inspection. Most plastics components are subject to at least a cursory inspection immediately after they are moulded, if only to check that the mould has been properly filled, and that there are no obvious defects. Many specifications include a clause which states 'the moulding shall be free from flaws and visible imperfections' or something similar, and in the extreme it may be necessary to examine components microscopically. Such an inspection can detect obvious faults, such as surface blemishes, gross deformations, inclusions and bubbles (in transparent or translucent materials) and many other defects. The weight of components is rarely of critical importance, except from a commercial point of view, although table tennis balls are made to close weight tolerances. Mensuration, the final member of this trio of elementary non-destructive tests, is dealt with in Section 5.2 of Chapter 5, as it is an essential preliminary to most of the other tests already described. It should be remembered, however, that in many cases it is an exceedingly important test in its own right.

15.3 HARDNESS

Hardness tests are dealt with in Section 9.2 of Chapter 9 and, strictly speaking, they are not 'non-destructive'. However a number of the micro-techniques involve very small indentations, and the specimens are often undamaged and sometimes unmarked after test.

15.4 DYNAMIC MECHANICAL TESTS

Dynamic mechanical tests generally involve subjecting a specimen to a changing stress or force, and measuring the resulting deformation or strain. For reasons of convenience, both experimentally and in subsequent calculation, the stress variation is usually sinusoidal, and the tests yield data concerning an elastic modulus of the specimen and its mechanical damping. Apart from being of immediate practical value, such data have proved to be of great value in studying the structure of high polymers, but for this purpose it is necessary to cover as wide a range of temperature as possible. Consequently most dynamic tests are designed with this requirement in mind. That is, the test specimen is so positioned that it can be placed in a cryostat or oven as a matter of routine. Nielsen[1] discusses the theoretical implications in some detail.

In practical terms, a dynamic mechanical test must consist of three essential components:

1. a mechanism which causes the specimens to experience a stress which varies sinusoidally with time, and with a periodicity which can be varied over a range to suit experimental requirements;
2. a method of measuring or controlling the frequency of the varying stress;
3. a method of measuring the amplitude of the resultant deformation.

There are many methods of satisfying these three requirements and the simplest is probably the torsional pendulum. This usually consists essentially of an upper, rigidly mounted clamp which supports a specimen usually in the form of a thin strip, which in turn supports an inertia bar or disc. (An alternative arrangement involving an inertia member supported by a 'thin' wire, and in turn supporting the specimen, is similar in most functional details. In both configurations the specimen is subjected to a small tensile stress which is ignored in the calculation of results). In use the specimen is maintained at a selected constant temperature, the inertia bar is displaced from its position of equilibrium by rotating it about the long axis of symmetry of the specimen, and the frequency and the amplitudes of the subsequent oscillations are recorded by any suitable means—a stopwatch and a protractor in simple cases, or some form of automatic recorder. The shear modulus of a material, as a specimen which is not very highly damped (i.e. for which the oscillations do not decrease in size too rapidly) is given by the equations (taken from A.S.T.M. D.2236[2]).

$$G = \frac{64\pi^2 ILf^2}{\mu bt} \quad \text{(specimens of rectangular cross-section)}$$

$$G = \frac{8\pi ILf^2}{r^4} \quad \text{(specimens of circular cross-section)}$$

where I = moment of inertia of the inertia member, g/cm^2
f = frequency of oscillation, Hz
L = length of specimen, cm

b = width of specimen, cm
t = thickness of specimen, cm
r = radius of specimen, cm
μ = a shape factor for specimens of rectangular cross-section
G = shear modulus, dyne/cm^2

Typical values of the shape factor μ, are given by Table 15.1 (taken from A.S.T.M. 2236).

Table 15.1 SHAPE FACTORS

Ratio of specimen width to thickness	*Shape factor, μ*	*Ratio of specimen width to thickness*	*Shape factor, μ*
1·00	2·249	3·50	4·373
1·20	2·658	4·00	4·493
1·40	2·990	4·50	4·586
1·60	3·250	5·00	4·662
1·80	3·479	6·00	4·773
2·00	3·659	7·00	4·853
2·25	3·842	8·00	4·913
2·50	3·990	10·00	4·997
2·75	4·110	20·00	5·165
3·00	4·213	40·00	5·232

The mechanical damping or loss of the material is expressed as the logarithmic decrement Δ, which is defined as the natural logarithm of the ratio of the amplitude of two successive oscillations.

In this type of test the periodicity of the oscillation can be varied by changing the dimensions of the specimen or the inertia bar. A.S.T.M. D.2236[2] describes the method briefly, whilst DIN.53445 'Testing of Plastics —Torsion Pendulum Test[3]' gives rather more details and helpful explanations. ISO/R.537 also refers[4].

Torsional pendulum techniques are described in papers by Nielsen[1, 5, 6], James et alia[7], Lord and his co-workers[8], Wheeler[9] and Lepie and Adicoff[10].

The torsional pendulum technique determines shear modulus, and the frequency of the oscillation *in any one test* is fixed by the modulus of the material and the geometry of the stressed and oscillating members. An alternative method involves the imposition of an oscillation of controlled frequency on the test specimen. The frequency is varied whilst the amplitude of the response of the specimen is monitored. If the experimental parameters are correctly selected the response will pass through a maximum value at some particular value of frequency (the natural frequency of the specimen under the test conditions) and from this frequency, and the dimensions of the specimen, the tensile (or other appropriate) modulus can be calculated. The mechanical damping is derived from the shape of the peak in the amplitude against frequency curve. A typical technique of this type involves a 'vibrating reed'—a thin, rectangular strip of the plastics material under test which is attached to a relatively massive vibrating member, such as a loud-speaker moving coil, which is driven by a variable frequency oscillator, with suitable amplification.

Preliminary tests are made with the vibrator alone, to ensure that its amplitude is constant at all frequencies in the test range. The test specimen is then attached in such a way that it will vibrate transversely and the amplitude of the movement of the free end is recorded continuously as the frequency is varied through the resonance value. This amplitude can be measured simply by means of a microscope with a calibrated eyepiece, or recorded via an electronic sensing device. Young's modulus is then given by the equation (taken from Nielsen[1]):

$$E = \frac{38{\cdot}24dL^4}{D^2} f_r^2$$

where E = Young's modulus, dyne/cm^2
d = Density of plastic specimen, g/cm^3
L = Free length of specimen, cm
D = Thickness of specimen, cm
f_r = Resonant frequency, Hz

A damping term may be calculated from the width of the resonance peak. Vibrating reed tests of this type are described by Newman[11], Atkinson and Eagling[12], Strella[13] and many others[14, 15]. As an alternative to the transverse oscillation of the vibrating reed, rod shaped specimens can be induced to resonate longitudinally; Crissman and McCammon[16] describe such a method, and give the following equation for the derivation of Young's modulus:

$$E = 4\, l^2 \rho f_0^2$$

where

E = Young's modulus
l = Length of specimen
ρ = Density of specimen
f_0 = Resonant frequency of the specimen

Similar techniques are described by Lee[17] and Learmonth and Pritchard[18], whilst Hansen and others[19] used a related method to study the dynamic tensile modulus of thin film specimens. Kline[20] used a horizontal specimen supported on two cotton thread stirrups, inducing the test piece to vibrate in transverse mode.

A useful summary of these and similar dynamic tests is given by Karas and Warburton[21].

The techniques so far described depend on measurements made on specimens which are vibrating at, or close to, their natural resonant frequency. Payne[22] describes a mechanically driven apparatus in which rubber specimens are compressed sinusoidally at non-resonant frequencies, whilst stress, strain and phase angle (a measure of the lag of strain behind stress) are measured. Maxwell[23] subjected specimens to stresses at non-resonant frequencies.

Finally, it is necessary to consider non-cyclic dynamic tests. Barrett and Gordon[24] describe a technique for the determination of the rebound resilience of a specimen. This property is determined repeatedly (at up to twenty times per minute) whilst the temperature of the specimen is increased (at up to 30 degC per min), and the data is correlated with the degree of cure of various polymers. Roberts[25] discusses various techniques whereby the effects on a specimen of a single non-destructive impact can be observed, over very short time scales.

15.5 BIREFRINGENCE, PHOTOELASTICITY AND DICHROISM

The velocity of light through a transparent solid medium is dependent on its physical properties. If the material is anisotropic an incident light ray will be split into two components polarised in mutually perpendicular directions and these will be propagated with different velocities. The material is then said to be birefringent (that is, to have two refractive indices) and the phenomenon is discussed in very much greater detail in Chapter IV of Reference 26. In plastics, the refractive index parallel to the polymer chain is usually different from the refractive index perpendicular to the chain, and thus birefringence is a measure of molecular orientation.

One of the earliest uses of plastics involved the exploitation of this property, and in the early part of this century[27, 28] it was found that if a transparent model was made of a component in a suitable 'optically active' material (celluloid) it would affect a beam of polarised light in such a way that the distribution of stresses within the model, resulting from an applied load, could be deduced. Nowadays many plastics materials are known to be optically active and although the volume of material used in this way is very small its importance is disproportionately great.

It is therefore necessary to evaluate materials for this use, to assess the degree of optical activity or to measure the stress–optical coefficient. However, this is a highly specialised field and the interested reader is referred to Reference 26 for a complete study of the subject, and to References 29–35 for work specifically related to plastics.

The birefringence of plastics materials, apart from being a useful phenomenon which materially assists in the evaluation of other, non-plastics components, is of some value in the study of plastics themselves. Many plastics processing techniques yield finished products which are unavoidably oriented —that is the individual links in the polymer chain tend to lie predominantly in one direction or plane—and other techniques deliberately encourage orientation as a means of increasing the strength of the product in a given direction or directions. The use of simple stress–optical techniques, such as the polarised light tester described by Pugh et alia[36] gives a quick general assessment of the residual stresses in transparent mouldings.

Nielsen[1] discusses the theory of birefringence as applied to oriented polymers, and also briefly mentions the related phenomenon of dichroism—the ability of an oriented polymer to absorb different amounts of polarised infrared radiation, depending whether the plane of polarisation is parallel or perpendicular to the direction of orientation.

15.6 ULTRASONICS

Ultrasonic techniques can be used in several ways in the testing and evaluation of plastics. In the simplest case a suitable piezoelectric crystal can be placed in contact with one surface of a component, and energised so that it emits a very approximately parallel beam of ultrasonic energy into the component. A similar transducer in contact with the opposite face of the component will receive the (attenuated) ultrasonic wave, and can indicate this fact with the aid of suitable electronics. If it is then arranged that the emitter and the receiver can scan the surface of the component, moving together relative to the test piece, there may be differences in the received signal. In the extreme, a laminar fault within the thickness of the specimen, with an area similar to or greater than that of the beam of ultra sound, will completely stop all transmission. Far less gross faults can also be detected, depending on the sophistication of the apparatus and the skill of the operator.

This basic method can be used with a simple frequency modulator to convert the received signal into an audible tone, from which the operator derives information based largely on previous experience. More information can be gained if the input signal is pulsed, and the start of each pulse is used to trigger the horizontal sweep of an oscilloscope beam, whilst the received signal is applied to the vertical deflection plates. If relevant parameters of the circuitry are known the unmodulated horizontal trace is related to the path length of the ultrasonic beam (a function of thickness, density and modulus) and the amplitude of the modulation, and its form, will be affected by the impedance of the material, and the presence of flaws, inclusions and voids. This technique is informative but it suffers from one major limitation in that it only indicates the presence of a fault—not the position. Further details will be found in reference 37.

The position of the faults is indicated in the reflection technique, for which emitter and receiver are mounted side by side on one face of the test piece. Each pulse of ultrasonic energy is transmitted through the thickness of the specimen to the opposite face where it is reflected back to the receiver. Flaws of various types give rise to reflections, and the energy thus reflected returns to the receiver before that reflected by the farther surface of the specimen. This fact is indicated on the oscilloscope trace, the length of the unmodulated sweep corresponding to path length, whilst the form of the modulation indicates the extent and type of the fault. This method is particularly suited to the study of adhesive bonds, and is described by Baumeister[38], whilst Miller and Boruff[39] give an elegant variation, in which the effects of bond quality on the resonant frequency and amplitude of the exciting crystal are studied.

The above test methods are primarily related to the finding of faults in specimens: ultrasonic techniques can also yield data of more fundamental interest. The basic equipment already described can effectively determine the speed at which a sound (or ultrasound) wave is propagated through a specimen, and hence the modulus can be derived from the equation (from Hitt and Ramsey[40]):

$$V_L = \left[\frac{E(1-\delta)}{\rho(1-\delta)\,(1-2\delta)}\right]^{\frac{1}{2}}$$

Where V_L = longitudinal wave velocity cm/s
E = Young's modulus g/cm^2 × gravitational constant
ρ = material density g/cm^3
δ = Poisson's ratio

This aspect of the subject is also considered by Dietz[41] who used a similar technique to study glass reinforced plastics, concluding that the wave velocity and attenuation were related to glass content and laminate quality and Zurbrick[42] who related wave velocity to modulus. Sofer, Dietz and Hauser[43] followed the cure of phenol-formaldehyde resins by a method involving the continuous measurement of wave velocity and attenuation.

Useful summaries of ultrasonic methods applied to plastics are given in References 40 and 44.

15.7 ELECTRICAL TEST METHODS

Electrical test methods are discussed in Chapter 10 of this text. However, it should be remembered that many electrical tests are excellent examples of non-destructive techniques for the evaluation of polymers.

15.8 X-RAY METHODS

X-ray test methods applicable to plastics fall into two categories—the conventional engineering inspection procedures and the analytical techniques for the investigation of crystalline structure. The former technique is far too expensive to be applied to the vast majority of plastics components, although Slocum and Robbins[45] do describe the X-ray inspection of fibre reinforced plastics items for use in missiles, an application where cost would be of little importance. X-ray diffraction techniques are used to study the structure of crystalline polymers, and Reynolds[46] described the use of the technique to investigate the crystallinity of carbon fibre within a composite. (See Chapter 4, Section 4.2.)

15.9 XERORADIOGRAPHY

Nemet, Black and Cox[47] describe the use of xeroradiography for inspecting plastics components. The technique is essentially similar to conventional X-ray inspection, but in place of the photographic film or plate a xeroradiographic plate, consisting of a layer of photoconducting amorphous selenium, applied to a sheet of a conductor (usually aluminium), is used. The plate is sensitised before use by exposure to a corona discharge, and the selenium layer becomes charged. This charge can be dissipated by light or more particularly X-rays.

Thus if a specimen is placed in contact with a prepared xeroradiographic plate and irradiated with X-rays, a charge image of the specimen will be set up on the plate. Application of a suitable, oppositely charged, powder makes the image visible, when it can be photographed by conventional methods. The image differs from a normal X-ray photograph in that the overall contrast is very low (allowing a wide range of specimen thicknesses to be accommodated with one exposure) whilst the local contrast, at a change of density or thickness or any discontinuity, is enhanced. The plates are re-usable.

REFERENCES

1. NIELSEN, L. E., *Mechanical Properties of Polymers,* Reinhold Publishing Corporation, New York (1962)
2. *A.S.T.M. D.2236-67T,* 'Tentative Method of Test for Dynamic Mechanical Properties of Plastics by Means of a Torsional Pendulum', American Society for Testing and Materials (Philadelphia)
3. *D.I.N. 53445:* 'Testing of Plastics—Torsional Pendulum Test', Deutscher Normen Ausschuss (Berlin)
4. *I.S.O. Recommendation R.537,* 'Testing of Plastics with the Torsional Pendulum', 1st edn, International Organisation for Standardisation (January 1967)
5. NIELSEN, L. E., 'Dynamic Mechanical Properties of High Polymers', *S.P.E. J.*, **16,** No. 5, 525 (May 1960)
6. NIELSEN, L. E., 'Some Instruments for Measuring the Dynamic Mechanical Properties of Plastics Materials', *Bull. Amer. Soc. Test. Mat.* **No. 165,** 48 (April 1950)
7. JAMES, D. I., NORMAN, R. H. and PAYNE, A. R., 'The Relation Between Coefficient of Friction and Dynamic Properties of Polyvinyl Chloride', *S.C.I. Monograph No. 5,* The Physical Properties of Polyvinyl Chloride, Society of Chemical Industry (London), 233 (1959)
8. LORD, P., PITHEY, E. R. and WETTON, R. E., 'Improved Instrument for the Continuous Determination of Dynamic Mechanical Properties of High Polymer Solids', *Lab. Pract.*, **10,** 884 (December 1961)
9. WHEELER, A., 'Evaluation of Flexible Materials by Means of an Automatic Torsion Pendulum', *Plast. & Polymers,* **37,** No. 131, 469 (October 1969)
10. LEPIE, A. H. and ADICOFF, A., 'Torsional Pendulum for Plastic Materials under Tensile Strain', *Rev. Sci. Instrum.*, **38,** No. 11, 1615 (November 1967)
11. NEWMAN, S., 'A Vibrating Reed Apparatus for Measuring the Dynamic Mechanical Properties of Polymers', *J. appl. Polymer Sci.*, **2,** No. 6, 333 (1959)
12. ATKINSON, E. B. and EAGLING, R. F., 'Some Applications of Dynamic Elastic Measurements in Polymer Systems', *S.C.I. Monograph No. 5,* The Physical Properties of Polymers, Society of Chemical Industry (London), 197 (1959)
13. STRELLA, S., 'Vibrating Reed Test for Plastics', *Bull. Amer. Soc. Test. Mat.* **No. 124,** 47 (May 1956)
14. SCHERR, H. J., 'Measuring the Amplitude of Vibration of a Reed', *Mater. Res. Stand.*, **6,** No. 12, 614 (December 1966)
15. FIELDING-RUSSELL, G. S. and WETTON, R. E., 'The Dynamic Mechanical Method in Plastics Testing, I, A Multi-Sample Vibrating Reed Apparatus', *Plast. & Polymers,* **38,** No. 135, 179 (June 1970)
16. CRISSMAN, J. M. and MCCAMMON, R. D., 'Apparatus for Measuring the Dynamic Mechanical Properties of Polymeric Materials Between 4° and 300°K', *J. acoust. Soc. Amer.*, **34,** No. 11, 1703 (November 1962)
17. LEE, TUNG-MING, 'Method of Determining Dynamic Properties of Viscolastic Solids Employing Forced Vibration', *J. Appl. Phys.*, **34,** No. 5, 1524 (May 1963)
18. LEARMONTH, G. S. and PRITCHARD, G., 'The Dynamic Behaviour of Undercured Polyester Resins', *S.P.E. J.*, **24,** No. 11, 47 (November 1968)
19. HANSEN, O. C. JR., FABRY, T. L., MARKER, L. and SWEETING, O. J., 'Apparatus for Measuring the Dynamic Tensile Modulus of Thin Polymeric Films', *J. Polymer Sci.*, **1,** No. 5, Part A, 1585 (1963)
20. KLINE, D. E., 'A Recording Apparatus for Measuring the Dynamic Mechanical Properties of Polymers', *J. Polymer Sci.*, **22,** No. 102, 449 (December 1956)
21. KARAS, G. C. and WARBURTON, B., 'Dynamic Mechanical Testing of Polymers', Part 1, *Brit. Plast.*, **34,** No. 3, 131 (March 1961); Part 2, **34,** No. 4, 189 (April 1961)
22. PAYNE, A. R., 'Sinusoidal-Strain Dynamic Testing of Rubber Products', *Mater. Res. Stand.*, **1,** No. 12, 942 (December 1961)
23. MAXWELL, B., 'An Apparatus for Measuring the Response of Polymeric Materials to an Oscillating Strain', *Bull. Amer. Soc. Test. Mat.* **No. 215,** 76 (July 1956)
24. BARRETT, R. M. and GORDON, M., 'Applications of Rebound Resilience to the Cure of Polyester Resins', *S.C.I. Monograph No. 5, The Physical Properties of Polymers,* Society of Chemical Industry, (London), 183 (1959)
25. ROBERTS, J., 'Method to Measure Dynamic Elastic Constants of Polymers', *S.C.I. Monograph No. 17, Techniques of Polymer Science,* Society of Chemical Industry, (London), 211 (1963)
26. JESSOP, H. T. and HARRIS, F. C., *Photoelasticity, Principles and Methods,* Cleaver-Hume Press Ltd., London (1949)

27. JESSOP, H. T., 'Photoelasticity', *Brit. Plast.*, **20,** No. 234, 513 (1948)
28. LEVEN, M. M. and SAMPSON, R. C., 'Recent Trends in Photoelastic Plastics', *Mod. Plast.*, **34,** No. 9, 151 (May 1957)
29. GURNEE, E. F., 'Theory of Orientation and Double Refraction in Polymers', *J. appl. Phys.*, **25,** No. 10, 1232 (October 1954)
30. MARSHALL, D. F., 'Dynamic Stress Optic Coefficient of Perspex', *Proc. Phys. Soc.*, Section B, **70,** No. 11, 1033 (November 1957)
31. CRAWFORD, S. M., 'The Relation Between Stress, Strain and Birefringence in Some High Polymers', *Proc. Phys. Soc.*, Section B, **66,** No. 10, 884 (October 1953)
32. KEEDY, D. A., VOLUNGIS, R. J. and KAWAI, H., 'Use of a Tensile Testing Machine for the Determination of Stress and Strain-Optical Coefficients', *Rev. Sci. Instrum.*, **32,** No. 4, 415 (April 1961)
33. ANDREWS, R. D. and RUDD, J. F., 'Photoelastic Properties of Polystyrene in the Glassy State, I, Effect of Molecular Orientation', *J. appl. Phys.*, **28,** No. 10, 1091 (October 1957)
34. RUDD, J. F. and GURNEE, E. F., 'Photoelastic Properties of Polystyrene in the Glassy State, II, Effect of Temperature', *J. appl. Phys.*, **28,** No. 10, 1096 (October 1957)
35. SAUNDERS, D. W., 'The Photo-elastic Properties of Cross-Linked Amorphous Polymers', *Trans. Faraday Soc.*, **52,** No. 10, 1414 and 1425 (October 1956); **53,** No. 6, 860 (June 1957)
36. PUGH, D. W., MCDONALD, W. F. and FUNK, W. V., 'Do-it-yourself Polarized Light Tester', *Mod. Plast.*, **37,** No. 6, 114 (February 1960)
37. SEAMAN, R. E., 'Ultrasonic Inspection by Pulsed Transmissions', *Brit. Plast.*, **29,** No. 7, 262 (July 1956)
38. BAUMEISTER, G. B., 'Production Testing of Bonding Materials with Ultrasonics', *Bull. Amer. Soc. Test. Mat.* **No. 204,** 50 (February 1955)
39. MILLER, N. B. and BORUFF, V. H., 'Adhesive Bonds Tested Ultrasonically', *Adhes. Age*, **6,** 32 (June 1963)
40. HITT, W. C. and RAMSEY, J. B., 'Ultrasonic Inspection of Plastics', *Rubb. & Plast. Age*, **44,** No. 4, 411 (April 1963)
41. DIETZ, A. G. H., 'Ultrasonic Evaluation of Reinforced Plastics', Journal of Engineering Mechanics Division, *Proc. Amer. Soc. civ. Engrs*, **87,** 31 (June 1961)
42. ZURBRICK, J. R., 'Non-Destructive Testing of Glass-Fiber—R.P.: Key to Composition Characterisation and Design Properties Prediction', *S.P.E. J.*, **24,** No. 9, 56 (September 1968)
43. SOFER, G. A., DIETZ, A. G. H. and HAUSER, E. A., 'Cure of Phenol-Formaldehyde Resin', *Industr. Engng Chem.*, **45,** No. 12, 2743 (December 1953)
44. HATFIELD, P., 'Ultrasonic Measurements in High Polymers', *Research*, **9,** 388 (October 1956)
45. SLOCUM, D. and ROBBINS, G., 'The Case for X-Ray Inspection', *Plast. Technol.*, **8,** No. 12, 29 (December 1962)
46. REYNOLDS, W. N., 'Non Destructive Testing of Bonded-Fibre Composites', *Plast. & Polymers*, **37,** No. 128, 155 (April 1969)
47. NEMET, A., BLACK, A. D. and COX, W. F., 'Xeroradiography for the Testing of Plastics and Light Materials', *Trans. Plast. Inst. Lond.*, **30,** No. 87, 192 (June 1962)

16

Optical Properties

16.1 INTRODUCTION

Optical properties of plastics are in general important for mainly psychophysical reasons. Exceptions are important though; for example, plastics lenses and windows both involve the basic properties of refractive index and light transmission. Nevertheless, vast quantities of plastics materials and articles or plastics-wrapped commodities are sold on account of their aesthetic appeal and properties such as gloss, reflectance, surface texture, haze, colour, in various combinations tend to overshadow the more fundamental characteristics. However, as with other aesthetic properties they are extremely difficult to measure accurately and correlate with visual experience —and even to define!

The subject is broad and will be dealt with broadly. Thus no attempt to describe the large variety of glossmeters which have been and are being used will be made, but attention confined to those established instruments and specification tests which are in common use. Nor will attention be directed at those instruments which are employed with success in other fields such as paints, paper and ceramics, unless such instruments are widely used for plastics also.

Finally, we shall not be concerned with those properties and techniques which are of increasing relevance to the investigation of polymer structure and to analytical work. These include infrared and ultraviolet absorption, light scattering both by visible light and by X-rays, and birefringence.

Definitions are given where relevant and some help in understanding optical terms is given by B.S. 233: 1953, 'Glossary of Terms used in Illumination and Photometry', and B.S. 1611: 1953, 'Glossary of Colour Terms used in Science and Industry'.

16.2 REFRACTIVE INDEX

The refractive index of transparent plastics materials is dealt with in A.S.T.M. D.542–50, 'Index of Refraction of Transparent Organic Plastics', which gives two alternative methods. The first is based on the standard Abbé refractometer, almost universally used for liquids. For solids the illuminating prism is removed and replaced by a small test specimen, $\frac{1}{4}$ in $\times \frac{1}{2}$ in being convenient, which is kept in contact with the fixed prism by a drop of a suitable liquid of refractive index at least 0·01 units greater than the test piece. A variety of liquids for different materials is suggested. The mating face of the specimen must be polished flat and also have one truly perpendicular edge. White light is suggested, the measurement is made at 23°C and an accuracy of 0·000 1 is stated to apply to such a measurement. DIN.53491, 'Testing of Plastics; Determination of Refractive Index and Dispersion', gives useful advice on practical details relevant to Abbé measurements.

The second, less preferred, method in A.S.T.M. D.542 is based on a standard microscope technique in which the apparent thickness of a uniformly thick specimen is determined by focussing on opposite faces alternately and measuring the traverse. The refractive index is obtained by dividing the true thickness by apparent thickness.

An extensive review of methods has been made by Wiley and Hobson[1], although this is not very recent (1948). Wiley and Garrett[2] describe methods for use in the range −70°C to +70°C based on a spectrometer, and other authors (McAlister, Villa and Salzberg[3]) have employed a commercial spectrometer in the infrared region. Adaptations to the Abbé refractometer enabling measurements to be made to 100°C are described by Black, Harvey and Ferris[4]. Difficulties in measuring polymeric films are underlined by the relative preponderance of papers in this field as typified by those of Billmeyer[5], Ellis[6] and Schael[7]. Faust[8] describes a method for inhomogeneous solids, e.g. fibres, using interference microscopy.

16.3 LIGHT TRANSMISSION AND HAZE

16.3.1 Introduction

The light transmission and haze of a material are two closely linked properties which together constitute transparency. The *transmission* is usually defined as the percentage of incident light which is transmitted without deviation, and the *haze* as the percentage of incident light which is transmitted with more than a certain angular deviation by forward scattering. Both properties vary with the geometry of the optical measuring system used which must be carefully specified.

To illustrate these properties, glazing material and diffusing material for lighting fixtures must both have a high light transmittance, but the former must also be free from haze and very transparent, whilst the latter must have maximum diffusion and minimum transparency so that a bright light source cannot be seen through it.

Film for many packaging purposes must also be transparent, but in addition it must be possible to resolve objects clearly through it and the term 'see-through clarity' has been coined to describe this subjective phenomenon which is usually investigated by using standard test charts (Snellen charts) under standard viewing conditions. This property should not be confused with haze and a hazy material may in fact give good resolution. Webber[9] and Miles and Thornton[10] deal with packaging aspects and both express the view that 'see-through clarity' is a function of very small angle forward light scatter as distinct from the rather large angle scatter utilised in haze measurements (a fraction of a degree compared with several degrees). Haze may be induced in plastics by weathering and Niegisch[11] has shown that this can be caused by the formation of minute internal voids resulting from photo induced molecular degradation.

16.3.2 Small Scale Methods

The A.S.T.M. method for plastics is D.1003–61, 'Haze and Luminous Transmittance of Transparent Plastics', and Procedure A therein employs a spherical hazemeter as shown in Figure 16.1, which is pivotable about a vertical axis through the specimen placed in contact with the entrance port. In the normal position the collimated incident light passes straight through

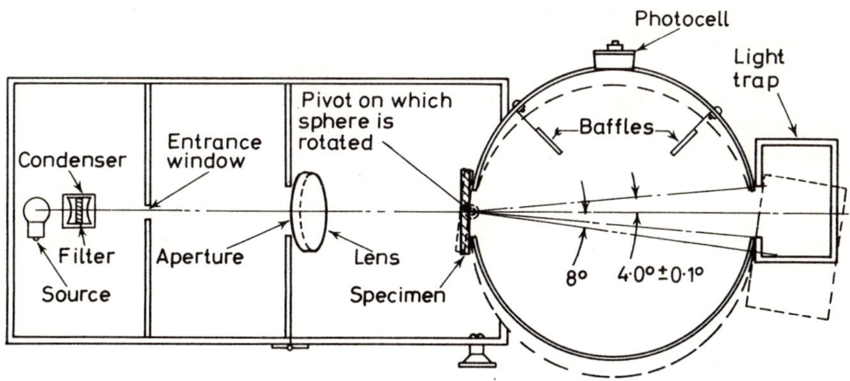

Figure 16.1. Integrating sphere hazemeter (A.S.T.M. D.1003)

the sphere, leaving through the exit port which is closed by an absorbent light trap. Any light which is scattered by the instrument alone (specimen removed) or instrument plus specimen (specimen in) is reflected from the region around the edge of the exit port and finally collected by the photocell after multiple reflections from the highly reflective walls of the sphere.

When the sphere is rotated slightly so that the incident light hits the opposite highly reflecting wall of the sphere (which forms a reflectance standard) adjacent to the exit port, a measurement with and without specimen gives a measure of the total transmittance.

The three properties of interest are:

$$\text{Total transmittance } T_t = \frac{T_2}{T_1}$$

$$\text{Diffuse transmittance } T_d = \frac{T_4 - T_3\left(\dfrac{T_2}{T_1}\right)}{T_1}$$

Haze $= \dfrac{T_d}{T_t}$

where T_1 = Photocell output, specimen out, reflectance standard in beam
T_2 = Photocell output, specimen in, reflectance standard in beam
T_3 = Photocell output, specimen out, reflectance standard out of beam
T_4 = Photocell output, specimen in, reflectance standard out of beam

and these four quantities represent respectively the incident light, the total light transmitted by the specimen, the light scattered by the instrument and the light scattered by instrument and specimen.

The A.S.T.M. D.1003 method is also invoked by A.S.T.M. D.1044–56, 'Resistance of Transparent Plastics to Surface Abrasion', to measure the diffuse transmittance of specimens after Taber abrasion tests, as a measure of the degree of abrasion (see Chapter 9).

The method given in B.S. 2782: 1970, 'Methods of Testing Plastics', Method 515A, 'Haze of Film', is essentially identical to the A.S.T.M. procedure, but oddly refers to haze only and not to light transmission. C.I.E. Source C is the light source used although the A.S.T.M. (but not the B.S.) permits the alternative Source A. (The three principle sources of light used in illumination and photometry are CIE Sources, A, B, and C which respectively correspond to incandescent light, noon sunlight and overcast sky daylight.)

These methods use very small specimens and the results obtained obviously do not apply to large sheets of corrugated roofing material, where additional scattering is involved.

16.3.3 Large Scale Methods

A light transmission test for large panels (2 ft square) appears in two British Standards (B.S. 4154: 1967, 'Specification for Corrugated Plastics Translucent Sheets made from Thermo-setting Polyester Resins (Glass Fibre Reinforced)' and B.S. 4203: 1967, 'Specification for Extruded Rigid PVC Corrugated Sheeting') and in A.S.T.M. D.1494–60, 'Diffuse Light Transmission Factor of Reinforced Plastics Panels' (see also Reference 12). Since it is a test on a finished product rather than a material it will not be described in detail. The A.S.T.M. version is shown in Figure 16.2. A photocell measures the transmitted light and the transmittance is given as the ratio of photocell output—specimen in/specimen out. In B.S. 4154, an additional test for light diffusion is given, based on a simple slit diffusion photometer, and an appropriate correction is then calculated and applied to the total light transmission. This refinement is not required in B.S. 4203, presumably because only transparent material is involved.

16.4 GLOSS

16.4.1 Introduction

According to Hunter[13] 'Gloss is defined as the degree to which a surface simulates a perfect mirror in its capacity to reflect incident light' and he goes

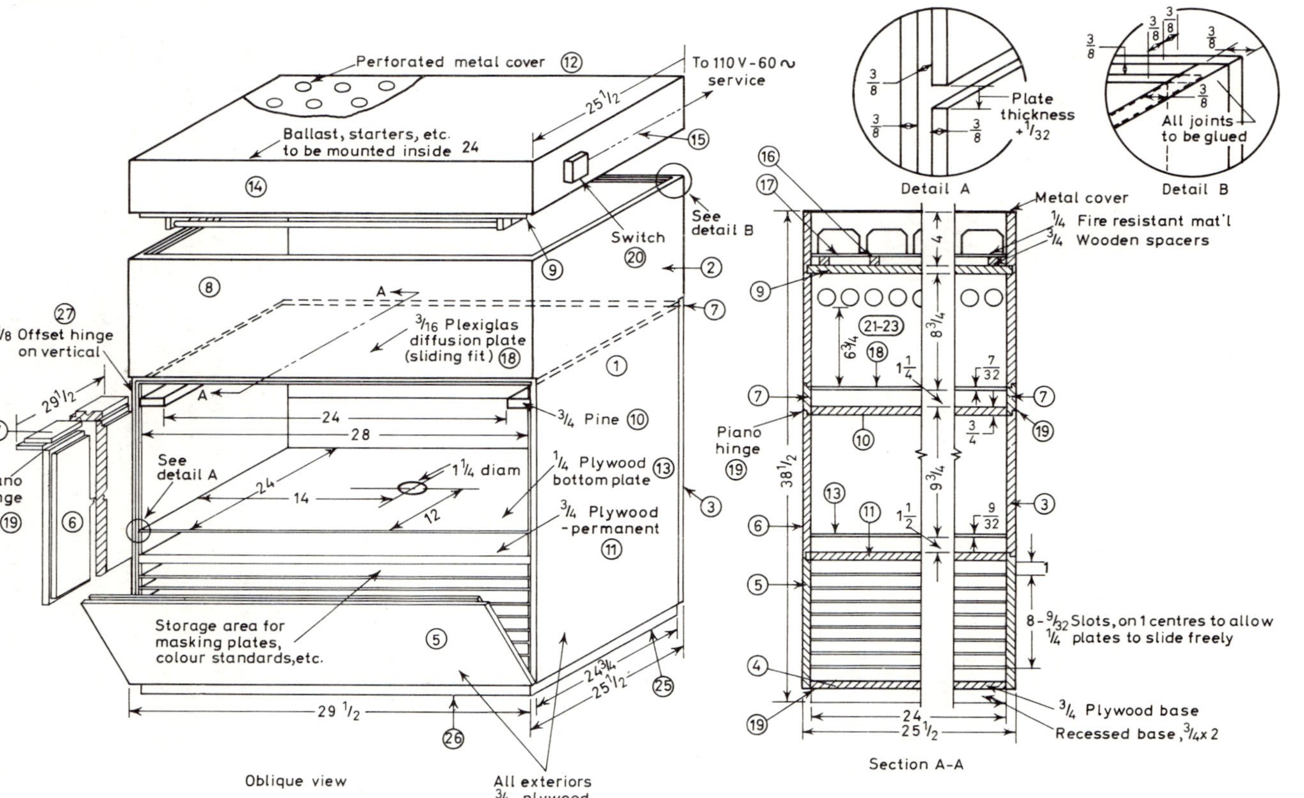

Figure 16.2. Transmissometer (A.S.T.M. D.1494). All dimensions in inches. Notes: 1. Illumination shall consist of 20 W fluorescent tubes which are assembled in three banks of four tubes each according to the following positions: daylight, deluxe cool white, blue, daylight. 2. All plywood to be exterior type. 3. All inside areas above diffusion plate to be painted white, below the plate a dull black

on to say that 'Gloss is determined by the surface's geometric selectivity in reflecting light'. In most practical glossmeters *specular gloss* is the property determined and this is usually taken to be the fraction of light flux reflected in the direction of mirror reflection (the specular direction) when a specimen is illuminated by a parallel light beam.

The standard methods of measuring gloss are numerous. Superficially many of the methods are similar, but small differences in optical geometry can cause large variations in results from one type of instrument to another. The greatest difficulty with gloss measurements lies in correlating measurements with visual impressions and Hammond[14], Dinsdale and Malkin[15], Hunter[16] and Elm[17] discuss the difficulties at some length.

Knittel[18] points out that an observer in assessing visual gloss turns the sample over and over using many angles of incidence and reflection and his judgement is therefore based on a whole series of observations.

Because an absolute standard for gloss does not exist, existing standards (usually optically flat black glass plates) are assigned values which depend on the refractive index of the glass employed and which may vary from one test to another.

16.4.2 Specification Tests

A.S.T.M. D.523–67, 'Specular Gloss', is the recommended standard for measurements on plastics (excepting films) and uses a 20 degree (to the vertical) incident and reflected light beam geometry for high gloss materials and 85 degree geometry for low gloss materials. An intermediate 60 degree angle is used for inter-comparative purposes and for deciding which of the other angles should be used with a given specimen.

Although the geometry of source and receiver is defined with close tolerances on the angles, the actual instrument to be used is not described in great detail and Figure 16.3 taken from the standard merely indicates (in plan) the general layout. The primary standard of gloss is a highly polished, plane, black glass surface with a refractive index of 1·567 to which is assigned the arbitrary value of 100 gloss units for each of the three geometries (the reader is warned that elsewhere and in other circumstances different values may be assigned to similar gloss standards).

The specification requires that any measurements should be in accordance with C.I.E. Source C values without actually stating that this source should be used, the inference being that the type of source is not very critical.

For films, A.S.T.M. D.2457–65T, 'Specular Gloss of Plastic Films', is based on the A.S.T.M. D.523 document for the 20 degree and 60 degree angle tests, but the third angle is 45 degrees, not 85 degrees, and the 45 degree test is based on a method for ceramics (A.S.T.M. C.346). The same black glass standard is used at 20 degrees and 60 degrees as in the A.S.T.M. D.523 test, but at 45 degrees the refractive index is 1·540 and the assigned gloss value is *54·5*. A note warns of the possibility of obtaining gloss values on clear films of more than 100 units with any one of the three geometries, because of reflection at both surfaces of the specimen. Three specimen mounting devices are described for ensuring the flatness so essential in gloss measurements.

No B.S. gloss test exists for solid plastics (as distinct from film) at the present time, but B.S. 3900, 'Methods of Test for Paints', Part 2: 1967, 'Gloss (Specular Reflection Value)', and B.S. 3962, 'Methods of Test for Clear Finishes for Wooden Furniture', Part 1: 1965, 'Test for Low-angle Glare', are both relevant to modern polymer based paints and varnishes.

The B.S. 3900 test is based on a well known British defence specification DEF 1053 Method No. 11, 'Gloss (Specular Reflection Value)', and employs

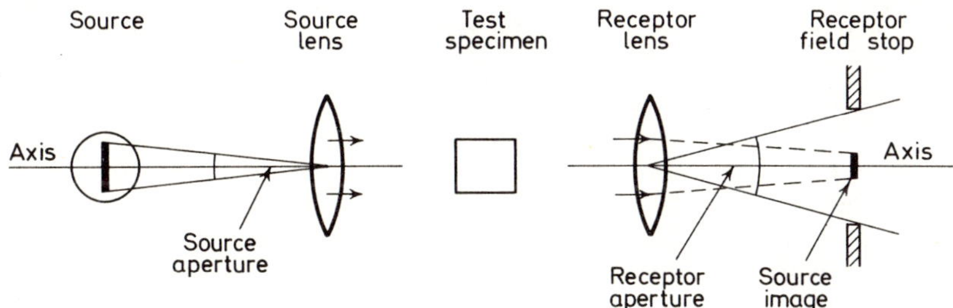

Figure 16.3. Generalised glossmeter showing apertures and source image for a collimated beam type instrument (A.S.T.M. D.523)

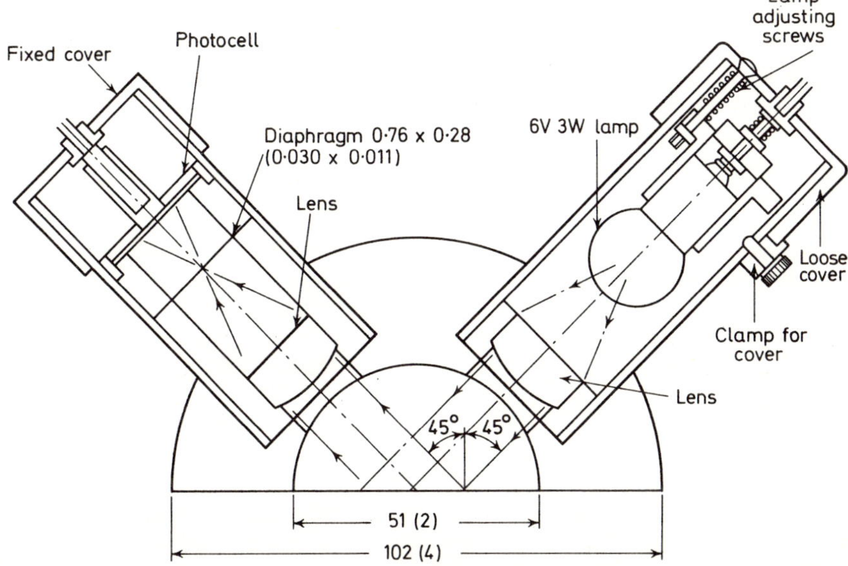

Figure 16.4. High-gloss head for paints (B.S. 3900). Dimensions in millimetres with inch equivalents in parentheses

the 45 degree geometry head marketed by Evans Electroselenium Ltd.[19] as their 'High gloss Head'. The optical arrangement is shown in Figure 16.4.

The standard is unique in quoting figures for the variation in gloss of blackened glass with refractive index as shown in Table 16.1.

The importance of flatness in the reflectance standard is underlined by the tolerance permitted, viz. plane to 2 fringes/cm. The standard itself is highly polished clear glass with its underside and edges roughened and coated with

Table 16.1 VARIATION OF GLOSS WITH REFRACTIVE INDEX OF BLACKENED GLASS (B.S. 3900)

Refractive index	1·4	1·5	1·523	1·55	1·6	1·7
Maximum specular reflection (μ of 1·523 assumed ≡ 100 units)	68	94	100	107	120	148

black paint. The refractive index is 1·523±0·002 and the gloss value assigned to it is 100 units. The head is placed on it and the photocell output or galvanometer input adjusted to 100 scale divisions. When placed on the test specimen, in this case coated plate glass, the gloss is given direct. Gloss values above 100 are evidently not impossible according to the operating procedure given.

The B.S. 3962 gloss head employs a 75°/75° (to the vertical) geometry and the Evans Electroselenium Ltd.[19] 'Variable-angle Glossmeter' conforms to the requirements shown in Figure 16.5. This test is used to assess low angle

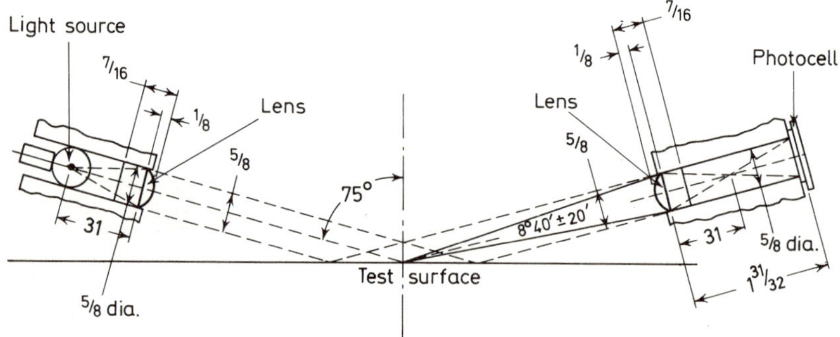

Figure 16.5. Apparatus for measuring low-angle glare of clear finishes for wood (B.S. 3962). Dimensions in inches

glare emanating from clear finishes used on school furniture, which may present a nuisance.

The gloss standard used is identical to that of B.S. 3900 and the procedure similar.

B.S. 2782: 1970 'Methods of Testing Plastics', Method 515B, 'Gloss (45°) of sheet', uses a 45° optical system and the reference standard is highly polished black glass of refractive index 1·52, to which a gloss value of 53 units is assigned. On this scale a perfect mirror gives a value of 1000. The gloss head is similar to that of B.S. 3900 with important but minor alterations to the aperture. For films a clamp is described, which enables specimens to be held completely taut and free from wrinkles. The light source specified is C.I.E. Source C.

16.5 COLOUR

The physical measurement of colour, as opposed to matching by eye, is not widely used as a control test for plastics, but since it is an important part of the general plastics scene, for aesthetic and other reasons a number of references will be given. For convenience these are in chronological order.

Willmer, in an interesting paper[20] deals with the physiology of perception of colour and in the same publication Vickerstaff and Walls[21] discuss matching and measurement of colour. Goodwin[22] discusses small colour differences and gives some examples in colour. Adams[23] is concerned with basic fundamentals.

Wright's report[24] on a symposium entitled 'Colorimetry: Its Errors and Accuracy' is of interest for its account of intercomparison measurements on standard tiles circulated to U.K. laboratories, the instruments including General Electric (Hardy) and Beckman spectrophotometers. The use of the General Electric instrument is also described by Morse in Reference 25 and a number of other U.S. instruments are described in the same reference. The Gardner 'Automatic Color Difference Meter' and its application to colour matching is described by Menard[26]. Foote has reviewed the methods of measuring colour and the commercial instruments available listing a score of the latter, and giving over 150 references[27]. See Reference 28 for general reading. Finally, B.S. 1611: 1953, 'Glossary of Colour Terms used in Science and Industry', is relevant.

16.6 MICROSCOPY

The use of the optical microscope, and even the humble pocket lens, as adjuncts in the testing of plastics should not be overlooked. The examination of tensile and impact specimens for edge notches or other stress raisers and the examination of the surfaces of an electric strength specimen for voids and bubbles before test are two examples where low power microscopy could be used with value.

Weber[29] and Morehead[30] deal with two important areas where low powered microscopy is extensively employed, respectively those of glass fibre laminates and films. The examination of materials and specimens containing fillers is covered by Wredden[31] in a long series of articles, and by Stephens and Langton[32], the former dealing with processing and composition faults, the latter being more concerned with the techniques of sample preparation. However, beyond this microscopy is more of an adjunct to analysis and is therefore mainly outside the scope of this handbook.

The use of the electron microscope is also beyond the scope of this book, but some references will be given for completeness, dealing specifically with the examination of polymers, viz. Menter[33], Andrews[34], Langton and Stephens[35] and Reding and Walter[36].

REFERENCES

1. WILEY, R. H. and HOBSON, P. H., 'Determination of Refractive Index of Polymers', *Anal. Chem.*, **20,** No. 6, 520 (June 1948)
2. WILEY, R. H. and GARRETT, R. R., 'Spectrometric Determination of Refractive Index at Low Temperatures', *J. Polymer Sci.*, **7,** Nos. 2 and 3, 121 (August–September 1951)
3. MCALISTER, E. D., VILLA, J. J. and SALZBERG, C. D., 'Rapid and Accurate Measurements of Refractive Index in the Infrared', *J. opt. Soc. Amer.*, **46,** No. 7, 485 (July 1956)
4. BLACK, E. P., HARVEY, W. T. and FERRIS, S. W., 'High Temperature Refractometry with an Abbé-type Instrument', *Anal. Chem.*, **26,** No. 6, 1089 (June 1954)

5. BILLMEYER, F. W., 'Measurements of the Refractive Index of Films', *J. appl. Phys.*, **18,** No. 5, 431 (May 1947)
6. ELLIS, R. H., 'Measuring Refractive Index of Plastic Films', *Rev. Sci. Instrum.*, **28,** No. 7, 557 (July 1957)
7. SCHAEL, G. W., 'Determination of Polyolefin Film Properties from Refractive Index Measurements', *J. appl. Polymer Sci.*, **8,** No. 6, 2717 (November 1964)
8. FAUST, R. C., 'Determination of the Refractive Indices of Inhomogeneous Solids by Interference Microscopy', *Proc. Phys. Soc.*, **67,** No. 410B, Part 2, 138 (February 1954)
9. WEBBER, A. C., 'Method for the Measurement of Transparency of Sheet Materials', *J. opt. Soc. Amer.*, **47,** No. 9, 785 (September 1957)
10. MILES, J. A. C. and THORNTON, A. E., 'See-through Clarity of Polythene', *Brit. Plast.*, **35,** No. 1, 26 (January 1962)
11. NIEGISCH, W. D., 'Voids as a Source of Haze in Weathered Poly (2, 2, 4, 4, Tetramethyl–1, 3–cyclobutylene Carbonate)', *J. appl. Polymer Sci.*, **13,** No. 11, 2483 (November 1969)
12. MEYER, R. W. and SMITH, D., 'The S.P.I. Method of Measuring Light Transmission of Reinforced Plastic Panels', *Reinf. Plast.*, **2,** No. 5, 12 (January 1958)
13. HUNTER, R. S., 'The Measurement of Color and Gloss', *Product Engng*, **27,** No. 2, 176 (February 1956)
14. HAMMOND, H. K., 'Gloss Measurement—Past, Present and Future', *Amer. Paint J.*, **41,** No. 37, 95 (May 1957)
15. DINSDALE, A. and MALKIN, F., 'The Measurement of Gloss with Special Reference to Ceramic Materials', *Trans. Brit. Ceram. Soc.*, **54,** No. 2, 94 (February 1955)
16. HUNTER, R. S., 'Gloss Evaluation of Materials', *Bull. Amer. Soc. Test. Mat.* **No. 186,** 48 (December 1952)
17. ELM, A. C., 'Reflections on Gloss', *Off. Dig. Fed. Paint Varn. Prod. Cl.*, **33,** No. 433, 163 (February 1961)
18. KNITTEL, R. R., 'Development of a Gloss Method for Transparent Plastic Films', *Mater. Res. Stand.*, **2,** No. 3, 180 (March 1962)
19. Evans Electroselenium Ltd., Colchester Road, Halstead, Essex.
20. WILLMER, E. N., 'The Physiology of the Perception of Surface Colours', *J.O.C.C.A.*, **36,** No. 399, 491 (September 1953)
21. VICKERSTAFF, T. and WALLS, I. S. M., 'Colour Matching and Colour Measurement', *J.O.C.C.A.*, **36,** No. 399, 507 (September 1953)
22. GOODWIN, W. J., 'Measurement and Specification of Color and Small Color Differences', *Mod. Plast.*, **32,** No. 10, 32 (June 1955)
23. ADAMS, J. M., 'Fundamentals of Colour', *Paint Manuf.*, **26,** No. 4, 105 (April 1956)
24. WRIGHT, W. D., 'A Challenge to Colorimetry', *Nature, Lond.*, **179,** No. 4552, 179 (January 1957)
25. MORSE, M. P., 'Colour Measurement with the General Electric Spectrophotometer', *Off. Dig.*, **28,** No. 383, 1278 (December 1956)
26. MENARD, D. F., 'Measuring Color Difference', *Mod. Packag.*, **32,** No. 7, 115 (March 1959)
27. FOOTE, P. V., 'Measurement of Colour and Colour Difference', *Rev. Curr. Lit. Paint All. Ind.*, **37,** No. 259, 1 (January 1964)
28. BILLMEYER, F. W. and SALTZMAN, M., *Principles of Colour Technology,* Interscience Publishers, London (1967)
29. WEBER, M. K., 'Microscopic Examination of Glass Fiber Reinforced Plastics', *Bull. Amer. Soc. Test. Mat. No. 208,* 49 (September 1955)
30. MOREHEAD, F. F., 'Modern Microscopy of Films and Fibers', *Bull. Amer. Soc. Test. Mat. No. 163,* 54 (January 1950)
31. WREDDEN, J. H., 'The Microscopic Examination of Plastics Materials', A series of articles starting in *Plastics, Lond.*, **9,** No. 98, 341 (July 1945) and finishing in **13,** No. 146, 388 (July 1949)
32. STEPHENS, M. and LANGTON, N. H., 'The Microscopic Examination of Rubbers, Plastics and Fillers', *Plastics, Lond.*, **25,** No. 274, 329 (August 1960)
33. MENTER, J. W., 'Electron Microscopy', *J. Roy. Inst. Chem.*, **86,** No. 11, 415 (November 1962)
34. ANDREWS, E. H., 'Microstructure of Melt Crystallised Polythene', *J. Polymer Sci.* B., **3,** No. 5, 353 (May 1965)
35. LANGTON, N. H. and STEPHENS, M., 'The Electron Microscopy of High Polymers', *Plastics, Lond.*, **23,** No. 253, 384 (October 1958) also *Plastics, Lond.*, **23,** No. 254, 422 (November 1958)
36. REDING, F. P. and WALTER, E. R., 'An Electron Microscope Study of the Growth and Structure of Spherulites in Polyethylene', *J. Polymer Sci.*, **38,** No. 133, 141 (July 1959)

17

Tests for Specific Types of Products

17.1 INTRODUCTION

In the preceding chapters, each property or group of properties of plastics materials has been taken in turn and has been discussed and described without, so far as possible, specific reference to any particular material or finished shape. The title of this monograph is 'A Handbook of Plastics Test Methods' and an endeavour has been made to keep to a logical interpretation of this phrase. The subject is vast enough as it stands, even at the relatively modest level of coverage attempted, without wandering into the realms of testing the many fabricated forms of plastics, describing the various specific tests that have been developed and/or standardised for each particular plastic or, even worse, embracing those quasi-tests (more correctly analytical methods) which exist for filler content, volatile matter, monomer content, anti-oxidant concentration, etc., etc. For the latter, reference may be made to, in particular, B.S. 2782: 1970, 'Methods of Testing Plastics', the product specifications given in Appendix 2 of Chapter 2 (above), A.S.T.M. Parts 26 and 27 and, as more textbooks, References 1–3. Similarly the B.S. product specifications and A.S.T.M. Parts 26 and 27, should be consulted for standard tests on fabricated forms such as finished mouldings and the technical literature contains many articles on ad hoc methods developed to deal with particular proving or quality control methods. Thus it is not intended to consider the evaluation of a polyethylene washing-up bowl or a fabric reinforced resin gear wheel. What has been attempted, however, is a description of how to measure the important properties of thermoplastics and thermosets of which polyethylene and fabric reinforced resin are respectively examples.

The reader must not, therefore, be surprised at the omission of such an old friend as 'acetone soluble matter'—which is specific to phenol-formaldehyde type materials. (Methods 401A, B and C of B.S. 2782: 1970 cover the subject adequately, including a comment on the limitations of the values obtained from such methods.) Likewise many tests specific to laminates, casting resins, plastics coated fabrics, etc., are omitted. There remain, however, a number of important tests which have not fallen logically into the previous chapters and the purpose of this chapter is to tie up as many 'loose ends' as possible.

17.2 EXPANDED, CELLULAR OR FOAM PLASTICS

In the light of the reservations made in Section 17.1 above, only four properties will be mentioned: closed cell count, permeability, water vapour permeability and resilience. Others, such as thermal conductivity, compression strength and flammability may be of equal or greater importance, but have either already been covered (at least in principle) and/or are described adequately in the relevant product standards. The British Standards may be found from Appendix 2 of Chapter 2, particularly B.S. 3667: 1963–6 and B.S. 4370: 1968 of Group I thereof, the A.S.T.M. standards from Parts 26 and 27 thereof (and also Reference 4), but mention should be made of the relevant DIN. standards, as the coverage of test methods is quite comprehensive:

DIN.7726 Definitions of foam materials; classification
DIN.16990 Rigid Cellular Plastics Based on Polyurethane; Classification and Designation
DIN.18164 Foamed Plastics as Insulating Building Materials; Dimensions, Properties and Testing
DIN.53420 Density
DIN.53421 Compression stength–(Rigid foam materials)
DIN.53422 Shear strength—(Rigid foam materials)
DIN.53423 Bending strength—(Rigid foam materials)
DIN.53424 Temperature of deflection under bending and compression load—(Rigid foam materials)
DIN.53425 Compression creep due to heating—(Rigid foam materials)
DIN.53426 (Draft) Determination of dynamic elastic modulus and loss factor
DIN.53428 Determination of the resistance to liquids, vapours, gases and solid materials
DIN.53429 (Draft) Determination of water vapour transmission (Rigid foam materials)
DIN.53570 Determination of linear dimensions (Flexible foam materials)
DIN.53571 Tensile strength (Flexible foam materials)
DIN.53572 Compression set (Flexible foam materials)
DIN.53574 Dynamic flexing test: Determination of the compression and shear fatigue (Flexible foam materials)
DIN.53575 Tear resistance (Flexible foam materials)
DIN.53576 Hardness number by compression (Flexible foam materials)
DIN.53577 Compression strength (Flexible foam materials)

Useful general articles on the evaluation of expanded plastics materials, and some guide to their properties, will be found in References 5–13 inclusive.

17.2.1 Closed Cell Content Porosity

For rigid cellular materials, a method for determining 'porosity' is given in A.S.T.M. D.1940–62T, based on the work of Remington and Pariser[14]. The apparatus is shown in Figure 17.1.

The apparatus in Figure 17.1 comprises:

1. The specimen chamber system consisting of the specimen chamber A, the left half of mercury manometer B and the left half of auxiliary manometer C. The total volume of this system is V_1 (between 100 and 125 ml).
2. The reference chamber system consisting of air chamber E, auxiliary bulb J (volume V_j), the right half of mercury manometer B, the right half of auxiliary manometer C and gas burette D. The total volume of this system is V_2 (between 100 and 125 ml).

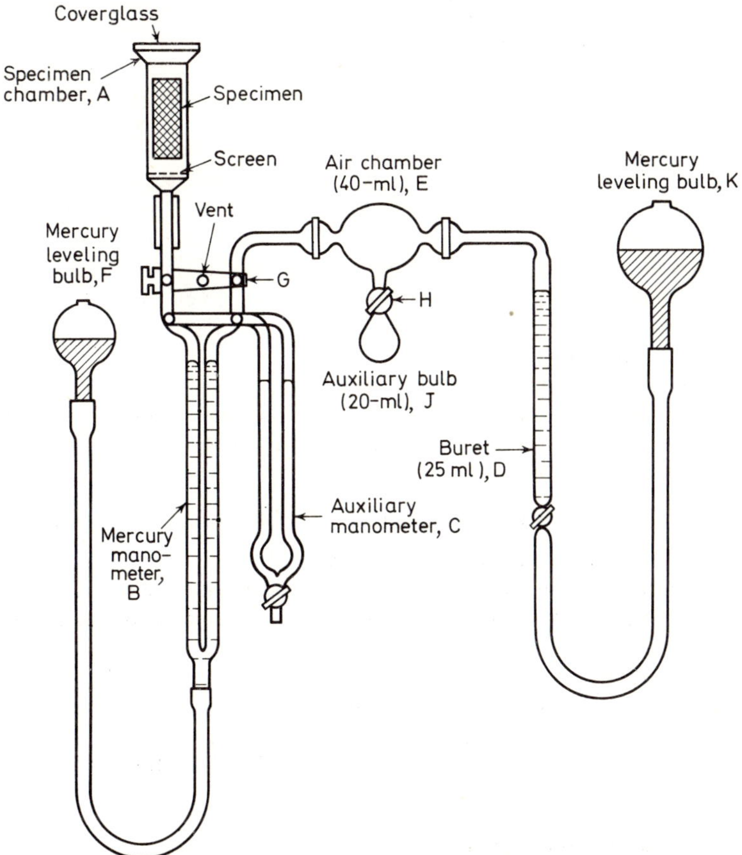

Figure 17.1. Remington-Pariser apparatus (A.S.T.M. D.1940)

The specimen chamber A is designed to contain a maximum specimen volume of approximately 3 in^3 (49 cm^3). The gas burette D, by means of mercury levelling bulb K, is used to match the volume of air on the specimen side of the apparatus. The calibrated auxiliary bulb J is required when measuring foams of greater than approximately 65% open cells. The mercury manometer B, the level of which is controlled by adjusting bulb F, connects chambers 1 and 2. It is used to apply a slight vacuum to the two systems so that differences of pressure may be detected. Auxiliary manometer C, containing dibutyl phthalate, is connected in parallel with the mercury manometer for sensitivity in detecting any pressure differential.

At least five specimens are tested, each carefully prepared and of about 2 in^3 (30–35 cm^3), either cylindrical or bar shaped. The dimensions and weights are accurately taken and the approximate cell size obtained by observing a square inch of surface, selected at random, under a low powered microscope (20 ×).

The whole apparatus must be kept constant to $\pm 1{\cdot}8$ degF, preferably at $73{\cdot}4 \pm 1{\cdot}8$°F (23 ± 1°C), and 50 ± 2% relative humidity. The specimens are preconditioned for at least one hour.

The apparatus is brought to balance by (a) closing the specimen chamber tightly and bringing the mercury level in manometer B to the zero position with stopcock G open to the atmosphere, (b) opening stopcock H (to include auxiliary bulb J), at the same time setting the mercury level in gas burette D at about 20 ml, (c) isolating the volumes of air at atmospheric pressure in both systems by closing stopcock G. The mercury level in manometer B is lowered by approximately 2 ml (by lowering F); if the levels in both arms remain equal, then V_1 equals V_2. (Procedures are given to correct the whole system if this equality is not observed.) The mercury level in the burette is used as reference R.

To check the accuracy of the apparatus a piece of solid metal of known volume, about 2 in^3 (33 cm^3), is placed in A and the chamber resealed. The systems are brought to balance as before, with auxiliary bulb J removed by closing stopcock H, and the burette level is read at balance, (R_1).

$$\text{Volume of metal} = R-(R_1-V_j)$$

This should agree within $\pm\frac{1}{2}$% of the true value.

To examine a foam specimen the same procedure is followed as with the metal piece; if the foam is predominantly closed cell, stopcock H must be closed.

Percentage open cells

If stopcock H is open for both R and R_1:

$$P = \frac{(v_1-\Delta V_1)}{v_1} \times 100 = \frac{(v_1+R_1-R)}{v_1} \times 100$$

If stopcock H is open for R_1:

$$P = \frac{(v_1-\Delta V_1)}{v_1} \times 100 = \frac{(v_1+R_1-V_j-R)}{v_1} \times 100$$

Percentage closed cells $= 100{\cdot}0-P-C_w$

where C_w = Percentage of cell wall $= \dfrac{D_f}{D_s} \times 100$

where D_f = Density of foam
D_s = Density of solid polymer
v_1 = Volume of specimen (by mensuration)
V_1 = Volume of air displaced

Methods are given for correcting the values for (i) the cutting of the specimens (to expose cells) and (ii) differing size specimens.

17.2.2 Permeability

The properties of flexible foams are very dependent on a similar function, conveniently measured as the air permeability. R. E. Jones and G. Fesman[15] have described an apparatus for the measurement of this property—see Figure 17.2.

The operation of the apparatus is obvious, a pressure drop of 0·5 ± 0·002 in water gauge across the specimen being recommended. In the article results

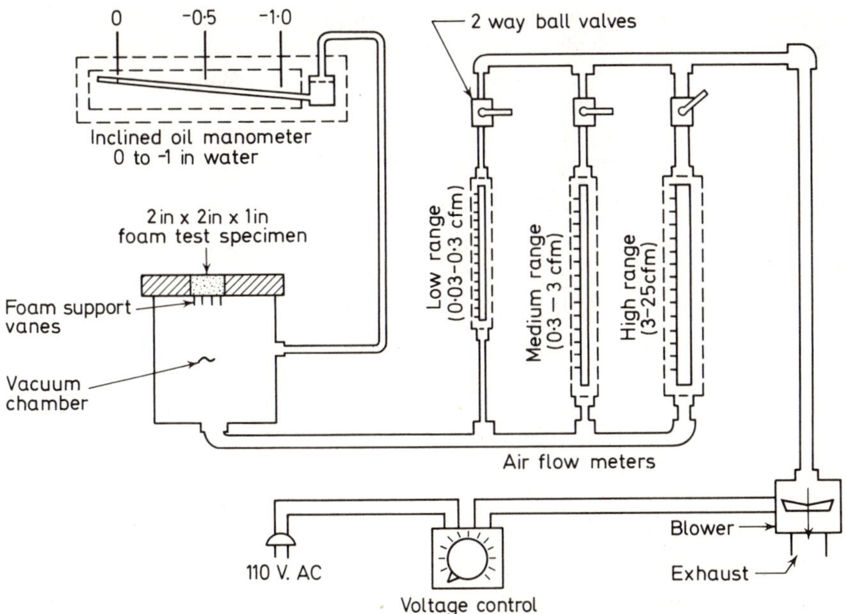

Figure 17.2. Air flow apparatus (Jones and Fesman)

are correlated with a number of process variables and those of ball rebound tests (see below, Section 17.2.4). Theoretical considerations of the permeability of open cell foamed materials are provided by Gent and Rusch[16].

17.2.3 Water Vapour Permeability

Water vapour transmission significantly affects the thermal efficiency of expanded plastics, probably the most important characteristic of the rigid ones at least, as shown by Levy[17]. A simple method is included, as Appendix C, in B.S. 3837: 1965, 'Expanded Polystyrene Board for Thermal Insulation Purposes'. The specimen is a right cylinder 2 ± 0·05 in (50·8 ± 1·3 cm) in height and diameter 0·03 in (0·8 mm) greater than the internal diameter of a light cylindrical can, or glass beaker of approximately $2\frac{1}{2}$ in (64 mm) internal diameter and $3\frac{1}{2}$ in (89 mm) height. The top of the can or beaker is slightly 'belled out', as shown in Figure 17.3.

The can is filled with granulated anhydrous calcium chloride to about $\frac{1}{4}$ in (6·3 mm) below the expected lower surface of the specimen (the weight of dessicant must be at least ten times the total expected take-up water during the test). The specimen is inserted into the can or beaker so that its top surface is flush with the top of the can or beaker. Melted wax (e.g. 90% micro-crystalline wax and 10% plasticiser) is run in to seal the space between the 'belled out' rim and the upper edge of the perimeter of the specimen, the temperature of the wax being insufficient to soften the specimen. The whole assembly is then weighed to 0·001 g and placed in an atmosphere of $38 \pm 0{\cdot}5$°C and 90 ± 2% relative humidity. Successive weighings are taken at 24 h intervals and the cumulative weight increase plotted against time, until at least three points (excluding the origin) lie on a

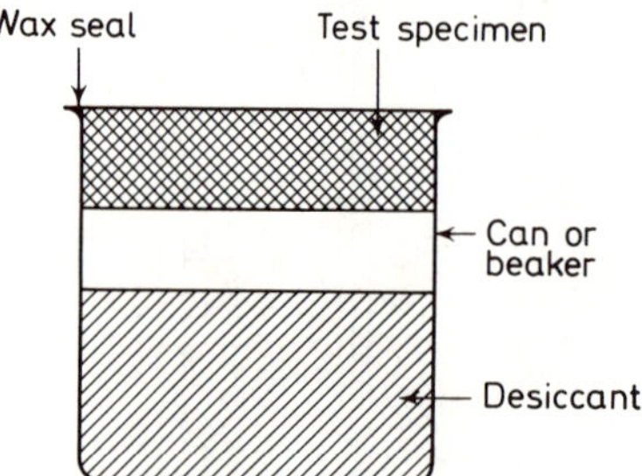

Figure 17.3. Water vapour transmission test (B.S. 3837)

straight line. (The assemblies are allowed to cool before weighing, but for the minimum of time necessary and this must be constant each time.)

$$\text{Permeability} = \frac{240x}{Ay}\ \text{g/m}^2/\ \text{day}$$

where x mg is the weight increase over y h and A is the exposed area in cm^2. Three specimens are examined and the results averaged.

A.S.T.M. C.355–64, 'Water Vapor Transmission of Thick Materials', contains a very similar method and also one where a dish contains water and the rate of permeation through the material into a defined atmosphere is measured.

17.2.4 Resilience of Flexible Foams

Flexible foamed materials can vary, according to their structure particularly, from 'dead' energy absorbing types through to extremely 'bouncy' resilient—with appropriate uses in each case. Resilience is defined as the ratio of the energy returned to the energy imparted (%). Jones et alia[18] describe a dynamic method based on ball rebound. $4 \times 4 \times 2$ in specimens are used, with a $\frac{5}{8}$ in diameter steel ball dropped from 16 in or 18 in. The ball is released by electromagnet down a calibrated transparent tube, the lower end of which just rests on the upper 4×4 in surface of the specimen. The height of rebound is determined and expressed as a percentage of the original fall height to obtain the resilience.

A method using a pendulum was described by Rosenthal and Addis[19] and is shown in Figures 17.4 and 17.5.

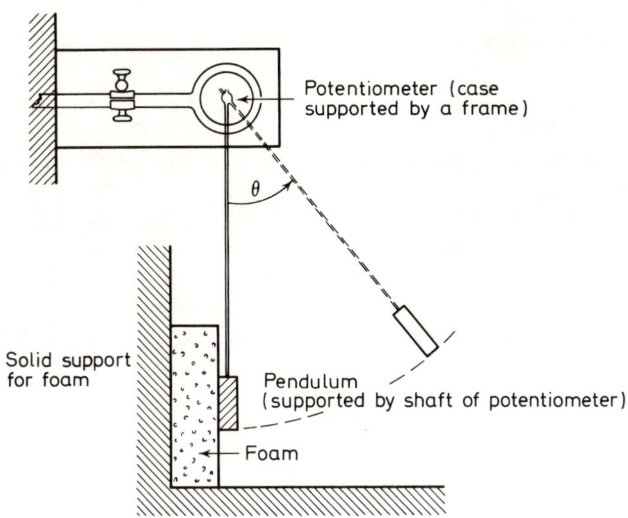

Figure 17.4. Mechanical layout of rebound pendulum (Rosenthal and Addis)

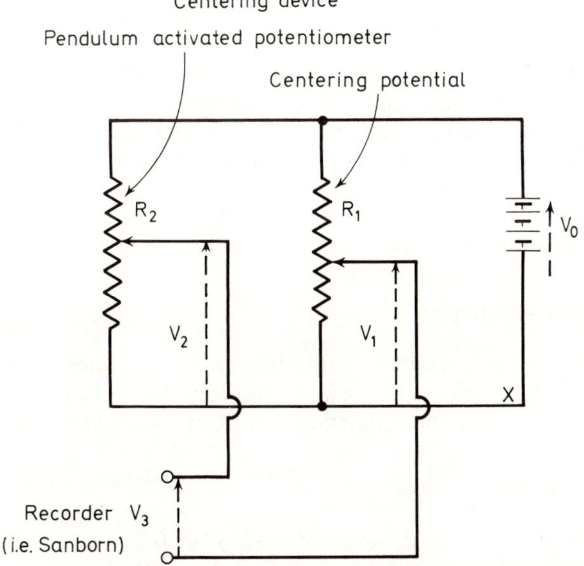

Figure 17.5. Electrical circuit of Figure 17.4

The pivot of the pendulum (Figure 17.4) was made into a precision ball bearing potentiometer and the output fed into a recorder. The electrical circuit (Figure 17.5) is so arranged that the recorder registers zero with the pendulum at rest and just touching the surface of the foam specimen (Figure 17.4) and that the voltage V_3 recorded is directly proportional to the angle of displacement θ; thus the energy of the pendulum before release and after successive indentations and rebounds may be computed.

17.3 PIPES AND TUBES

Probably the most important property of a plastics pipe or tube, that is its long term bursting strength, has been considered in Chapter 14; short term quality control tests follow a similar pattern and reference should be made to the appropriate products specifications (e.g. for the British Standards see Appendix 2 of Chapter 2 (above)) wherein are described also such tests as for reversion, softening point and impact strength; none of these involves any new principle that has not already been described for the material in the relevant preceding chapters. For the transport of potable water, however, there remains the very important matter of toxicity hazard. This subject is mentioned in Section 17.4 (below) in relation to sheet and film and, as the principles are the same, reference should be made to that section for dealing with the problem as applied to pipes.

There is a vast amount of technical literature on the properties and testing of plastics pipes and tubes; by way of example the reader is referred to References 20–24.

17.4 FILM AND SHEET

Under this heading it is proposed to consider permeability, impact resistance, tear strength and, briefly, examination for toxicological hazards of flexible film and sheet. The general principles relating to the last mentioned subject apply equally to any packaging contacting foodstuffs or beverages or any products likely to be ingested (such as children's toys!). Highly important topics such as mechanical and thermal properties generally, electrical properties, light transmission, haze and glass, have either been described specifically or covered in general terms, in the relevant chapters above—7 to 13 inclusive. Some useful references to the testing of film and sheet will be found in References 25–35.

17.4.1 Permeability

There is already in existence a Plastics Institute[36] publication devoted specifically to this subject; in view of this and the availability of another useful reference[37], only the important standard tests will be described.

WATER VAPOUR PERMEABILITY (W.V.P.)

Basically, two different methods are described in B.S. 2782: 1970, 'Methods of Testing of Plastics', the so-called dish and sachet methods, though each exists in two variants:

1. At 'Temperate' conditions of 25±0·5°C and relative humidity 75±2% (Methods 513A and 513C)
2. At 'Tropical' conditions of 38±0·5° and relative humidity 90±2% (Methods 513B and 513D)

In the dish methods (513A and 513B) shallow aluminium dishes are used,

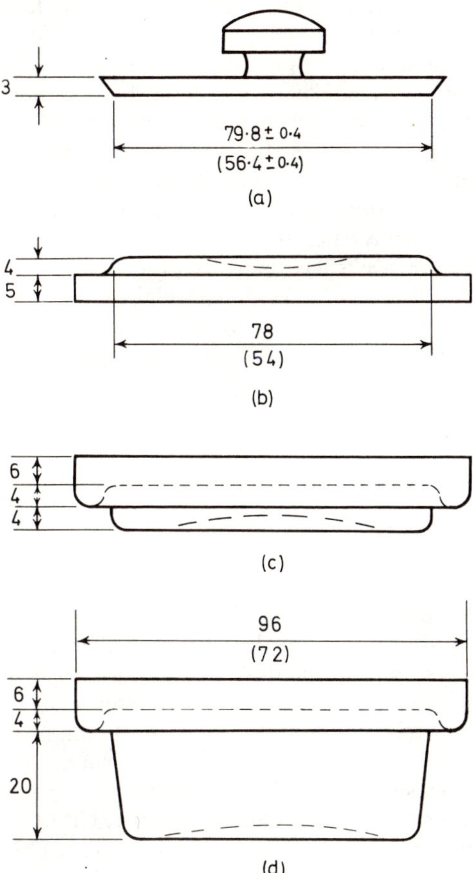

Figure 17.6. Water vapour permeability dishes and templates (B.S. 2782). (a) Waxing template, (b) lid, (c) shallow dish for materials of normal permeability, (d) deep dish for materials of permeability greater than 100 g/m²/day tapered to nest. Dimensions are shown for test areas of 50 cm² and 25 cm², the latter in brackets. Dimensions in millimetres. Figures are inside dimensions, except for overall diameters of dishes, which are outside dimensions. Recommended material: aluminium sheet about 1 mm thick

so designed that the exposed area of film test specimen is accurately defined at either 25 cm^2 or 50 cm^2, edge transmission of water vapour being prevented by a wax seal (a suitable composition is 90% microcrystalline wax and 10% plasticiser; it must change in weight by less than 0·001 g when a sample of surface area 50 cm^2 is exposed for 24 h to an atmosphere of type 2 above).

To carry out the test, calcium chloride or similar dessicant is placed in the dish to within 1–2 mm of the supporting ring on which the test specimen is then placed centrally (projecting half way over the annular recess of the dish). The template, waxed on its edges *only* to facilitate removal later, is next placed centrally over the dish and specimen and, without delay, molten wax is run into the annular recess until it is level with the upper surface of the template. The waxing template is then removed carefully.

Sufficient test assemblies are prepared to give a total test specimen area of not less than 100 cm^2 and placed in a humidity cabinet controlled to give conditions 1 or 2 (as required) over the specimen surface. Successive weighings are taken at suitable intervals and the cumulative weight increase of each assembly, in mg, is plotted against total time, ensuring that the test does not last for such a duration that the relative humidity in the dish rises above 2% (for anhydrous calcium chloride as dessicant this will be ensured if the total weight increase is less than 1·5 g for 15 g of dessicant). Weighings are made to an accuracy of 1% of the weight change between successive weighings or 0·001 g, whichever is the larger. When three, but preferably four, points of weight plotted against time lie on a straight line, the test is discontinued.

A best straight line is drawn for each assembly and from the slope the permeability, in grammes per square metre per 24 h, is calculated as $240x/Ay$ where x mg is the weight increase over a period of y h for an exposed area of A cm^2. The average value is reported to two significant figures but not closer than to the nearest 0·2 units, and the thickness of the test specimen must be stated. Permeability is not invariably proportional to thickness.

The sachet methods (513C and 513D) are suitable for thermoplastics film which can be heat sealed and when the transmission is known to be considerably less through the seal than through the film.

The technique is similar to the dish method already described except that the dessicant is heat sealed into a sachet of the film made from two pieces of film each approximately 100 mm (4 in) square. Using a suitable heat sealer, three edges of the two pieces are heat sealed to form an open ended bag, the inner edges of the seals being 76 ± 1 mm ($3 \pm 0{\cdot}05$ in) apart. A bag of blotting paper or filter paper, is suitably folded so that it will fit into the sealed bag just described and about 15 g of anhydrous calcium chloride poured into the paper bag. The latter is folded over and sealed with paperclips, after which it is placed in the sealed bag and the fourth side thereof is heat sealed to form a sachet of nominal area 116 cm^2 (18 in^2)—i.e. two sides each of 58 cm^2 (9 in^2).

Two such sachet assemblies are prepared and the procedure described above for Methods 513A and B is followed, taking care that both surfaces of each sachet are freely exposed.

All four methods are stated to be suitable for film materials of permeability not less than 1 g/m^2/day.

In A.S.T.M. E.96–66, 'Water Vapor Transmission of Materials in Sheet Form', the dish method is used for all six alternative procedures described. In only three of these, however, (A, C and E) is the dessicant placed in the dish, the first in an atmosphere of 23·0°C (73·4°F) and the second of 32·2°C (90°F), both at a relative humidity of 50%. In Procedure E the temperature is 37·8°C (100°F) and the relative humidity 90%. In Procedures B, BW and

D water is placed in the dish with the temperatures respectively 23·0°C (73·4°F), 23·0°C (73·4°F) and 32·2°C (90°F). The external relative humidity is 50% in each case and in Procedure BW the dish is inverted so that the water contacts the film specimen.

The A.S.T.M. method defines, in addition to 'water vapour transmission' (w.v.t.) as identical to w.v.p. in B.S. 2782 above, 'water vapour permeance' as w.v.t. divided by the vapour pressure differential across the specimen surfaces, and 'water vapour permeability' which is the permeance reduced to unit thickness (see above for comment on calculation of results from Methods 513A and B of B.S. 2782). Water vapour permeance is quoted in 'Perms', one metric perm being 1 g per 24 h per m^2 per mm Hg.

GAS PERMEABILITY

Method 514A of B.S. 2782: 1965 is described as 'Gas Transmission Rate ('Permeability')' and is for determining the rate of transmission of gases through film. The latter is defined as the volume (at n.t.p.) of gas in ml/m^2 passing in 24 h through a test specimen of the film when the difference in pressure between the two faces is one atmosphere and the temperature is 23 ± 2°C. Two procedures are specified: (1) for air, where the pressure is two atmospheres on one side of the specimen and one atmosphere on the other, and (2) for air or other gases, where the pressure is one atmosphere on one side and less than 0·3 mbar (0·2 mm Hg) on the other. The apparatus is shown in Figure 17.7.

The lower part of the (metal) cell contains a cavity of internal diameter 50–70 mm over which the test specimen is clamped. A filter of rigid porous material, e.g. sintered glass or sintered bronze, is fitted in the cavity, to support the film specimen and allow gas to flow to the capillary tube at the bottom. The upper surface of the filter is level with the top of the cavity so that the film lies flat when clamped. The volume of the capillary over its working length must not be more than 5% of the net volume of the cavity and the cross-section of the enlarged side tube must be at least 100 times that of the capillary tube.

In Procedure I the total volume of the mercury reservoir and the enlarged side tube should be at least 100 times the volume of the cavity; this is achieved if necessary by the addition of the glass vessel shown on the r.h.s. of the figure. The amount of mercury in the reservoir is such that when it is tipped into the capillary and side tubes the level is near the bottom of the enlarged portion.

In the test, the upper part of the cell is removed, stopcock B opened to the air and stopcock A opened, and the lower part of the apparatus tilted so that mercury flows into the capillary tube and sides to such a level that the total movement of mercury during the subsequent test will take place within the working length of the capillary. The apparatus is then set upright again, a test specimen placed over the cavity and filter and the (elastomeric) gasket placed over it as shown. The upper part of the cell is then clamped down uniformly to obtain a gas tight seal and stopcock B is closed. The air pressure in the upper part of the cell is then raised to two atmospheres and the changes in height of mercury (changes in pressure) in the capillary tube are recorded at appropriate intervals. According to the absolute values of the gas transmission rate, so the accuracy of measurement and intervals of time are altered (as specified), at least six readings of height must be taken that have a linear relationship with time and the rate of change of the mercury height in the capillary tube must not be less than 10 mm per 24 h.

In Procedure II, the section of the apparatus to the right of joint J is removed. With the mercury in the reservoir, after the specimen has been fitted, the upper part of the cell is clamped to give a vacuum tight seal. The

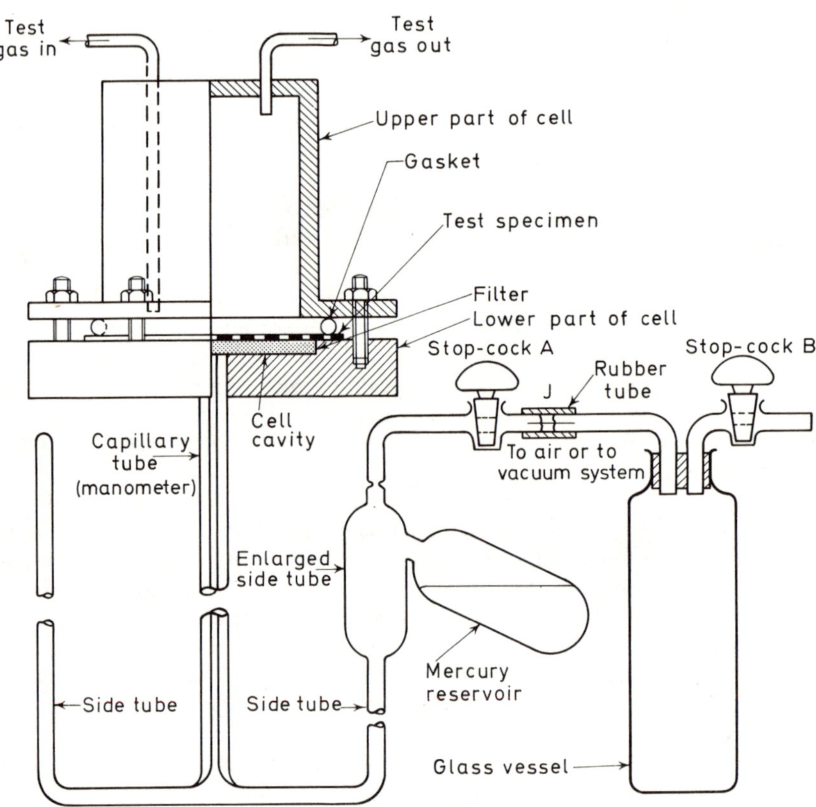

Figure 17.7. Apparatus for gas transmission rate (B.S. 2782)

chamber in the upper part of the cell and the connecting line are flushed with the gas under test and the cell and manometer system are evacuated to a pressure not greater than 0·3 mbar (0·2 mm of mercury). Stopcock A is then closed and the apparatus tilted so that the mercury flows into the capillary and side tubes; after this the apparatus is returned to the upright position and the level of the mercury in the capillary noted. The movement thereof is noted as in Procedure I.

Gas Transmission Rate:

$$\text{For Procedure I} \quad = \frac{(273p)\,(aH+V)\,(24\times 10^4)}{ATP}$$

$$\text{For Procedure II} = \frac{(273p)\ (24 \times 10^4)}{ATP}$$

where V is the volume in ml between the test specimen and the mercury level in the capillary tube as determined by calculating the volume in ml between the test specimen, when clamped in position over the cell cavity, and the level in the capillary that the mercury will occupy half way through the test and subtracting from this the volume of the filter. The latter is calculated from its weight and the true density of the material from which it is made, and the depth of the cell cavity is taken to be the thickness of the filter. (It is assumed that the change of volume of the air in the capillary is much less than V.)

a is the area in cm^2 of the manometer capillary,
H is the atmospheric pressure in cm of mercury,
p is the rate of change of pressure, in cm Hg/h, in the cell cavity and capillary tube,
A is the area in cm^2 of the test specimen above the cavity,
T is the temperature in K of the gas during test and
P is the pressure difference across the sample at the start of the test in cm Hg.

The gas transmission rate is obtained as the mean of the results from at least three test specimens and is expressed as ml/(m^2 24 h atm); the average thickness of the specimen must be stated in the report—see the comment made above with reference to water vapour 'permeability'. Individual results differing by up to 20% from the mean can be tolerated; this variance of test results justifies the approximate formula given (however, see James[38]).

A.S.T.M. D.1434–66, 'Gas Transmission Rate of Plastic Film and Sheeting', essentially follows the procedures above, with filter paper in place of the porous filter, but allows for measurement of the mercury by cathetometer or by electrical resistance change of a calibrated wire running through the tube into the mercury.

17.4.2 Impact Resistance

The impact resistance of film can be measured in a manner somewhat similar to that described in Section 8.10.3 of Chapter 8. Method 306F of B.S. 2782: 1970 is termed 'Impact Resistance of Flexible Film with Falling Dart' and requires tests on at least 60 specimens, each of which in turn is held in a carefully specified vacuum operated specimen holder with a pressure difference of 750 mbar (560 mm Hg; 22 in Hg). The impactor is a 'dart' as shown in Figure 17.8.

Specimens of thickness greater than 0·1 mm (0·005 in) are clamped by a metal clamping ring into the holder in addition to applying the vacuum: the 'free' test area of film in the specimen holder is a circle of diameter 127 mm (5·00 in).

The dart is released from 660 mm (26 in) electromagnetically with added circular weights (a range between 1 g and 550 g is suggested). A set of ten specimens is tested at each of at least six equally spaced values of (total) dart weight such that in each of at least three sets of specimens there are some broken and some unbroken specimens, in at least one set all are broken and in at least one set none are broken.

Impact Resistance (M_{50})

$$= m_{100} - D\,(S_1 - \tfrac{1}{2})$$

where m_{100} is the lowest dart weight in grammes at which all ten specimens broke

D is the difference in grammes between two successive dart weights

S_1 is the sum of the fractions of specimens that have broken in each set, starting with a weight corresponding to no failures, up to m_{100}.

A formula is given for calculating the standard deviation and a description is also included of a graphical method for determining M_{50}.

The same principle is used in A.S.T.M. D.1709–67, 'Impact Resistance of Polyethylene Film by the Free Falling Dart Method'. Two methods are

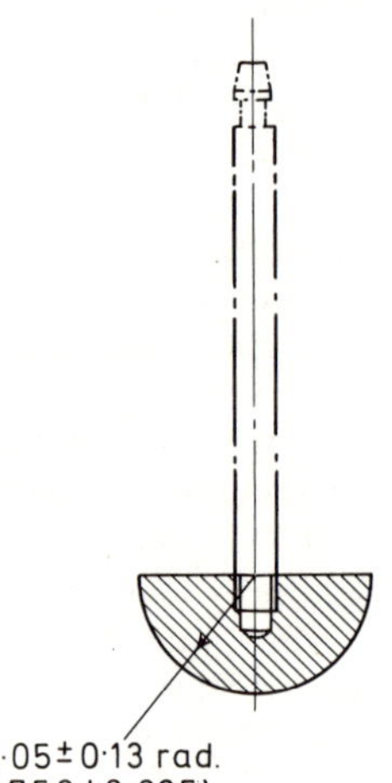

Figure 17.8. Dart for impact resistance of film (B.S. 2782). Head of dart is of cotton fabric filled laminated phenolic rod, highly polished. Dimensions in millimetres with inch equivalents in parentheses

given, differing in the size and weight of the darts and of the additional weights. Figure 17.9 shows the general layout of the equipment:

This type of test has been subjected to critical study of the influence of relevant variables by Cohen et alia[39].

17.4.3 Tear Strength

The subject of tear strength has been discussed by Hulse[40], in another Plastics Institute Monograph, and Scott[41] with particular reference to elastomers. The former deals inter alia in some measure with the theory—in so far as it has yet been worked out—of tear failure and the latter describes a whole range of test pieces and the different methods which have been standardised for elastomeric materials.

In B.S. 2782: 1970, two methods are described. In Method 308A, 'Tear Strength of Flexible Unsupported Polyvinyl Chloride Sheet', the specimen

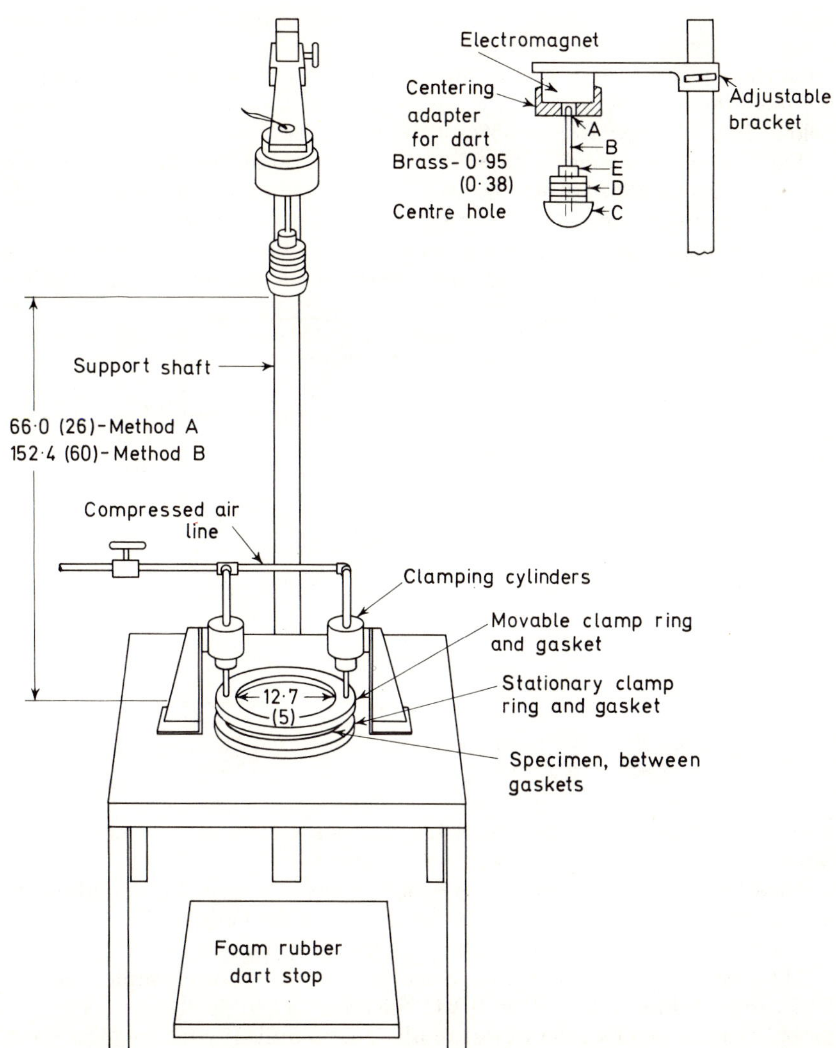

Figure 17.9. Free falling dart impact test for polyethylene film (A.S.T.M. D.1709). A. Steel shaft tip 0·64 (0·25) o.d. by 1·27 (0·50) long. B. Dart shaft: Method A—aluminium 0·65 (0·25) o.d. by 12·70 (5) long; ¼-20 thd (N.C.) 1·27 (0·50) long on bottom; No. 5–40 thd, (N.F.) for steel tip. Method B—same as in Method A except dart shaft is 11.43 (4.50) long. C. Hemispherical head: Method A—3.81 (1.500 ± 0·005) in diameter. Method B—5·08 (2·000 ± 0·005) in diameter. D. Removable weights. E. Collar and screw. Dimensions in centimetres with inch equivalents in parentheses

used is shown in Figure 17.10 and must be punched out using a single stroke of a press with a sharp cutting tool.

(a) for sheet up to and including 0·18 mm (0·007 in) thick, fifteen specimens are used; (b) for sheet above this, nine are used—directional effects are examined if appropriate. The specimens are tested in a machine with grip separation rate of 275 ± 25 mm/min (11 ± 1 in/min) and the load range of the machine must be such that the tearing load of a group of specimens (see below) falls between 15 and 85% of the maximum scale reading.

The test is conducted at 23 ± 1°C and the average thickness of the test specimen is calculated from the weight per unit area and the density. In case (a), three groups of five test specimens are rested and, for (b), three groups of three. The components of each group (five or three specimens as appropriate) are superimposed, held together in the grips of the testing

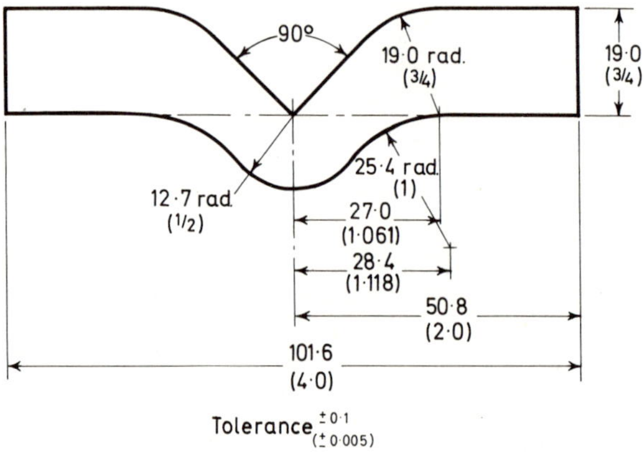

Figure 17.10. Tear strength specimen (B.S. 2782). Dimensions in millimetres with inch equivalents in parentheses

machine (see Chapter 8, Section 8.4.3) and the maximum load noted when the machine is operated as described above.

The tear strength of each group of specimens is obtained by dividing the maximum load by the total thickness. The tear strength of the material is the average of the values of the three groups of specimens.

This test is falling into disuse—it has been omitted for example from the 1967 issue of B.S. 1763, 'Thin PVC Sheeting (Flexible, Unsupported)'—probably because of scatter of test results due to difficulty in making a clean cut at the right angle of the crescent shape.

In B.S. 2739: 1967, 'Thick PVC Sheeting (Flexible, Unsupported)', three 'trouser type' specimens are used (see below, Figure 17.14); each is 7 × 2 in (179 × 51 mm) and each has a single cut 3 in (76 mm) long made centrally down the longitudinal axis, to give two 1 in (25 mm) wide 'legs'. Each specimen in turn is then tested as follows:

One leg is inserted into each jaw, 3 in (76 mm) apart, of a tensile testing machine capable of being driven at a constant rate of jaw separation of 11 ± 1 in/min (280 ± 25 mm/min). The tearing load is autographically recorded and the horizontal median of the load plateaux is divided by the gravimetric thickness of the test specimen to give tear strength in lbf/in thickness. The mean tear strength of the three specimens is calculated. The test is performed at 23 ± 1°C.

Method 308B of B.S. 2782: 1970 is 'Resistance between Propagation of Thin Flexible Sheet', the Elmendorf tear test, which has been described by Stackhouse[42]. The apparatus is shown in Figure 17.11 and the test specimen in Figure 17.12.

Once again the slit is made with a sharp blade. Five specimens are tested (in each direction if necessary) singly.

The test apparatus consists of a stationary jaw (right hand clamp in Figure 17.11) and a moveable one (left hand clamp in Figure 17.11), the latter carried on a pendulum—here shown as a sector of a wheel, free to

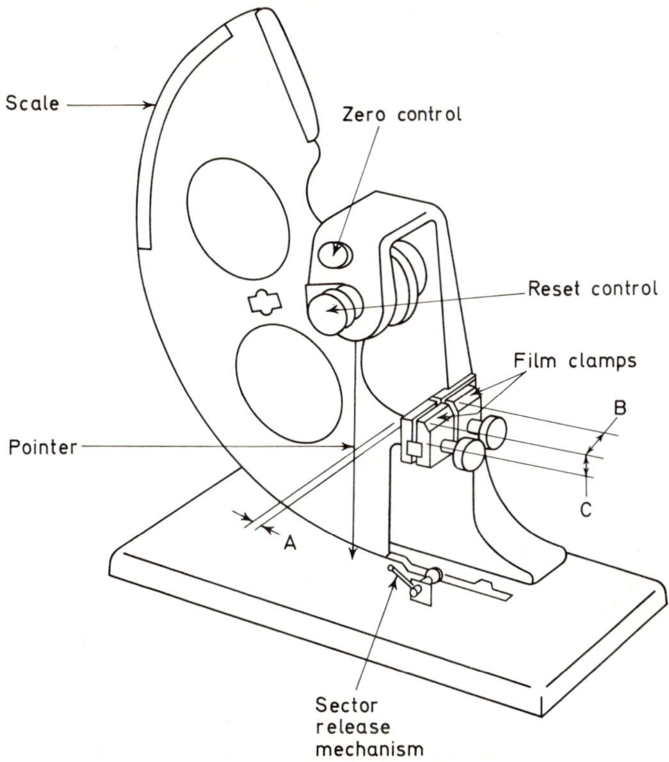

Figure 17.11. Elmendorf tear testing machine (B.S. 2782)

swing on a substantially frictionless bearing. The dimensions B and C of the clamping surface of each of the jaws are not less than 25 mm (1 in) and not less than 13 mm (0·5 in) respectively. Dimension A is between 9 mm and 13 mm (0·35 in and 0·5 in).

To start the test the two jaws are separated by an interval of 2·5 mm (0·10 in) and so aligned that the test specimen clamped in them lies in a plane perpendicular to the plane of oscillation of the pendulum, with the edges of the jaw gripping the test specimen in a horizontal line, a perpendicular to which though the axis of suspension of the pendulum is 102 mm (4 in) long and makes an angle of 27·5° with the plane of the test specimen. The pendulum carries a circumferential scale graduated 0–100 so as to read the work done in tearing the specimen (when the pendulum is released from its raised position).

Thickness measurements are taken under a specified pressure on each specimen around the area to be torn and the specimen mounted, one 'tongue' in each jaw. The pointer is placed against its (upper) stop and the pendulum is released. The scale reading is noted but is rejected if the line of tear deviates by more than 5° from the line of the slit. If the energy required to tear the test specimen is more than 60% of the maximum scale reading, an additional weight is added and the test repeated. If the energy is again more than 60%, the specimen cannot be tested by this method. The results are expressed as gf/mil or gf/micron. (This is the method now specified in B.S. 1763: 1967.)

A.S.T.M. D.1004–66, 'Tear Resistance of Plastic Film and Sheeting', uses a specimen similar to that of Figure 17.10 but a rate of grip separation

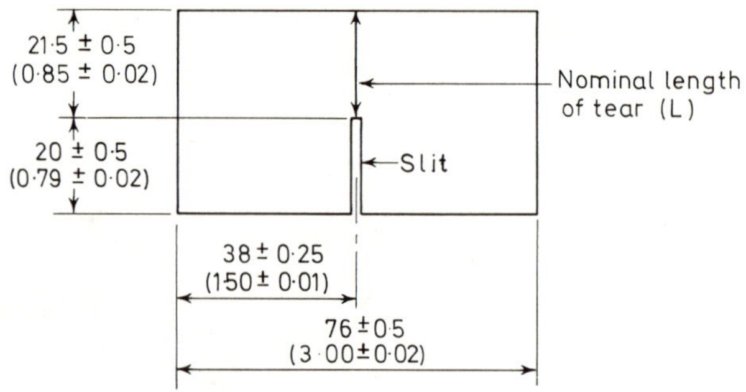

Figure 17.12. Elmendorf tear test specimen. Dimensions in millimetres with inch equivalents in parentheses

of only 2 in/min. Also, at least ten specimens in each direction must be tested—singly. Results are normally expressed in pounds—i.e. not reduced to unit thickness.

A.S.T.M. D.1922–67, 'Propagation Tear Resistance of Plastic Film and Thin Sheeting', is of the Elmendorf type, using a specimen as shown in Figure 17.13.

The specimen is cut out and the slit is made by a sharp spring loaded knife or equivalent. The particular apparatus specified (see the A.S.T.M. manual for details) gives a rate of tearing between 760 and 4600 cm (300–1800 in) per minute.

A 'trouser' type of specimen is employed in A.S.T.M. D.1938–67, 'Resistance to Tear Propagation in Plastic Film and Thin Sheeting by a Single-Tear Method' (see Figure 17.14).

The 5 cm slit is made by a sharp razor blade or equivalent and the thickness below the slit (i.e. the 2·5 cm length) is measured in several places. Five specimens (at least) are examined in each sheet direction, at a rate of grip separation of 25 cm (10 in) per minute, after the specimens in turn have been mounted in the grips by securing tongue A in one grip and tongue B in the other, the grips being initially 5 cm (2 in) apart. The major axis of the specimen must coincide with an imaginary line joining the centre lines of the grips. Advice is given on how to express the results depending on whether the specimen tears in a reasonably steady fashion or yields an initial tear load and then a maximum tear load.

The test specimen used in DIN.53363 is trapezoid in shape, with the slit cut centrally in the shorter of the two parallel sides. Stressing is carried out with the two non-parallel ends of the trapezium gripped in the chucks, so that these ends of the specimen are perpendicular to the direction of the stressing, i.e. the slit is opened out.

Reinhart and Mandel[43] have compared results of crescent (Figure 17.10)

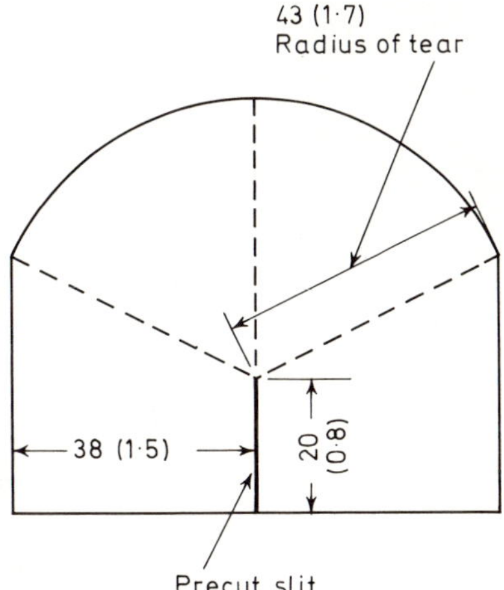

Figure 17.13. Constant radius test specimen for tear resistance test (A.S.T.M. D.1922). Dimensions in millimetres with inch equivalents in parentheses

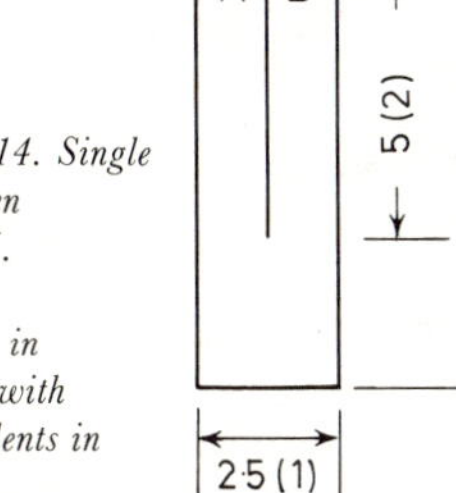

Figure 17.14. Single tear specimen (A.S.T.M. D.1938). Dimensions in centimetres with inch equivalents in parentheses

tear, Elmendorf tear and tensile tests on three vinyl films and one of polyethylene and concluded that the three tests measured different properties. Grimminger and Jacobshagen[44] investigated another range of four tear tests, not including the Elmendorf, and favoured a specimen with a trapezoid tear shape.

A.S.T.M. D.2582–67, 'Resistance to Puncture-Propagation of Tear in Thin Plastics Sheeting', is stated to assess the resistance of film and thin sheeting to dynamic tearing. The apparatus is shown in Figure 17.15.

The carriage, carrying a sharp pointed probe, drops normally from a height of 50·8 cm (20 in) to cut into the specimen, approximately 20 cm (8 in) long in the direction of the tear. Two sets of specimens are used, one each cut from the machine and transverse directions, and sufficient in number to give a minimum of five tear measurements in each direction.

The thickness of the specimens is recorded and then they are mounted in turn in the apparatus, as shown in the figure. By trial and error a carriage weight is selected (six alternatives are available) which produces a minimum tear length of 4·0 cm but does not 'bottom' onto the base of the apparatus. 'Zeroing' is next carried out by lowering the selected carriage until the probe

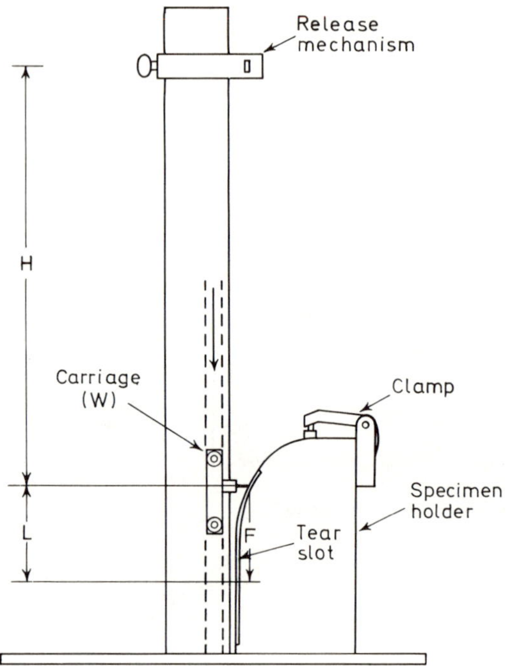

Figure 17.15. Apparatus for resistance to puncture propagation of tear of thin plastic film (A.S.T.M. D.2582)

point touches, but does not indent, the film specimen and then the tear length indicating rod is adjusted to '0'.

The specimens are tested with this carriage and this zero position, the tear length being measured in each case.

$$\text{Tear Resistance } F = \frac{W \times H}{L} + W \text{kgf}$$

where W = weight of carriage, kg
H = height of carriage before release, cm
and L = length of tear, cm

Average values and standard deviations are calculated for each film direction.

17.4.4 Toxicity

The wide use of plastics materials, often printed, and very likely containing one or several components such as plasticisers, stabilisers, lubricants and colouring matter, as food packaging is a matter of some concern to the responsibly minded user. Quite literally, fractions of a part per million of certain compounds may present toxic hazards, especially if their effects are cumulative, so that even if the basic polymer is harmless—as it often is—residual polymerisation catalysts or breakdown products derived therefrom and emulsion stabilisers may present additional hazards. However, unless the packaging is actually going to be ingested the presence of a potentially dangerous element or radical, such as lead, arsenic or phosphate for instance, need not necessarily be a reason for rejection of the packaging unless it is in such a chemically combined form that it is still toxically hazardous and it is extractable by the foodstuffs material in significant quantity ('significant' in this context may be a very low figure indeed—see below).

Legislation on the subject varies from country to country and the reader is referred to References 47 and 48 for information on this matter. In the U.K., at least, the basic principles is to establish the inherent toxicological hazard of every compound by pathological trials with rodents and mammals (results cannot be extrapolated from one member of a homologous series to another and, worse still, results are not necessarily applicable to all species of the same type of animal! For instance, monkeys and humans may react differently to the same materials). Having thus obtained some safe limit of concentration which may be tolerated, in the original approach[45] the packaging was then extracted for each ingredient either with the materials it was known it would contact, or if unknown with a representative selection of all types; the extracts were then analysed for the 'hazardous' ingredients. The concentration of each was then equated with its pathologically determined 'toxicity factor' and the 'toxicity quotients' thus obtained were summated. More recently[46], the concept has been changed to one of allowing only the use of 'safe' materials in the polymerisation process and subsequent fabricating operations.

It will be seen that basically the problem is a biological one followed by refined analysis. As such it is beyond the scope of this monograph; further information may be obtained from references 45–55 inclusive.

17.5 LAMINATES

It may seem strange, not to say unrealistic, to dismiss the testing of plastics laminates in a few lines when one considers their importance and potential in the industry. However, generally speaking their evaluation does not involve any new principles or ideas that have not already been covered (see the introduction to this chapter); many of the standard tests for laminates have in fact been described in the relevant preceding chapters. True, there are certain methods unique to laminates, for example A.S.T.M. D.2733–68T, 'Interlaminar Shear Strength of Structural Reinforced Plastics at Elevated Temperatures', but even in this instance unless the data are specifically required for design purposes (which is unlikely), the more routine measurement of cross-breaking strength (Chapter 8, Section 8.7) serves as a good guide to the coherence between laminated layers.

There is, nevertheless, an out-of-the-ordinary technique developed particularly for the evaluation of the raw materials of filament wound structures. It involves the making of NOL (Naval Ordnance Laboratory) rings, which is described in A.S.T.M. D.2291–67, 'Fabrication of Ring Test Specimens for Reinforced Plastics'. Figure 17.16 gives a suggested apparatus and Figure 17.17 a suitable mould.

Three alternative rings are specified, each 146 mm (5·750 in) diameter by 6.35 mm (0·250 in) wide; the wall thicknesses are in two cases 1·52 mm (0·060 in) and the other 3·18 mm (0·125 in). The difference between the

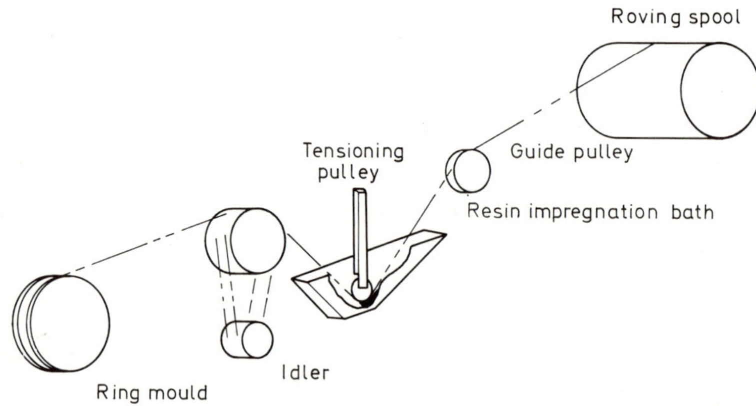

Figure 17.16. Wet roving ring winding apparatus (A.S.T.M. D.2291)

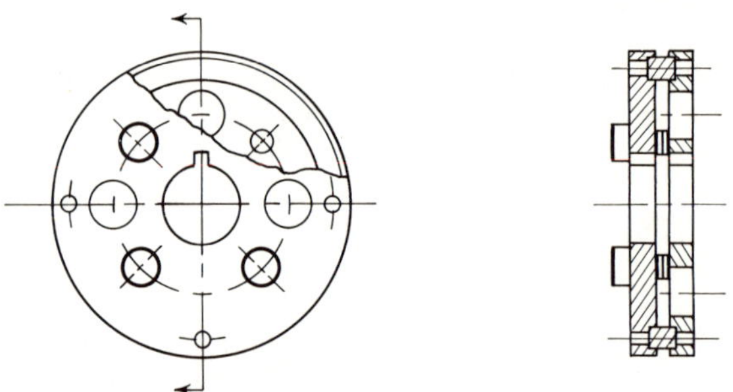

Figure 17.17. Single ring mould (A.S.T.M. D.2291)

two apparently identical rings is that one is made over thick (and is the larger one) and machined down, whereas the other is moulded to the correct wall thickness directly.

Tests that may be applied to these rings include split disc ring tensile strength (A.S.T.M. D.2290–64T)—see Figure 17.18—ring compression strength, flexural strength (on a ring segment) and short beam interlaminar shear strength (A.S.T.M. D.2344–67).

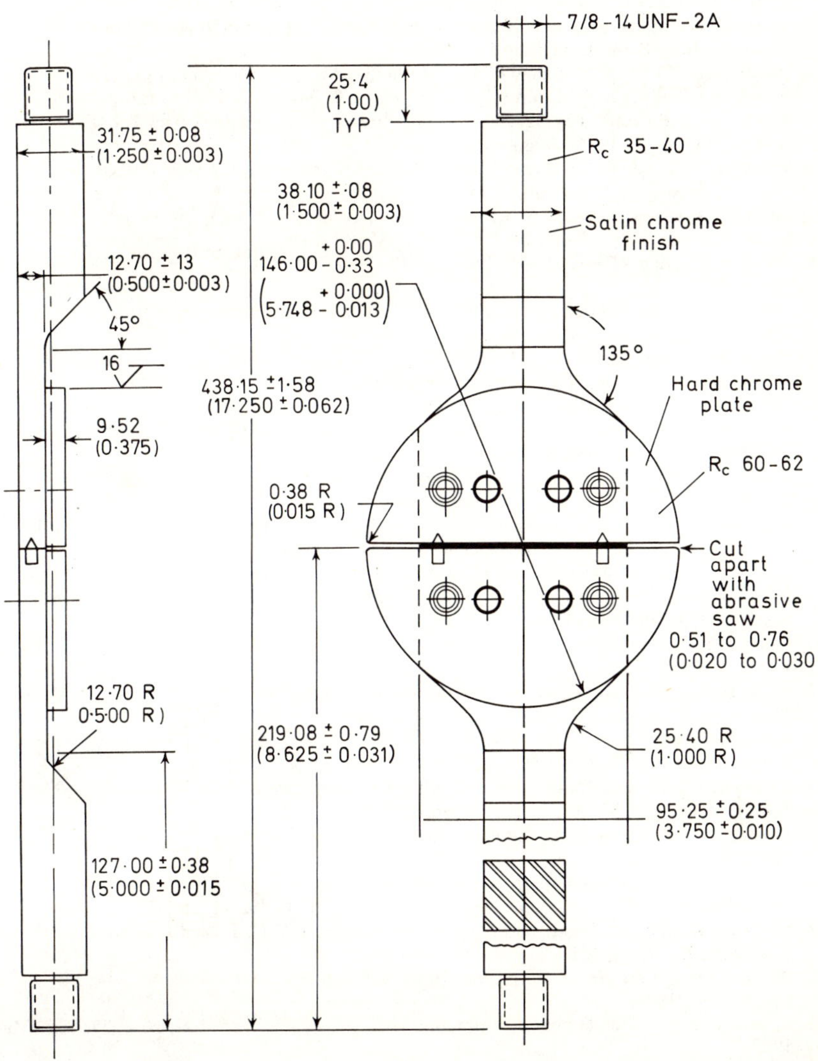

Figure 17.18. Split disc tension test fixture (A.S.T.M. D.2290). Dimensions in millimetres with inch equivalents in parentheses. Material AISI 4140

REFERENCES

1. HASLAM, J., WILLIS, H. A. and SQUIRREL, D. C. M., *Identification and Analysis of Plastics,* 2nd edn, Iliffe Books Ltd., London (in preparation)
2. KLINE, G. M. (Ed.), *Analytical Chemistry of Polymers,* Part 1—Analysis of Monomers and Polymeric Materials (1959) and Part 3—Identification Procedures and Chemical Analysis (1962), Interscience Publishers, New York
3. KRAUSE, A. and LANGE, A., *Introduction to the Chemical Analysis of Plastics* (Translated by J. Haslam), Iliffe Books Ltd., London (1969)
4. HILADO, C. J., 'The Development of Test Methods for Rigid Cellular Plastics in the American Society for Testing and Materials', *J. Cell. Plast.,* **3,** No. 11, 502 (November 1967)
5. BUIST, J. M., 'Polymer Testing and its Contribution to Developments in Industry', *Trans. I. R. I.,* **33,** No. 4, T126 (August 1957)
6. JONES, R. E., 'Rigid Urethane Foam—Process Variables and Test Procedures', *Plast. Technol.,* **7,** No. 10, 27 (October 1961)
7. PHILLIPS, T. L. and LANNON, D. A., 'The Assessment of the Physical Properties of Rigid Expanded Plastics', *Brit. Plast.,* **34,** No. 5, 236 (May 1961)
8. 'Physical Testing of Foams—I', *Rubb. & Plast. Weekly,* **141,** No. 25, 976 (December 16 1961)
9. 'Physical Testing of Foams—II', *Rubb. & Plast. Weekly,* **141,** No. 26, 1014 (December 23 1961)
10. BUIST, J. M. and LOWE, A., 'The Properties of Polyurethanes and their Applications', *Trans. Plast. Inst. Lond.,* **27,** No. 67, 13 (February 1959)
11. 'Properties of Rigid Cellular Plastics', *Des. Compon. Engng,* 34, (August 11 1966)
12. 'Properties of Cellular Polymers', *Kirk-Othmer Encyclopedia of Chemical Technology,* 2nd edn, 961, Interscience Publishers, New York, Volume 9 (1966)
13. GERSTIN, H., 'How to Evaluate Rigid Plastic Foams', *Product Engng,* **36,** No. 13, 59 (June 21 1965)
14. REMINGTON, W. J. and PARISER, R., 'A New Apparatus for Determining the Cell Structure of Cellular Materials', *Rubb. World,* **138,** 261 (May 1958)
15. JONES, R. E. and FESMAN, G., 'Air Flow Measurement and its Relations to Cell Structure, Physical Properties, and Processability for Flexible Urethane Foam', *J. Cell. Plast.,* **1,** No. 1, 200 (January 1965)
16. GENT, A. N. and RUSCH, K. C., 'Permeability of Open-cell Foamed Materials', *J. Cell. Plast.,* **2,** No. 1, 46 (January 1966)
17. LEVY, M. M., 'Moisture Vapor Transmission and its Effect on Thermal Efficiency of Foam Plastics', *J. Cell. Plast.,* **2,** No. 1, 37 (January 1966)
18. JONES, R. E., HERSCH, P., STIER, G. G. and DOMBROW, R. A., 'Measuring Resilience of Flexible Urethane Foams', *Plast. Technol.,* **5,** No. 9, 55 (September 1959)
19. ROSENTHAL, L. A. and ADDIS, G. I., 'A Modified Rebound Pendulum for Evaluating Flexible Foams', *Rubb. Age,* **85,** No. 5, 790 (August 1959)
20. KARAS, G. C., 'Equipment for the Assessment of Thermoplastics Pipes', *Internat. Plast. Engng,* **1,** No. 1, 18 (February 1961)
21. CANN, J. M., 'The Assessment of the Impact Resistance of Rigid P.V.C. Pipe', Part 1, *Brit. Plast.,* **36,** No. 9, 516 (September 1963); Part 2, *Brit. Plast.,* **36,** No. 10, 579 (October 1963)
22. RICHARD, K., GAUBE, E. and DEIDRICH, G., 'Testing of P.V.C. Pipes for Impact Strength', *Kunststoffe,* **51,** No. 8, 431 (August 1961)
23. 'Burst Testing Equipment for Plastics Pipes and Fittings', *Internat. Plast. Engng,* **3,** No. 1, 17 (January 1963)
24. LLOYD, P. F. V., 'The Testing of Pipes and Fittings', *Plastics, Lond.,* **32,** No. 355, 565 (May 1967)
25. FRANK, K. (Ed.), 'Testing of Plastics Foils and Films', *Prufungbuch fur Kautschuk und Kunststoffe,* Chapter 5, 84, Berliner Unior, Stuttgart (1955)
26. 'German Equipment for Testing Flexible Materials', *Brit. Plast.,* **29,** No. 7, 259 (July 1956)
27. DAVIS, E. G., KAREL, M. and PROCTOR, B. E., 'Film Strengths in Heat Processing', *Mod. Packag.,* **33,** No. 4, 135 (December 1959)
28. ROUGEAUX, J., 'Contribution to the Study of the Mechanical Properties of Thin Films of Plastics Materials. Bursting Properties', *Industr. Plast. mod.,* **12,** No. 12, 30 (December 1960)
29. GRIMMINGER, H., 'Electronic Tearing Apparatus for Investigating Shock Behaviour of Plastics Sheeting', *Kunststoffe,* **50,** No. 9, 491 (September 1960)
and:
GRIMMINGER, H., 'Measurements on Plastics Films with an Electronic Tensile Tester',

Kunststoffe, **50,** No. 11, 618 (November 1960)

30. GLYDE, B. S., HOLMES-WALKER, W. A. and JEFFS, K. D., 'An Impact Fatigue Test for Polythene Film', *Brit. Plast.,* **34,** No. 8, 432 (August 1961)
31. DALTON, W. K., 'Testing of Flexible Vinyl Sheet', *Brit. Plast.,* **35,** No. 3, 136 (March 1962)
32. PATTERSON, D. JR. and WINN, E. B., 'A New Tester for Puncture and Tear Resistance', *Mater. Res. Stand.,* **2,** No. 5, 396 (May 1962)
33. BERGMAN, H. A. and CALDERWOOD, R. H., 'Pressure-Deflection and Burst Characteristics', *Mod. Plast.,* **41,** No. 11, 143 (July 1964)
34. BAIRD, M. E. and LANNON, D. A., 'Physical Testing of Plastics Films', *Brit. Plast.,* **37,** No. 2, 82 (February 1964)
35. PATTERSON, G. D. JR., 'An Interlaboratory Study of Cutting Plastic Film Tension Specimens', *Mater. Res. Stand.,* **4,** No. 4, 159 (April 1964)
36. HENNESEY, B. J., MEAD, J. A. and STENNING, T. C., *The Permeability of Plastics Films,* The Plastics Institute, London (1966)
37. STANNETT, V. and YASUDA, H., 'The Measurement of Gas and Vapor Permeation and Diffusion in Polymers', *Testing of Polymers* (Edited by J. V. Schmitz), Volume 1, Chapter 13, Interscience Publishers, New York (1965)
38. JAMES, D. I., 'Gas Transmission Through Polymer Films. A Discussion of the Approximations in the B. S. Method', *Research Report No. 179,* Rubber and Plastics Research Association of Great Britain (Shawbury) (1969)
39. COHEN, S. M., FERRIS, T. V., MONT, G. E. and MARTINS, J. G., 'A Versatile Plastic Sheet Impact Tester', *Mater. Res. Stand.,* **9,** No. 5, 21 (May 1962)
40. RITCHIE, P. D. (Ed.), *Physics of Plastics,* Chapter 3, Iliffe Books Ltd., London, 150 (1965)
41. SCOTT, J. R., *Physical Testing of Rubbers,* 111, Maclaren and Sons Ltd., London (1965)
42. STACKHOUSE, N., 'The Elmendorf Tear Test for Thin Unsupported P.V.C. Sheet', *Brit. Plast.,* **34,** No. 1, 34 (January 1961)
43. REINHART, F. W. and MANDEL, J., 'Comparison of Methods of Measuring the Tensile and Tear Properties of Plastics Films', *Bull. Amer. Soc. Test. Mat.* **No. 209,** 50 (October 1955)
44. GRIMMINGER, H. and JACOBSHAGEN, E., 'Tear Propagation of Film', *Gummi, Asbest, Kunststoffe,* **20,** No. 12, 1271 (December 1967)
45. Second Report of the Toxicity Sub-Committee of the Main Technical Committee with Methods of Analysis of Representative Extractants, The British Plastics Federation Publication No. 45 (London) (1962) also: *Inform. Bull.,* British Industrial Biological Research Association, **4,** No. 1, 5 (January 1965)
46. 'Plastics for Food Contact Applications. A Code of Practice for Safety in Use', The British Plastics Federation (with the cooperation of the British Industrial Biological Research Association) (London) (1969)
47. HOPF, P. P., 'Toxicity of Plastics', *Trans. Plast. Inst. Lond.,* **29,** No. 79, 2 (February 1961)
48. RODEYNS, A., 'Toxicity in Relation to the Use of Plastics Materials in Europe', *Brit. Plast.,* **35,** No. 2, 90 (February 1962)
49. PHILLIPS, I. and MARKS, G. C., 'Plastics in Foodstuffs Applications, Part, 1, Toxicity Testing and Regulations', *Brit. Plast.,* **34,** No. 6, 319 (June 1961)
50. PHILLIPS, I. and MARKS, G. C., 'Plastics in Foodstuffs Applications, Part 2, Development of Non-Toxic Materials and Toxicological Studies', *Brit. Plast.,* **34,** No. 7, 385 (July 1961)
51. NIKLAS, H. and MEYER, W., 'The Migration of Lead from Lead-Stabilised P.V.C. Pipes', *Kunststoffe,* **51,** No. 1, 2 (January 1961)
52. 'Toxicity Problems in Plastics Packaging', *Rubb. & Plast. Weekly,* **145,** No. 6, 178 (August 10 1963)
53. GARLOCK, E. A., 'Analysis of Packaging Materials for Extractable Substances', *The Analyzer,* **4,** No. 4, 8 (Nobember 1963)
54. NIKLAS, H. and MEYER, W., 'The Migration of Lead from Lead-Stabilised P.V.C. Drinking Water Pipes', *Kunststoffe, 54,* No. 2, 82 (February 1964)
55. LEFAUX, R., *Practical Toxicology of Plastics,* English Editor P. P. Hopf, Iliffe Books Ltd., London (1968)
56. Symposium for Filament-Wound Reinforced Plastics, *Special Technical Publication No. 327,* 66 et seq., American Society for Testing and Materials, Philadelphia (1963)
57. ERICKSON, P. W. SR., SILVER, I. and PERRY, H. A. JR., 'NOL Ring Test for Glass Roving-Reinforced Plastic', *Plast. Technol.,* **4,** No. 11, 1017 (November 1958)
58. BARNET, F. R., 'NOL Rings for Composite Materials Research and Development', *Trans. Plast. Inst. Lond.,* **33,** No. 107, 177 (October 1965)
59. ALGRA, E. A. H. and VAN DER BEEK, M. H. B., 'Standards and Test Methods for Filament-wound Reinforced Plastics', *Plast. & Polymers,* **36,** No. 123, 229 (June 1968)

Index